Lecture Notes in Computer Science 9133

Commenced Publication in 1973
Founding and Former Series Editors:
Gerhard Goos, Juris Hartmanis, and Jan van Leeuwen

More information about this series at http://www.springer.com/series/7407

Ferdinando Cicalese · Ely Porat
Ugo Vaccaro (Eds.)

Combinatorial Pattern Matching

26th Annual Symposium, CPM 2015
Ischia Island, Italy, June 29 – July 1, 2015
Proceedings

 Springer

Editors
Ferdinando Cicalese
Department of Computer Science
University of Verona
Verona
Italy

Ugo Vaccaro
Department of Computer Science
University of Salerno
Fisciano
Italy

Ely Porat
Department of Computer Science
Bar-Ilan University
Ramat Gan
Israel

ISSN 0302-9743   ISSN 1611-3349   (electronic)
Lecture Notes in Computer Science
ISBN 978-3-319-19928-3   ISBN 978-3-319-19929-0   (eBook)
DOI 10.1007/978-3-319-19929-0

Library of Congress Control Number: 2015940414

LNCS Sublibrary: SL1 – Theoretical Computer Science and General Issues

Springer Cham Heidelberg New York Dordrecht London

Printed on acid-free paper

Springer International Publishing AG Switzerland is part of Springer Science+Business Media
(www.springer.com)

Preface

This volume contains the papers presented at the 26th Annual Symposium on Combinatorial Pattern Matching (CPM 2015) held during June 29 – July 1, 2015, in Ischia, Italy.

The conference program included 34 contributed papers and three invited talks by Sorin Istrail from Brown University, USA, Rasmus Pagh from IT University of Copenhagen, Denmark, and Wojciech Szpankowski from Purdue University, USA. The contributed papers were selected out of 83 submissions from 35 countries, corresponding to an acceptance ratio of 40.9 %. Each submission received at least three reviews. We thank the members of the Program Committee and all the additional external reviewers for their hard and invaluable work that resulted in an excellent scientific program. Their names are listed on the following pages.

The objective of the annual CPM meetings is to provide an international forum for research in combinatorial pattern matching and related applications. It addresses issues of searching and matching strings and more complicated patterns such as trees, regular expressions, graphs, point sets, and arrays. The goal is to derive combinatorial properties of such structures and to exploit these properties in order to achieve a superior performance for the corresponding computational problems. The meeting also deals with problems in computational biology, data compression and data mining, coding, information retrieval, natural language processing, and pattern recognition.

The Annual Symposium on Combinatorial Pattern Matching started in 1990, and has since taken place every year. Previous CPM meetings were held in Paris, London (UK), Tucson, Padova, Asilomar, Helsinki, Laguna Beach, Aarhus, Piscataway, Warwick, Montreal, Jerusalem, Fukuoka, Morelia, Istanbul, Jeju Island, Barcelona, London (Ontario, Canada), Pisa, Lille, New York, Palermo, Helsinki, Bad Herrenalb, and Moscow. Starting from the third meeting, proceedings of all meetings have been published in the LNCS series, as volumes 644, 684, 807, 937, 1075, 1264, 1448, 1645, 1848, 2089, 2373, 2676, 3109, 3537, 4009, 4580, 5029, 5577, 6129, 6661, 7354, 7922, and 8486, respectively.

Selected papers from the first meeting appeared in volume 92 of *Theoretical Computer Science*, from the 11th meeting in volume 2 of the *Journal of Discrete Algorithms*, from the 12th meeting in volume 146 of *Discrete Applied Mathematics*, from the 14th meeting in volume 3 of the *Journal of Discrete Algorithms*, from the 15th meeting in volume 368 of *Theoretical Computer Science*, from the 16th meeting in volume 5 of the *Journal of Discrete Algorithms*, from the 19th meeting in volume 410 of *Theoretical Computer Science*, from the 20th meeting in volume 9 of the *Journal of Discrete Algorithms*, from the 21st meeting in volume 213 of *Information and Computation*, from the 22nd meeting in volume 483 of *Theoretical Computer Science*, and from the 23rd meeting in volume 25 of the *Journal of Discrete Algorithms*. A special issue of *Algorithmica* is planned for the extended versions of a selection of the papers presented at this year's meeting.

The whole submission and review process was carried out with the help of the EasyChair conference system. We thank the CPM Steering Committee for supporting Ischia as the site for CPM 2015, and for their advice and help in different issues. We thank the Italian Chapter of the European Association for Theoretical Computer Science for its scientific endorsement. We thank Dr. Aniello Castiglione for his invaluable help in the local organization.

The conference was sponsored by the Department of Computer Science of the University of Salerno and the Department of Computer Science of the University of Verona, which we thank for their financial support.

April 2015

Ferdinando Cicalese
Ely Porat
Ugo Vaccaro

Organization

Program Committee

Francine Blanchet-Sadri	University of North Carolina, USA
Timothy M. Chan	University of Waterloo, Canada
Ferdinando Cicalese	University of Verona, Italy (Co-chair)
Raphael Clifford	University of Bristol, UK
Paolo Ferragina	University of Pisa, Italy
Travis Gagie	University of Helsinki, Finland
Leszek A. Gasieniec	University of Liverpool, UK
Raffaele Giancarlo	University of Palermo, Italy
Roberto Grossi	University of Pisa, Italy
John Iacono	New York University, USA
Tsvi Kopelowitz	University of Michigan, USA
Gregory Kucherov	Université Paris-Est Marne-la-Vallée, France
Eduardo Laber	PUC-Rio, Brazil
Gad M. Landau	University of Haifa, Israel
Jesper Larsson	Malmö University, Sweden
Noa Lewenstein	Netanya College, Israel
Inge Li Gørtz	Technical University of Denmark, Denmark
Stefano Lonardi	University of California, Riverside, USA
Veli Makinen	University of Helsinki, Finland
Ian Munro	University of Waterloo, Canada
Gonzalo Navarro	University of Chile, Chile
Ely Porat	Bar-Ilan University, Israel (Co-chair)
Simon Puglisi	University of Helsinki, Finland
Kunihiko Sadakane	University of Tokyo, Japan
Marie-France Sagot	Inria Grenoble Rhône-Alpes, Lyon, France
Tatiana Starikovskaya	Higher School of Economics, Russia
Jens Stoye	Bielefeld University, Germany
Oren Weimann	University of Haifa, Israel

Additional Reviewers

Amit, Mika	Boucher, Christina
Backofen, Rolf	Bowe, Alexander
Badkobeh, Golnaz	Boytsov, Leonid
Belazzougui, Djamal	Bucher, Philipp
Blondin Massé, Alexandre	Bulteau, Laurent
Bodnar, Michelle	Cannon, Sarah

Chikhi, Rayan
Christiansen, Anders Roy
Claude, Francisco
Cording, Patrick Hagge
Cunial, Fabio
Doerr, Daniel
El-Zein, Hicham
Elzein, Hicham
Farruggia, Andrea
Farrugia, Ashley
Feijao, Pedro
Fertin, Guillaume
Fici, Gabriele
Fischer, Johannes
Fontaine, Allyx
Fox, Nathan
Ganguly, Arnab
Gawrychowski, Pawel
Grant, Oliver
Gutin, Gregory
Hamel, Sylvie
Hamilton, David
He, Meng
Hermelin, Danny
Hoener Zu Siederdissen, Christian
Hon, Wing-Kai
I, Tomohiro
Jahn, Katharina
Kociumaka, Tomasz
Konow, Roberto
Kurpicz, Florian
Kuszner, Lukasz
Kärkkäinen, Juha
Köppl, Dominik
Lewenstein, Moshe
Li, Jing
Lombardy, Sylvain
Manea, Florin
Manzini, Giovanni
Meidanis, Joao
Mercas, Robert
Mitzenmacher, Michael
Nekrich, Yakov
Nicholson, Patrick K.
Nishimoto, Takaaki

Niskanen, Reino
Ono, Hirotaka
Ordóñez Pereira, Alberto
Ounit, Rachid
Pacheco, Eduardo
Park, Heejin
Patil, Manish
Pedersen, Christian Nørgaard Storm
Pissis, Solon
Radoszewski, Jakub
Raman, Rajeev
Rampersad, Narad
Roscigno, Gianluca
Rozenberg, Liat
Różański, Michał
Sach, Benjamin
Sadakane, Kunihiko
Salikhov, Kamil
Salmela, Leena
Seki, Shinnosuke
Shah, Rahul
Sheinwald, Dafna
Simmons, Sean
Simpson, Olivia
Sirèn, Jouni
Skjoldjensen, Frederik Rye
Sun, He
Takeda, Masayuki
Thachuk, Chris
Theodoridis, Evangelos
Tiskin, Alexander
To, Thu-Hien
Tomescu, Alexandru I.
Tsur, Dekel
Uznański, Przemysław
Valenzuela, Daniel
van Iersel, Leo
Venturini, Rossano
Vialette, Stéphane
Vildhøj, Hjalte Wedel
Vind, Søren
Vuillon, Laurent
Walen, Tomasz
Wittler, Roland
Zhou, Gelin

Invited Talks

On Humans, Plants and Disease: Algorithmic Strategies for Haplotype Assembly Problems

Sorin Istrail

Department of Computer Science, Brown University, USA
sorin_istrail@brown.edu

This talk is about a set of computational problems about haplotypes reconstruction from genome sequencing data for diploid organisms, such as humans, and for polyploid organisms, such as plants. Polyploidy is a fundamental area of molecular biology with powerful methods of Nobel prize recognition: polyploidy inducement for cell reprogramming, mosaicism for aneuploid chromosome content as the constitutional make-up of the mammalian brain, and the polyploidy design for highly sought after agricultural crops and animal products. On the medical side, polyploidy refers to changes in the number of whole sets of chromosomes of an organism, and aneuploidy refers to changes in number of specific chromosomes or of parts of them. We will present an algorithmic framework, HapCOMPASS for these problems that is based on graph theory. The software tools implementing our algorithms (available from the Istrail Lab) are already in use, by many users, and recognized as among the leading tools in the areas of human genome haplotype assembly, plant polyploidy haplotype assembly, and tumor haplotype assembly. We will also present a number of unresolved computational problems whose solutions would advance our understanding of human biology, plant biology and human disease.

To introduce the application areas, and a hint at the type of combinatorial problems of that biological import, a short primer follows. Humans, like most species whose cells have nuclei, are diploid, meaning they have two sets of chromosomesone set inherited from each parent. In the genome era, the genome sequencing technologies are generating big data-bases of empirical patterns of genetic variation within and across species. A SNP (single nucleotide polymorphism) is a DNA sequence variation occuring commonly (e.g. 3 %) at a fixed site on the genome within a population in which a single nucleotide A, C, G or T differs between individuals of a species, or between the mother-father chromosomes of a single individual. The different nucleotide bases at the SNP site are called alleles. SNPs account for large majority of genetic variation of species. For humans, there are about 10 million SNPs, so conceptually the SNPs variation of any individual is captured by two allele vectors (each with 10 million components), one inherited from mother and one from the father. Our approach to haplotype assembly is based on graph theoretical modeling of sequencing reads linking SNPs and assembling whole haplotypes based on such basic read-SNPs linkings.

S. Istrail—Work in collaboration with Derek Aguiar (Princeton University) and Wendy Wong (INOVA Translational Medicine Institute).

Polyploid organisms have more than two sets of chromosomes. Although this phenomenon is particularly common in plants (e.g., seedless watermelon is 3x, wheat 6x, strawberries 10x), it is also present in animals (e.g. fish could have 12x and up to 400 haplotypes), and in humans (e.g., some mammalian liver cells or heart cells or bone marrow cells are polyploid). While polyploidy refers to numerical change in the whole set of chromosomes, aneuploidy refers to organisms in which a part of the set of chromosomes (e.g. a particular chromosome or a segment of a chromosome) is under- or over- represented. Polyploidy and aneuploidy phenomena are recognized as disease mechanisms. Examples for polyploidy: triploidy birth conceptions end in miscarriages, although mixoploidy, when both diploid and triploid cells are present, could lead to survival; triploidy, as a result of either digyny (the extra haploid set is from the mother by failure of one meiotic division during oogenesis) or diandry (mostly caused by reduplication of paternal haploid set from a single sperm or dispermic fertilization of the egg) could have parent-of-origin (genomic imprinting) medical consequences: diandry predominate among preterm labor miscarrieges while digyny predominates into survival into fetal period, although with a poor grown fetus and very small placenta). Examples for aneuploidy: trisomy in the the Down syndrome, cells with one chromosome missing while others with an extra copy of the chromosome, cells with unpredictably many chromosomes of a given type; mosaicism (when two or more populations of cells with different genotypes derived from a single individual) aneuploidy occurs in virtually all cancer cells.

Analytic Pattern Matching: From DNA to Twitter

Philippe Jacquet[1] and Wojciech Szpankowski[2]

[1] Alcatel-Lucent Bell Labs, Nozay, France
philippe.jacquet@alcatel-lucent.com
[2] Department of Computer Science, Purdue University, USA
spa@cs.purdue.edu

Repeated patterns and related phenomena in words are known to play a central role in many facets of computer science, telecommunications, coding, data compression, data mining, and molecular biology. One of the most fundamental questions arising in such studies is the frequency of pattern occurrences in a given string known as the text. Applications of these results include gene finding in biology, executing and analyzing tree-like protocols for multiaccess systems, discovering repeated strings in Lempel–Ziv schemes and other data compression algorithms, evaluating string complexity and its randomness, synchronization codes, user searching in wireless communications, and detecting the signatures of an attacker in intrusion detection.

This talk is based on our yet unpublished book *"Analytic Pattern Matching: From DNA to Twitter"*, Cambridge, 2015. After a brief motivation, we review several pattern matching problems (e.g., exact string matching, constrained pattern matching, generalized pattern matching, and subsequence pattern matching), and then we discuss a few applications (e.g., spike trains of neuronal data, Google search, Lempel-Ziv'77 and Lempel-Ziv'78 data compression schemes, and string complexity used in Twitter classification). Finally, we illustrate our approach to solve these problems using tools of analytic combinatorics, which we discuss in some depth.

The basic *pattern matching* problem is to find for a given (or random) pattern w or set of patterns $\mathcal{W}$ and a text X how many times $\mathcal{W}$ occurs in the text X and how long it takes for $\mathcal{W}$ to occur in X for the first time. There are many variations of this basic pattern matching setting which is known as *exact string matching*. In *generalized string matching* certain words from $\mathcal{W}$ are expected to occur in the text while other words are *forbidden* and cannot appear in the text. In some applications, especially in constrained coding and neural data spikes, one puts restrictions on the text (e.g., only text without the patterns 000 and 0000 is permissible), leading to *constrained string matching*. Finally, in the most general case, patterns from the set $\mathcal{W}$ do not need to occur as strings (i.e., consecutively) but rather as subsequences; this leads to *subsequence pattern matching*, also known as *hidden pattern matching*.

The approach we advocate to study these problems is the analysis of pattern matching problems through a formal description by means of regular languages.

This work was supported in part by NSF Science and Technology Center on Science of Information Grant CCF-0939370, NSF Grants DMS-0800568 and CCF-0830140, NSA Grant H98230-11-1-0141.

Basically, such a description of the *contexts* of one, two, or more occurrences of a pattern gives access to the expectation, the variance, and higher moments. A systematic translation into the *generating functions* of a complex variable is available by methods of analytic combinatorics deriving from the original Chomsky–Schützenberger theorem. The structure of the implied generating functions at a pole, an algebraic singularity, or a saddle point provides the necessary asymptotic information. In fact, there is an important phenomenon, that of *asymptotic simplification*, in which the essentials of combinatorial-probabilistic features are reflected by the singular forms of generating functions.

On Multiseed Lossless Filtration

Rasmus Pagh

IT University of Copenhagen, Denmark

Abstract. In *approximate string matching* a string $x \in \Sigma^n$ is given and preprocessed in order to support k-approximate match queries: we seek all substrings of x that differ from a query string q in at most k positions. This problem is motivated for example by biological sequence analysis where approximate occurrences of a sequence q are of interest.

Filtration is an approach to approximate string matching that aims to be efficient when most substrings of x have distance to q considerably larger than k. In these approaches a *seed* is used to extract multisets of subsequences S_x and S_q from x and q, respectively, such that every k-approximate match gives rise to at least one element in $S_q \cap S_x$. (Elements in S_x are annotated with the substring position(s) they correspond to.) Thus, computing $S_q \cap S_x$ (for example using an index data structure for S_x to look up each element of S_q) gives a set of *candidate* positions for k-approximate matches. The filter is *efficient* if it generates few candidates that do not correspond to k-approximate matches. It is known that filtering can be particularly effective in high-entropy strings such as biological sequences.

In this talk we consider so-called *multiseed* methods where several sequences of sets S_x^i, S_q^i, $i = 1, 2, \ldots$ are extracted from x and q, and candidate matches are found in $\bigcup_i S_q^i \cap S_x^i$. Multiseed methods can yield better filtering efficiency, at the expense of a higher candidate generation cost. While some filtration methods allow a nonzero error probability, we focus on *lossless* methods that are guaranteed to report all k-approximate matches. We present a randomized construction of a set of roughly 2^k seeds for which a substring x' having $k + t$ mismatches with q becomes a candidate match $\Theta(2^{-t})$ times in expectation. Since the method is lossless, every x' with at most k mismatches becomes a candidate at least once. This filtering efficiency is better than previous methods with the same number of seeds for $k > 3$. Finally, we use a general transformation to present a new, improved trade-off between the number of seeds and the filtering efficiency.

The research leading to these results has received funding from the European Research Council under the European Union's Seventh Framework Programme (FP7/2007-2013) / ERC grant agreement no. 614331.

Contents

On the Hardness of Optimal Vertex Relabeling and Restricted Vertex Relabeling

Amihood Amir[1,2]($\boxtimes$) and Benny Porat[1]

[1] Department of Computer Science, Bar-Ilan University, 52900 Ramat-gan, Israel
amir@cs.biu.ac.il, bennyporat@gmail.com
[2] Department of Computer Science, Johns Hopkins University,
Baltimore, MD 21218, USA

Abstract. Vertex Relabeling is a variant of the graph relabeling problem. In this problem, the input is a graph and two vertex labelings, and the question is to determine how close are the labelings. The distance measure is the minimum number of label swaps necessary to transform the graph from one labeling to the other, where a swap is the interchange of the labels of two adjacent nodes. We are interested in the complexity of determining the swap distance. The problem has been recently explored for various restricted classes of graphs, but its complexity in general graphs has not been established.

We show that the problem is $\mathcal{NP}$-hard. In addition we consider restricted versions of the problem where a node can only participate in a bounded number of swaps. We show that the problem is $\mathcal{NP}$-hard under these restrictions as well.

1 Introduction

Graph labeling is a well-studied subject in computer science and mathematics and has widespread applications in many other disciplines. Here we explore a variant of graph labeling called the *Vertex Relabeling Problem*. In this problem, the input is a graph and two vertex labelings, and the question is to determine how close are the labelings. The distance measure is the minimum number of label swaps necessary to transform the graph from one labeling to the other, where a swap is the interchange of the labels of two adjacent nodes. We are interested in the complexity of determining the swap distance. Some instances of this problem were explored by Kantaburta [28] and later by Agnarsson et al. [1].

The graph labeling field has a rich and long history. It was first introduced in the late 1960's. In the intervening years dozens of graph labelings techniques and variation have been studied. For a comprehensive survey of the topic see Gallian's excellent dynamic survey [24].

The Vertex Relabeling Problem is not only interesting in its own right but also has applications in several area such as BioInformatics, networks and VLSI.

A. Amir—Partly supported by ISF grant 571/14.

B. Porat—Partly supported by a Bar Ilan University President Fellowship. This work is part of Benny Porat's Ph.D. thesis.

© Springer International Publishing Switzerland 2015
F. Cicalese et al. (Eds.): CPM 2015, LNCS 9133, pp. 1–12, 2015.
DOI: 10.1007/978-3-319-19929-0_1

New application for such work are constantly emerging, and sometimes in unexpected contexts. For instance the Vertex Relabeling Problem can be used to model a *wormhole routing* in processor networks [22]. Perhaps the most famous special case of this problem is the so-called *15-Puzzle* [27]. The 15-Puzzle consists of 15 tiles numbered from 1 to 15 that are placed on a 4×4 board leaving one position empty. The goal is to reposition the tiles of an arbitrary arrangement into increasing order from left-to-right and from top-to-bottom by swapping an adjacent tile with the open hole. In [28] a generalized version of this puzzle called the $(n \times n)$ Puzzle was used to show that a variant of the *Vertex Relabeling Problem With Privileged Labels* is $\mathcal{NP}$-hard. Other well known problems, for example, the Pancake Flipping Problem [23,25,26], can also be viewed as special cases of the vertex relabeling problem.

Another special case of the vertex relabeling problem appears in pattern matching. The *Pattern Matching with Swaps* problem (the *Swap Matching* problem, for short), defined by Muthukrishnan [34], requires finding all swapped occurrences of a pattern of length m in a text of length n. The pattern is said to *match* the text at a given location i if adjacent pattern characters can be swapped, if necessary, so as to make the pattern identical to the substring of the text starting at location i. All the swaps are constrained to be disjoint, i.e., each character is involved in at most one swap. Muthukrishnan asked whether all swap matches of a pattern in a text can be found in time $o(nm)$.

This question led to a flurry of activity. Amir et al. [7] obtained the first non-trivial results on this problem. They showed that the case when the size of the alphabet set Σ exceeds 2 can be reduced to the case when it is exactly 2 with a time overhead of $O(\log^2 \sigma)$, where $\sigma = min\{|\Sigma|, m\}$. (The reduction overhead was reduced to $O(\log \sigma)$ in the journal version [8].) They then showed how to solve the problem for alphabet sets of size 2 in time $O(nm^{1/3} \log m)$, which was the best deterministic time bound known to date. Amir et al. [12] also gave certain special cases for which $O(n\text{polylog}(m))$ time can be obtained. However, these cases are rather restrictive. In a Technical Report [19] Cole and Hariharan provide a randomized algorithm that solves the swap matching problem over a binary alphabet in time $O(n \log n)$. Finally, Amir et al. [9] showed an algorithm for swap matching over a general alphabet whose running time is $O(n \log m \log \sigma)$. The question of measuring the swap distance, i.e. counting the minimum number of swaps, which concerns us in vertex relabeling, was also considered [13]. It was shown that this too can be done in time $O(n \log m \log \sigma)$. In the literature, mismatches were considered in conjunction with other forms of inexactness [16–18,20]. Similarly, swaps were considered in conjunction with other edit operations. It was shown [10] that the swap and mismatch edit distance can be computed in time $O(n\sqrt{m \log m})$. Algorithms for approximate swap and mismatch appeared in [21,31]. This time is the same as the best-known time for computing pattern matching with mismatches alone. It should be noted that the Swap Matching problem also led to the *pattern matching with rearrangements* paradigm [3,5,6,11,29]. Most of the pattern matching work is carried out in the traditional string matching model, where both the pattern and the text are strings (one dimensional arrays). The function matching work [4] considers both pattern and text as two dimensional arrays. However, there are applications where a pattern is sought in non-linear structures. These applications are derived both from searching in hypertext, or comparing non-linear structures, such as folded proteins. Manber and Wu [32] pioneered the study of pattern matching in hypertext and defined a hypertext model for pattern matching. This led to much activity [2, 14, 15, 30, 35].

The vertex relabeling problem is the extension of the swap matching problem to graphs and thus is a natural step in the direction of pattern matching research over non-linear structures.

The Contribution of this Paper: The main contributions of this paper are **conceptual** rather than technical. The value of this paper lies in the fact that we juxtapose two research efforts in two disparate areas and this leads to a crisper understanding of issues in both realms. This cross-fertilization of ideas led to some of the following new insights:

1. We define the vertex relabeling problem as a *distance measure*. To our understanding, this is the first time it is so defined in the vertex relabeling literature. We prove that finding the optimal number of swaps necessary to relabel is $\mathcal{NP}$-hard. We show that in the pattern matching case this problem is polynomially computable.
2. We define the *restricted* version of graph relabeling, which bounds the number of swaps each vertices can participate in. We show that, unlike in the pattern matching case where this problem is solved in almost linear time, in the graph domain the problem is $\mathcal{NP}$-hard. The only exception is the restriction to a single swap, which is linear time computable in the vertex relabeling case as well.

2   Definitions

2.1   Pattern Matching

At the core of the swap matching problem is the constraint that no character is allowed to participate in more then 1 swap. Following the vertex relabeling problem, we generalize this constraint and allow each character to participate in up to k swaps. We call this problem the *k-Swap Matching* problem. Of course the $1-$Swap Matching problem is exactly the well known Swap Matching problem. Formally, the k-swap matching problem is defined as follows:

Definition 1. *Let $S = S[0], \ldots, S[n-1]$ be a string over alphabet Σ. A swap permutation for S is a permutation $\pi : \{0, \ldots, n-1\} \to \{0, \ldots, n-1\}$ such that for all i, $\pi(i) \in \{i-1, i, i+1\}$ (only adjacent characters are swapped).*

For a given series of swap permutations $\pi_1, \pi_2, \ldots, \pi_k$ (we denote $f = \pi_1 \circ \pi_2 \ldots \circ \pi_k$) and string S. We denote $f(S) = S[f(0)], S[f(1)], \ldots, S[f(n-1)]$. We call $f(S)$ a k-swapped version of S. For pattern $P = P[0], \ldots, P[m-1]$ and text $T = T[0], \ldots, T[n-1]$, we say that P has a k-swapped match at location *i if there exists a k-swapped version P' of P such that P' has an exact match with T starting at location i, i.e. $P'[j] = T[i+j]$ for $j = 0, \ldots, m-1$.*

The Pattern Matching with *k*-Swaps Problem *is the following:*

INPUT: *Text string $T = T[0], \ldots, T[n-1]$ and pattern string*
    $P = P[0], \ldots, P[m-1]$ *over alphabet Σ.*
OUTPUT: *All locations i where P has a k-swapped match in T.*

If there is no limit on the number of swaps each location can be involved in, then we call the problem the Unbounded Swap Matching Problem.

2.2   Graphs

Definition 2. *Let $G = (V, E)$ be an undirected, connected, simple graph, let $L_V, L'_V : V \to \Sigma$ be two vertex labelings of the vertices of G. We call the operation of exchanging a pair of labels of adjacent vertices the* label swap operation. *The* Vertex Relabeling up to k Problem *is to transform G from L_V into L'_V using the swap operation with one constraint: Each vertex can participate in at most k label swap operation.*

If there is no limit on the number of swaps each vertex can participate in, then we call the problem the Vertex Relabeling Problem.

Example: In this example, G can be transformed from L_v to L'_v by swapping

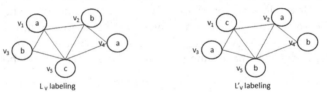

the labels of v_2 and v_4, v_3 and v_5, and then v_3 and v_1. This means that it was a relabeling up to 2, since v_3 participated in two swaps.

3   Unbounded Swap Matching and Vertex Relabeling

Each of these problems can be regarded as either a *decision* problem, where we need to decide if there *exists* a swap matching or a vertex relabeling, or as an *optimization* problem, where we seek the *swap distance*, i.e. the smallest number of swaps necessary for the transformation.

3.1   Decision Problem

For the decision problem is it easy to see that the Unbounded Swap Matching can be solved iff the two string have the same *histogram*, i.e. the same alphabet symbols and the same frequency of occurrence of each symbol. For the Vertex Relabeling problem we also show (the details will appear in the journal version) that there exists a vertex relabeling iff the histograms of the two labelings are equal. This give us a trivial linear time solution for the decision problem for fixed finite label sets, and an almost linear time algorithm for unbounded label sets.

3.2   The Distance Problem

Vertex Relabeling: The distance version of the vertex relabeling problem, is whether one vertex label can be transformed to the other using at most t swap operations. Formally:

Definition 3. *Let $G = (V, E)$ be an undirected, connected, simple graph, let $L_V, L'_V : V \rightarrow \Sigma$ be two vertex labelings of the vertices of G. The* Vertex Rela-*beling with distance t Problem is that of determining whether G's labeling can be transformed from L_V into L'_V using at most t swap operations.*

Note that in Definition 3 the number of label swap operation per vertex is not bound.

To the best of our knowledge it was not known in the vertex relabeling literature whether vertex relabeling with distance t is efficiently computable. Using pattern matching results we can easily prove the following.

Theorem 1. *The Vertex Relabeling Problem with at most t swap, is $\mathcal{NP}$-complete.*

Proof. We reduce the *interchange distance on strings* problem to the vertex relabeling with distance t problem. In the interchange distance on strings problem we are given two string $x = x[0], x[1], \ldots, x[m-1]$ and $y = y[0], y[1], \ldots y[m-1]$. The question is whether x can be transformed to y using at most t interchange operations. When an interchange operation, exchanges the values of two indices i and j, i.e. exchange $x[i]$ with $x[j]$. and x_j. Formally:

Definition 4. *Let $x, y \in \Sigma^m$ be two strings that have the same histogram, and let $s = s_1, \ldots, s_k$ be a sequence of interchanges that transforms x to y, where s_j interchanges elements in positions i_j, i'_j. We define $cost(s) = k$. The interchange distance problem is to compute $d(x, y) = \min\{cost(s)|\ s\ \ transforms\ x\ to\ y\}$.*

Amir et al., prove that the interchange distance problem is $\mathcal{NP}$-complete [11]. We show a reduction from the interchange distance problem to the vertex relabeling problem.

The Reduction: The input of the reduction is an instance of the interchange distance problem, and the output is an instance for vertex relabeling problem.

Given two string $x = x[0], x[1], \ldots, x[m-1]$ and $y = y[0], y[1], \ldots, y[m-1]$ construct $G = (V, E)$, L_v and L'_v in the following manner. Build a vertex for each index of the string, so $V = \{v_0, v_1, \ldots, v_{m-1}\}$,. An interchange operation exchanges the value of two indices. To simulate an interchange by a swap we need to construct graph edges that allow a swap between any two nodes. This is accomplished by constructing a complete graph on the m vertices, and labeling it by x and by y. In other words, $L_V(v_i) = x[i]$ and $L'_V(v_i) = y[i]$ for $0 \le i \le m-1$.

Claim. x can be transformed to y using at most t interchange operations iff L_V can be transformed to L'_V using at most t *swap* operations.

Proof. Assume that there exist a sequence $s = s_1, s_2, \ldots s_t$ of interchange operations that transform x to y. We build a sequence $s' = s'_1, s'_2, \ldots s'_t$ of swap operation that transforms L_v into L'_v. s_j interchanges the contents of positions i_j, i'_j. Then the swap s'_j swaps the labels of $v[i_j]$ and $v[i'_j]$.

We show that $s' = s'_1, s'_2, \ldots s'_t$ transforms L_v to L'_v

Assume, to the contrary, that there exists a $v[i] \in V$ such that after applying the sequence of swap operations, the label of $v[i]$ is σ, and $\sigma \neq L'_V(v[i])$. let $s'^\sigma = s'_{i_1}, s'_{i_2}, \ldots s'_{i_k}$ be the sequence of swap operations that moves the label σ. This sequence of swaps takes the σ label form some starting vertex $v[j]$ and moves it to the vertex $v[i]$. Consider, now, the corresponding sequence of interchanges operation $s^\sigma = s_{i_1}, s_{i_2}, \ldots, s_{i_k}$. $L(v_j) = x[j] = \sigma$, hence, this sequence of interchanges takes σ, and moves it until index i. Conclude that after the sequence of interchanges, σ is in the ith position. But $\sigma \neq L'_V(v_i) = y[i]$.   □

Swap Matching Distance: The following observation is key to the efficient swap matching algorithm.

Observation 1. *Two equal adjacent characters need not be swapped.*

The observation allows us to treat each occurrence of the same symbol as a different character, all we need to do is maintain the order. Therefore, if a string has an alphabet letter σ that appears multiple times, we can convert each occurrence of σ to σ_i where i is a running counter. Formally:

Definition 5. *Given a string $s = s[0], s[1], \ldots s[m-1]$ we define the unique version of s and denote it by s_u, to be $s_u[i] = s[i]_j$ for $0 \leq i \leq m-1$ where $j = |\{k \mid s_k = s_i \text{ and } 1 \leq k \leq i\}|$.*

Example: The string $s = a\ b\ a\ c\ c\ a$ we will convert to $s_u = a_1\ b_1\ a_2, c_1\ c_2\ a_3$. Given two strings as an input for the swap matching distance problem, we can convert them easily to their unique versions, where all their characters are different. The following theorem, a version of which appears as early as 1882 [33] is key to our algorithm.

Theorem 2. *Given two string $s = s[0], s[1], \ldots, s[m-1]$ and $t = t[0], t[1], \ldots, t[m-1]$ where $s[i] \neq s[j]$ and $t[i] \neq t[j]$ for all $0 \leq i, j \leq m-1$, the minimum number of swaps needed in order to transform s to t is the number of inversions between s and t, where an inversion is a pair of indices (i, j) such that $i > j$ and $s[i]$ appears before $s[j]$ in t.*

Proof
Minimality: Assume, to the contrary, that there is a sequences of swaps of length k that transforms s to t where k is smaller than the number of inversions. This means that there must exist two character $s[i]$ and $s[j]$ such that $i > j$ and $s[i]$ appears before $s[j]$ in t that don't swap. Hence, at the end of the sequence of swaps, $s[i]$ appears after $s[j]$, but this not the case on the string t. Contradiction.

Correctness: For all $0 \leq i \leq m-1$, we denote the position of $s[i]$ after the sequence of swaps by i'.

Observe that after the sequence of swaps, for all $0 \le i \le m - 1$, there is no $0 \le j \le m - 1$ such that $i' > j'$ and $s[i]$ appears before $s[j]$ in t. The observation says that for each character $s[i]$, all the characters $s[j]$ that appear after $s[i]$ in t after the sequence of swaps, actually occur after s_i. Applying the observation inductively from the first character of t until the last, gives us that for all $0 \le i \le m - 1$, there is no $0 \le j \le m - 1$ such that $i' < j'$ and $s[i]$ appears after $s[j]$ in t.

This sequence of swaps transforms s to t. □

Now we are ready for the algorithm:

Algorithm – Swap Matching Distance(s, t)

1. **Initialization:** Given two strings s and t build the corresponding unique strings $s_u = s[0], s[1], ..., s[m - 1]$ and $t_u = t[0], t[1], ...t[m - 1]$.
2. $C = 0$.
3. For $i = 0$ to $m - 1$
   - $C+ = |\{s[j] \mid 0 \le j < i \text{ and } s[i] \text{ appears before } s[j] \text{ in } t\}|$
4. C is the minimum number of swaps needed to transform s to t.

end Algorithm

Time: Using such data structures as, AVL trees, for example, this algorithm can be implemented in time $O(m \log m)$.

4 (1)-Swap Matching and Vertex Relabeling up to 1

4.1 Swap Matching

As previously mentioned, Swap Matching is a well studied problem. The best know result solves it in time $O(n \log m \log \Sigma)$, for a length m pattern "sliding" across a length n text. It is easy to see that when $n = m$ it can be trivially solved in linear time.

4.2 Vertex Relabeling up to 1

We show a two way reduction between Vertex Relabeling and the Graph Perfect Matching problem.

Definition 6. *Given a graph $G = (V, E)$, a matching M in G is a set of pairwise non-adjacent edges; that is, no two edges share a common vertex.*

Definition 7. *A perfect matching is a matching which matches all vertices of the graph. That is, every vertex of the graph is incident to exactly one edge of the matching.*

The Perfect Matching on Bipartite Graph problem is, given a bipartite graph to find if their is a perfect matching in that graph.

We will show two reductions:

- Given an algorithm for the Perfect Matching on bipartite graphs problem, we can use that algorithm to solve the Vertex Relabeling problem.
- Given an algorithm for the Vertex Relabeling up to 1 problem we can use that algorithm to solve the Perfect Matching on bipartite graphs problem.

4.3   Perfect Matching On Bipartite Graphs ⇒ Vertex Relabeling up to 1

Given an undirected, connected, simple graph $G = (V, E)$, with two vertex labelings L_V and L'_V of the vertices of G. Construct $G' = (V' = (X, Y), E')$ in the following way.

For each pair of labels a, b, and for each $u, v \in V$ such that $(L_V(u) = L'_V(v) = a) \wedge (L'_V(u) = L_V(v) = b)$. We put u in X and v in Y. Also we build an edge between u and v.

Example:

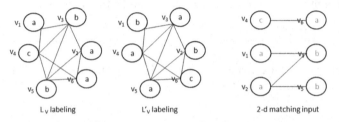

L$_V$ labeling   L$'_V$ labeling   2-d matching input

Theorem 3. *G' has a Perfect Matching iff G has a Vertex Relabeling up to 1 from L_V to L'_V.*

Proof. Immediate form the definitions of the problems.   □

Construction Time: Although the construction considers all pairs of labels, it still only takes linear time because the number of pairs is bounded by $|V|$.

4.4   Vertex Relabeling up to 1 ⇒ Perfect Matching on Bipartite Graph

Given a bipartite graph $G = (V = (X, Y), E)$. We construct $G' = (V', E')$ and two vertex labeling L_V and L'_V in the following way. $V' = X \cup Y$ and $E' = E$ $L_V(v) = 0$ for $v \in X$ and $L_V(v) = 1$ for $v \in Y$ $L'_V(v) = 1$ for $v \in X$ and $L'_V(v) = 0$ for $v \in Y$. See Fig. 1 for an example.

Theorem 4. *G has a Perfect Matching iff G' has a Vertex Relabeling up to 1 from L_V to L'_V.*

Proof. Immediate from the definitions of the problems.   □

Algorithm Time: The best known algorithm for Perfect Matching in bipartite graphs run in time $O(\sqrt{V}E)$ or in $O(V^{2.376})$. Hence, due to the two side reduction, those solution also apply to the Vertex relabeling up to 1 problem.

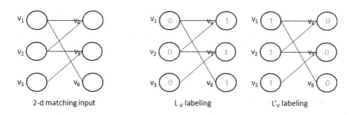

Fig. 1. Example of reduction from Perfect Matching to Vertex Relabeling.

5   2-Swap Matching and Vertex Relabeling up to 2

The 2-Swap matching problem can be easily solved in polynomial time. However the Vertex Relabeling up to 2 is $\mathcal{NP}$-complete.. We present a reduction from $3D-$Matching.

Definition 8. *Let X, Y, and Z be finite, disjoint sets, and let T be a subset of $X \times Y \times Z$. That is, T consists of triples (x, y, z) such that $x \in X$, $y \in Y$, and $z \in Z$. $M \subseteq T$ is a 3-dimensional matching if the following holds: for any two distinct triples $(x_1, y_1, z_1) \in M$ and $(x_2, y_2, z_2) \in M$, we have $x_1 \neq x_2$, $y_1 \neq y_2$, and $_z1 \neq z_2$. The 3D$-$matching problem is that of deciding if there exist $M \subseteq T$ such that $|M| = k$.*

5.1   The Reduction

We show that a polynomial-time algorithm that solves the Vertex Relabeling up to 2 problem can be used to solve the $3D-$Matching problem.

Let X, Y, and Z be finite, disjoint sets, $|X| = |Y| = |Z| = m$, and let T be a subset of $X \times Y \times Z$. Construct $G = (V, E)$ in the following way:
$V = X \cup Y \cup Z$ and $E = \{(x, y), (y, z) | \forall (x, y, z) \in T\}$.

Define L_V and L'_V in the following way:

$L_V \Rightarrow L_V(x) = 0$ for $x \in X$, $L_V(y) = 0$ for $y \in Y$ and $L_V(z) = 1$ for $z \in Z$.
$L'_V \Rightarrow L'_V(x) = 1$ for $x \in X$, $L'_V(y) = 0$ for $y \in Y$ and $L'_V(z) = 0$ for $z \in Z$.

Example:

Theorem 5. *Given X, Y, Z and T, there exist $3D-$Matching of size m iff graph G can be vertex relabeled up to 2 for labels L_V and L'_V.*

Proof. $\Leftarrow$: Assume that there exist a sequence of swap operations $S = s_1, s_2, \ldots$ that relabels L_V to L'_V. We will build a $M \subset T$ such that M will be a perfect Matching for X, Y, Z. $|Z| = |X|$, $L_V(x) = L'_V(z) = 1$ and $L_V(z) = L'_V(x) = 0$,

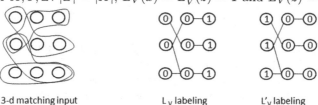

hence for all $z_i \in Z$ there exist $x_j \in X$ such that z_i got the label of x_j. For any $x_j \in X$ and $z_i \in Z$ $(x_j, z_i) \notin E$. Hence there exist some y_k such that x_j swaps with y_k and y_k swaps with z_i. We put each such triplet in M.

Lemma 1. $M \subset T$ *is a perfect Matching for* X, Y, Z.

Proof. From the building process of M we know that for any $x_j \in X$ and $z_i \in Z$ M covers them exactly once. We need to prove this also for all $y_k \in Y$. Assume that there exist $y_k \in Y$ that appear in two triplets $m_1 = (x_1, y_k, z_1)$ and $m_2 = (x_2, y_k, z_2)$ such that $m_1, m_2 \in M$. The triplets construction means that y_k swaps 4 times. A contradiction. Assume that there exist $y_k \in Y$ such that for any (x, y, z) inM $y_k \neq y$. $|M| = |Y| = n$ and we prove that each $y_k \in Y$ appears at most in 1 triplet in M. This is a contradiction to the size of M. M is a perfect Matching.

$\Rightarrow$: Assume that there exist $M \subset T$ such that M is a perfect matching. Construct a sequence s of swap operations, such that S relabels G from L_V to L'_V. For any triplet $(x, y, z) \in M$ we execute two swap operations in S, $(z, y), (y, x)$. $\square$

Lemma 2.

1. *There are now vertices that participate in more then 2 swap operation in S.*
2. *S relabels from L_V to L'_V.*

Proof

1. M is a perfect matching. Hence for any $x \in X$ x appears in exactly 1 triplet in M. From that triplet x has exactly 1 swap operation. The same applies to all $z \in Z$. For all $y \in Y$. y appears in exactly one triplet in M and from that triplet it has exactly 2 swap operations. Overall there is no vertex that participates in more then two swap operations.
2. Each triplet $(x, y, z) \in M$ produces two swap operations, $(z, y), (y, x)$. Starting with the labeling L_V, will finish with label 0 to z and y and label 1 for x which is exactly L'_V. $\square$

6   Conclusions and Open Problems

To fully understand the behavior of swaps, many more issues need to be explored. Unlike graphs where graph sub-isomorphism is already $\mathcal{NP}$-hard, in pattern matching one usually considers a small pattern of length m sliding across a large text of length n. We have pointed out that the k-swap matching on strings of equal length can be done in linear time. Can we use previous knowledge and find, in a text of length n, all locations with a k-swap match in time shorter than $o(nm)$?

In the vertex relabeling arena, $\mathcal{NP}$-hardness is only the beginning of the story. The next step is providing approximation algorithms for these swap-distance problems. Another intriguing question is considering these graph problems as on-line problems, where the graphs are input together a vertex at a time.

References

1. Agnarsson, G., Greenlaw, R., Kantabutra, S.: On the graph relabeling problem. Thai J. Math. **8**(1), 21–42 (2010)
2. Akutsu, T.: A linear time pattern matching algorithm between a string and a tree. In: Proceedings of 4th Symposium on Combinatorial Pattern Matching (CPM), pp. 1–10, Padova (1993)
3. Amir, A., Aumann, Y., Indyk, P., Levy, A., Porat, E.: Efficient computations of ℓ_1 and ℓ_∞ rearrangement distances. In: Ziviani, N., Baeza-Yates, R. (eds.) SPIRE 2007. LNCS, vol. 4726, pp. 39–49. Springer, Heidelberg (2007)
4. Amir, A., Aumann, A., Lewenstein, M., Porat, E.: Function matching. SIAM J. Comput. **35**(5), 1007–1022 (2006)
5. Amir, A., Aumann, Y., Benson, G., Levy, A., Lipsky, O., Porat, E., Skiena, S., Vishne, U.: Pattern matching with address errors: rearrangement distances. In: Proceedings of 17th ACM-SIAM Symposium on Discrete Algorithms (SODA), pp. 1221–1229 (2006)
6. Amir, A., Aumann, Y., Kapah, O., Levy, A., Porat, E.: Approximate string matching with address bit errors. In: Ferragina, P., Landau, G.M. (eds.) CPM 2008. LNCS, vol. 5029, pp. 118–129. Springer, Heidelberg (2008)
7. Amir, A., Aumann, Y., Landau, G., Lewenstein, M., Lewenstein, N.: Pattern matching with swaps. In: Proceedings of 38th IEEE FOCS, pp. 144–153 (1997)
8. Amir, A., Aumann, Y., Landau, G., Lewenstein, M., Lewenstein, N.: Pattern matching with swaps. J. Algorithms **37**(2), 247–266 (2000). (Preliminary version appeared at FOCS 1997)
9. Amir, A., Cole, R., Hariharan, R., Lewenstein, M., Porat, E.: Overlap matching. Inf. Comput. **181**(1), 57–74 (2003)
10. Amir, A., Eisenberg, E., Porat, E.: Swap and mismatch edit distance. Algorithmica **45**(1), 109–120 (2006)
11. Amir, A., Hartman, T., Kapah, O., Levy, A., Porat, E.: On the cost of interchange rearrangement in strings. SIAM J. Comp. **39**(4), 1444–1461 (2009)
12. Amir, A., Landau, G.M., Lewenstein, M., Lewenstein, N.: Efficient special cases of pattern matching with swaps. Inf. Process. Lett. **68**(3), 125–132 (1998)
13. Amir, A., Lewenstein, M., Porat, E.: Approximate swapped matching. Inf. Process. Lett. **83**(1), 33–39 (2002)
14. Amir, A., Lewenstein, N., Lewenstein, M.: Pattern matching in hypertext. J. Algorithms **35**, 82–99 (2000)
15. Amir, A., Navarro, G.: Parameterized matching of non-linear structures. Inf. Process. Lett. **109**(15), 864–867 (2009)
16. Clifford, R., Efremenko, K., Porat, E., Rothschild, A.: From coding theory to efficient pattern matching. In: Proceedings of 20th Annual ACM-SIAM Symposium on Discrete Algorithms (SODA), pp. 778–784 (2009)
17. Clifford, R., Efremenko, K., Porat, E., Rothschild, A.: k-Mismatch with don't cares. In: Arge, L., Hoffmann, M., Welzl, E. (eds.) ESA 2007. LNCS, vol. 4698, pp. 151–162. Springer, Heidelberg (2007)
18. Clifford, R., Porat, E.: A Filtering algorithm for k-mismatch with don't cares. In: Ziviani, N., Baeza-Yates, R. (eds.) SPIRE 2007. LNCS, vol. 4726, pp. 130–136. Springer, Heidelberg (2007)
19. Cole, R., Hariharan, R.: Randomized swap matching in $o(m \log m \log |\sigma|)$ time. Technical report TR1999-789, New York University, Courant Institute, September 1999

20. Cole, R., Hariharan, R., Lewenstein, M., Porat, E.: A faster implementation of the goemans-williamson clustering algorithm. In: Proceedings of 12th ACM-SIAM Symposium on Discrete Algorithms (SODA), pp. 17–25 (2001)
21. Dombb, Y., Lipsky, O., Porat, B., Porat, E., Tsur, A.: Approximate swap and mismatch edit distance. In: Ziviani, N., Baeza-Yates, R. (eds.) SPIRE 2007. LNCS, vol. 4726, pp. 149–163. Springer, Heidelberg (2007)
22. Duato, J.: A theory of fault-tolerant routing in wormhole networks. IEEE Trans. Parallel Distrib. Syst. **8**(8), 790–802 (1997)
23. Dweighter, H.: Problem e2569. Am. Math. Mon. **82**, 1010 (1975)
24. Gallian, J.A.: A dynamic survey of graph labeling. Electronic Journal of Combinatorics **18**, 1–219 (2011)
25. Gates, W.H., Papadimittiou, C.H.: Bounds for sorting by prefix reversal. Discrete Math. **27**, 47–57 (1979)
26. Heydari, M.H., Sudborough, I.H.: On the diameter of the pancake network. J. Algorithms **25**(1), 67–94 (1997)
27. Johnson, W.W., Woolsay, W.E.: Notes on the '15 puzzle'. Am. J. Math. **2**(4), 397–404 (1879)
28. Kantabutra, S.: The complexity of label relocation problems on graphs. In: Proceedings of 8th Asian Symposium of Computer Mathematics, National University of Singapore (2007)
29. Kapah, O., Landau, G.M., Levy, A., Oz, N.: Interchange rearrangement: the element-cost model. In: Amir, A., Turpin, A., Moffat, A. (eds.) SPIRE 2008. LNCS, vol. 5280, pp. 224–235. Springer, Heidelberg (2008)
30. Kim, D.K., Park, K.: String matching in hypertext. In: Proceedings of 6th Symposium on Combinatorial Pattern Matching (CPM 1995) (1995)
31. Lipsky, O., Porat, B., Porat, E., Shalom, B.R., Tzur, A.: Approximate string matching with swap and mismatch. In: Tokuyama, T. (ed.) ISAAC 2007. LNCS, vol. 4835, pp. 869–880. Springer, Heidelberg (2007)
32. Manber, U., Wu, S.: Approximate string matching with arbitrary cost for text and hypertext. In: Proceedings International Workshop on Structural and Syntactic Pattern Recognition, pp. 22–33 (1992)
33. Muir, T.: A treatise of the thoery of determinants with graduated sets of exercises. Macmillan and Co., London (1882)
34. Muthukrishnan, S.: New results and open problems related to non-standard stringology. In: Galil, Zvi, Ukkonen, Esko (eds.) CPM 1995. LNCS, vol. 937, pp. 298–317. Springer, Heidelberg (1995)
35. Navarro, G.: Improved approximate pattern matching on hypertext. Theoret. Comput. Sci. **237**, 455–463 (2000)

A Framework for Space-Efficient String Kernels

Djamal Belazzougui[1,2] and Fabio Cunial[1,2($\boxtimes$)]

[1] Department of Computer Science, University of Helsinki, Helsinki, Finland
[2] Helsinki Institute for Information Technology, Helsinki, Finland
fabio.cunial@cs.helsinki.fi

Abstract. String kernels are typically used to compare genome-scale sequences whose length makes alignment impractical, yet their computation is based on data structures that are either space-inefficient, or incur large slowdowns. We show that a number of exact string kernels, like the k-mer kernel, the substrings kernels, a number of length-weighted kernels, the minimal absent words kernel, and kernels with Markovian corrections, can all be computed in $O(nd)$ time and in $o(n)$ bits of space in addition to the input, using just a `rangeDistinct` data structure on the Burrows-Wheeler transform of the input strings that takes $O(d)$ time per element in its output. The same bounds hold for a number of measures of compositional complexity based on multiple values of k, like the k-mer profile and the k-th order empirical entropy, and for calibrating the value of k using the data.

1 Introduction

Given two strings T^1 and T^2, a *kernel* is a function that simultaneously converts T^1 and T^2 into vectors $\mathbf{T}^1$ and $\mathbf{T}^2$ in $\mathbb{R}^n$ for some $n > 0$, and computes a similarity or a distance measure between $\mathbf{T}^1$ and $\mathbf{T}^2$, *without building and storing* $\mathbf{T}^i$ *explicitly* [14]. Kernels are often the method of choice for comparing extremely long strings, like genomes, read sets, and metagenomic samples, whose size makes alignment infeasible, yet their computation is typically based on space-inefficient data structures, like (truncated) suffix trees, or on space-efficient data structures with $O(\log^\epsilon n)$ slowdowns, like compressed suffix trees (see e.g. [1,9] and references therein). The (possibly infinite) dimensions of $\mathbf{T}^i$ are, for example, all strings of a specific family on the alphabet of T^1 and T^2, and the value assigned to vector $\mathbf{T}^i$ along dimension W corresponds to the number of occurrences of string W in T^i, often rescaled and corrected in domain-specific ways. $\mathbf{T}^i$ is often called *composition vector*, and a large number of its components can be zero in practice. In this paper we focus on space- and time-efficient algorithms for computing the *cosine of the angle between two composition vectors* $\mathbf{T}^1$ and $\mathbf{T}^2$, i.e. on computing the kernel $\kappa(\mathbf{T}^1, \mathbf{T}^2) = N/\sqrt{D^1 D^2} \in [-1..1]$, where $N = \sum_W \mathbf{T}^1[W]\mathbf{T}^2[W]$ and $D^i = \sum_W \mathbf{T}^i[W]^2$. This measure of similarity can be converted into a distance $d(\mathbf{T}^1, \mathbf{T}^2) = (1 - \kappa(\mathbf{T}^1, \mathbf{T}^2))/2 \in [0..1]$, and the

---

This work was partially supported by Academy of Finland under grant 284598 (Center of Excellence in Cancer Genetics Research).

F. Cicalese et al. (Eds.): CPM 2015, LNCS 9133, pp. 13–25, 2015.
DOI: 10.1007/978-3-319-19929-0_2

algorithms we describe can be applied to compute norms of vector $\mathbf{T}^1 - \mathbf{T}^2$, like the p-norm and the infinity norm. When $\mathbf{T}^1$ and $\mathbf{T}^2$ are bitvectors, we are more interested in interpreting them as sets and in computing the Jaccard distance $J(\mathbf{T}^1, \mathbf{T}^2) = ||\mathbf{T}^1 \wedge \mathbf{T}^2||/||\mathbf{T}^1 \vee \mathbf{T}^2|| = ||\mathbf{T}^1 \wedge \mathbf{T}^2||/(||\mathbf{T}^1|| + ||\mathbf{T}^2|| - ||\mathbf{T}^1 \wedge \mathbf{T}^2||)$, where $\wedge$ and $\vee$ are the bitwise AND and OR operators, and where $||\cdot||$ measures the number of ones in a bitvector.

Given a data structure that supports `rangeDistinct` queries on the Burrows-Wheeler transform of each string in input, we show that a number of popular string kernels, like the k-mer kernel, the substrings kernels, a number of length-weighted kernels, the minimal absent words kernel, and kernels with Markovian corrections, can all be computed in $O(nd)$ time and in $o(n)$ bits of space in addition to the input, *all in a single pass over the BWTs of the input strings*, where d is the time taken by the `rangeDistinct` query per element in its output. The same bounds hold for computing a number of measures of compositional complexity *for multiple values of k at the same time*, like the k-mer profile and the k-th order empirical entropy, and for choosing the value of k used in k-mer kernels from the data. All these algorithms become $O(n)$ using the `rangeDistinct` data structure described in [4], and concatenating this setup to the BWT construction algorithm described in [3], we can compute all such kernels and complexity measures *from the input strings* in randomized $O(n)$ time and in $O(n \log \sigma)$ bits of space in addition to the input. Finally, we show that measures of expectation based on Markov models are related to the left and right extensions of maximal repeats.

2   Preliminaries

2.1   Strings

Let $\Sigma = [1..\sigma]$ be an integer alphabet, let $\# = 0$, $\#_1 = -1$ and $\#_2 = -2$ be distinct separators not in Σ, and let $T = [1..\sigma]^{n-1}\#$ be a string. We assume $\sigma \in o(\sqrt{n}/\log n)$ throughout the paper. A *k-mer* is any string $W \in [1..\sigma]$ of length $k > 0$. We denote by $f_T(W)$ the number of (possibly overlapping) occurrences of a string W in the circular version of T, and we use the shorthand $p_T(W) = f_T(W)/(n - |W|)$ to denote an approximation of the *empirical probability* of observing W in T, assuming that all positions of T except the last $|W|$ ones are equally probable starting positions for W. A *repeat* W is a string that satisfies $f_T(W) > 1$. We denote by $\Sigma_T^\ell(W)$ the set of characters $\{a \in [0..\sigma] : f_T(aW) > 0\}$ and by $\Sigma_T^r(W)$ the set of characters $\{b \in [0..\sigma] : f_T(Wb) > 0\}$. A repeat W is *right-maximal* (respectively, *left-maximal*) iff $|\Sigma_T^r(W)| > 1$ (respectively, iff $|\Sigma_T^\ell(W)| > 1$). It is well known that T can have at most $n - 1$ right-maximal substrings and at most $n - 1$ left-maximal substrings. A *maximal repeat* of T is a repeat that is both left- and right-maximal.

For reasons of space we assume the reader to be familiar with the notion of *suffix tree* ST_T of a string T, and with the notion of *generalized suffix tree* of two strings, which we do not define here. We denote by $\ell(v)$ the string label of a node v in a suffix tree. It is well known that a substring W of T is right-maximal iff $W = \ell(v)$ for some internal node v of ST_T. We assume the reader to be familiar with the notion of *suffix link* connecting a node v with $\ell(v) = aW$ for some $a \in [0..\sigma]$ to a node w with $\ell(w) = W$: we say that $w = \mathtt{suffixLink}(v)$ in this case. Here we just recall that suffix links and internal nodes of ST_T form a tree, called the *suffix-link tree* of T and denoted by SLT_T, and that inverting the direction of all suffix links yields the so-called *explicit Weiner links*. Given an internal node v and a symbol $a \in [0..\sigma]$, it might happen that string $a\ell(v)$ does occur in T, but that it is not right-maximal, i.e. it is not the label of any internal node of ST_T: all such left extensions of internal nodes that end in the middle of an edge are called *implicit Weiner links*. An internal node v of ST_T can have more than one outgoing Weiner link, and all such Weiner links have distinct labels: in this case, $\ell(v)$ is a maximal repeat. It is known that the number of suffix links (or, equivalently, of explicit Weiner links) is upper-bounded by $2n-2$, and that the number of implicit Weiner links can be upper-bounded by $2n - 2$ as well.

2.2   Enumerating Right-Maximal Substrings and Maximal Repeats

For reasons of space we assume the reader to be familiar with the notion and uses of the Burrows-Wheeler transform of T, including the C array, the $\mathtt{rank}$ function, and backward searching. In this paper we use BWT_T to denote the BWT of T, we use $\mathtt{range}(W) = [\mathtt{sp}(W)..\mathtt{ep}(W)]$ to denote the lexicographic interval of a string W in a BWT that is implicit from the context, and we use $\Sigma_{i,j}$ to denote the set of distinct characters that occur inside interval $[i..j]$ of a string that is implicit from the context. We also denote by $\mathtt{rangeDistinct}(i,j)$ the function that returns the set of tuples $\{(c, \mathtt{rank}(c, p_c), \mathtt{rank}(c, q_c)) : c \in \Sigma_{i,j}\}$, *in any order*, where p_c and q_c are the first and the last occurrence of c inside interval $[i..j]$, respectively. Here we focus on a specific application of BWT_T: enumerating all the right-maximal substrings of T, or equivalently all the internal nodes of ST_T. In particular, we use the algorithm described in [3] (Sect. 4.1), which we sketch here for completeness.

Given a substring W of T, let $b_1 < b_2 < \cdots < b_k$ be the sorted sequence of all the distinct characters in $\Sigma_T^r(W)$, and let $a_1, a_2, \ldots, a_h$ be the list of all the characters in $\Sigma_T^\ell(W)$, not necessarily sorted. Assume that we represent a substring W of T as a pair $\mathtt{repr}(W) = (\mathtt{chars}[1..k], \mathtt{first}[1..k+1])$, where $\mathtt{chars}[i] = b_i$, $\mathtt{range}(Wb_i) = [\mathtt{first}[i]..\mathtt{first}[i+1]-1]$ for $i \in [1..k]$, and $\mathtt{range}()$ refers to BWT_T. Note that $\mathtt{range}(W) = [\mathtt{first}[1]..\mathtt{first}[k+1]-1]$, since it coincides with the concatenation of the intervals of the right extensions of W in lexicographic order. If W is not right-maximal, array $\mathtt{chars}$ in $\mathtt{repr}(W)$ has length one. Given a data structure that supports $\mathtt{rangeDistinct}$ queries on BWT_T, and given the C array of T, there is an algorithm that converts $\mathtt{repr}(W)$ into the sequence $a_1, \ldots, a_h$ and into the corresponding sequence $\mathtt{repr}(a_1 W), \ldots, \mathtt{repr}(a_h W)$, in $O(de)$ time and $O(\sigma^2 \log n)$ bits of space in addition to the input and the output [3], where d is the time taken by the $\mathtt{rangeDistinct}$ operation per element in its output, and e is the number of distinct strings $a_i W b_j$ that occur in the circular version of T, where $i \in [1..h]$ and $j \in [1..k]$. We encapsulate this algorithm into a function that we call `extendLeft`.

If $a_i W$ is right-maximal, i.e. if array `chars` in `repr`$(a_i W)$ has length greater than one, we push pair $(\text{repr}(a_i W), |W|+1)$ onto a stack S. In the next iteration we pop the representation of a string from the stack and we repeat the process, until the stack itself becomes empty. This process is equivalent to following all the explicit Weiner links from the node v of ST_T with $\ell(v) = W$, not necessarily in lexicographic order. Thus, running the algorithm from a stack initialized with $\text{repr}(\varepsilon)$ is equivalent to performing a preorder depth-first traversal of the suffix-link tree of T (with children explored in arbitrary order), which guarantees to enumerate all the right-maximal substrings of T. Every operation performed by the algorithm can be charged to a distinct node or Weiner link of ST_T, thus the algorithm runs in $O(nd)$ time. The depth of the stack is $O(\log n)$ rather than $O(n)$, since at every iteration we push the pair $(\text{repr}(a_i W), |a_i W|)$ with largest $\text{range}(a_i W)$ first. Every suffix-link tree level in the stack contains at most σ pairs, and each pair takes at most $\sigma \log n$ bits of space, thus the total space used by the stack is $O(\sigma^2 \log^2 n)$ bits. The following theorem follows from our assumption that $\sigma \in o(\sqrt{n}/\log n)$:

Theorem 1 ([3]). *Let $T \in [1..\sigma]^{n-1}\#$ be a string. Given a data structure that supports `rangeDistinct` queries on BWT_T, we can enumerate all the right-maximal substrings W of T, and for each of them we can return $|W|$, $\text{repr}(W)$, the sequence $a_1, a_2, \ldots, a_h$ of all characters in $\Sigma_T^\ell(W)$ (not necessarily sorted), and the sequence $\text{repr}(a_1 W), \ldots, \text{repr}(a_h W)$, in $O(nd)$ time and in $o(n)$ bits of space in addition to the input and the output, where d is the time taken by the `rangeDistinct` operation per element in its output.*

Theorem 1 does not specify the order in which the right-maximal substrings must be enumerated, nor the order in which the left extensions of a right-maximal substring must be returned. The algorithm we just described can be adapted to return all the maximal repeats of T, with the same bounds, by outputting a right-maximal string W iff $|\text{rangeDistinct}(\text{sp}(W), \text{ep}(W))| > 1$. A version of the same algorithm can also enumerate all the internal nodes of the *generalized suffix tree* of two string T^1 and T^2, using BWT_{T^1} and BWT_{T^2}: in this case, a string W is represented as a quadruple $\text{repr}'(W) = (\text{chars}_1[1..k_1], \text{first}_1[1..k_1+1], \text{chars}_2[1..k_2], \text{first}_2[1..k_2+1])$, and we assume that $\text{first}_i[1] = 0$ iff W does not occur in T^i. We call `extendLeft`$'$ the function that maps $\text{repr}'(W)$ to the list of its left extensions $\text{repr}'(a_i W)$.

Theorem 2 ([3]). *Let $T^1 \in [1..\sigma]^{n_1-1}\#_1$ and $T^2 \in [1..\sigma]^{n_2-1}\#_2$ be two strings. Given two data structures that support `rangeDistinct` queries on BWT_{T^1} and on BWT_{T^2}, respectively, we can enumerate all the right-maximal substrings W of $T = T^1 T^2$, and for each of them we can return $|W|$, $\text{repr}'(W)$, the sequence $a_1, a_2, \ldots, a_h$ of all characters in $\Sigma_{T^1 T^2}^\ell(W)$ (not necessarily sorted), and the sequence $\text{repr}'(a_1 W), \ldots, \text{repr}'(a_h W)$, in $O(nd)$ time and in $o(n)$ bits of space in addition to the input and the output, where $n = n_1 + n_2$ and d is the time taken by the `rangeDistinct` operation per element in its output.*

For reasons of space, we assume throughout the paper that d is the time per element in the output of a `rangeDistinct` data structure that is implicit from the context. We also replace T^i by i in subscripts, or we waive subscripts completely whenever they are clear from the context.

3   Kernels and Complexity Measures on k-mers

Given a string $T \in [1..\sigma]^{n-1}\#$ and a length $k > 0$, let vector $\mathbf{T}_k[1..\sigma^k]$ be such that $\mathbf{T}_k[W] = f_T(W)$ for every $W \in [1..\sigma]^k$. The k-mer complexity $\mathsf{C}_k(T)$ of string T is the number of nonzero components of $\mathbf{T}_k$. The k-mer kernel of two strings T^1 and T^2 is $\kappa(\mathbf{T}_k^1, \mathbf{T}_k^2)$. Recall that Theorems 1 and 2 enumerate all nodes of a suffix tree in no specific order. In this section we describe algorithms to compute $\mathsf{C}_k(T)$ and $\kappa(\mathbf{T}_k^1, \mathbf{T}_k^2)$ in a way that does not depend on the order in which the nodes of a suffix tree are enumerated: we can thus implement such algorithms on top of Theorems 1 and 2. The main idea behind our approach is a telescoping strategy that works by adding and subtracting terms in a sum, as described below:

Theorem 3. *Let $T \in [1..\sigma]^{n-1}\#$ be a string. Given an integer k and a data structure that supports* `rangeDistinct` *queries on* BWT_T*, we can compute $\mathsf{C}_k(T)$ in $O(nd)$ time and in $o(n)$ bits of space in addition to the input.*

Proof. A k-mer of T can either be the label of a node of ST_T, or it could end in the middle of an edge (u, v) of ST. In the latter case, we assume that the k-mer is represented by its locus v, which might be a leaf. Let $\mathsf{C}_k(T)$ be initialized to $n - k$, i.e. to the number of leaves that correspond to suffixes of T of length at least $k + 1$. We enumerate the internal nodes of ST using Theorem 1, and every time we enumerate a node v we proceed as follows: if $|\ell(v)| < k$ we leave $\mathsf{C}_k(T)$ unaltered, otherwise we increment $\mathsf{C}_k(T)$ by one and we decrement $\mathsf{C}_k(T)$ by the number of children of v in ST, which is the length of array `chars` in $\mathtt{repr}(\ell(v))$. In this way, every internal node v of ST that is located at string depth at least k and that is not the locus of a k-mer is both added to $\mathsf{C}_k(T)$ (when the algorithm visits v) and subtracted from $\mathsf{C}_k(T)$ (when the algorithm visits $\mathtt{parent}(v)$). Leaves at depth at least $k + 1$ that are not the locus of a k-mer are added by the initialization of $\mathsf{C}_k(T)$, and they are subtracted during the enumeration. Conversely, every locus v of a k-mer of T (including leaves) is just added to $\mathsf{C}_k(T)$, since $|\ell(\mathtt{parent}(v))| < k$.

We can apply the same telescoping strategy to compute $\kappa(\mathbf{T}_k^1, \mathbf{T}_k^2)$:

Theorem 4. *Let $T^1 \in [1..\sigma]^{n_1-1}\#_1$ and $T^2 \in [1..\sigma]^{n_2-1}\#_2$ be strings. Given an integer k and two data structures that support* `rangeDistinct` *queries on* BWT_{T^1} *and on* BWT_{T^2}*, respectively, we can compute $\kappa(\mathbf{T}_k^1, \mathbf{T}_k^2)$ in $O(nd)$ time and in $o(n)$ bits of space in addition to the input, where $n = n_1 + n_2$.*

Proof. Recall that $\kappa(\mathbf{T}_k^1, \mathbf{T}_k^2) = N/\sqrt{D^1 D^2}$, where $N = \sum_W \mathbf{T}_k^1[W]\mathbf{T}_k^2[W]$, $D^i = \sum_W \mathbf{T}_k^i[W]^2$, and $W \in [1..\sigma]^k$. We initially set $N = 0$ and $D^i = n_i - k$, since these are the contributions of all the leaves at depth at least $k + 1$ in the generalized suffix tree of T^1 and T^2. Then, we enumerate every internal node u of the generalized suffix tree, using Theorem 2: if $|\ell(u)| < k$ we keep all variables unchanged, otherwise we set N to $N + f_1(\ell(u)) \cdot f_2(\ell(u)) - \sum_v f_1(\ell(v)) \cdot f_2(\ell(v))$ and we set D^i to $D^i + f_i(\ell(u))^2 - \sum_v f_i(\ell(v))^2$, where v ranges over all children of u in the generalized suffix tree. Clearly $f_i(\ell(u)) = \mathtt{first}_i[k_i + 1] - \mathtt{first}_i[1]$ where k_i is the size of array $\mathtt{chars}_i$ in $\mathtt{repr'}(\ell(u))$, and $f_i(\ell(v)) = f_i(\ell(u)b_j) = \mathtt{first}_i[j + 1] - \mathtt{first}_i[j]$ for some $j \in [1..k_i]$. In analogy to Theorem 3, the contribution of the loci of the distinct k-mers of T^1, of T^2, or of both, is added to the three temporary variables and never subtracted, while the contribution of every other node u at depth at least k in the generalized suffix tree is both added (when the algorithm visits u, or when N and D^i are initialized) and subtracted (when the algorithm visits $\mathtt{parent}(u)$).

An even more specific notion of compositional complexity is $\mathsf{C}_{k,f}(T)$, the number of distinct k-mers that occur exactly f times in T. In the *k-mer profiling* problem [6,7] we are given a string T, an interval $[k_1..k_2]$ of lengths and an interval $[f_1..f_2]$ of frequencies, and we are asked to compute the matrix $\mathtt{profile}[k_1..k_2, f_1..f_2]$ defined as follows: $\mathtt{profile}[i, j] = \mathsf{C}_{i,j}(T)$ if $j < f_2$, and $\mathtt{profile}[i, j] = \sum_{h \geq j} \mathsf{C}_{i,h}(T)$ if $j = f_2$. Note that the jth column of $\mathtt{profile}$ can have nonzero cells only if f_j is the frequency of some internal node of ST_T. In practice $\mathtt{profile}$ is often computed by running a k-mer extraction algorithm $k_2 - k_1 + 1$ times, and by scanning the output of all such runs (see e.g. [6] and references therein). The following lemma shows that we can compute $\mathtt{profile}$ in just one pass over the BWT of the input string, and in linear time in the size of $\mathtt{profile}$:

Theorem 5. *Let $T \in [1..\sigma]^{n-1}\#$ be a string. Given ranges $[k_1..k_2]$ and $[f_1..f_2]$, and given a data structure that supports $\mathtt{rangeDistinct}$ queries on BWT_T, we can compute matrix $\mathtt{profile}[k_1..k_2, f_1..f_2]$ in $O(nd + (k_2 - k_1)(f_2 - f_1))$ time and in $o(n)$ bits of space in addition to the input and the output.*

Proof. We use Theorem 1 again. Assume that, for every internal node u of ST_T with string depth at least k_1 and with frequency at least f_1, and for every $k \in [k_1.. \min\{|\ell(u)|, k_2\}]$, we increment $\mathtt{profile}[k, \min\{f(u), f_2\}]$ by one and we decrement $\mathtt{profile}[k, \min\{f(v), f_2\}]$ by one for every child v of u in ST such that $f(v) \geq f_1$. This would take $O(n^2)$ total updates to $\mathtt{profile}$. However, we can perform all of these updates in batch, as follows: for every node u of ST with $f(u) \geq f_1$ and with $|\ell(u)| \geq k_1$, we just increment $\mathtt{profile}[\min\{|\ell(u)|, k_2\}, \min\{f(u), f_2\}]$ by one, and we just decrement $\mathtt{profile}[\min\{|\ell(u)|, k_2\}, \min\{f(v), f_2\}]$ by one for every child v of u in ST such that $f(v) \geq f_1$. After having traversed all the internal nodes of ST, we scan $\mathtt{profile}$ as follows: for every $j \in [f_1..f_2]$, we traverse all values of i in the decreasing order $k_2 - 1, \ldots, k_1$, and we set $\mathtt{profile}[i, j] = \mathtt{profile}[i, j] + \mathtt{profile}[i + 1, j]$. If $f_1 = 1$, at the end of this process the first column of $\mathtt{profile}$ contains negative numbers, since Theorem 1 does not enumerate the leaves of ST. Thus, before returning, we add to profile$[i, 1]$ the number of leaves with string depth at least $k_i + 1$, i.e. value $n - k_i$, for all $i \in [k_1..k_2]$.

A similar algorithm allows computing $\kappa(\mathbf{T}_k^1, \mathbf{T}_k^2)$ for all k in a user-specified range $[k_1..k_2]$ in $O(nd + k_2 - k_1)$ time. Matrix profile can be used to determine a range of values of k to be used in k-mer kernels. The smallest number in this range is typically the value of k that maximizes the number of distinct k-mers that occur at least twice in T [15]. The largest number in the range is typically determined using some measure of expectation: we cover this computation in Sect. 5.

A related notion of compositional complexity is the *k-th order empirical entropy* of T, defined as $\mathsf{H}_k(T) = (1/|T|) \cdot \sum_W \sum_{a \in \Sigma^r(W)} f_T(Wa) \cdot \log(f_T(W)/f_T(Wa))$, where W ranges over all strings in $[1..\sigma]^k$. Clearly only the internal nodes of ST_T contribute to some $\mathsf{H}_k(T)$ [9], thus our methods allow computing $\mathsf{H}_k(T)$ for a user-specified *range of lengths* $[k_1..k_2]$ in $O(nd + k_2 - k_1)$ time, using just one pass over BWT_T.

4   Kernels and Complexity Measures on All Substrings

Given a string $T \in [1..\sigma]^{n-1}\#$, consider the infinite-dimensional vector $\mathbf{T}_\infty$ indexed by all distinct substrings $W \in [1..\sigma]^+$, such that $\mathbf{T}_\infty[W] = f_T(W)$. The *substring complexity* $\mathsf{C}_\infty(T)$ of T is the number of nonzero components of $\mathbf{T}_\infty$. The *substring kernel* of two strings T^1 and T^2 is the cosine of composition vectors $\mathbf{T}_\infty^1$ and $\mathbf{T}_\infty^2$. Computing substring complexity and substring kernel amounts to applying the same telescoping strategy described in Theorems 3 and 4, but with different contributions:

Corollary 1. *Let $T \in [1..\sigma]^{n-1}\#$ be a string. Given a data structure that supports* rangeDistinct *queries on* BWT_T, *we can compute* $\mathsf{C}_\infty(T)$ *in $O(nd)$ time and in $o(n)$ bits of space in addition to the input.*

Proof. The substring complexity of T coincides with the number of characters in $[1..\sigma]$ that occur on all edges of ST_T. We can thus proceed as in Theorem 3, initializing $\mathsf{C}_\infty(T)$ to $(n-1)n/2$, or equivalently to the sum of the lengths of all suffixes of $T[1..n-1]$. Whenever we visit a node v of ST, we add to $\mathsf{C}_\infty(T)$ the quantity $|\ell(v)|$, and we subtract from $\mathsf{C}_\infty(T)$ the quantity $|\ell(v)| \cdot |\mathtt{children}(v)|$. The net effect of all such operations coincides with summing the lengths of all edges of ST, discarding all occurrences of character $\#$. Note that $|\ell(u)|$ is provided by Theorem 1, and $|\mathtt{children}(v)|$ is the size of array chars in $\mathtt{repr}(\ell(v))$.

Corollary 2. *Let $T^1 \in [1..\sigma]^{n_1-1}\#_1$ and $T^2 \in [1..\sigma]^{n_2-1}\#_2$ be strings. Given data structures that support* rangeDistinct *queries on* BWT_{T^1} *and on* BWT_{T^2}, *respectively, we can compute* $\kappa(\mathbf{T}_\infty^1, \mathbf{T}_\infty^2)$ *in $O(nd)$ time and in $o(n)$ bits of space in addition to the input, where $n = n_1 + n_2$.*

Proof. We proceed as in Theorem 4, setting again $N = 0$ and $D^i = (n_i - 1)n_i/2$ at the beginning of the algorithm. When we visit a node u of the generalized suffix tree of T^1 and T^2, we set N to $N + |\ell(u)| \cdot (f_1(\ell(u))f_2(\ell(u)) - \sum_v f_1(\ell(v))f_2(\ell(v)))$ and we set D^i to $D^i + |\ell(u)| \cdot (f_i(\ell(u))^2 - \sum_v f_i(\ell(v))^2)$, where v ranges over all children of u in the generalized suffix tree.

In a substring kernel it is common to weight a substring W by a user-specified function of its length: typical choices are $\epsilon^{|W|}$ for a given constant ϵ, or indicators that select only substrings within a specific range of lengths [16]. We denote by $\mathbf{T}^i_{\infty,g}$ a weighted version of the infinite-dimensional vector $\mathbf{T}^i_\infty$ such that $\mathbf{T}^i_{\infty,g}[W] = g(|W|) \cdot \mathbf{T}^i_\infty[W]$, where g is any user-specified function. We assume that the number of bits required to represent the output of g with sufficient precision is $O(\log n)$. It is easy to adapt Corollary 2 to support this type of composition vector:

Corollary 3. *Let $T^1 \in [1..\sigma]^{n_1-1}\#_1$ and $T^2 \in [1..\sigma]^{n_2-1}\#_2$ be strings. Given a function $g(k)$ that can be evaluated in constant time, and given data structures that support* `rangeDistinct` *queries on* BWT_{T^1} *and on* BWT_{T^2}, *respectively, we can compute* $\kappa(\mathbf{T}^1_{\infty,g}, \mathbf{T}^2_{\infty,g})$ *in $O(nd)$ time and in $o(n)$ bits of space in addition to the input, where $n = n_1 + n_2$.*

Proof. We modify Corollary 2 as follows. Assume that we are processing an internal node v of the generalized suffix tree, let $\ell(v) = W$, and assume that we have computed `repr'`(aW) for all the left extensions aW of W. In addition to pushing `repr'`(aW) onto the stack, we also push value `prefixSum`$(aW) = \sum_{i=1}^{|W|+1} g(i)^2$ with it, where `prefixSum`$(aW) = $ `prefixSum`$(W) + g(|W| + 1)^2$. When we pop `repr'`(aW), we compute its contributions to N and D^i as described in Corollary 2, but replacing $|aW|$ by `prefixSum`(aW). We initialize D^i to $\sum_{j=1}^{n_i-1} g(j)^2$.

Corollary 3 can clearly support distinct weight functions for T^1 and T^2. For some functions, like $\epsilon^{|W|}$, prefix sums can be computed in closed form [16], thus there is no need to push `prefixSum` values on the stack. Another frequent weighting scheme for a string W associates a score $q(c)$ to every character c of W, and it weights W by e.g. $q(W) = \prod_{i=1}^{|W|} q(W[i])$. In this case we could just push `prefixSum`$(V) = \sum_{i=1}^{|V|} \prod_{j=1}^{i} q(V[j])^2$ onto the stack, where $V = aW$ and `prefixSum`$(V) = q(a)^2 \cdot (1 + $ `prefixSum`$(W))$. A similar weighting scheme can be used for k-mers as well. Let $\mathbf{T}_{k,q}$ be a version of $\mathbf{T}_k$ such that $\mathbf{T}_{k,q}[W] = f_T(W) - (|T| - |W|)q(W)$ for every $W \in [1..\sigma]^k$, and consider the following distances defined in [13]:

$$D_2^s(\mathbf{T}^1_{k,q}, \mathbf{T}^2_{k,q}) = \sum_W \mathbf{T}^1_{k,q}[W]\mathbf{T}^2_{k,q}[W]/\sqrt{(\mathbf{T}^1_{k,q}[W])^2 + (\mathbf{T}^2_{k,q}[W])^2}$$

$$D_2^*(\mathbf{T}^1_{k,q}, \mathbf{T}^2_{k,q}) = \sum_W \mathbf{T}^1_{k,q}[W]\mathbf{T}^2_{k,q}[W]/\left(\sqrt{(n_1 - k)(n_2 - k)} \cdot q(W)\right)$$

where W ranges over all strings in $[1..\sigma]^k$. We can compute such distances using just a minor modification to Theorem 4:

Corollary 4. *Let $T^1 \in [1..\sigma]^{n_1-1}\#_1$ and $T^2 \in [1..\sigma]^{n_2-1}\#_2$ be strings. Given an integer k and data structures that support* rangeDistinct *queries on* BWT_{T^1} *and on* BWT_{T^2}, *respectively, we can compute $D_2^s(\mathbf{T}_{k,p}^1, \mathbf{T}_{k,p}^2)$ and $D_2^*(\mathbf{T}_{k,p}^1, \mathbf{T}_{k,p}^2)$ in $O(nd)$ time and in $\lambda \log \sigma + o(n)$ bits of space in addition to the input, where $n = n_1 + n_2$ and λ is the length of the longest repeat in $T^1 T^2$.*

Proof. We proceed as in Theorem 4, pushing on the stack value $q(W, k) = \prod_{j=1}^{k} q(W[j])$ in addition to repr'(W), and maintaining a separate stack of characters to represent the string we are processing during the depth-first traversal of the generalized suffix-link tree. We set $q(aW, k) = q(a) \cdot q(W, k)/q(b)$, where b is the kth character from the top of the character stack when we are processing W.

An orthogonal way to measure the similarity between T^1 and T^2 consists in comparing the repertoire of all strings that *do not appear* in T^1 and in T^2. Given a string T and two frequency thresholds $\tau_1 < \tau_2$, a string W is a *minimal rare word* of T if $\tau_1 \le f_T(W) < \tau_2$ and if $f_T(V) \ge \tau_2$ for every proper substring V of W. Setting $\tau_1 = 0$ and $\tau_2 = 1$ gives the well-known *minimal absent words* (see e.g. [5,10] and references therein), whose total number can be $\Theta(\sigma n)$ [8]. Setting $\tau_1 = 1$ and $\tau_2 = 2$ gives the so-called *minimal unique substrings* (see e.g. [11] and references therein), whose total number is $O(n)$, like the number of strings obtained by any other setting of $\tau_1 \ge 1$. In what follows we focus on minimal absent words, but our algorithms can be generalized to other settings of the thresholds.

To decide whether aWb is a minimal absent word of T, where a and b are characters, it clearly suffices to check whether $f_T(aWb) = 0$ and whether both $f_T(aW) \ge 1$ and $f_T(Wb) \ge 1$. It is well known that only a maximal repeat of T can be the infix W of a minimal absent word aWb, and this applies to any setting of τ_1 and τ_2. To enumerate all the minimal absent words, for example to count their total number $\mathsf{C}_-(T)$, we can thus iterate over all nodes of ST_T associated with maximal repeats, as described below:

Theorem 6. *Let $T \in [1..\sigma]^{n-1}\#$ be a string. Given a data structure that supports* rangeDistinct *queries on* BWT_T, *we can compute $\mathsf{C}_-(T)$ in $O(nd)$ time and in $o(n)$ bits of space in addition to the input.*

Proof. For clarity, we first describe how to *enumerate* all the distinct minimal absent words of T: we specialize this algorithm to counting at the end of the proof. We use Theorem 1 to enumerate all nodes v of ST_T associated with maximal repeats, as described in Sect. 2.2. Let $\{a_1, \ldots, a_h\}$ be the set of distinct left extensions of string $\ell(v)$ in T returned by operation extendLeft(repr($\ell(v)$)), let extensions$[1..\sigma + 1, 0..\sigma]$ be a boolean matrix initialized to all zeros, and let leftExtensions$[1..\sigma + 1]$ be an array initialized to all zeros. Let h' be a pointer initialized to one. Operation extendLeft allows following all the Weiner links from v, not necessarily in lexicographic order: for every string $a_i\ell(v)$ obtained in this way, we set leftExtensions$[h'] = a_i$, we enumerate its right extensions $\{c_1, \ldots, c_{k'}\}$ using array chars of repr($a_i\ell(v)$), we set extensions$[h', c_j] = 1$

for all $j \in [1..k']$, and we finally increment h' by one. Note that only the columns of extensions that correspond to the right extensions of $\ell(v)$ are updated by this procedure. Then, we enumerate all the right extensions $\{b_1, \ldots, b_k\}$ of $\ell(v)$ using array chars of $\mathrm{repr}(\ell(v))$, and for every such extension b_j we report all pairs (a_i, b_j) such that $a_i = \mathrm{chars}[x]$, $x \in [1..h']$, and $\mathrm{extensions}[x, b_j] = 0$. This process takes time proportional to the number of Weiner links from v, plus the number of children of v, plus the number of Weiner links from v multiplied by σ. When applied to all nodes of ST, this takes in total $O(n\sigma)$ time, which is optimal in the size of the output. The matrices and vectors used by this process can be reset to all zeros after processing each node: the total time spent in such reinitializations in $O(n)$.

If we just need $\mathsf{C}_-(T)$, rather than storing the temporary matrices extensions and leftExtensions, we store just a number area which we initialize to hk before processing node v. Whenever we observe a right extension c_j of a string $a_i\ell(v)$, we decrease area by one. Before moving to the next node, we increment $\mathsf{C}_-(T)$ by area.

Let $\mathbf{T}_-$ be the infinite-dimensional vector indexed by all distinct substrings $W \in [1..\sigma]^+$, such that $\mathbf{T}_-[W] = 1$ iff W is a minimal absent word of T. Theorem 6 can be adapted to compute the Jaccard distance between the composition vectors of two strings:

Corollary 5. *Let $T^1 \in [1..\sigma]^{n_1-1}\#_1$ and $T^2 \in [1..\sigma]^{n_2-1}\#_2$ be strings. Given data structures that support* rangeDistinct *queries on* BWT_{T^1} *and on* BWT_{T^2}, *respectively, we can compute $J(\mathbf{T}_-^1, \mathbf{T}_-^2)$ in $O(nd)$ time and in $o(n)$ bits of space in addition to the input, where $n = n_1 + n_2$.*

Proof. We apply the strategy of Theorem 6 to the internal nodes of the generalized suffix tree of T^1 and T^2 whose label is a maximal repeat of T^1 and a maximal repeat of T^2: such strings are clearly maximal repeats of T^1T^2 as well. We enumerate such nodes as described in Sect. 2.2. We keep a global variable intersection and a bitvector sharedRight$[1..\sigma]$. For every node v that corresponds to a maximal repeat of T^1 and of T^2, we merge the sorted arrays chars$_1$ and chars$_2$ of $\mathrm{repr}'(\ell(v))$, we set sharedRight$[c] = 1$ for every character c that belongs to the intersection of the two arrays, and we cumulate in a variable k' the number of ones in sharedRight. Then, we scan every left extension a_i provided by extendLeft$'$, we determine in constant time whether it occurs in both T^1 and T^2, and if so we increment a variable h' by one. Finally, we initialize a variable area to $h'k'$, and we process again every left extension a_i provided by extendLeft$'$: if $a_i\ell(v)$ occurs in both T^1 and T^2, we compute the union of arrays chars$_1$ and chars$_2$ of $\mathrm{repr}'(a_i\ell(v))$, and for every character c in the union such that sharedRight$[c] = 1$, we decrement area by one. At the end of this process, we add area to the global variable intersection. To compute $\|\mathbf{T}_-^1 \vee \mathbf{T}_-^2\|$ we apply Theorem 6 to T^1 and T^2 separately.

It is easy to extend Corollary 5 to compute $\kappa(\mathbf{T}_-^1, \mathbf{T}_-^2)$, as well as to support weighting schemes based on the length and on the characters of minimal absent words.

5   Markovian Corrections

In some applications it is desirable to assign to component $W \in [1..\sigma]^k$ of composition vector $\mathbf{T}_\infty$ an estimate of the *statistical significance* of observing $f_T(W)$ occurrences of W in T: intuitively, strings whose frequency departs from its expected value are more likely to carry "information", and they should be weighted more [12]. Assume that T is generated by a Markov random process of order $k - 2$ or smaller, that produces strings on alphabet $[1..\sigma]$ according to a probability distribution $\mathbb{P}$. It is well known that the probability of observing W in a string generated by such a random process is $\mathbb{P}(W) = \mathbb{P}(W[1..k-1]) \cdot \mathbb{P}(W[2..k])/\mathbb{P}(W[2..k-1])$. We can estimate $\mathbb{P}(W)$ using the empirical probability $p_T(W)$, obtaining the following approximation for $\mathbb{P}(W)$: $\tilde{p}_T(W) = p_T(W[1..k-1]) \cdot p_T(W[2..k])/p_T(W[2..k-1])$ if $p_T(W[2..k-1]) \neq 0$, and $\tilde{p}_T(W) = 0$ otherwise. We can thus estimate the significance of the event that substring W has empirical probability $p_T(W)$ in string T using the following score: $z_T(W) = (p_T(W) - \tilde{p}_T(W))/\tilde{p}_T(W)$ if $\tilde{p}_T(W) \neq 0$, and $z_T(W) = 0$ if $\tilde{p}_T(W) = 0$ [12]. After elementary manipulations [2], $z_T(W)$ becomes:

$$z_T(W) = g(n, k) \cdot \frac{f_T(W) \cdot f_T(W[2..k-1])}{f_T(W[1..k-1]) \cdot f_T(W[2..k])} - 1$$

$$g(x, y) = (x - y + 2)^2 / (x - y + 1)(x - y + 3)$$

Since $g(x, y) \in [1..1.125]$, we temporarily assume $g(x, y) = 1$ in what follows, removing this assumption later.

Let $\mathbf{T}_z$ be a version of the infinite-dimensional vector $\mathbf{T}_\infty$ in which $\mathbf{T}_z[W] = z_T(W)$. Among all strings that occur in T, only strings aWb such that a and b are characters in $[0..\sigma]$ and such that W is a maximal repeat of T can have $\mathbf{T}_z[aWb] \neq 0$. Similarly, among all strings that *do not occur* in T, only the minimal absent words of T have a nonzero component in $\mathbf{T}_z$: specifically, $\mathbf{T}_z[aWb] = -1$ for all minimal absent words aWb of T, where a and b are characters in $[0..\sigma]$ [2]. Given two strings T^1 and T^2, we can thus compute $\kappa(\mathbf{T}_z^1, \mathbf{T}_z^2)$ using the same strategy as in Corollary 5:

Theorem 7. *Let $T^1 \in [1..\sigma]^{n_1-1}\#_1$ and $T^2 \in [1..\sigma]^{n_2-1}\#_2$ be strings. Given data structures that support* rangeDistinct *queries on* BWT_{T^1} *and on* BWT_{T^2}, *respectively, and assuming $g(x, y) = 1$ for all settings of x and y, we can compute $\kappa(\mathbf{T}_z^1, \mathbf{T}_z^2)$ in $O(nd)$ time and in $o(n)$ bits of space in addition to the input, where $n = n_1 + n_2$.*

Proof. We focus here on computing component N of $\kappa(\mathbf{T}_z^1, \mathbf{T}_z^2)$: computing D^i follows a similar algorithm on BWT_{T^i}. We keep again a bitvector sharedRight$[1..\sigma]$, and we enumerate all the internal nodes of the generalized suffix tree of T^1 and T^2 whose label is a maximal repeat of T^1 and a maximal repeat of T^2, as described in Corollary 5. For every such node v, we merge the sorted arrays chars$_1$ and chars$_2$ of repr$'(\ell(v))$, we set sharedRight$[c] = 1$ for every character c that belongs to the intersection of the two arrays, and we cumulate in a variable k' the number of ones in sharedRight. Then, we scan every left extension a_i provided by `extendLeft'`, we determine in constant time whether it occurs in both T^1 and T^2, and if so we increment a variable h' by one. Finally, we initialize a variable `area` to $h'k'$, and we process again every left extension a_i provided by `extendLeft'`. If $a_i\ell(v)$ occurs in both T^1 and T^2, we merge arrays `chars`$_1$ and `chars`$_2$ of `repr'`$(a_i\ell(v))$: for every character b in the intersection of `chars`$_1$ and `chars`$_2$, we add to N value $z_1(a_i\ell(v)b) \cdot z_2(a_i\ell(v)b)$, retrieving the corresponding frequencies from `repr'`$(a_i\ell(v))$ and from `repr'`$(\ell(v))$, and we decrement `area` by one. For every character b that occurs only in `chars`$_1$, we test whether `sharedRight`$[b] = 1$: if so, a_iWb is a minimal absent word of T^2 that occurs in T^1, thus we decrement `area` by one and we add to N value $-z_1(a_i\ell(v)b)$. We proceed symmetrically if b occurs only in `chars`$_2$. At the end of this process, `area` counts the number of minimal absent words with infix $\ell(v)$ that are shared by T^1 and T^2: thus, we add `area` to N.

It is easy to remove the assumption that $g(x,y)$ is always equal to one. There are only two differences from the previous case. First, the score of the substrings W of T^i that have a maximal repeat of T^i as an infix changes, but $g(n_i, |W|)$ can be immediately computed from $|W|$, which is provided by the enumeration algorithm. Second, the score of all substrings W of T^i that do not have a maximal repeat as an infix changes from zero to $g(n_i, |W|) - 1$: we can take all such contributions into account by pushing prefix-sums to the stack, as in Corollary 3. For example, to compute component N of $\kappa(\mathbf{T}_z^1, \mathbf{T}_z^2)$, we can first assume that *all* substring W that occur both in T^1 and in T^2 have score $g(n_i, |W|) - 1$, by pushing on the stack the prefix-sums described in [2] and by enumerating only nodes v of the generalized suffix tree of T^1 and T^2 such that $\ell(v)$ occurs both in T^1 and in T^2. Then, we can run a similar algorithm as in Theorem 7, subtracting quantity $(g(n_1, |W| + 2) - 1) \cdot (g(n_2, |W| + 2) - 1)$ from the contribution to N of every string a_iWb that occurs both in T^1 and in T^2.

Finally, recall that in Sect. 3 we mentioned the problem of determining an *upper bound* on the values of k to be used in k-mer kernels. Let $\mathbf{T}_k$ be the composition vector indexed by all strings in $[1..\sigma]^k$ such that $\mathbf{T}_k[W] = p_T(W)$, and let $\tilde{\mathbf{T}}_k$ be a similar composition vector with $\tilde{\mathbf{T}}_k[W] = \tilde{p}_T(W)$, where $\tilde{p}_T(W)$ is defined as in the beginning of this section. It makes sense to disregard values of k for which $\mathbf{T}_k$ and $\tilde{\mathbf{T}}_k$ are very similar, and more formally whose Kullback-Leibler divergence $\mathsf{KL}(\mathbf{T}_k, \tilde{\mathbf{T}}_k) = \sum_W \mathbf{T}_k[W] \cdot (\log(\mathbf{T}_k[W]) - \log(\tilde{\mathbf{T}}_k[W]))$ is small, where W ranges over all strings in $[1..\sigma]^k$. Thus, we could use as an upper bound on k the minimum value k^* such that $\sum_{k'=k^*}^{\infty} \mathsf{KL}(\mathbf{T}_{k'}, \tilde{\mathbf{T}}_{k'}) < \tau$ for some user-specified threshold τ [15]. Note again that only strings aWb such that a and b are characters in $[0..\sigma]$ and W is a maximal repeat of T contribute to $\mathsf{KL}(\mathbf{T}_{|W|+2}, \tilde{\mathbf{T}}_{|W|+2})$. We can thus adapt Theorem 7 to compute the KL divergence for a user-specified *range of lengths* $[k_1..k_2]$, using just one pass over BWT_T, in $O(nd)$ time and in $o(n)$ bits of space in addition to the input and the output. The same approach can be used to compute the *KL-divergence kernel* $\kappa(\mathbf{T}_{KL}^1, \mathbf{T}_{KL}^2)$, where $\mathbf{T}_{KL}^i[W] = \mathsf{KL}_{T^i}(W)$ and $\mathsf{KL}_{T^i}(W) = \sum_{a,b \in \Sigma} p_{T^i}(aWb) \cdot (\log(p_{T^i}(aWb)) - \log(\tilde{p}_{T^i}(aWb)))$.

References

1. Apostolico, A.: Maximal words in sequence comparisons based on subword composition. In: Elomaa, T., Mannila, H., Orponen, P. (eds.) Ukkonen Festschrift 2010. LNCS, vol. 6060, pp. 34–44. Springer, Heidelberg (2010)
2. Apostolico, A., Denas, O.: Fast algorithms for computing sequence distances by exhaustive substring composition. Algorithms Mol. Biol. **3**(1), 13 (2008)
3. Belazzougui, D.: Linear time construction of compressed text indices in compact space. In Symposium on Theory of Computing, STOC 2014, New York, NY, USA, 31 May–03 June, pp. 148–193 (2014)
4. Belazzougui, D., Navarro, G., Valenzuela, D.: Improved compressed indexes for full-text document retrieval. J. Discret. Algorithms **18**, 3–13 (2013)
5. Chairungsee, S., Crochemore, M.: Using minimal absent words to build phylogeny. Theoret. Comput. Sci. **450**, 109–116 (2012)
6. Chikhi, R., Medvedev, P.: Informed and automated k-mer size selection for genome assembly. Bioinformatics **30**(1), 31–37 (2014)
7. Chor, B., Horn, D., Goldman, N., Levy, Y., Massingham, T., et al.: Genomic DNA k-mer spectra: models and modalities. Genome Biol. **10**(10), R108 (2009)
8. Crochemore, M., Mignosi, F., Restivo, A.: Automata and forbidden words. Inf. Process. Lett. **67**(3), 111–117 (1998)
9. Gog, S.: Compressed suffix trees: design, construction, and applications. Ph.D. thesis, University of Ulm, Germany (2011)
10. Herold, J., Kurtz, S., Giegerich, R.: Efficient computation of absent words in genomic sequences. BMC Bioinform. **9**(1), 167 (2008)
11. İleri, A.M., Külekci, M.O., Xu, B.: Shortest unique substring query revisited. In: Kulikov, A.S., Kuznetsov, S.O., Pevzner, P. (eds.) CPM 2014. LNCS, vol. 8486, pp. 172–181. Springer, Heidelberg (2014)
12. Qi, J., Wang, B., Hao, B.-I.: Whole proteome prokaryote phylogeny without sequence alignment: a k-string composition approach. J. Mol. Evol. **58**(1), 1–11 (2004)
13. Reinert, G., Chew, D., Sun, F., Waterman, M.S.: Alignment-free sequence comparison (I): statistics and power. J. Comput. Biol. **16**(12), 1615–1634 (2009)
14. Shawe-Taylor, J., Cristianini, N.: Kernel Methods for Pattern Analysis. Cambridge University Press, Cambridge (2004)
15. Sims, G.E., Jun, S.-R., Wu, G.A., Kim, S.-H.: Alignment-free genome comparison with feature frequency profiles (FFP) and optimal resolutions. Proc. Natl. Acad. Sci. **106**(8), 2677–2682 (2009)
16. Smola, A.J., Vishwanathan, S.V.N.: Fast kernels for string and tree matching. In: Becker, S., Thrun, S., Obermayer, K. (eds.) Advances in Neural Information Processing Systems 15, pp. 585–592. MIT Press, Cambridge (2003)

Composite Repetition-Aware Data Structures

Djamal Belazzougui[1,2]([✉]), Fabio Cunial[1,2], Travis Gagie[1,2],
Nicola Prezza[3], and Mathieu Raffinot[4]

[1] Department of Computer Science, University of Helsinki, Helsinki, Finland
djamal.belazzougui@cs.helsinki.fi
[2] Helsinki Institute for Information Technology, Helsinki, Finland
[3] Department of Mathematics and Computer Science,
University of Udine, Udine, Italy
[4] LIAFA, Paris Diderot University, Paris 7, France

Abstract. In highly repetitive strings, like collections of genomes from
the same species, distinct measures of repetition all grow sublinearly in
the length of the text, and indexes targeted to such strings typically
depend only on one of these measures. We describe two data structures whose size depends on multiple measures of repetition at once, and
that provide competitive tradeoffs between the time for counting and
reporting all the exact occurrences of a pattern, and the space taken by
the structure. The key component of our constructions is the run-length
encoded BWT (RLBWT), which takes space proportional to the number
of BWT runs: rather than augmenting RLBWT with suffix array samples, we combine it with data structures from LZ77 indexes, which take
space proportional to the number of LZ77 factors, and with the compact
directed acyclic word graph (CDAWG), which takes space proportional
to the number of extensions of maximal repeats. The combination of
CDAWG and RLBWT enables also a new representation of the suffix
tree, whose size depends again on the number of extensions of maximal
repeats, and that is powerful enough to support matching statistics and
constant-space traversal.

1 Introduction

The space taken by compressed data structures for highly-repetitive strings is
typically a function of a specific measure of repetition, for example the number z
of factors in a Lempel-Ziv parsing [1,11], or the number r of runs in a Burrows-
Wheeler transform [14]. For many such compressed data structures, computing
all the occurrences of a pattern in the indexed string is a bottleneck. In this
paper we explore the advantages of *combining data structures that depend on
distinct measures of repetition*. Specifically, we describe a data structure that
takes approximately $O(z+r)$ words of space, and that reports all the occurrences

Travis Gagie—Supported by the Academy of Finland.
This work was partially supported by Academy of Finland under grant 284598 (Center of Excellence in Cancer Genetics Research).

F. Cicalese et al. (Eds.): CPM 2015, LNCS 9133, pp. 26–39, 2015.
DOI: 10.1007/978-3-319-19929-0_3

of a pattern of length m in $O(m(\log \log n + \log z) + \mathsf{pocc} \log^\epsilon z + \mathsf{socc} \log \log n)$ time, where n is the length of the string and pocc and socc are the number of primary and of secondary occurrences, respectively (see Sect. 2.2 for definitions). This compares favorably to the $O(m^2 h + (m + \mathsf{occ}) \log z)$ reporting time of LZ77 indexes [11], where h is the height of the parse tree. It also compares favorably in space to solutions based on run-length encoded BWT (RLBWT) and suffix array samples [14], which take $O(n/k + r)$ words of space to achieve $O(m \log \log n + k \cdot \mathsf{occ} \log \log n)$ reporting time, where k is a sampling rate.

We also introduce a new measure of the repetitiveness of a string, the number e of right extensions of maximal repeats, which is related to the number of arcs in the compact directed acyclic word-graph (CDAWG) and which is an upper bound on r and z. We show a data structure whose size depends on e and that reports all the occ occurrences of a pattern of length m in a string of length n in $O(m \log \log n + \mathsf{occ})$ time. The main component of our constructions is the RLBWT, which we use to count the number of occurrences of a pattern, and which we combine with the CDAWG and with data structures from LZ indexes, rather than with suffix array samples, for reporting. Similar combinations have already appeared in the literature, but their space has been related to statistical compressibility rather than to the number of repetitions: for example, an FM-index has already been combined with an LZ78 self-index to achieve faster search or reporting [1,7], but the size of the resulting data structure depends on k-th order empirical entropy. Bounds in terms of k-th order empirical entropy have redundancy terms that depend exponentially on k, so they cannot capture compressibility based on long repetitions.

Combining the RLBWT with the CDAWG enables also a new representation of the suffix tree, which takes space proportional to $e + e^\ell$ (where e^ℓ is the number of left extensions of maximal repeats) and which supports a number of operations in $O(\log \log n)$ time. Among other properties, this new representation allows computing the matching statistics of a pattern of length m in $O(m \log \log n)$ time. Our constructions are targeted to highly-repetitive strings, like large databases of similar genomes, in which all the measures of repetition on which our data structures depend grow sublinearly in the size of the database (see Fig. 1 for an example). In a future paper we will provide a full experimental comparison of our results against other data structures for pattern matching in highly-repetitive strings.

2 Preliminaries

Let $\Sigma = [1..\sigma]$ be an integer alphabet, let $\# = 0 \notin \Sigma$ be a separator, and let $T = [1..\sigma]^{n-1}\#$ be a string. We denote the reverse of T by $\overline{T}$. Given a substring W of T, let $\mathcal{P}_T(W)$ be the set of all starting positions of W in the circular version of T. A *repeat* W is a string that satisfies $|\mathcal{P}_T(W)| > 1$. We denote by $\Sigma^\ell_T(W)$ the set of characters $\{a \in [0..\sigma] : |\mathcal{P}_T(aW)| > 0\}$ and by $\Sigma^r_T(W)$ the set of characters $\{b \in [0..\sigma] : |\mathcal{P}_T(Wb)| > 0\}$. A repeat W is *right-maximal* (respectively, *left-maximal*) iff $|\Sigma^\ell_T(W)| > 1$ (respectively, iff $|\Sigma^r_T(W)| > 1$).

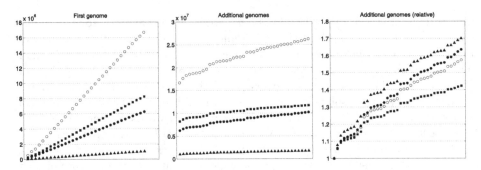

Fig. 1. Growth of the number of maximal repeats $|\mathcal{M}_T|$ (black circles), of $|\mathcal{E}_T^r \cup \mathcal{F}_T^r|$ (white circles, e in the introduction), of the number of runs in BWT $|\mathcal{R}_T|$ (squares, r in the introduction), and of $|\mathcal{Z}_T|$ (triangles, z in the introduction) in a concatenation T of 39 highly similar *Saccharomyces cerevisiae* genomes [8] (see Sect. 2 for definitions). Left: growth inside the first genome of the database. Center: growth after the addition of each genome (one sample per genome). Right: the same as the plot in the center, but with each curve normalized by its first sample. $|\mathcal{E}_T^\ell \cup \mathcal{F}_T^\ell|$, $|\mathcal{R}_{\overline{T}}|$ and $|\mathcal{Z}_{\overline{T}}|$ are not shown since they behave approximately as their symmetrical counterparts.

It is well known that T can have at most $n-1$ right-maximal substrings and at most $n-1$ left-maximal substrings. A *maximal repeat* of T is a repeat that is both left- and right-maximal: we call $\mathcal{M}_T$ the set of all maximal repeats of T. A maximal repeat W can be seen as a set of right-maximal substrings of T, and specifically as the set of all right-maximal strings $W[i..|W|]$ for $i \in [1..k]$ that are not left-maximal, and such that $W[k+1..|W|]$ is left-maximal.

For reasons of space we assume the reader to be familiar with the notion of *suffix tree* $\mathsf{ST}_T = (V, E)$ of T, which we do not define here. We denote by $\ell(\gamma)$, or equivalently by $\ell(u, v)$, the label of edge $\gamma = (u, v) \in E$, and we denote by $\ell(v)$ the string label of node $v \in V$. It is well known that a substring W of T is right-maximal (respectively, left-maximal) iff $W = \ell(v)$ for some internal node v of ST_T (respectively, iff $W = \overline{\ell(v)}$ for some internal node v of $\mathsf{ST}_{\overline{T}}$). We assume the reader to be familiar with the notion of *suffix link* connecting a node v with $\ell(v) = aW$ for some $a \in [0..\sigma]$ to a node w with $\ell(w) = W$: we say that $w = \mathtt{suffixLink}(v)$ in this case. Here we just recall that inverting the direction of all suffix links yields the so-called *explicit Weiner links*. Given an internal node v and a symbol $a \in [0..\sigma]$, it might happen that string $a\ell(v)$ does occur in T, but that it is not right-maximal, i.e. it is not the label of any internal node: all such left extensions of internal nodes that end in the middle of an edge are called *implicit Weiner links*. An internal node can have more than one outgoing Weiner link, and all such Weiner links have distinct labels.

The *compact directed acyclic word graph* of a string T (denoted by CDAWG_T in what follows) is the minimal compact automaton representing the set of suffixes of a given string [3,6]. It can be seen as the minimization of ST_T, in which all leaves are merged to the same node (the sink) that represents T itself, and in which all nodes except the sink are in one-to-one correspondence with the maximal repeats of T [16]. Since a maximal repeat corresponds to a set of right-maximal substrings, CDAWG_T can be built by putting in the same equivalence class all nodes of ST_T that belong to the same maximal unary path of explicit Weiner links.

For reasons of space we assume the reader to be familiar with the notion and uses of the Burrows-Wheeler transform of T, including the C array and backward searching. In this paper we use BWT_T to denote the BWT of T, and we use $\mathtt{range}(W) = [\mathtt{sp}(W)..\mathtt{ep}(W)]$ to denote the lexicographic interval of a string W in a BWT that is implicit from the context. We say that $\mathsf{BWT}_T[i..j]$ is a run iff $\mathsf{BWT}_T[k] = c \in [0..\sigma]$ for all $k \in [i..j]$, and moreover if any substring $\mathsf{BWT}_T[i'..j']$ such that $i' \leq i$, $j' \geq j$, and either $i' \neq i$ or $j' \neq j$, contains at least two distinct characters. It is well known that repetitions in T tend to be converted into runs of BWT_T. We denote by $\mathcal{R}_T$ the set of all triplets (c, i, j) such that $\mathsf{BWT}_T[i..j]$ is a run of character c, and we use r_T and $\bar{r}_T$ as shorthands for $|\mathcal{R}_T|$ and $|\mathcal{R}_{\bar{T}}|$, respectively.

The $LZ77$ $factorization$ of T [20] is the greedy decomposition $T_1 T_2 \cdots T_z$ of T obtained as follows. Assume that T is virtually preceded by the σ distinct characters in its alphabet, and assume that $T_1 T_2 \cdots T_i$ has already been computed for some prefix of length k of T: then, T_{i+1} is the longest prefix of $T[k + 1..n]$ such that there is a $j \leq k$ that satisfies $T[j..j + |T_{i+1}| - 1] = T_{i+1}$. We denote by $\mathcal{Z}_T$ the set of pairs (T_i, p_i) for all $i \in [1..z]$, where p_i is the starting position of T_i in T, and we use z_T as a shorthand for $|\mathcal{Z}_T|$. From now on, we drop subscripts whenever the string T they specify is clear from the context.

2.1   Relationships Among Maximal Repeats, Runs in BWT, and LZ Factors

Clearly $|\mathcal{R}|$ can be as small as two, e.g. in string $0^{n-1}\#$, and as large as $\Theta(n)$, e.g. in the string of length n that contains exactly n distinct characters, or in a de Bruijn string of order $k > 1$ on a binary alphabet: this string of length $\sigma^k + k - 1$ contains all the distinct k-mers, thus the interval of every $(k - 1)$-mer in BWT_T contains exactly σ distinct characters, and the number of runs in BWT_T is thus at least $\sigma^{k-1}(k-1)$. It is known that $|\mathcal{Z}|$ is $O(n/\log_\sigma n)$ [12], and it can be constant, e.g. in $0^{n-1}\#$. Conversely, $|\mathcal{M}|$ can be zero, e.g. in a string of length n that contains exactly n distinct characters, and it can be $\Theta(n)$ in the worst case, e.g. in string $0^{n-1}\#$. When maximal repeats exist, the number of $right$ $extensions$ of $maximal$ $repeats$ $\sum_{W \in \mathcal{M}} |\Sigma^r(W)|$ is $\Omega(\log n)$, and this lower bound is matched by Fibonacci strings and by Thue-Morse strings of length n, whose CDAWG contains $O(\log n)$ nodes [15,17]. Both $|\mathcal{M}|/|\mathcal{R}|$ and $|\mathcal{M}|/|\mathcal{Z}|$ can be $\Theta(n)$, for example in the already mentioned $0^{n-1}\#$. $|\mathcal{R}|/|\mathcal{Z}|$ can be $\Theta(\log n)$, e.g. in the already mentioned de Bruijn string T of order k, which has $\Theta(n/\log_\sigma n)$ LZ factors. However, $|\mathcal{M}|$, $|\mathcal{R}|$ and $|\mathcal{Z}|$ can all grow at the same asymptotic rate in the same family of strings. Consider e.g. string $T = 0^1 10^2 1 \cdots 0^x 1\#$ of length $x(x + 3)/2 + 1$. Clearly $|\mathcal{Z}| = x + 3$, and $|\mathcal{M}| = 3(x - 1)$ since the maximal repeats of T are only the substrings $0^i 1$ for $i \in [1..x - 1]$, 0^j for $j \in [1..x - 1]$, and $0^{k-1}10^k$ for $k \in [2..x-1]$. Replacing $\#$ with a new block $0^{x+1}1\#$ in string T creates two new runs for every $x > 1$, thus $|\mathcal{R}| = 2x$ for $x > 1$.

Recall that a substring W of T is a maximal repeat iff $W = \ell(v)$ for some internal node v of $\mathsf{ST}_T = (V, E)$, and moreover if there are at least two Weiner links from v. Since the set of all left-maximal substrings of T is closed under the prefix operation, there is a bijection between $\mathcal{M}$ and the nodes that lie on the paths of ST_T that start from the root and that end at nodes labeled by maximal repeats defined as follows:

Definition 1. *A maximal repeat W of a string $T \in [1..\sigma]^{n-1}\#$ is* rightmost *if no string WV with $V \in [0..\sigma]^+$ is left-maximal in T.*

We denote the set of rightmost maximal repeats of T by $\mathcal{M}_T^r$. We also denote by $\mathcal{E}_T^r$ the set of edges of ST_T that connect pairs of nodes labeled by maximal repeats, and we denote by $\mathcal{F}_T^r$ the set of edges (v, w) in ST_T such that $\ell(v) \in \mathcal{M}_T$ and $\ell(w) \notin \mathcal{M}_T$. We use $\mathcal{M}_T^\ell$, $\mathcal{E}_T^\ell$ and $\mathcal{F}_T^\ell$ to denote symmetrical concepts in $\mathsf{ST}_{\overline{T}}$, and we use e_T and e_T^ℓ as shorthands for $|\mathcal{E}_T^r| + |\mathcal{F}_T^r|$ and for $|\mathcal{E}_T^\ell| + |\mathcal{F}_T^\ell|$, respectively. Clearly $\mathcal{E}^r$ and $\mathcal{F}^r$ are the image of explicit and implicit Weiner links of $\mathsf{ST}_{\overline{T}}$:

Lemma 1. *Let $\mathsf{ST}_T = (V, E)$. There is a bijection between $\mathcal{E}_T^r$ and the set of all explicit Weiner links from nodes of $\mathsf{ST}_{\overline{T}}$ that correspond to maximal repeats of T. There is a bijection between $\mathcal{F}_T^r$ and the set of all implicit Weiner links from nodes of $\mathsf{ST}_{\overline{T}}$ that correspond to maximal repeats of T.*

The proof of Lemma 1 is provided in the appendix. It is clear that the set of suffix tree edges $\mathcal{E}_T^r \cup \mathcal{F}_T^r$ is in one-to-one correspondence with the set of all arcs of CDAWG_T. This set of edges is also related to runs in BWT_T:

Theorem 1. $|[0..\sigma] \setminus \cup_{W \in \mathcal{M}_T^r} \Sigma_T^\ell(W)| + \sum_{W \in \mathcal{M}_T^r} |\Sigma_T^\ell(W)| - |\mathcal{M}_T^r| + 1 \leq |\mathcal{R}_T| \leq |\mathcal{F}_T^r|$.

Proof. The root of ST_T is a maximal repeat, thus the destinations of all edges in $\mathcal{F}^r$ partition all leaves of ST_T into disjoint subtrees, or equivalently they partition the entire BWT_T in disjoint blocks. Since every such block is the interval in BWT_T of some string that is not left-maximal, all characters of BWT_T in the same block are identical, thus the number of runs in BWT_T cannot be bigger than $|\mathcal{F}^r|$.

The interval of a string $W \in \mathcal{M}^r$ in BWT_T contains exactly $|\Sigma^\ell(W)|$ distinct characters, and at most one of them is identical to the character that precedes the largest suffix of T smaller than W in lexicographic order (note that such suffix might not be prefixed by any string in $\mathcal{M}^r$). Thus, the number of runs in BWT_T is at least $\sum_{W \in \mathcal{M}^r} |\Sigma^\ell(W)| - |\mathcal{M}^r| + 1$. Factor $[0..\sigma] \setminus \cup_{W \in \mathcal{M}^r} \Sigma_T^\ell(W)$ in the claim takes into account symbols of T that never occur to the left of strings in $\mathcal{M}^r$. $\square$

A symmetrical argument holds for $\mathcal{R}_{\overline{T}}$. The set of arcs in CDAWG_T is also related to the LZ factorization of T:

Theorem 2. $|\mathcal{Z}_T| \leq |\mathcal{E}_T^r \cup \mathcal{F}_T^r|$

Proof. Let $T = T_1 T_2 \ldots T_z$ be the LZ factorization of T, and let $p_1, p_2, \ldots, p_z$ be the sequence such that p_i is the starting position of factor T_i in T. Every factor is a right-maximal substring of T, but it is not necessarily left-maximal: let W_i be a suffix of $T[1..p_i - 1]$ such that $W_i T_i$ is both right-maximal and left-maximal, and assume that we assign T_i to the edge (v, w) in $\mathcal{E}_T^r \cup \mathcal{F}_T^r$ such that $\ell(v) = W_i T_i$, $v = \mathtt{parent}(w)$, and the first character of T_{i+1} equals the first character of $\ell(v, w)$. Assume that there is some $j > i$ for which we assign T_j to the same maximal repeat $W_i T_i$. Then, the first character of T_{j+1} must be different from the first character of T_{i+1}, otherwise factor T_j would have been longer. It follows that every LZ factor can be assigned to a distinct element of $\mathcal{E}_T^r \cup \mathcal{F}_T^r$. □

The gap between r and e, and between z and e, is apparent from Fig. 1 (center). However, all these measures seem to grow at the same relative rate in practice (right panel).

2.2   Repetition-Aware Data Structures

Given a string $T \in [1..\sigma]^{n-1}\#$, we call *run-length encoded BWT* any representation of BWT_T that takes $O(|\mathcal{R}_T|)$ words of space, and that supports rank and select operations: see for example [13,14,18]. Let $\mathcal{R}_T$ be a set of triplets (c, i, j) such that $\mathsf{BWT}_T[i..j]$ is a run of character c. It is easy to implement rank in $O(\log \log n)$ time, by encoding $\mathcal{R}_T$ as $\sigma + 1$ predecessor data structures [19], each of which stores the second component of all triplets with the same first component. For every such second component i, we also store in an array the sum of all occurrences of c up to i, exclusive. To implement select in $O(\log \log n)$ time, we can similarly encode $\mathcal{R}_T$ as $\sigma + 1$ predecessor data structures, each of which stores value $\mathtt{rank}_c(\mathsf{BWT}_T, i-1)$ for all triplets (c, i, j) with the same value of c. We also store the value of i for every such triplet. We denote the run-length encoded BWT of T by RLBWT_T.

   For reasons of space we assume the reader to be familiar with LZ77-indexes: see e.g. [9,10]. Here we just recall that a *primary occurrence* of a pattern P in a string $T \in [1..\sigma]^{n-1}\#$ is one that crosses a phrase boundary in the LZ77 factorization $T_1 T_2 \cdots T_z$ of T. All other occurrences are called *secondary*. Once we have determined all primary occurrences, locating secondary occurrences reduces to two-sided range reporting and takes $O(\mathrm{occ} \log \log n)$ time with a data structure that takes $O(z)$ words of space [10]. To locate primary occurrences, we can use a data structure for four-sided range reporting on a $z \times z$ grid, with a marker at (x, y) if the xth LZ factor in lexicographic order is preceded in the text by the lexicographically yth reversed prefix ending at a phrase boundary. This data structure takes $O(z)$ words of space, and it returns all the phrase boundaries immediately followed by a factor in the specified range, and immediately preceded by a reversed prefix in the specified range, in $O((1 + k) \log^\epsilon z)$ time, where k is the number of phrase boundaries reported [4].

3   Combining Runs in BWT and LZ Factors

In this section we describe how to combine data structures whose size depends on the number of LZ factors of a string $T \in [1..\sigma]^{n-1}\#$, and data structures whose size depends on the number of runs in BWT_T, to report all the occurrences of a pattern in T. To do so, we first need to solve the following subproblem. Let $\mathsf{ST}_T = (V, E)$ be the suffix tree of T, and let $V' = \{v_1, v_2, \ldots, v_k\} \subseteq V$ be a subset of the nodes of ST_T. Consider the list of node labels $L = \ell(v_1), \ell(v_2), \ldots, \ell(v_k)$, sorted in lexicographic order. Given a string $W \in [0..\sigma]^*$, we want to implement function $\mathbb{I}(W, V')$ that returns the (possibly empty) interval of W in L. The following lemma describes how to do this in $O(k)$ words of space:

Lemma 2. *Let $T \in [1..\sigma]^{n-1}\#$ be a string, and let V' be a subset of k nodes of its suffix tree, represented as intervals in BWT_T. Given the interval $[i..j]$ of a string $W \in [0..\sigma]^*$ in BWT_T, there is a data structure that takes $O(k)$ words of space and that computes $\mathbb{I}(W, V')$ in $O(\log k)$ time.*

Proof. Let $F[1..n]$ be a bitvector such that $F[i] = 1$ iff there is a node $v' \in V'$ such that $\mathtt{range}(v') = [i..j]$. Similarly, let $L[1..n]$ be a bitvector such that $L[j] = 1$ iff there is a node $v' \in V'$ such that $\mathtt{range}(v') = [i..j]$. Let α and β be the number of ones in F and L, respectively. We store in array $\mathtt{first}[1..\alpha]$ (respectively, $\mathtt{last}[1..\beta]$) the sorted positions of the ones in F (respectively, in L), using $O(k)$ words of space. Let $F'[1..\alpha]$ be the array such that $F'[i]$ equals the number of intervals $[p..q]$ such that p is the ith one in F and $[p..q] = \mathtt{range}(v')$ for a node $v' \in V'$. Similarly, let $L'[1..\beta]$ be the array such that $L'[i]$ equals the number of intervals $[p..q]$ such that q is the ith one in L and $[p..q] = \mathtt{range}(v')$ for a node $v' \in V'$. We represent F' and L' as prefix-sum arrays $\mathtt{first}'[1..\alpha]$ and $\mathtt{last}'[1..\beta]$ using $O(k)$ words of space, i.e. $\mathtt{first}'[i] = \sum_{h=1}^{i} F'[h]$ and $\mathtt{last}'[i] = \sum_{h=1}^{i} L'[h]$.

Let $\mathbb{I}(W, V') = [x..y]$. Given the interval $[i..j]$ of a string W in BWT_T, we find the corresponding interval $[i'..j']$ in array $\mathtt{first}$ in $O(\log \alpha)$ time, using binary search on $\mathtt{first}'$. Specifically, $i' = \min\{h \in [1..\alpha] : \mathtt{first}'[h] \geq i\}$ and $j' = \max\{h \in [1..\alpha] : \mathtt{first}'[h] \leq j\}$. If $j' < i'$ then W is not the prefix of a label of a node in V'. Otherwise, since all nodes $v' \in V'$ whose BWT interval starts inside $[i+1..j]$ are right extensions of W, we set $y = \sum_{h=1}^{j'} F'[h] = \mathtt{first}'[j']$ in constant time. If $\mathtt{first}[i'] \neq i$, i.e. if no interval of a node $v' \in V'$ starts at position i in BWT_T, then we can just set $x = 1 + \sum_{h=1}^{i'-1} F'[h] = 1 + \mathtt{first}'[i'-1]$ in constant time and stop.

Otherwise, it could happen that just a (possibly empty) subset of all the nodes in V' whose interval starts at position i in BWT_T correspond to W or to right extensions of W: the intervals of such nodes necessarily end inside $[i..j]$. All the other intervals that start at position i could correspond instead to *prefixes* of W, and they necessarily end after position j in BWT_T. Thus, let $[i''..j'']$ be the interval in $\mathtt{last}$ that corresponds to $[i..j]$: specifically, let $i'' = \min\{h \in [1..\beta] : \mathtt{last}[h] \geq i\}$ and $j'' = \max\{h \in [1..\beta] : \mathtt{last}[h] \leq j\}$. To determine the number of intervals that start at position i in BWT_T and that correspond to prefixes of W, it suffices to compute the difference δ between the number of starting positions and the number of ending positions inside interval $[i..j]$, as follows: $\delta = \sum_{h=1}^{j'} F'[h] - \sum_{h=1}^{i'-1} F'[h] - \sum_{h=1}^{j''} L'[h] + \sum_{h=1}^{i''-1} L'[h]$. Then, $x = \sum_{h=1}^{i'-1} F'[h] + \delta + 1$. All such sums can be computed in constant time using the prefix-sum representations of F' ad L'.

If the interval of some node in V' starts at i and ends after j in BWT_T, then no interval can end at j and start before i, so δ is nonnegative. $\qquad\square$

Consider now a factorization of T such that all factors are right-maximal substrings of T, and let V' be the set of nodes of ST_T that correspond to the distinct factors. To locate all the occurrences of a pattern that cross or end at a boundary between two factors, we just need an implementation of function $\mathbb{I}(W, V')$ and a pair of RLBWTs:

Lemma 3. *Let $T \in [1..\sigma]^{n-1}\#$ be a string, and let $T = T_1 T_2 \cdots T_z$ be a factorization of T in which all factors are right-maximal substrings. There is a data structure that takes $O(z + r_T + \bar{r}_T)$ words of space and that reports all the occ occurrences of a pattern $P \in [0..\sigma]^m$ that cross or end at a boundary between two factors of T, in $O(m(\log\log n + \log z) + \mathrm{occ}\log^{\epsilon} z)$ time.*

Proof. Let $p_1, p_2, \ldots, p_z$ be the sequence such that p_i is the starting position of factor T_i in T. The same occurrence of P in T can cover up to m boundaries between two factors, thus we organize the computation as follows. We consider every possible way to place *the rightmost boundary between two factors* in P, i.e. every possible split of P into two parts $P[1..k-1]$ and $P[k..m]$ for $k \in [1..m]$, such that $P[k..m]$ is either a factor or a proper prefix of a factor. For every such k, we use four-sided range reporting queries to list all the occurrences of P in T that conform to this split, as described in Sect. 2.2. The four-sided range reporting data structure represents the mapping between the lexicographic rank of a factor W among all the distinct factors of T, and the lexicographic ranks of all the reversed prefixes $\overline{T[1..p_i - 1]}$ such that $T_i = W$, among all the reversed prefixes of T that end at the last position of a factor. As described in Sect. 2.2, this data structure takes $O(z)$ words of space.

We encode sequence $p_1, p_2, \ldots, p_z$ implicitly, as follows: we use a bitvector $\mathtt{last}[1..n]$ such that $\mathtt{last}[i] = 1$ iff $\mathsf{SA}_{\overline{T}}[i] = n - p_j + 2$ for some $j \in [1..z]$, i.e. iff $\mathsf{SA}_{\overline{T}}[i]$ is the last position of a factor. We represent such bitvector as a predecessor data structure with partial ranks, using $O(z)$ words of space [19]. Then, we build the data structure described in Lemma 2, where V' is the set of loci in ST_T of all factors of T. This data structure takes $O(z)$ words of space, and together with $\mathtt{last}$, RLBWT_T and $\mathsf{RLBWT}_{\overline{T}}$, it is the output of our construction.

Given a pattern $P \in [0..\sigma]^m$, we first perform a backward search in RLBWT_T to determine the number of occurrences of P in T: if this number is zero, we stop. During this backward search, we store in a table the interval $[i_k..j_k]$ of $P[k..m]$ in BWT_T for every $k \in [2..m]$. Then, we compute the interval $[i'_{k-1}..j'_{k-1}]$ of $\overline{P[1..k-1]}$ in $\mathsf{BWT}_{\overline{T}}$ for every $k \in [2..m]$, using backward search in $\mathsf{RLBWT}_{\overline{T}}$: if $\mathtt{rank}_1(\mathtt{last}, j'_{k-1}) - \mathtt{rank}_1(\mathtt{last}, i'_{k-1} - 1) = 0$, then $P[1..k-1]$ never ends at the last position of a factor, and we can discard this value of k. Otherwise, we convert $[i'_{k-1}..j'_{k-1}]$ to the interval $[\text{rank}_1(\text{last}, i'_{k-1}) + 1..\text{rank}_1(\text{last}, j'_{k-1})]$ of all the reversed prefixes of T that end at the last position of a factor. Rank operations on last can be implemented in $O(\log \log n)$ time using predecessor queries. We get the lexicographic interval of $P[k..m]$ in the list of all the distinct factors of T using operation $\mathbb{I}(P[k..m], V')$, in $O(\log z)$ time. We use such intervals to query the four-sided range reporting data structure. $\square$

The algorithm described in Lemma 3 can be engineered in a number of ways in practice. Here we just apply it to the LZ factorization of T to find all the primary occurrences of P in T, and we use the strategy described in Sect. 2.2 to compute secondary occurrences, obtaining the key result of this section:

Theorem 3. *Let $T \in [1..\sigma]^{n-1}\#$ be a string, and let $T = T_1 T_2 \ldots T_z$ be its LZ factorization. There is a data structure that takes $O(z + r_T + \bar{r}_T)$ words of space and that reports all the pocc primary occurrences and all the socc secondary occurrences of a pattern $P \in [0..\sigma]^m$ in $O(m(\log \log n + \log z) + \text{pocc} \log^\epsilon z + \text{socc} \log \log n)$ time.*

4   Combining Runs in BWT and Maximal Repeats

An alternative way to compute all the occurrences of a pattern in a string T consists in combining RLBWT_T with CDAWG_T, using an amount of space proportional to the number of right extensions of the maximal repeats of T:

Theorem 4. *Let $T \in [1..\sigma]^{n-1}\#$ be a string. There is a data structure that takes $O(e_T)$ words of space (or alternatively, $O(e^\ell_T)$ words of space) and that reports all the occ occurrences of a pattern $P \in [0..\sigma]^m$ in $O(m \log \log n + \text{occ})$ time.*

Proof. We build RLBWT_T and CDAWG_T. For every node v in the CDAWG, we store $|\ell(v)|$ in a variable $v.\text{length}$. Recall that an arc (v, w) of the CDAWG means that maximal repeat $\ell(w)$ can be obtained by extending maximal repeat $\ell(v)$ to the right and to the left. Thus, for every arc $\gamma = (v, w)$ of CDAWG_T, we store the first character of $\ell(\gamma)$ in a variable $\gamma.\text{char}$, and we store the length of the right extension implied by γ in a variable $\gamma.\text{right}$. The length $\gamma.\text{left}$ of the left extension implied by γ can be computed by $w.\text{length} - v.\text{length} - \gamma.\text{right}$. Clearly arcs of CDAWG_T that correspond to edges of ST_T in set $\mathcal{E}^r_T$ induce no left extension. For every arc of CDAWG_T that connects a maximal repeat W to the sink, we store just $\gamma.\text{char}$ and the starting position $\gamma.\text{pos}$ of string $W \cdot \gamma.\text{char}$ in T. The total space used by the CDAWG is clearly $O(e)$ words, and by Theorem 1 the space used by RLBWT_T is $O(|\mathcal{F}^r_T|)$ words. An alternative construction could use $\text{CDAWG}_{\overline{T}}$ and $\text{RLBWT}_{\overline{T}}$.

We use the RLBWT to count the number of occurrences of P in T in $O(m \log \log n)$ time: if this number is greater than zero, we use the CDAWG to report all the occ occurrences of P in T in $O(\text{occ})$ time, using the technique sketched in [5]. Specifically, since we know that P occurs in T, we perform a blind search for P in the CDAWG, as is typically done with Patricia trees. We keep a variable i, initialized to zero, that stores the length of the prefix of P that we have matched so far, and we keep a variable j, initialized to one, that stores the starting position of P inside the last maximal repeat encountered during the search. For every node v in the CDAWG, we choose the arc γ such that γ.char $= P[i+1]$ in constant time using hashing, we increment i by γ.right, and we increment j by γ.left. If the search leads to the sink by an arc γ, we report γ.pos $+ j$ and we stop. If the search leads to a node v that is associated with the maximal repeat W, we determine all the occurrences of W in T by performing a depth-first traversal of all the nodes in the CDAWG that are reachable from v, updating variables i and j as described above, and reporting γ.pos $+ j$ for every arc γ that leads to the sink. The total number of nodes and arcs reachable from v is clearly $O(\mathsf{occ})$.   □

The combination of CDAWG_T and RLBWT_T can also be used to implement a repetition-aware representation of ST_T. We will apply the following property to support operations on ST_T:

Property 1. A maximal repeat $W = [1..\sigma]^m$ of T is the equivalence class of all the right-maximal strings $\{W[1..m], \ldots, W[k..m]\}$ such that $W[k+1..m]$ is left-maximal, and $W[i..m]$ is not left-maximal for all $i \in [2..k]$. Equivalently, the node v' of CDAWG_T with $\ell(v') = W$ is the equivalence class of the nodes $\{v_1, \ldots, v_k\}$ of ST_T such that $\ell(v_i) = W[i..m]$ for all $i \in [1..k]$, and such that $v_k, v_{k-1}, \ldots, v_1$ is a maximal unary path of Weiner links.

Thus, the set of right-maximal strings that belong to the equivalence class of a maximal repeat can be represented by a single integer k, and a right-maximal string can be identified by the maximal repeat W it belongs to, and by the length of the corresponding suffix of W. In BWT_T, the right-maximal strings in the same equivalence class enjoy the following additional properties:

Property 2. Let $\{W[1..m], \ldots, W[k..m]\}$ be the right-maximal strings that belong to the equivalence class of maximal repeat $W \in [1..\sigma]^m$, and let $\mathtt{range}(W[i..m]) = [p_i..q_i]$ for $i \in [1..k]$. Then:

1. $|q_i - p_i + 1| = |q_j - p_j + 1|$ for all i and j in $[1..k]$.
2. $\mathsf{BWT}_T[p_i..q_i] = W[i-1]^{q_i-p_i+1}$ for $i \in [2..k]$. Conversely, $\mathsf{BWT}_T[p_1..q_1]$ contains at least two distinct characters.
3. $p_{i-1} = C[c] + \mathtt{rank}_c(\mathsf{BWT}_T, p_i)$ and $q_{i-1} = p_{i-1} + q_i - p_i$ for $i \in [2..k]$, where $c = W[i-1] = \mathsf{BWT}_T[p_i]$.
4. $p_{i+1} = \mathtt{select}_c(\mathsf{BWT}_T, p_i - C[c])$ and $q_{i+1} = p_{i+1} + q_i - p_i$ for $i \in [1..k-1]$, where $c = W[i]$ is the character that satisfies $C[c] < p_i \leq C[c+1]$. This can be computed in $O(\log \log n)$ time using a predecessor data structure that uses $O(\sigma)$ words of space [19].
5. Let $c \in [0..\sigma]$, and let $\mathtt{range}(W[i..m]c) = [x_i..y_i]$ for $i \in [1..k]$. Then, $x_i = p_i + x_1 - p_1$ and $y_i = p_i + y_1 - p_1$.

Table 1. Time complexities of two representations of ST_T: with intervals in BWT_T (row 1) and without intervals in BWT_T (row 2).

	stringDepth	isAncestor	parent	child	suffixLink	weinerLink	edgeChar	nLeaves
	locateLeaf		nextSibling	firstChild				
1	$O(1)$	$O(1)$	$O(\log\log n)$	$O(1)$	$O(\log\log n)$	$O(\log\log n)$	$O(\log\log n)$	$O(1)$
2	$O(1)$		$O(\log\log n)$	$O(1)$	$O(1)$			

The final property we will exploit relates the equivalence class of a maximal repeat to the equivalence classes of its in-neighbors in the CDAWG:

Property 3. Let w be a node in $CDAWG_T$ with $\ell(w) = W \in [1..\sigma]^m$, and let $\mathcal{S}_w = \{W[1..m], \ldots, W[k..m]\}$ be the right-maximal strings that belong to the equivalence class of node w. Let $\{v^1, \ldots, v^t\}$ be the in-neighbors of w in $CDAWG_T$, and let $\{V^1, \ldots, V^t\}$ be their labels. Then, $\mathcal{S}_w$ is partitioned into t disjoint sets $\mathcal{S}_w^1, \ldots, \mathcal{S}_w^t$ such that $\mathcal{S}_w^i = \{W[x^i+1..m], W[x^i+2..m], \ldots, W[x^i + |\mathcal{S}_{v^i}|..m]\}$, and the right-maximal string $V^i[p..|V^i|]$ labels the parent of the locus of the right-maximal string $W[x^i + p..m]$ in ST_T.

Proof. It is clear that the parent in ST_T of every right-maximal string in the equivalence class of node w belongs to the equivalence class of an in-neighbor of w: we focus here just on showing that the in-neighbors of w induce a partition on the equivalence class of w. Assume that the character that labels arc $\gamma = (v^i, w)$ in the CDAWG is c. Since arc γ exists, we can factorize W as $X^i V^i Y^i$, where $Y^i[1] = c$, and we know that no prefix of $V^i Y^i$ longer than V^i is right-maximal, and that no suffix of W longer than $|V^i Y^i|$ is left-maximal. Consider any suffix $V^i[p..|V^i|]$ of V^i that belongs to the equivalence class of V^i: if $p > 1$, then $W[|X^i|+p..m]$ is not left-maximal, thus $W[|X^i|+p..m]$ belongs to the equivalence class of W. Its prefix $V^i[p..|V^i|]$ is right-maximal, and no longer prefix is right-maximal. Indeed, assume that string $V^i[p..|V^i|]Z^i$ is right-maximal for some prefix Z^i of Y^i. Since $V^i[p..|V^i|]$ is not left-maximal, then string $V^i[p..|V^i|]Z^i$ is not left-maximal either, and this implies that $V^i Z^i$ is right-maximal, contradicting the hypothesis. Thus, string $V^i[p..|V^i|]$ labels the parent of the locus of string $W[|X^i| + p..m]$ in ST_T. If $p = 1$ and $V^i Y^i$ is not left-maximal, the same argument applies. If $V^i Y^i$ is left-maximal, then $W = V^i Y^i$, and since no right-maximal prefix of W longer than V^i exists, we have that V^i labels the parent of the locus of W in ST_T.   □

Combining Properties 1, 2 and 3, we obtain the following results:

Theorem 5. *Let $T \in [1..\sigma]^{n-1}\#$ be a string. There are two implementations of ST_T that take $O(e_T + e_T^\ell)$ words of space each, and that support the operations in Table 1 with the specified time complexities.*

Proof. We build $RLBWT_T$ and $CDAWG_T$, and we annotate the latter as described in Theorem 4, with the only difference that arcs that connect a maximal repeat to the sink are annotated with character and length like all other arcs. We store in every node v of the CDAWG the number $v.\texttt{size}$ of right-maximal strings that belong to its equivalence class, the interval $[v.\texttt{first}..v.\texttt{last}]$ of $\ell(v)$ in BWT_T, a linear-space predecessor data structure [19] on the boundaries induced on the equivalence class of v by its in-neighbors (see Observation 3), and pointers to the in-neighbor that corresponds to the interval associated with each boundary. Finally, we add to the CDAWG all suffix links (v, w) from ST_T such that both v and w are maximal repeats, and the corresponding explicit Weiner links.

We represent a node v of ST_T as a tuple $\texttt{id}(v) = (v', |\ell(v)|, i, j)$, where v' is the node in CDAWG_T that corresponds to the equivalence class of v, and $[i..j]$ is the interval of $\ell(v)$ in BWT_T. Thus, operation $\texttt{stringDepth}$ can be implemented in constant time, and if v is a leaf, the second component of $\texttt{id}(v)$ is its starting position in T. Operation $\texttt{isAncestor}$ can be implemented by testing the containment of the corresponding intervals in BWT_T. To implement operation $\texttt{suffixLink}$, we first check whether $|\ell(v)| = v'.\texttt{length} - v'.\texttt{size} + 1$: if so, we take the suffix link (v', w') from v' and we return $(w', w'.\texttt{length}, w'.\texttt{first}, w'.\texttt{last})$. Otherwise, we return $(v', |\ell(v)| - 1, i', j')$, where $[i'..j']$ is computed as described in point 2 of Property 2. To implement $\texttt{weinerLink}$ for some character c, we first check whether $|\ell(v)| = v'.\texttt{length}$: if so, we take the Weiner link (v', w') from v' labeled by character c (if any), and we return $(w', w'.\texttt{length} - w'.\texttt{size} + 1, i', j')$, where $[i'..j']$ is computed by taking a backward step with character c from $[v'.\texttt{first}..v'.\texttt{last}]$. Otherwise, we check whether $\mathsf{BWT}_T[i] = c$: if so, we return $(v', |\ell(v)| + 1, i', j')$, where $[i'..j']$ is computed as described in point 2 of Property 2.

To implement $\texttt{child}$ for some character c, we follow the arc $\gamma = (v', w')$ in the CDAWG labeled by c (see Observation 3), and we return tuple $(w', |\ell(v)| + \gamma.\texttt{right}, i', j')$, where $[i'..j']$ is computed as described in point 2 of Property 2. To implement $\texttt{parent}$ we exploit Property 2, i.e. we determine the partition of the equivalence class of v' that contains v by searching the predecessor of value $|\ell(v)|$ in the set of boundaries of v': this can be done in $O(\log \log n)$ time [19]. Let $\gamma = (u', v')$ be the arc that connects to v' the in-neighbor u' associated with the partition that contains v: we return tuple $(u', |\ell(v)| - \gamma.\texttt{right}, i', j')$, where $i' = i - v'.\texttt{first} + u'.\texttt{first}$ and $j' = j + u'.\texttt{last} - v'.\texttt{last}$ as described in point 2 of Property 2. Operation $\texttt{nextSibling}$ can be implemented in the same way.

We read the label of an edge γ of ST_T in $O(\log \log n)$ time per character (operation $\texttt{edgeChar}$), by storing $\mathsf{RLBWT}_{\overline{T}}$ and the interval in $\mathsf{BWT}_{\overline{T}}$ of the reverse of the maximal repeat that corresponds to every node of the CDAWG. By removing from $\texttt{id}(v)$ the interval of $\ell(v)$ in BWT_T, we can implement $\texttt{stringDepth}$, $\texttt{child}$, $\texttt{firstChild}$ and $\texttt{suffixLink}$ in constant time, and $\texttt{parent}$ and $\texttt{nextSibling}$ in $O(\log \log n)$ time. □

Corollary 1. *Let $T \in [1..\sigma]^{n-1}\#$ be a string. There is an implementation of ST_T that takes $O(e_T + e_T^\ell)$ words of space, that computes the matching statistics of a pattern $S \in [1..\sigma]^m$ with respect to T in $O(m \log \log n)$ time, and that can be traversed in $O(n \log \log n)$ time and in a constant number of words of space.*

Proof. We combine the implementation in the first row of Table 1 with the folklore algorithm for matching statistics, that issues $\texttt{suffixLink}$ and $\texttt{child}$ operations on ST_T, and that reads the label of some edges of ST_T. For traversal, we combine the implementation in the second row of Table 1 with the folklore algorithm that issues just firstChild, parent and nextSibling operations. □

By storing $\mathsf{RLBWT}_{\overline{T}}$ in addition to RLBWT_T, and by adding to $\mathrm{id}(v)$ the interval of $\overline{\ell(v)}$ in $\mathsf{BWT}_{\overline{T}}$, we can also implement a bidirectional index on T like those described in [2], that supports the left and right extension of a string with any character in $O(\log \log n)$ time and that takes $O(e + e^\ell)$ words of space.

References

1. Arroyuelo, D., Navarro, G., Sadakane, K.: Stronger Lempel-Ziv based compressed text indexing. Algorithmica **62**(1–2), 54–101 (2012)
2. Belazzougui, D., Cunial, F., Kärkkäinen, J., Mäkinen, V.: Versatile succinct representations of the bidirectional burrows-wheeler transform. In: Bodlaender, H.L., Italiano, G.F. (eds.) ESA 2013. LNCS, vol. 8125, pp. 133–144. Springer, Heidelberg (2013)
3. Blumer, A., Blumer, J., Haussler, D., McConnell, R., Ehrenfeucht, A.: Complete inverted files for efficient text retrieval and analysis. J. ACM **34**(3), 578–595 (1987)
4. Chan, T.M., Larsen, K.G., Pătraşcu, M.: Orthogonal range searching on the RAM, revisited. In: Proceedings of SoCG, pp. 1–10 (2011)
5. Crochemore, M., Hancart, C.: Automata for matching patterns. In: Rozenberg, G., Salomaa, A. (eds.) Handbook of Formal Languages, vol. 2, pp. 399–462. Springer, Heidelberg (1997)
6. Crochemore, M., Vérin, R.: Direct construction of compact directed acyclic word graphs. In: Apostolico, A., Hein, J. (eds.) Proceedings of CPM. LNCS, vol. 1264, pp. 116–129. Springer, Heidelberg (1997)
7. Ferragina, P., Manzini, G.: Indexing compressed text. J. ACM **52**(4), 552–581 (2005)
8. P. Ferragina and G. Navarro. Pizza&Chili repetitive corpus. http://pizzachili.dcc. uchile.cl/repcorpus.html. Accessed on 25 January 2015
9. Gagie, T., Gawrychowski, P., Kärkkäinen, J., Nekrich, Y., Puglisi, S.J.: LZ77-based self-indexing with faster pattern matching. In: Pardo, A., Viola, A. (eds.) LATIN 2014. LNCS, vol. 8392, pp. 731–742. Springer, Heidelberg (2014)
10. Kärkkäinen, J., Ukkonen, E.: Lempel-Ziv parsing and sublinear-size index structures for string matching. In: Proceedings of WSP, pp. 141–155 (1996)
11. Kreft, S., Navarro, G.: On compressing and indexing repetitive sequences. Theor. Comput. Sci. **483**, 115–133 (2013)
12. Lempel, A., Ziv, J.: On the complexity of finite sequences. IEEE Trans. Info. Theory **22**(1), 75–81 (1976)
13. Mäkinen, V., Navarro, G.: Succinct suffix arrays based on run-length encoding. In: Apostolico, A., Crochemore, M., Park, K. (eds.) CPM 2005. LNCS, vol. 3537, pp. 45–56. Springer, Heidelberg (2005)
14. Mäkinen, V., Navarro, G., Sirén, J., Välimäki, N.: Storage and retrieval of highly repetitive sequence collections. J. Comput. Biol. **17**(3), 281–308 (2010)
15. Radoszewski, J., Rytter, W.: On the structure of compacted subword graphs of ThueMorse words and their applications. J. Discret. Algorithms **11**, 15–24 (2012)
16. Raffinot, M.: On maximal repeats in strings. Inform. Process. Lett. **80**(3), 165–169 (2001)

17. Rytter, W.: The structure of subword graphs and suffix trees of Fibonacci words. Theoret. Comput. Sci. **363**(2), 211–223 (2006)
18. Sirén, J., Välimäki, N., Mäkinen, V., Navarro, G.: Run-length compressed indexes are superior for highly repetitive sequence collections. In: Amir, A., Turpin, A., Moffat, A. (eds.) SPIRE 2008. LNCS, vol. 5280, pp. 164–175. Springer, Heidelberg (2008)
19. Willard, D.E.: Log-logarithmic worst-case range queries are possible in space $\Theta(N)$. Inform. Process. Lett. **17**(2), 81–84 (1983)
20. Ziv, J., Lempel, A.: A universal algorithm for sequential data compression. IEEE Trans. Info. Theory **23**(3), 337–343 (1977)

Efficient Construction of a Compressed de Bruijn Graph for Pan-Genome Analysis

Timo Beller[(✉)] and Enno Ohlebusch

Institute of Theoretical Computer Science, University of Ulm,
89069 Ulm, Germany
{Timo.Beller,Enno.Ohlebusch}@uni-ulm.de

Abstract. Recently, Marcus et al. (Bioinformatics 2014) proposed to use a compressed de Bruijn graph of maximal exact matches to describe the relationship between the genomes of many individuals/strains of the same or closely related species. They devised an $O(n \log g)$ time algorithm called splitMEM that constructs this graph directly (i.e., without using the uncompressed de Bruijn graph) based on a suffix tree, where n is the total length of the genomes and g is the length of the longest genome. In this paper, we present an algorithm that outperforms their algorithm in theory and in practice. More precisely, our algorithm has a better worst-case time complexity of $O(n \log \sigma)$, where σ is the size of the alphabet ($\sigma = 4$ for DNA). Moreover, experiments show that it is much faster than splitMEM while using only a fraction of the space required by splitMEM.

1 Introduction

Today, next generation sequencers produce vast amounts of DNA sequence information and it is often the case that multiple genomes of the same or closely related species are available. An example is the 1000 Genomes Project, which started 2008. Its goal was to sequence the genomes of at least 1000 humans from all over the world and to produce a catalog of all variations (SNPs, indels, etc.) in the human population. The genomic sequences together with this catalog is called the "pan-genome" of the population. There are several approaches that try to capture variations between many individuals/strains in a population graph; see e.g. [11,20,22]. These works all require a multi-alignment as input. By contrast, Marcus et al. [14] use a compressed de Bruijn graph of maximal exact matches (MEMs) as a graphical representation of the relationship between genomes; see Sect. 3 for a definition of de Bruijn graphs. They describe an $O(n \log g)$ time algorithm that directly computes the compressed de Bruijn graph based on a suffix tree, where n is the total length of the genomes and g is the length of the longest genome. Marcus et al. write in [14, Sect. 4]: "Future work remains to improve splitMEM and further unify the family of sequence indices. Although ..., most desired are techniques to reduce the space consumption ..." In this paper, we present such a technique. To be more precise, we will develop an $O(n \log \sigma)$ time algorithm that computes the compressed de Bruijn

F. Cicalese et al. (Eds.): CPM 2015, LNCS 9133, pp. 40–51, 2015.
DOI: 10.1007/978-3-319-19929-0_4

i	SA	LCP	B	BWT	$S_{SA[i]}$
1	15	-1	0	G	$
2	12	0	1	T	ACG$
3	8	3	0	T	ACGTACG$
4	4	7	1	T	ACGTACGTACG$
5	1	2	0	$	ACTACGTACGTACG$
6	13	0	0	A	CG$
7	9	2	0	A	CGTACG$
8	5	6	0	A	CGTACGTACG$
9	2	1	0	A	CTACGTACGTACG$
10	14	0	0	C	G$
11	10	1	0	C	GTACG$
12	6	5	0	C	GTACGTACG$
13	11	0	0	G	TACG$
14	7	4	0	G	TACGTACG$
15	3	8	0	C	TACGTACGTACG$
16		-1			

Fig. 1. The suffix array of the string ACTACGTACGTACG$.

graph directly based on an FM-index of the genomes, where σ is the size of the underlying alphabet.

Closely related is the *contracted* de Bruijn graph introduced by Cazaux et al. [5]. A node in the contracted de Bruijn graph is not necessarily a substring of one of the genomic sequences (see the remark following Definition 3 in [5]). Thus the contracted de Bruijn graph, which can be constructed in linear time from the suffix tree [5], is not useful for our purposes. Nevertheless, it is worth mentioning that Cazaux et al. write in the full version of their paper (a technical report): "Other topics for future research include transforming compressed indexes, such as a FM-index" into a contracted de Bruijn graph. Maybe our new method can be applied to their problem as well, but we did not investigate this yet.

As discussed in [14, Sect. 1.4], techniques such as Bloom filters or succinct representations of de Bruijn graphs cannot directly be extended to pan-genome analysis.

2 Preliminaries

Let Σ be an ordered alphabet of size σ whose smallest element is the sentinel character $. In the following, S is a string of length n on Σ having the sentinel character at the end (and nowhere else). In pan-genome analysis, S is the concatenation of multiple genomic sequences, where the different sequences are separated by special symbols (in practice, we use one separator symbol and treat the different occurrences of it as if they were different characters; see Sect. 5). For $1 \leq i \leq n$, $S[i]$ denotes the *character at position* i in S. For $i \leq j$, $S[i..j]$ denotes the *substring* of S starting with the character at position i and ending with the character at position j. Furthermore, S_i denotes the i-th suffix $S[i..n]$ of S. The *suffix array* SA of the string S is an array of integers in the range 1 to n specifying the lexicographic ordering of the n suffixes of S, that is, it satisfies $S_{SA[1]} < S_{SA[2]} < \cdots < S_{SA[n]}$; see Fig. 1 for an example. A suffix array can be constructed in linear time; see e.g. the overview article [19]. For every substring ω of S, the ω-interval is the suffix array interval $[i..j]$ so that ω is a prefix of $S_{SA[k]}$ if and only if $i \leq k \leq j$.

The Burrows-Wheeler transform [4] converts S into the string $\mathsf{BWT}[1..n]$ defined by $\mathsf{BWT}[i] = S[\mathsf{SA}[i] - 1]$ for all i with $\mathsf{SA}[i] \neq 1$ and $\mathsf{BWT}[i] = \$$ otherwise; see Fig. 1. Several semi-external and external memory algorithms are known that construct the BWT directly (i.e., without constructing the suffix array); see e.g. [3, 6, 13, 18]. The *wavelet tree* [10] of the BWT supports one backward search step in $O(\log \sigma)$ time [7]: Given the ω-interval and $c \in \Sigma$, it returns the $c\omega$-interval if $c\omega$ is a substring of S (otherwise it returns an empty interval). This crucially depends on the fact that a bit vector B can be preprocessed in linear time so that an arbitrary $rank_1(B, i)$ query (asks for the number of ones in B up to and including position i) can be answered in constant time [12]. Backward search can be generalized on the wavelet tree as follows: Given an ω-interval $[lb..rb]$, a slight modification of the procedure $getIntervals([lb..rb])$ described in [2] returns the list $[(c, [i..j]) \mid c\omega$ is a substring of S and $[i..j]$ is the $c\omega$-interval], where the first component of an element $(c, [i..j])$ must be a character. The worst-case time complexity of the procedure $getIntervals$ is $O(z + z \log(\sigma/z))$, where z is the number of elements in the output list; see [8, Lemma 3].

The suffix array SA is often enhanced with the so-called LCP-array containing the lengths of longest common prefixes between consecutive suffixes in SA; see Fig. 1. Formally, the LCP-array is an array so that $\mathsf{LCP}[1] = -1 = \mathsf{LCP}[n+1]$ and $\mathsf{LCP}[i] = |\mathsf{lcp}(S_{SA[i-1]}, S_{SA[i]})|$ for $2 \leq i \leq n$, where $\mathsf{lcp}(u, v)$ denotes the longest common prefix between two strings u and v. The LCP-array can be computed in linear time from the suffix array and its inverse, but it is also possible to construct it directly from the wavelet tree of the BWT in $O(n \log \sigma)$ time with the help of the procedure $getIntervals$ [2].

A substring ω of S is a *repeat* if it occurs at least twice in S. Let ω be a repeat of length ℓ and let $[i..j]$ be the ω-interval. The repeat ω is *left-maximal* if $|\{\mathsf{BWT}[x] \mid i \leq x \leq j\}| \geq 2$, i.e., the set $\{S[\mathsf{SA}[x] - 1] \mid i \leq x \leq j\}$ of all characters that precede at least one of the suffixes $S_{SA[i]}, \ldots, S_{SA[j]}$ is not singleton (where $S[0] := \$$). Analogously, the repeat ω is *right-maximal* if $|\{S[\mathsf{SA}[x] + \ell] \mid i \leq x \leq j\}| \geq 2$.

A detailed explanation of the techniques used here can be found in [16].

3   Construction of a Compressed de Bruijn Graph

Given S and $k > 0$, the de Bruijn graph representation of S contains a node for each distinct length k substring of S, called a k-mer. Two nodes u and v are connected by a directed edge (u, v) if $u = S[i..i + k - 1]$ and $v = S[i + 1..i + k]$; see Fig. 2 for an example. Clearly, the graph contains at most n nodes and n

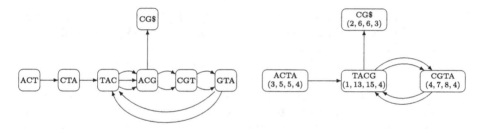

Fig. 2. The de Bruijn graph for $k = 3$ and the string ACTACGTACGTACG\$ is shown on the left, while its compressed counterpart is shown on the right.

edges. By construction, adjacent nodes will overlap by $k - 1$ characters, and the graph can include multiple edges connecting the same pair of nodes or self-loops representing overlapping repeats. For every node, except for the start node (containing the first k characters of S) and the stop node (containing the last k characters of S), the in-degree coincides with the out-degree. A de Bruijn graph can be "compressed" by merging non-branching chains of nodes into a single node with a longer string. More precisely, if node u is the only predecessor of node v and v is the only successor of u (but there may be multiple edges (u, v)), then u and v can be merged into a single node that has the predecessors of u and the successors of v. After maximally compressing the graph, every node (apart from possibly the start node) has at least two different predecessors or its single predecessor has at least two different successors and every node (apart from the stop node) has at least two different successors or its single successor has at least two different predecessors; see Fig. 2. Of course, the compressed de Bruijn graph can be built from its uncompressed counterpart (a much larger graph), but this is disadvantageous because of the huge space consumption. That's why we build it directly from an FM-index (the wavelet tree of the BWT) of S, using Lemma 1 (the simple proof is omitted).

Lemma 1. *Let ω be a node in the compressed de Bruijn graph. If ω is not the start node, then it has at least two different predecessors if and only if the length k prefix of ω is a left-maximal repeat. It has at least two different successors if and only if the length k suffix of ω is a right-maximal repeat.*

The general idea behind our algorithm is as follows. Compute all right-maximal k-mers. For each such k-mer v, compute all cv-intervals, where $c \in \Sigma$. Then, for each $u = cv$, compute all bu-intervals, where $b \in \Sigma$, etc. In other words, we start with all right-maximal k-mers and extend them as long as possible (and in all possible ways with the procedure *getIntervals*), character by character, to the left. According to Lemma 1, the left-extension of a string ω must stop if (i) the length k prefix of ω is a left-maximal repeat (this is the case if the procedure *getIntervals* applied to the ω-interval returns a non-singleton list). It must also stop if (ii) the length k prefix v of $c\omega$ is a right-maximal repeat for some $c \in \Sigma$. This is because by Lemma 1 there is a node uv, $u \in \Sigma^*$, in the compressed de Bruijn graph with at least two different successors (the length k suffix v of uv is a right-maximal repeat). Consequently, there must be a directed edge (uv, ω) in the compressed de Bruijn graph. In the following, we will explain the different phases of the algorithm in detail.

First, we compute all right-maximal k-mers and their suffix array intervals. Moreover, we number them in lexicographic order. This number will serve as a unique identifier. Let u be a right-maximal k-mer and consider the u-interval $[lb..rb]$ in the suffix array. Note that (1) $\mathsf{LCP}[lb] < k$ and (2) $\mathsf{LCP}[rb + 1] < k$. Since u is right-maximal, u is the longest common prefix of all suffixes in the interval $[lb..rb]$. This implies (3) $\mathsf{LCP}[i] \geq k$ for all i with $lb + 1 \leq i \leq rb$ and (4) $\mathsf{LCP}[i] = k$ for at least one i with $lb + 1 \leq i \leq rb$ (in the terminology of Abouelhoda et al. [1], $[lb..rb]$ is an lcp-interval of lcp-value k). With this knowledge, it is not difficult to verify that lines 7 to 18 (ignore lines 11 to 12 for a moment) of Algorithm 1 compute all suffix array intervals of right-maximal k-mers, including their identifiers. Furthermore, on lines 10 and 15 the boundaries of the k-mer intervals are marked by setting the entries of a bit vector B at these positions to 1 (the purpose of the bit vector B will be explained below); see Fig. 1. We would like to stress, however, that all right-maximal k-mers can be determined without the entire LCP-array. In order to verify whether or not an interval satisfies properties (1)–(4), it is sufficient to compute all entries $\leq k$ in the LCP-array (the others have a value $> k$). Since the algorithm presented in [2] calculates entries in the LCP-array in ascending order, we use it for our purpose; see [24] for related work. We initialize an array L with values 2 and set $L[1] = 0$ and $L[n + 1] = 0$. Two bits are enough to encode the case "$< k$" by 0, the case "$= k$" by 1, and the case "$> k$" by 2 (so initially all entries in the LCP-array are marked as being $> k$, except for $L[1]$ and $L[n + 1]$, which are marked as being $< k$). Then, for ℓ from 0 to $k - 1$, the algorithm of Beller et al. [2] calculates all indices p with entries $\mathsf{LCP}[p] = \ell$ and sets $L[p] = 0$. Furthermore, it continues to calculates all indices q with entries $\mathsf{LCP}[q] = k$ and sets $L[q] = 1$. Now the array L contains all the information that is needed to compute right-maximal k-mers. For each such k-mer u, Algorithm 1 inserts a node $node = (id, lb, rb, len)$ into the compressed de Bruijn graph G (line 11), where id is its identifier (lexicographic rank), $[lb..rb]$ is its suffix array interval, and $len = k$ is its length. Furthermore, on line 12, a pointer $node$ to the node (id, lb, rb, len) is added to an initially empty queue Q (one could also use another data structure like a stack to administer these pointers). The identifier $node.id$ of a node will never change, but its attributes $node.lb$, $node.rb$, and $node.len$ will change when a left-extension is possible. As long as the queue Q is not empty, the algorithm removes a pointer $node$ from Q and in a repeat-loop computes $list = getIntervals([node.lb..node.rb])$. During the repeat-loop, the interval $[node.lb..node.rb]$ is the suffix array interval of some string ω of length $node.len$. In the body of the repeat-loop, a flag $extendable$ is set to false. The procedure call $getIntervals([node.lb..node.rb])$ then returns the list $list$ of all $c\omega$-intervals. At this point, the algorithm tests whether or not the length k prefix of $c\omega$ is a right-maximal repeat. It is not difficult to see that the length k prefix of $c\omega$ is a right-maximal repeat if and only if the $c\omega$-interval $[i..j]$ is a subinterval of a right-maximal k-mer interval. Here, the bit vector B comes into play. At the beginning of Algorithm 1, all suffix array intervals of right-maximal k-mers have been computed and their boundaries have been marked in B. It is crucial to note that these intervals are disjoint. Thus, the $c\omega$-interval $[i..j]$ is not a subinterval of a right-maximal k-mer interval if and only if $rank_1(B, i)$ is even and $B[i] = 0$ (the proof of this fact is omitted due to lack of space). Now, the algorithm proceeds by case analysis:

1. If the length k prefix of $c\omega$ is a right-maximal repeat, there must be a node v that ends with the length k prefix of $c\omega$ (note that $c\omega[1..k]$ and ω have a suffix-prefix-overlap of $k - 1$ characters), and this node v will be detected by a computation that starts with the k-mer $c\omega[1..k]$. Consequently, an edge $(v.id, node.id)$ is added to the compressed de Bruijn graph, and the computation stops here. If the length k prefix of $c\omega$ is not a right-maximal repeat, one of the following two cases occurs:

2. If $list$ contains just one element $(c, [i..j])$, then ω is not left maximal. In this case, the algorithms sets $extendable$ to true, $node.lb$ to i, $node.rb$ to j, and increments $node.len$ by one. Now $node$ represents the $c\omega$-interval $[i..j]$ and the repeat-loop continues with this interval.

3. Otherwise, ω is left maximal. In this case, the attributes of $node$ will not change any more. So a node $newNode = (counter, i, j, k)$ is inserted into G. For the correctness of the algorithm, it is important to note that the interval $[i..j]$ is the $c\omega[1..k]$-interval (the proof of this fact is omitted due to lack of space). An edge $(counter, node.id)$ is added to G (note that $c\omega[1..k]$ and ω have a suffix-prefix-overlap of $k - 1$ characters) and the pointer $newNode$ is added to Q.

As an example, we apply Algorithm 1 to $k = 3$ and the LCP-array and the BWT of the string ACTACGTACGTACG\$; see Fig. 1. There is only one right maximal k-mer, ACG, so a node $(id, lb, rb, len) = (1, 2, 4, 3)$ is inserted to G on line 11 and a pointer to it is added to the queue Q. Furthermore, the stop node $(2, 1, 1, 1)$ of G is inserted on line 20 and a pointer to it is enqueued to Q. In the while-loop, the pointer to node $(1, 2, 4, 3)$ is dequeued and the procedure call $getIntervals([2..4])$ returns a list that contains just one interval, the TACG-interval $[13..15]$. Since $rank_1(B, 13) = 2$ is even and $B[13] = 0$, Case 2 applies. So $extendable$ is set to true and node 1 is modified to $(1, 13, 15, 4)$. In the next iteration of the repeat-loop, $getIntervals([13..15])$ returns the list $[(C, [9..9]), (G, [11..12])]$, where $[9..9]$ is the CTACG-interval and $[11..12]$ is the GTACG-interval. It is readily verified that Case 3 applies in both cases, so a node $(3, 9, 9, 3)$ corresponding to the CTA-interval and a node $(4, 11, 12, 3)$ corresponding to the GTA-interval is added to G and pointers to them are added to Q. Furthermore, an edge $(3, 1)$ and two edges $(4, 1)$ (the multiplicity is 2) are added to G. Next, the pointer to node $(2, 1, 1, 1)$ is dequeued and the procedure call $getIntervals([1..1])$ returns a list that contains just one interval, the G\$-interval $[10..10]$. Case 2 applies, so node 2 is modified to $(2, 10, 10, 2)$. In the second iteration of the repeat-loop, $getIntervals([10..10])$ returns the CG\$-interval $[6..6]$. Again Case 2 applies and node 2 is modified

---

Algorithm 1. FM-index based construction of a compressed de Bruijn graph

---

1: **function** CREATE-COMPRESSED-GRAPH(k, LCP, BWT)
2:     $open \leftarrow$ false
3:     $counter \leftarrow 1$
4:     create an empty graph G
5:     create an empty queue Q
6:     initialize a bit vector B of length n with zeros
7:     **for** $i \leftarrow 1$ **to** n **do**   ▷ Note that LCP ends with -1
8:         **if** LCP$[i] < k$ **and** $open$ **then**
9:             $open \leftarrow$ false
10:             $B[i-1] \leftarrow 1$
11:             insert a node $node = (counter, lb, i-1, k)$ into G
12:             $enqueue(Q, node)$ ▷ Add a pointer to node $(counter, lb, i-1, k)$
13:             $counter \leftarrow counter + 1$
14:         **else if** LCP$[i] = k$ **and not** $open$ **then**
15:             $B[lb] \leftarrow 1$
16:             $open \leftarrow$ true
17:         **if** LCP$[i] < k$ **then**
18:             $lb \leftarrow i$

19:     process bit vector B so that $rank_1$ queries can be answered in constant time
20:     insert a node $stopNode = (counter, 1, 1, 1)$ into G
21:     $enqueue(Q, stopNode)$ ▷ Add a pointer to the stop node
22:     $counter \leftarrow counter + 1$
23:     **while** Q is not empty **do**
24:         $node \leftarrow dequeue(Q)$
25:         **repeat**
26:             $extendable \leftarrow$ false
27:             $list \leftarrow getIntervals([node.lb..node.rb])$
28:             **for each** $(c, [i..j])$ in $list$ **do**
29:                 $ones \leftarrow rank_1(B, i)$
30:                 **if** $ones$ is even **and** $B[i] = 0$ **then**
31:                     $number \leftarrow \bot$
32:                 **else**
33:                     $number \leftarrow (ones + 1)/2$
34:                 **if** $number \neq \bot$ **then**   ▷ Case 1
35:                     add an edge $(number, node.id)$ with multiplicity $j - i + 1$ to G
36:                 **else if** $c \neq \$$ **then**
37:                     **if** $list$ contains just one element **then**   ▷ Case 2
38:                         $extendable \leftarrow$ true
39:                         $node.len \leftarrow node.len + 1$
40:                         $node.lb \leftarrow i$
41:                         $node.rb \leftarrow j$
42:                     **else**   ▷ Case 3
43:                         insert a node $newNode = (counter, i, j, k)$ into G
44:                         add edge $(counter, node.id)$ with multiplicity $j - i + 1$ to G
45:                         $enqueue(Q, newNode)$ ▷ Add a pointer to $(counter, i, j, k)$
46:                         $counter \leftarrow counter + 1$
47:         **until not** $extendable$

Algorithm 2. Adding the sorted lists of positions and edges to the nodes.

```
 1: function CREATE-COMPRESSED-GRAPH2(G)
 2:     initialize an array A of length n with ⊥
 3:     m ← number of nodes in G
 4:     for i ← 1 to m do
 5:         node ← node with identifier i
 6:         for j ← node.lb to node.rb do
 7:             A[SA[j]] ← i
 8:     id ← A[1]
 9:     node ← node with identifier id                  ▷ start node
10:     append(node.posList, 1)
11:     for j ← 2 to n do
12:         i ← A[j]
13:         if i ≠ ⊥ then
14:             append(node.adjList, i)                 ▷ add the edge (node.id, i)
15:             node ← node with identifier i
16:         append(node.posList, j)                     ▷ add j at the back of node.posList
```

---

to $(2, 6, 6, 3)$. In the third iteration of the repeat-loop, $getIntervals([6..6])$ returns the ACG\$-interval $[2..2]$. This time, Case 1 applies because $number$ gets the value 1. Consequently, an edge $(1, 2)$ is added to G. The computation continues until the queue Q is empty; the final compressed de Bruijn graph is shown in Fig. 2.

We claim that Algorithm 1 has a worst-case time complexity of $O(n \log \sigma)$. This can be proven by considering the three mutually distinct cases 1–3. In Cases 1 and 3, an edge is added to G. Since G contains at most n edges, this implies that Cases 1 and 3 can occur at most n times. Case 2 can also occur at most n times because there are at most n left-extensions. In summary, at most $2n$ intervals are generated by the procedure $getIntervals$. Since this procedure takes $O(\log \sigma)$ time for each generated interval, the claim follows.

4   A Different Representation of the Graph

Algorithm 1 computes a compressed de Bruijn graph, in which a node is represented by the quadruple (id, lb, rb, len), where $[lb..rb]$ is the suffix array interval and len is the length of the corresponding string ω. Marcus et al. [14], store the positions $SA[lb], \ldots, SA[rb]$ at which ω occurs in S (together with the corresponding outgoing edges) in ascending order. To model this, a node now has five components $(id, lb, rb, len, posList, adjList)$, where $posList$ is the sorted list of positions and $adjList$ is the corresponding adjacency list (depending on the application, redundant information should be removed). In this way, the walk through the graph G that gives S is induced by the adjacency lists (if node v is visited for the i-th time, then its successor is the node that can be found at position i in the adjacency list of v). The representation suggested in [14] can be obtained in three different ways as follows:

i	1	2	3	4	5	6	7	8	9	10	11	12	13	14	15
$A[i]$	3	⊥	1	⊥	4	⊥	1	⊥	4	⊥	1	⊥	2	⊥	⊥

Fig. 3. The array A.

node	id	lb	rb	len	posList	adjList
TACG	1	13	15	4	$[3, 7, 11]$	$[4, 4, 2]$
CG$	2	6	6	3	$[13]$	$[\,]$
ACTA	3	5	5	4	$[1]$	$[1]$
CGTA	4	7	8	4	$[5, 9]$	$[1, 1]$

Fig. 4. Representation of the graph G.

A1 Use a comparison based sorting algorithm as in [14].
A2 Use a non-comparison based sorting algorithm; see Algorithm 2.
A3 Use a backward search to track the suffixes of S in G.

Marcus et al. [14] can afford a comparison-based sorting algorithm because their core algorithm already has a worst-case time complexity of $O(n \log n)$. If we use Algorithm 2 instead, the sort takes only $O(n)$ time. Algorithm 2 applied to the compressed de Bruijn graph of Fig. 2 first yields the array depicted in Fig. 3 and then scans this array from left to right. The identifier i of the start node is $A[1]$, so 1 is added to its *posList*. Let j be the second position with $A[j] \neq \bot$ that is encountered during the scan. The successor of the start node is the node $i = A[j]$, so the algorithm appends i to its (empty) *adjList*, etc. Upon termination, we have the situation depicted in Fig. 4.

Possibility 3, called Algorithm 3 in what follows, takes $O(n \log \sigma)$ time but it does not need the suffix array! This is an important advantage over the other algorithms because memory-usage is the bottleneck in pan-genome analysis. In fact, Algorithm 3 can be obtained by a modification of Algorithm 1. By means of the bit vector B, it is possible to detect (from right to left because a backward search is used) every node—together with its identifier—that ends with a right-maximal k-mer. The remaining difficulty is to modify lines 7 to 18 in such a way that every other node—together with its identifier—can be detected in a similar fashion. For space reasons, a detailed description of Algorithm 3 must be omitted here. It will be presented in the full version of this paper.

5   Implementation Details and Experimental Results

In pan-genome analysis, $S = S^1 \# S^2 \# \ldots S^{m-1} \# S^m \$$ is the concatenation of multiple genomic sequences $S^1, \ldots, S^m$, separated by a special symbol $\#$. To avoid that $\#$ may be part of a repeat, we treat (in contrast to splitMEM) the different occurrences of $\#$ as if they were different characters. Assuming that $\#$ is the second smallest character, this can be achieved as follows. We have seen that all right-maximal k-mers can be determined without the entire LCP-array if the algorithm in [2] is used. If there are $m - 1$ occurrences of $\#$ in total and

Table 1. The first column describes the input file, e.g. 40 *E.coli* (199) means the first forty *E.coli* genomes listed in the supplementary material of [14] (containing 199 million base pairs). The 7 genomes are human genomes and chr1 denotes their first chromosome. The second column contains the values of k and, in parentheses, the size of the compressed de Bruijn graph in bytes per base pair; a minus indicates that only the construction of the needed data structures was measured (which does not depend on k). The remaining columns show the runtimes in seconds and, in parentheses, the maximum main memory usage in bytes per base pair (including the construction). A minus indicates that the respective algorithm was not able to solve its task on our machine equipped with 128 GB of RAM.

file (Mbp)	k (size)	A1	A2	A3	splitMEM
40 *E.coli*	-	**38** (5.00)	**38** (5.00)	127 (**1.32**)	94 (315)
(199)	25 (1.50)	**55** (6.06)	57 (9.18)	190 (**2.87**)	2,170 (572)
	100 (0.65)	**58** (5.00)	65 (7.89)	207 (**1.63**)	1,684 (572)
	1000 (0.06)	**76** (5.00)	81 (7.08)	190 (**1.49**)	1,671 (572)
62 *E.coli*	-	**64** (5.00)	**64** (5.00)	201 (**1.24**)	134 (316)
(310)	25 (1.57)	**86** (6.06)	93 (9.38)	295 (**2.83**)	-
	100 (0.68)	**92** (5.00)	105 (8.19)	331 (**1.68**)	-
	1000 (0.06)	134 (5.00)	**123** (7.33)	305 (**1.53**)	-
7 *chr1*	-	**399** (5.00)	**399** (5.00)	1,163 (**1.24**)	-
(1,736)	25 (3.10)	**601** (7.70)	646 (11.44)	1,910 (**4.45**)	-
	100 (1.59)	**549** (5.88)	598 (9.70)	1,628 (**2.76**)	-
	1000 (1.50)	**606** (5.86)	621 (9.57)	1,655 (**2.66**)	-
7 *genomes*	-	-	-	**22,038** (**1.24**)	-
(21,201)	25 (3.34)	-	-	**33,247** (**4.84**)	-
	100 (1.16)	-	-	**29,641** (**2.22**)	-
	1000 (1.01)	-	-	**29,962** (**2.04**)	-

this algorithm starts with $m - 1$ singleton intervals $[i..i]$, $2 \leq i \leq m$, instead of the #-interval $[2..m]$, then the different occurrences of # are treated as if they were different characters.

We implemented the three algorithms A1–A3 using Simon Gog's library sdsl [9]. Software and test data are available at http://www.uni-ulm.de/in/theo/research/seqana.html. Both A1 and A2 require at least $n \log n$ bits because the suffix array is needed in main memory during the conversion of nodes of the form (id, lb, rb, len) to nodes with the components $(id, lb, rb, len, posList, adjList)$. Hence Yuta Mori's algorithm divsufsort can be used to construct the suffix array without increasing the memory requirements. Since A3 does not need the suffix array, we used a variant of the semi-external algorithm described in [3] to construct the BWT. The experiments were conducted on a 64 bit Ubuntu 14.04.1 LTS (Kernel 3.13) system equipped with two ten-core Intel Xeon processors E5-2680v2 with 2.8 GHz and 128GB of RAM (but no parallelism was used).

All programs were compiled with g++ (version 4.8.2) using the provided make-file. As test files we used the *E.coli* genomes listed in the supplementary material of [14]. Additionally, we used 5 different assemblies of the human reference genome (UCSC Genome Browser assembly IDs: hg16, hg17, hg18, hg19, and hg38) as well as the maternal and paternal haplotype of individual NA12878 (Utah female) of the 1000 Genomes Project [21]. The experimental results (Table 1) show that our algorithms are more than an order of magnitude faster than splitMEM while using significantly less space (two orders of magnitude).

6   Future Work

Future work will focus on even more space-efficient solutions to the problem, e.g. by using run-length encoded wavelet trees [23] (or similar compression techniques) or by matching the genomic sequences against the FM-index of just one of them [17]. Another possibility is to use index data structures that are optimized for repetitive text collections; see e.g. [15] and the references therein.

Acknowledgments. This work was supported by the DFG (OH 53/6-1).

References

1. Abouelhoda, M.I., Kurtz, S., Ohlebusch, E.: Replacing suffix trees with enhanced suffix arrays. J. Discrete Algorithms **2**, 53–86 (2004)
2. Beller, T., Gog, S., Ohlebusch, E., Schnattinger, T.: Computing the longest common prefix array based on the Burrows-Wheeler transform. J. Discrete Algorithms **18**, 22–31 (2013)
3. Beller, T., Zwerger, M., Gog, S., Ohlebusch, E.: Space-efficient construction of the Burrows-Wheeler transform. In: Kurland, O., Lewenstein, M., Porat, E. (eds.) SPIRE 2013. LNCS, vol. 8214, pp. 5–16. Springer, Heidelberg (2013)
4. Burrows, M., Wheeler, D.J.: A block-sorting lossless data compression algorithm. Research Report 124, Digital Systems Research Center (1994)
5. Cazaux, B., Lecroq, T., Rivals, E.: From indexing data structures to de Bruijn graphs. In: Kulikov, A.S., Kuznetsov, S.O., Pevzner, P. (eds.) CPM 2014. LNCS, vol. 8486, pp. 89–99. Springer, Heidelberg (2014)
6. Ferragina, P., Gagie, T., Manzini, G.: Lightweight data indexing and compression in external memory. In: López-Ortiz, A. (ed.) LATIN 2010. LNCS, vol. 6034, pp. 697–710. Springer, Heidelberg (2010)
7. Ferragina, P., Manzini, G.: Opportunistic data structures with applications. In: Proceedings of the 41st Annual IEEE Symposium on Foundations of Computer Science, pp. 390–398 (2000)
8. Gagie, T., Navarro, G., Puglisi, S.J.: New algorithms on wavelet trees and applications to information retrieval. Theoret. Comput. Sci. **426–427**, 25–41 (2012)
9. Gog, S., Beller, T., Moffat, A., Petri, M.: From theory to practice: plug and play with succinct data structures. In: Gudmundsson, J., Katajainen, J. (eds.) SEA 2014. LNCS, vol. 8504, pp. 326–337. Springer, Heidelberg (2014)

10. Grossi, R., Gupta, A., Vitter, J.S.: High-order entropy-compressed text indexes. In: Proceedings of 14th Annual ACM-SIAM Symposium on Discrete Algorithms, pp. 841–850 (2003)
11. Huang, L., Popic, V., Batzoglou, S.: Short read alignment with populations of genomes. Bioinformatics **29**(13), i361–i370 (2013)
12. Jacobson, G.: Space-efficient static trees and graphs. In: Proceedings of the 30th Annual IEEE Symposium on Foundations of Computer Science, pp. 549–554 (1989)
13. Kärkkäinen, J.: Fast BWT in small space by blockwise suffix sorting. Theoret. Comput. Sci. **387**(3), 249–257 (2007)
14. Marcus, S., Lee, H., Schatz, M.C.: SplitMEM: a graphical algorithm for pangenome analysis with suffix skips. Bioinformatics **30**(24), 3476–3483 (2014)
15. Navarro, G., Ordóñez, A.: Faster compressed suffix trees for repetitive text collections. In: Gudmundsson, J., Katajainen, J. (eds.) SEA 2014. LNCS, vol. 8504, pp. 424–435. Springer, Heidelberg (2014)
16. Ohlebusch, E.: Bioinformatics Algorithms: Sequence Analysis, Genome Rearrangements, and Phylogenetic Reconstruction. Oldenbusch Verlag, Germany (2013)
17. Ohlebusch, E., Gog, S., Kügel, A.: Computing matching statistics and maximal exact matches on compressed full-text indexes. In: Chavez, E., Lonardi, S. (eds.) SPIRE 2010. LNCS, vol. 6393, pp. 347–358. Springer, Heidelberg (2010)
18. Okanohara, D., Sadakane, K.: A linear-time Burrows-Wheeler transform using induced sorting. In: Karlgren, J., Tarhio, J., Hyyrö, H. (eds.) SPIRE 2009. LNCS, vol. 5721, pp. 90–101. Springer, Heidelberg (2009)
19. Puglisi, S.J., Smyth, W.F., Turpin, A.: A taxonomy of suffix array construction algorithms. ACM Comput. Surv. **39**(2), 1–31 (2007). Article 4
20. Rahn, R., Weese, D., Reinert, K.: Journaled string tree-a scalable data structure for analyzing thousands of similar genomes on your laptop. Bioinformatics **30**(24), 3499–3505 (2014)
21. Rozowsky, J., Abyzov, A., Wang, J., Alves, P., Raha, D., Harmanci, A., Leng, J., Bjornson, R., Kong, Y., Kitabayashi, N., Bhardwaj, N., Rubin, M., Snyder, M., Gerstein, M.: AlleleSeq: Analysis of allele-specific expression and binding in a network framework. Mol. Syst. Biol. **7**, 522 (2011)
22. Schneeberger, K., Hagmann, J., Ossowski, S., Warthmann, N., Gesing, S., Kohlbacher, O., Weigel, D.: Simultaneous alignment of short reads against multiple genomes. Genome Biol. **10**(9), R98 (2009)
23. Sirén, J., Välimäki, N., Mäkinen, V., Navarro, G.: Run-length compressed indexes are superior for highly repetitive sequence collections. In: Amir, A., Turpin, A., Moffat, A. (eds.) SPIRE 2008. LNCS, vol. 5280, pp. 164–175. Springer, Heidelberg (2008)
24. Välimäki, N., Rivals, E.: Scalable and versatile k-mer indexing for high-throughput sequencing data. In: Cai, Z., Eulenstein, O., Janies, D., Schwartz, D. (eds.) ISBRA 2013. LNCS, vol. 7875, pp. 237–248. Springer, Heidelberg (2013)

Longest Common Extensions in Trees

Philip Bille[1], Paweł Gawrychowski[2], Inge Li Gørtz[1], Gad M. Landau[3,4],
and Oren Weimann[3(✉)]

[1] DTU Informatics, Copenhagen, Denmark
{phbi,inge}@dtu.dk
[2] University of Warsaw, Warsaw, Poland
gawry@mimuw.edu.pl
[3] University of Haifa, Haifa, Israel
{landau,oren}@cs.haifa.ac.il
[4] NYU, New York, USA

Abstract. The longest common extension (LCE) of two indices in a string is the length of the longest identical substrings starting at these two indices. The LCE problem asks to preprocess a string into a compact data structure that supports fast LCE queries.

In this paper we generalize the LCE problem to trees and suggest a few applications of LCE in trees to tries and XML databases. Given a labeled and rooted tree T of size n, the goal is to preprocess T into a compact data structure that support the following LCE queries between subpaths and subtrees in T. Let v_1, v_2, w_1, and w_2 be nodes of T such that w_1 and w_2 are descendants of v_1 and v_2 respectively.

- $\mathrm{LCE}_{PP}(v_1, w_1, v_2, w_2)$: (path-path LCE) return the longest common prefix of the paths $v_1 \rightsquigarrow w_1$ and $v_2 \rightsquigarrow w_2$.
- $\mathrm{LCE}_{PT}(v_1, w_1, v_2)$: (path-tree LCE) return maximal path-path LCE of the path $v_1 \rightsquigarrow w_1$ and any path from v_2 to a descendant leaf.
- $\mathrm{LCE}_{TT}(v_1, v_2)$: (tree-tree LCE) return a maximal path-path LCE of any pair of paths from v_1 and v_2 to descendant leaves.

We present the first non-trivial bounds for supporting these queries. For LCE_{PP} queries, we present a linear-space solution with $O(\log^* n)$ query time. For LCE_{PT} queries, we present a linear-space solution with $O((\log \log n)^2)$ query time, and complement this with a lower bound showing that any path-tree LCE structure of size $O(n \operatorname{polylog}(n))$ must necessarily use $\Omega(\log \log n)$ time to answer queries. For LCE_{TT} queries, we present a time-space trade-off, that given any parameter τ, $1 \leq \tau \leq n$, leads to an $O(n\tau)$ space and $O(n/\tau)$ query-time solution. This is complemented with a reduction to the set intersection problem implying that a fast linear space solution is not likely to exist.

P. Bille and I. L. Gørtz—Partially supported by the Danish Research Council (DFF – 4005-00267, DFF – 1323-00178) and the Advanced Technology Foundation.

P. Gawrychowski—Work done while the author held a post-doctoral position at the Warsaw Center of Mathematics and Computer Science.

G. M. Landau—Partially supported by the National Science Foundation Award 0904246, Israel Science Foundation grant 571/14, Grant No. 2008217 from the United States-Israel Binational Science Foundation (BSF) and DFG.

O. Weimann—Partially supported by the Israel Science Foundation grant 794/13.

© Springer International Publishing Switzerland 2015
F. Cicalese et al. (Eds.): CPM 2015, LNCS 9133, pp. 52–64, 2015.
DOI: 10.1007/978-3-319-19929-0_5

1   Introduction

Given a string S, the *longest common extension* (LCE) of two indices is the length of the longest identical substring starting at these indices. The *longest common extension problem* (LCE problem) is to preprocess S into a compact data structure supporting fast LCE queries. The LCE problem is a basic primitive in a wide range of string matching problems such as approximate string matching, finding exact and approximate tandem repeats, and finding palindromes [2,9,15,18–20]. The classic textbook solution to the LCE problem on strings combines a suffix tree with a nearest common ancestor (NCA) data structure leading to a linear space and constant query-time solution [14].

In this paper we study generalizations of the LCE problem to trees. The goal is to preprocess an edge-labeled, rooted tree T to support the various LCE queries between paths in T. Here a path starts at a node v and ends at a descendant of v, and the LCEs are on the strings obtained by concatenating the characters on the edges of the path from top to bottom (each edge contains a single character). We consider path-path LCE queries between two specified paths in T, path-tree LCE queries defined as the maximal path-path LCE of a given path and *any* path starting at a given node, and tree-tree LCE queries defined as the maximal path-path LCE between *any* pair of paths starting from two given nodes. We next define these problems formally.

Tree LCE Problems. Let T be an edge-labeled, rooted tree with n nodes. We denote the subtree rooted at a node v by $T(v)$, and given nodes v and w such that w is in $T(v)$ the path going down from v to w is denoted $v \rightsquigarrow w$. A *path prefix* of $v \rightsquigarrow w$ is any subpath $v \rightsquigarrow u$ such that u is on the path $v \rightsquigarrow w$. Two paths $v_1 \rightsquigarrow w_1$ and $v_2 \rightsquigarrow w_2$ *match* if concatenating the labels of all edges in the paths gives the same string. Given nodes v_1, w_1 such that $w_1 \in T(v_1)$ and nodes v_2, w_2 such that $w_2 \in T(v_2)$ define the following queries:

- $\text{LCE}_{PP}(v_1, w_1, v_2, w_2)$: (path-path LCE) return the longest common matching prefix of the paths $v_1 \rightsquigarrow w_1$ and $v_2 \rightsquigarrow w_2$.
- $\text{LCE}_{PT}(v_1, w_1, v_2)$: (path-tree LCE) return the maximal path-path LCE of the path $v_1 \rightsquigarrow w_1$ and any path from v_2 to a descendant leaf.
- $\text{LCE}_{TT}(v_1, v_2)$: (tree-tree LCE) return a maximal path-path LCE of any pair of paths from v_1 and v_2 to descendant leaves.

We assume that the output of the queries is reported compactly as the endpoint(s) of the LCE. This allows us to report the shared path in constant time per edge. Furthermore, we will assume w.l.o.g. that for each node v in T, all the edge-labels to children of v are distinct. If this is not the case, then we can merge all identical edges of a node to its children in linear time, without affecting the result of all the above LCE queries.

We note that the direction of the paths in T is important for the LCE queries. In the above LCE queries, the paths start from a node and go downwards. If we instead consider paths from a node going upwards towards the root of T, the problem is easier and can be solved in linear space and constant query-time by combining Breslauer's suffix tree of a tree [7] with a nearest common ancestor (NCA) data structure [16].

Our Results. First consider the LCE_{PP} and LCE_{PT} problems. To answer an $LCE_{PP}(v_1, w_1, v_2, w_2)$ query, a straightforward solution is to traverse both paths in parallel-top down. Similarly, to answer an $LCE_{PT}(v_1, w_1, v_2)$ query we can traverse $v_1 \rightsquigarrow w_1$ top-down while traversing the matching path from v_2 (recall that all edges to a child are distinct and hence the longest matching path is unique). This approach leads to a linear-space solution with $O(h)$ query-time to both problems, where h is the height of T. Note that for worst-case trees we have that $h = \Omega(n)$.

We show the following results. For LCE_{PP} we give a linear $O(n)$ space and $O(\log^* n)$ query-time solution. For LCE_{PT} we give a linear $O(n)$ space and $O((\log \log n)^2)$ query-time solution, and complement this with a lower bound stating that any LCE_{PT} solution using $O(n \text{ polylog}(n))$ space must necessarily have $\Omega(\log \log n)$ query time.

Next consider the LCE_{TT} problem. Here, the simple top down traversal does not work and it seems that substantially different ideas are needed. We first show a reduction from the *set-intersection problem*, i.e., preprocessing a family of sets of total size n to support disjointness queries between any pairs of sets. In particular, the reduction implies that a fast linear space solution is not likely assuming a widely believed conjecture on the complexity of the set-intersection problem. We complement this result with a time-space trade-off that achieves $O(n\tau)$ space and $O(n/\tau)$ query time for any parameter $1 \le \tau \le n$.

All results assume the standard word RAM model with word size $\Theta(\log n)$. We also assume the alphabet is either sorted or is linear-time sortable.

Applications. We suggest a few immediate applications of LCE in trees. Consider a set of strings $\mathcal{S} = \{S_1, \ldots, S_k\}$ of total length $\sum_{i=1}^{k} |S_i| = N$ and let T be the *trie* of $\mathcal{S}$ of size n, i.e., T is the labeled, rooted tree obtained by merging shared prefixes in $\mathcal{S}$ maximally. If we want to support LCE queries between suffixes of strings in $\mathcal{S}$, the standard approach is to build a generalized suffix tree for the strings and combine it with an NCA data structure. This leads to a solution using $O(N)$ space and $O(1)$ query time. We can instead implement the LCE query between the suffixes of strings in $\mathcal{S}$ as an LCE_{PP} on the trie T. With our data structure for LCE_{PP}, this leads to a solution using $O(n)$ space and $O(\log^* n)$ query time. In general, n can be significantly smaller than N, depending on the amount of shared prefixes in $\mathcal{S}$. Hence, this solution provides a more space-efficient representation of $\mathcal{S}$ at the expense of a tiny increase in query time. An LCE_{PT} query on T corresponds to computing a maximal LCE of a suffix of a string in $\mathcal{S}$ with suffixes of strings in $\mathcal{S}$ sharing a common prefix. An LCE_{TT} query on T corresponds to computing a maximal LCE over pairs of suffixes of strings in $\mathcal{S}$ that share a common prefix. To the best of our knowledge these queries are novel one-to-many and many-to-many LCE queries. Since tries are a basic data structure for storing strings we expect these queries to be of interest in a number of applications.

Another interesting application is using LCE in trees as a query primitive for XML data. XML documents can be viewed as a labeled tree and typical queries (e.g., XPath queries) involve traversing and identifying paths in the tree. The LCE queries provide simple and natural primitives for comparing paths and subtrees without explicit traversal. For instance, our solution for LCE_{PT} queries can be used to quickly identify the "best match" of a given path in a subtree.

2 Preliminaries

Given a node v and an integer $d \geq 0$, the *level ancestor* of v at depth d, denoted $LA(v, d)$ is the ancestor of v at depth d. We explicitly compute and store the depth of every node v, denoted depth(v). Given a pair of nodes v and w the *nearest common ancestor* of v and w, denoted $NCA(v, w)$, is the common ancestor of v and w of greatest depth. Both LA and NCA queries can be supported in constant time with a linear space data structures, see e.g., [1,4–6,10,11,13,16].

Finally, the *suffix tree of a tree* [7,17,24] is the compressed trie of all suffixes of leaf-to-root paths in T. The suffix tree of a tree uses $O(n)$ space and can be constructed in $O(n \log \log n)$ time for general alphabets [24]. Note that the suffix tree of a tree combined with NCA can support LCE queries in constant time for paths going upwards. Since we consider paths going downwards, we will only use the suffix tree to check (in constant time) if two paths are completely identical.

We also need the following three primitives. *Range minimum queries:* A list of n numbers $a_1, a_2, \ldots a_n$ can be augmented with $2n + o(n)$ bits of additional data in $O(n)$ time, so that for any $i \leq j$ the position of the smallest number among $a_i, a_{i+1}, \ldots, a_j$ can be found in $O(1)$ time [11]. *Predecessor queries:* Given a sorted collection of n integers from $[0, U)$, a structure of size $O(n)$ answering predecessor queries in $O(\log \log U)$ time can be constructed in time $O(n)$ [25], where a predecessor query locates, for a given x, the largest $y \leq x$ such that $y \in S$. Finally, *Perfect hashing:* given a collection S of n integers a perfect hash table can be constructed in expected $O(n)$ time [12], where a perfect hash table checks, for a given x, if $x \in S$, and if so returns its associated data in $O(1)$ time. The last result can be made deterministic at the expense of increasing the preprocessing time to $O(n \log \log n)$ [23], but then we need one additional step in our solution for the path-tree LCE as to ensure $O(n)$ total construction time.

3 Difference Covers for Trees

In this section we introduce a generalization of difference covers from strings to trees. This will be used to decrease the space of our data structures. We believe it is of independent interest.

Lemma 1. *For any tree T with n nodes and a parameter x, it is possible to mark $2n/x$ nodes of T, so that for any two nodes $u, v \in T$ at (possibly different) depths at least x^2, there exists $d \leq x^2$ such that the d-th ancestors of both u and v are marked. Furthermore, such d can be calculated in $O(1)$ time and the set of marked nodes can be determined in $O(n)$ time.*

Proof. We distinguish between two types of marked nodes. Whether a node v is marked or not depends only on its depth. The marked nodes are determined as follows:

Type I. For every $i = 0, 1, \ldots, x - 1$, let V_i be the set of nodes at depth leaving a remainder of i when divided by x. Because $\bigcup_i V_i = T$ and all $V_i's$ are disjoint, there exists $r_1 \in [0, x - 1]$ such that $|V_{r_1}| \leq n/x$. Then v is a type I marked node iff $\mathrm{depth}(v) = r_1 \mod x$.

Type II. For every $i = 0, 1, \ldots, x - 1$, let V_i be the set of nodes v such that $\lfloor \mathrm{depth}(v)/x \rfloor$ leaves a remainder of i when divided by x. By the same argument as above, there exists $r_2 \in [0, x - 1]$ such that $|V_{r_2}| \leq n/x$. Then v is a type II marked node iff $\lfloor \mathrm{depth}(v)/x \rfloor = r_2 \mod x$.

Now, given two nodes u and v at depths at least x^2, we need to show that there exists an appropriate $d \leq x^2$. Let $\mathrm{depth}(u) = t_1 \mod x$ and choose $d_1 = t_1 + x - r_1$. Then the d_1-th ancestor of u is a type I marked node, because its depth is $\mathrm{depth}(u) - d_1 = \mathrm{depth}(u) - (t_1 + x - r_1) = \mathrm{depth}(u) - t_1 - x + r_1$, which leaves a remainder of r_1 when divided by x. Our d will be of the form $d_1 + d_2 x$. Observe that regardless of the value of d_2, we can be sure that the d-th ancestor of u is a type I marked node. Let v' be the d_1-th ancestor of v, $\lfloor \mathrm{depth}(v')/x \rfloor = t_2 \mod x$ and choose $d_2 = t_2 + x - r_2$. The $(d_2 x)$-th ancestor of v' is a type II marked node, because $\lfloor (\mathrm{depth}(v') - d_2 x)/x \rfloor = \lfloor \mathrm{depth}(v')/x \rfloor - t_2 - x + r_2$, which leaves a remainder of r_2 when divided by x. Therefore, choosing $d = d_1 + d_2 x$ guarantees that $d \leq x - 1 + x(x - 1) < x^2$, so the d-th ancestors of u and v are both defined, the d-th ancestor of u is a type I marked node, and the d-th ancestor of v is a type II marked node.

The total number of marked nodes is clearly at most $2n/x$, and the values of r and r' can be determined by a single traversal of T. To determine d, we only need to additionally know $\mathrm{depth}(u)$ and $\mathrm{depth}(v)$ and perform a few simple arithmetical operations. □

Remark. Our difference cover has the following useful property: whether a node v is marked or not depends only on the value of $\mathrm{depth}(v)$ (mod x^2). Hence, if a node at depth at least x^2 is marked then so is its (x^2)-th ancestor. Similarly, if a node is marked, so are all of its descendants at distance x^2.

4   Path-Path LCE

In this section we prove the following theorem.

Theorem 1. *For a tree T with n nodes, a data structure of size $O(n)$ can be constructed in $O(n)$ time to answer path-path LCE queries in $O(\log^* n)$ time.*

We start with a simple preliminary $O(n \log n)$-space $O(1)$-query data structure which will serve as a starting point for the more complicated final implementation. We note that a data structure with similar guarantees to Lemma 2 is also implied from [3].

Lemma 2. *For a tree T with n nodes, a data structure of size $O(n \log n)$ can be constructed in $O(n \log n)$ time to answer path-path LCE queries in $O(1)$ time.*

Proof. The structure consists of $\log n$ separate parts, each of size $O(n)$. The k-th part answers in $O(1)$ time path-path LCE queries such that both paths are of the same length 2^k. This is enough to answer a general path-path LCE query in the same time complexity, because we can first truncate the longer path so that both paths are of the same length ℓ, then calculate k such that $2^k \le \ell < 2^{k+1}$. Then we have two cases:

1. The prefixes of length 2^k of both paths are different. Then replacing the paths by their prefixes of length 2^k does not change the answer.
2. The prefixes of length 2^k of both paths are the same. Then replacing the paths by their suffixes of length 2^k does not change the answer.

We can check if the prefixes are the same and then (with level ancestor queries) reduce the query so that both paths are of the same length 2^k, all in $O(1)$ time.

Consider all paths of length 2^k in the tree. There are at most n of them, because every node u creates at most one new path $\mathrm{LA}(v, \mathrm{depth}(v) - 2^k) \rightsquigarrow v$. We lexicographically sort all such paths and store the longest common extension of every two neighbours on the sorted list. Additionally, we augment the longest common extensiones with a range minimum query structure, and keep at every v the position of the path $\mathrm{LA}(v, \mathrm{depth}(v) - 2^k) \rightsquigarrow v$ (if any) on the sorted list. This allows us to answer $\mathrm{LCE}_{PP}(\mathrm{LA}(u, \mathrm{depth}(u) - 2^k), u, \mathrm{LA}(v, \mathrm{depth}(v) - 2^k), v)$ in $O(1)$ time: we lookup the positions of $\mathrm{LA}(u, \mathrm{depth}(u) - 2^k) \rightsquigarrow u$ and $\mathrm{LA}(v, \mathrm{depth}(v) - 2^k) \rightsquigarrow v$ on the sorted list and use the range minimum query structure to calculate their longest common prefix, all in $O(1)$ time. The total space usage is $O(n)$, because every node stores one number and additionally we have a list of at most n numbers augmented with a range minimum query structure.

To construct the structure efficiently, we need to lexicographically sort all paths of length k. This can be done in $O(n)$ time for every k after observing that every path of length 2^{k+1} can be conceptually divided into two paths of length 2^k. Therefore, if we have already lexicographically sorted all paths of length 2^k, we can lexicographically sort all paths of length 2^{k+1} by sorting pairs of numbers from $[1, n]$, which are the positions of the prefix and the suffix of a longer path on the sorted list of all paths of length 2^k. With radix sorting, this takes $O(n)$ time. Then we need to compute the longest common extension of ever two neighbours on the sorted list, which can be done in $O(1)$ time by using the already constructed structure for paths of length 2^k. Consequently, the total construction time is $O(n \log n)$. □

To decrease the space usage of the structure from Lemma 2, we use the difference covers developed in Lemma 1. Intuitively, the first step is to apply the lemma with $x = \log n$ and preprocess only paths of length $2^k \log^2 n$ ending at the marked nodes. Because we have only $O(n/\log n)$ marked nodes, this requires $O(n)$ space. Then, given two paths of length ℓ, we can either immediately return their LCE using the preprocessed data, or reduce the query to computing the LCE of two paths of length at most $\log^2 n$. Using the same reasoning again with $x = \log(\log^2 n)$, we can reduce the length even further to at most $\log^2(\log^2 n)$ and so on. After $O(\log^* n)$ such reduction steps, we guarantee that the paths are of length $O(1)$, and the answer can be found naively. Formally, every step is implemented using the following lemma.

Lemma 3. *For a tree T with n nodes and a parameter b, a data structure of size $O(n)$ can be constructed in $O(n)$ time, so that given two paths of length at most b ending at $u \in T$ and $v \in T$ in $O(1)$ time we can either compute the path-path LCE or reduce the query so that the paths are of length at most $\log^2 b$.*

Proof. We apply Lemma 1 with $x = \log b$. Then, for every $k = 0, 1, \ldots, \log b$ separately, we consider all paths of length $2^k \log^2 b$ ending at marked nodes. As in the proof of Lemma 2, we lexicographically sort all such paths and store the longest common extension of every two neighbours on the sorted list augmented with a range minimum query structure. Because we have only $O(n/\log b)$ marked nodes, the space decreases to $O(n)$. Furthermore, because the length of the paths is of the form $2^k \log^2 b$ (as opposed to the more natural choice of 2^k), all lists can be constructed in $O(n)$ total time by radix sorting, as a path of length $2^{k+1} \log^2 b$ ending at a marked node can be decomposed into two paths of length $2^k \log^b$ ending at marked nodes, because if a node is marked, so is its (x^2)-th ancestor.

Consider two paths of the same length $\ell \leq b$ ending at $u \in T$ and $v \in T$. We need to either determine their LCE, or reduce the query to determining the LCE of two paths of length at most $\log^2 b$. If $\ell \leq \log^2 b$, there is nothing to do. Otherwise, first we check if the prefixes of length $\log^2 b$ of both paths are different in $O(1)$ time. If so, we replace the paths with their prefixes of such length and we are done. Otherwise, if $\ell \leq 2\log^2 b$ we replace the paths with their suffixes of length $\ell - \log^2 b \leq \log^2 b$ and we are done. The remaining case is that the prefixes of length $\log^2 b$ are identical and $\ell > 2\log^2 b$. In such case, we can calculate k such that $2^k \log^2 b \leq \ell - \log^2 b < 2^{k+1} \log^2 b$. Having such k, we cover the suffixes of length $\ell - \log^2 b$ with two (potentially overlapping) paths of length exactly $2^k \log^2 b$. More formally, we create two pairs of paths:

1. $\mathrm{LA}(u, \mathrm{depth}(u) - 2^k \log^2 b) \rightsquigarrow u$ and $\mathrm{LA}(v, \mathrm{depth}(v) - 2^k \log^2 b) \rightsquigarrow v$,
2. $\mathrm{LA}(u, \mathrm{depth}(u) - \ell + \log^2 b) \rightsquigarrow \mathrm{LA}(u, \mathrm{depth}(u) - \ell + \log^2 b + 2^k \log^2 b)$ and $\mathrm{LA}(v, \mathrm{depth}(v) - \ell + \log^2 b) \rightsquigarrow \mathrm{LA}(v, \mathrm{depth}(v) - \ell + \log^2 b + 2^k \log^2 b)$.

If the paths from the first pair are different, it is enough to compute their LCE. If they are identical, it is enough to compute the LCE of the paths from the second pair. Because we can distinguish between these two cases in $O(1)$ time, we focus on computing the LCE of two paths of length $2^k \log^2 b$ ending at some u' and v'. The important additional property guaranteed by how we have defined the pairs is that the paths of length $\log^2 b$ ending at $\mathrm{LA}(u', \mathrm{depth}(u') - 2^k \log^2 b)$ and $\mathrm{LA}(v', \mathrm{depth}(v') - 2^k \log^2 b)$ are the same. Now by the properties of the difference cover we can calculate in $O(1)$ time $d \leq \log^2 b$ such that the d-th ancestors of u' and v' are marked. We conceptually slide both paths up by d, so that they both end at these marked nodes. Because of the additional property, either the paths of length $2^k \log^2 b$ ending at $\text{LA}(u', \text{depth}('u) - d)$ and $\text{LA}(v', \text{depth}(v') - d)$ are identical, or their first mismatch actually corresponds to the LCE of the original paths ending at u' and v'. These two cases can be distinguished in $O(1)$ time. Then we either use the preprocessed data to calculate the LCE in $O(1)$ time, or we are left with the suffixes of length d of the paths ending at u' and v'. But because $d \leq \log^2 b$, also in the latter case we are done.   □

We apply Lemma 3 with $b = n, \log^2 n, \log^2(\log^2 n), \ldots$ terminating when $b \leq 4$. The total number of applications is just $O(\log^* n)$, because $\log^2(\log^2 z) = 4\log^2(\log z) \leq \log z$ for z large enough[1]. Therefore, the total space usage becomes $O(n \log^* n)$ and, by iteratively applying the reduction step, for any two paths of length at most n ending at given u and v we can in $O(\log^* n)$ time either compute their LCE, or reduce the query to computing the LCE of two paths of length $O(1)$, which can be computed naively in additional $O(1)$ time.

To prove Theorem 1, we need to decrease the space usage from $O(n \log^* n)$ down to $O(n)$. To this end, we create a smaller tree T' on $O(n/b)$ nodes, where $b = \log^* n$ is the parameter of the difference cover, as follows. Every marked node $u \in T$ becomes a node of T'. The parent of $u \in T$ in T' is the node corresponding in T' to the (b^2)-th ancestor of u in T, which is always marked. Additionally, we add one artificial node, which serves as the root of the whole T', and make it the parent of all marked nodes at depth (in T) less than b^2. Now edges of T' correspond to paths of length b^2 in T (except for the edges outgoing from the root; we will not be using them). We need to assign unique names to these paths, so that the names of two paths are equal iff the paths are the same. This can be done by traversing the suffix tree of T in $O(n)$ time. Finally, T' is preprocessed by applying Lemma 3 $O(\log^* n)$ times as described above. Because its size of T' is just $O(n/b)$, the total space usage preprocessing time is just $O(n)$ now.

To compute the LCE of two paths of length ℓ ending at $u \in T$ and $v \in T$, we first compare their prefixes of length b^2. If they are identical, by the properties of the difference cover we can calculate $d \leq b^2$ such that the d-th ancestors of both u and v, denoted u' and v', are marked, hence exist in T'. Consequently, if the prefixes of length $\ell - d$ of the paths are different, we can calculate their first mismatch by computing the first mismatch of the paths of length $\lfloor(\ell - d)/b^2\rfloor$ ending at $u' \in T'$ and $v' \in T'$. This follows because every edge of T' corresponds to a path of length b^2 in T, so a path of length $\lfloor(\ell - d)/b^2\rfloor$ in T' corresponds to a path of length belonging to $[\ell - d - b^2, \ell - d]$ in T, and we have already verified that the first mismatch is outside of the prefix of length b^2 of the original paths. Hence the first mismatch of the corresponding paths in T' allows us to narrow down where the first mismatch of the original paths in T occurs up to b^2 consecutive edges. All in all, in $O(1)$ time plus a single path-path LCE query in T' we can reduce the original query to a query concerning two paths of length at most b^2.

---

[1] This follows from $\lim_{z\to\infty} \frac{\log^2(\log^2 z)}{\log z} = \lim_{z\to\infty} \frac{4\log(\log^2 z)}{\ln z} = \lim_{z\to\infty} \frac{8}{\ln z} = 0$.

The final step is to show that T can be preprocessed in $O(n)$ time and space, so that the LCE of any two paths of length at most b^2 can be calculated in $O(b)$ time. We assign unique names to all paths of length b in T, which can be again done by traversing the suffix tree of T in $O(n)$ time. More precisely, every $u \in T$ such that depth(u) $\geq b$ stores a single number, which is the name of the path of length b ending at u. To calculate the LCE of two paths of length at most b^2 ending at $u \in T$ and $v \in T$, we proceed as follows. We traverse both paths in parallel top-down moving by b edges at once. Using the preprocessed names, we can check if the first mismatch occurs on these b consecutive edges, and if so terminate. Therefore, after at most b steps we are left with two paths of length at most b, such that computing their LCE allows us to answer the original query. But this can be calculated by naively traversing both paths in parallel top-down. The total query time is $O(b)$.

To summarize, the total space and preprocessing time is $O(n)$ and the query time remains $O(\log^* n)$, which proves Theorem 1.

5   Path-Tree LCE

In this section we prove the following theorem.

Theorem 2. *For a tree T with n nodes, a data structure of size $O(n)$ can be constructed in $O(n)$ time to answer path-tree LCE queries in $O((\log \log n)^2)$ time.*

The idea is to apply the difference covers recursively with the following lemma.

Lemma 4. *For a tree T with n nodes and a parameter b, a data structure of size $O(n)$ can be constructed in $O(n \log n)$ time, so that given a path of length $\ell \leq b$ ending at $u \in T$ and a subtree rooted at $v \in T$ we can reduce the query in $O(\log \log n)$ time so that the path is of length at most $b^{4/5}$.*

Proof. The first part of the structure is designed so that we can detect in $O(1)$ time if the path-tree LCE is of length at most $b^{4/5}$. We consider all paths of length exactly $b^{4/5}$ in the tree. We assign names to every such path, so that testing if two paths are identical can be done by looking at their names. Then, for every node w we gather all paths of length $b^{4/5}$ starting at w (i.e., $w \rightsquigarrow v$, where $w = LA(v, \text{depth}(v) - b^{4/5})$) and store their names in a perfect hash table, where every name is linked to the corresponding node w. This allows us to check if the answer is at least $b^{4/5}$ by first looking up the name of the prefix of length $b^{4/5}$ of the path, and then querying the perfect hash table kept at v. If the name does not occur there, the answer is less than $b^{4/5}$ and we are done. Otherwise, we can move by $b^{4/5}$ down, i.e., decrease ℓ by $b^{4/5}$ and replace v with its descendant of distance $b^{4/5}$.

The second part of the structure is designed to work with the marked nodes. We apply Lemma 1 with $x = b^{2/5}$ and consider *canonical paths* of length $i \cdot x^2$ in the tree, where $i = 1, 2, \ldots, \sqrt{x}$, ending at marked nodes. The total number of such paths is $O(n/\sqrt{x})$, because every marked node is the endpoint of at most

$\sqrt{x}$ of them. We lexicographically sort all canonical paths and store the longest common extension of every two neighbours on the global sorted list augmented with a range minimum query structure. Also, for every marked node v and every $i = 1, 2, \ldots, x$, we save the position of the path $LA(v, \text{depth}(v) - i \cdot x^2) \leadsto v$ on the global sorted list. Additionally, at every node u we gather all canonical paths starting there, i.e., $u \leadsto v$ such that $LA(v, \text{depth}(v) - i \cdot x^2) = u$ for some $i = 1, 2, \ldots, x$, sort them lexicographically and store on the local sorted list of u. Every such path is represented by a pair (u, i). The local sorted list is augmented with a predecessor structure storing the positions on the global sorted list.

Because we have previously decreased ℓ and replaced v, now by the properties of the difference cover we can find $d \leq x^2$ such that the $(\ell+d)$-th ancestor of u and the d-th ancestor of v are marked, and then increase ℓ by d and replace v by its d-th ancestor. Consequently, from now on we assume that both $LA(u, \text{depth}(u) - \ell)$ and v are marked.

Now we can use the second part of the structure. If $\ell \leq b^{4/5}$, there is nothing to do. Otherwise, the prefix of length $\lfloor \ell/x^2 \rfloor \cdot x^2$ of the path is a canonical path (because $\ell \leq \sqrt{x} \cdot x^2$), so we know its position on the global sorted list. We query the predecessor structure stored at v with that position to get the lexicographical predecessor and successor of the prefix among all canonical paths starting at v. This allows us to calculate the longest common extension p of the prefix and all canonical paths starting at v by taking the maximum of the longest common extension of the prefix and its predecessor, and the prefix and its successor. Now, because canonical paths are all paths of the form $i \cdot x^2$, the length of the path-tree LCE cannot exceed $p + x^2$. Furthermore, with a level ancestor query we can find v' such that the paths $LA(u, \text{depth}(u) - \ell) \leadsto LA(u, \text{depth}(u) - \ell + p)$ and $v \leadsto v'$ are identical. Then, to answer the original query, it is enough to calculate the path-tree LCE for $LA(u, \text{depth}(u) - \ell + p) \leadsto LA(u, \text{depth}(u) - \ell + \min(\ell, p + x^2))$ and the subtree rooted at v'. Therefore, in $O(\log \log n)$ time we can reduce the query so that the path is of length at most $x^2 = b^{4/5}$ as claimed.

To achieve $O(n)$ construction time, we need to assign names to all paths of length $b^{4/5}$ in the tree, which can be done in $O(n)$ by traversing the suffix tree of T. We would also like to lexicographically sort all canonical paths, but this seems difficult to achieve in $O(n)$. Therefore, we change the lexicographical order as follows: we assign names to all canonical paths of length exactly x^2, so that different paths get different names and identical paths get identical names (again, this can be done in $O(n)$ time by traversing the suffix tree). Then we treat every canonical path of length $i \cdot x^2$ as a sequence of consisting of i names, and sort these sequences lexicographically in $O(n)$ time with radix sort. Even though this is not the lexicographical order, the canonical paths are only used to approximate the answer up to an additive error of x^2, and hence such modification is still correct.

$\square$

We apply Lemma 4 with $b = n, n^{4/5}, n^{(4/5)^2}, \ldots, 1$. The total number of applications is $O(\log \log n)$. Therefore, the total space usage becomes $O(n \log \log n)$, and by applying the reduction step iteratively, for any path of length n ending at

u and a subtree rooted at v we can compute the path-tree LCE in $O((\log \log n)^2)$ time. The total construction time is $O(n \log \log n)$.

To prove Theorem 2, we need to decrease the space usage and the construction time. The idea is similar to the one from Sect. 4: we create a smaller tree T' on $O(n/b)$ nodes, where $b = \log \log n$ is the parameter of the difference cover. The edges of T' correspond to paths of length b^2 in T. We preprocess T' as described above, but because its size is now just $O(n/b)$, the preprocessing time and space become $O(n)$.

To compute the path-tree LCE for a given path of length ℓ ending at u and a subtree rooted at v, we first check if the answer is at least b^2. This can be done in $O(\log \log n)$ time by preprocessing all paths of length b^2 in T, as done inside Lemma 4 for paths of length $b^{4/5}$. If so, we can decrease ℓ and replace v with its descendant, so that both $\mathrm{LA}(u, \mathrm{depth}(u) - \ell)$ and v are marked, hence exist in T'. Then we use the structure constructed for T' to reduce the query, so that the path is of length at most b^2. Therefore, it is enough how to answer a query, where a path is of length at most b^2, in $O(\log \log n)$ time after $O(n)$ time and space preprocessing.

The final step is to preprocess T in $O(n)$ time and space, so that the path-tree LCE of a path of length at most b^2 and any subtree can be computed in $O(b)$ time. We assign unique names to all paths of length b in T. Then, for every u we gather the names of all paths $u \rightsquigarrow v$ of length b in a perfect hash table. To calculate the path-tree LCE, we traverse the path top-down while tracing the corresponding node in the subtree. Initially, we move by b edges by using the perfect hash tables. This allows us to proceed as long as the remaining part of the LCE is at least b. Then, we traverse the remaining part consisting of at most b edges naively. In total, this takes $O(b)$ time. The space is clearly $O(n)$ and the preprocessing requires constructing the perfect hash tables, which can be done in $O(n)$ time.

5.1   Lower Bound

In this section, we prove that any path-tree LCE structure of size $O(n \operatorname{polylog}(n))$ must necessarily use $\Omega(\log \log n)$ time to answer queries. As shown by Pătraşcu and Thorup [22], for $U = n^2$ any predecessor structure consisting of $O(n \operatorname{polylog}(n))$ words needs $\Omega(\log \log n)$ time to answer queries, assuming that the word size is $\Theta(\log n)$. In the full version of this paper, we show the following reduction, which implies the aforementioned lower bound.

Theorem 3. *For any $\epsilon > 0$, given an LCE_{PT} structure that uses $s(n) = \Omega(n)$ space and answers queries in $q(n) = \Omega(1)$ time we can build a predecessor structure using $O(s(2U^\epsilon + n \log |U|))$ space and $O(q(2U^\epsilon + n \log |U|))$ query time for any $S \subseteq [0, U]$ of size n.*

By applying the reduction with $U = n^2$ and $\epsilon = 1/2$, we get that an LCE_{PT} structure using $O(n \operatorname{polylog}(n))$ space and answering queries in $o(\log \log n)$ time implies a predecessor structure using $O(n \operatorname{polylog}(n))$ space and answering queries in $o(\log \log(n))$ time, which is not possible.

6   Tree-Tree LCE

The *set intersection problem* is defined as follows. Given a family $\mathcal{S} = \{S_1, \ldots, S_k\}$ of sets of total size $n = \sum_{i=1}^{k} |S_i|$ the goal is to preprocess $\mathcal{S}$ to answer queries: given two sets S_i and S_j determine if $S_i \cap S_j = \emptyset$. The set intersection problem is widely believed to require superlinear space in order to support fast queries. A folklore conjecture states that for sets of size polylogarithmic in k, supporting queries in constant time requires $\tilde{\Omega}(k^2)$ space [21] (see also [8]).

We now consider the LCE$_{TT}$ problem. In the full version of this paper we show that the problem is set intersection hard and give a time-space trade-off as stated by the following theorems:

Theorem 4. *Let T be a tree with n nodes. Given an* LCE$_{TT}$ *data structure that uses $s(n)$ space and answers queries in $q(n)$ time we can build a set intersection data structure using $O(s(n))$ space and $O(q(n))$ query time, for input sets containing $O(n)$ elements.*

Theorem 5. *For a tree T with n nodes and a parameter τ, $1 \leq \tau \leq n$, a data structure of size $O(n\tau)$ can be constructed in $O(n\tau)$ time to answer tree-tree LCE queries in $O(n/\tau)$ time.*

References

1. Alstrup, S., Holm, J.: Improved algorithms for finding level ancestors in dynamic trees. In: Welzl, E., Montanari, U., Rolim, J.D.P. (eds.) ICALP 2000. LNCS, vol. 1853, pp. 73–84. Springer, Heidelberg (2000)
2. Amir, A., Lewenstein, M., Porat, E.: Faster algorithms for string matching with k mismatches. J. Algorithms **50**(2), 257–275 (2004)
3. Bannai, H., Gawrychowski, P., Inenaga, S., Takeda, M.: Converting SLP to LZ78 in almost Linear Time. In: Fischer, J., Sanders, P. (eds.) CPM 2013. LNCS, vol. 7922, pp. 38–49. Springer, Heidelberg (2013)
4. Bender, M.A., Farach-Colton, M.: The LCA problem revisited. In: Gonnet, G.H., Viola, A. (eds.) LATIN 2000. LNCS, vol. 1776. Springer, Heidelberg (2000)
5. Bender, M.A., Farach-Colton, M.: The level ancestor problem simplified. Theoret. Comput. Sci. **321**(1), 5–12 (2004)
6. Berkman, O., Vishkin, U.: Finding level-ancestors in trees. J. Comput. Syst. Sci. **48**(2), 214–230 (1994)
7. Breslauer, D.: The suffix tree of a tree and minimizing sequential transducers. Theoret. Comput. Sci. **191**(1–2), 131–144 (1998)
8. Cohen, H., Porat, E.: Fast set intersection and two-patterns matching. Theor. Comput. Sci. **411**(40–42), 3795–3800 (2010)
9. Cole, R., Hariharan, R.: Approximate string matching: a simpler faster algorithm. SIAM J. Comput. **31**(6), 1761–1782 (2002)
10. Dietz, P.F.: Finding level-ancestors in dynamic trees. In: Dehne, F., Sack, J.-R., Santoro, N. (eds.) WADS '91. LNCS, vol. 519, pp. 32–40. Springer, Heidelberg (1991)

11. Fischer, J., Heun, V.: Space-efficient preprocessing schemes for range minimum queries on static arrays. SIAM J. Comput. **40**(2), 465–492 (2011)
12. Fredman, M.L., Komlos, J., Szemeredi, E.: Storing a sparse table with $O(1)$ worst case access time. In Proceedings of 23rd FOCS, pp. 165–169, November 1982
13. Geary, R.F., Raman, R., Raman, V.: Succinct ordinal trees with level-ancestor queries. ACM Trans. Algorithms **2**(4), 510–534 (2006)
14. Gusfield, D.: Algorithms on Strings, Trees, and Sequences: Computer Science and Computational Biology. Cambridge University Press, New York (1997)
15. Gusfield, D., Stoye, J.: Linear time algorithms for finding and representing all the tandem repeats in a string. J. Comput. Syst. Sci. **69**(4), 525–546 (2004)
16. Harel, D., Tarjan, R.E.: Fast algorithms for finding nearest common ancestors. SIAM J. Comput. **13**(2), 338–355 (1984)
17. Kosaraju, S.R.: Efficient tree pattern matching. In: Proceedings of 30th FOCS, pp. 178–183 (1989)
18. Landau, G.M., Myers, E.W., Schmidt, J.P.: Incremental string comparison. SIAM J. Comput. **27**(2), 557–582 (1998)
19. Landau, G.M., Vishkin, U.: Fast parallel and serial approximate string matching. J. Algorithms **10**, 157–169 (1989)
20. Main, M.G., Lorentz, R.J.: An $O(n \log n)$ algorithm for finding all repetitions in a string. J. Algorithms **5**(3), 422–432 (1984)
21. Pătraşcu, M., Roditty, L.: Distance oracles beyond the Thorup-Zwick bound. In: Proceedings of 51st IEEE FOCS, pp. 815–823 (To appear, 2010)
22. Pătraşcu, M., Thorup, M.: Time-space trade-offs for predecessor search. In: Proceedings of 38th STOC, pp. 232–240 (2006)
23. Ružić, M.: Uniform algorithms for deterministic construction of efficient dictionaries. In: Albers, S., Radzik, T. (eds.) ESA 2004. LNCS, vol. 3221, pp. 592–603. Springer, Heidelberg (2004)
24. Shibuya, T.: Constructing the suffix tree of a tree with a large alphabet. In: Aggarwal, A.K., Pandu Rangan, C. (eds.) ISAAC 1999. LNCS, vol. 1741, pp. 225–236. Springer, Heidelberg (1999)
25. van Emde Boas, P., Kaas, R., Zijlstra, E.: Design and implementation of an efficient priority queue. Math. Syst. Theory **10**, 99–127 (1977)

Longest Common Extensions in Sublinear Space

Philip Bille[1], Inge Li Gørtz[1], Mathias Bæk Tejs Knudsen[2],
Moshe Lewenstein[3], and Hjalte Wedel Vildhøj[1(✉)]

[1] Technical University of Denmark, DTU Compute, Lyngby, Denmark
hwv@hwv.dk
[2] Department of Computer Science, University of Copenhagen,
Copenhagen, Denmark
[3] Bar Ilan University, Ramat Gan, Israel

Abstract. The *longest common extension problem* (LCE problem) is to construct a data structure for an input string T of length n that supports $\mathrm{LCE}(i,j)$ queries. Such a query returns the length of the longest common prefix of the suffixes starting at positions i and j in T. This classic problem has a well-known solution that uses $\mathcal{O}(n)$ space and $\mathcal{O}(1)$ query time. In this paper we show that for any trade-off parameter $1 \leq \tau \leq n$, the problem can be solved in $\mathcal{O}(\frac{n}{\tau})$ space and $\mathcal{O}(\tau)$ query time. This significantly improves the previously best known time-space trade-offs, and almost matches the best known time-space product lower bound.

1 Introduction

Given a string T, the *longest common extension* of suffix i and j, denoted $\mathrm{LCE}(i,j)$, is the length of the longest common prefix of the suffixes of T starting at position i and j. The *longest common extension problem* (LCE problem) is to preprocess T into a compact data structure supporting fast longest common extension queries.

The LCE problem is a basic primitive that appears as a central subproblem in a wide range of string matching problems such as approximate string matching and its variations [1,4,11,13,16], computing exact or approximate repetitions [6,12,14], and computing palindromes [10,15]. In many cases the LCE problem is the computational bottleneck.

Here we study the time-space trade-offs for the LCE problem, that is, the space used by the preprocessed data structure vs. the worst-case time used by

P. Bille— Supported by the Danish Research Council and the Danish Research Council under the Sapere Aude Program (DFF 4005-00267).
I. L. Gørtz— Research partly supported by Mikkel Thorup's Advanced Grant from the Danish Council for Independent Research under the Sapere Aude research career programme and the FNU project AlgoDisc - Discrete Mathematics, Algorithms, and Data Structures.
H. W. Vildhøj— This research was supported by a Grant from the GIF, the German-Israeli Foundation for Scientific Research and Development, and by a BSF grant 2010437.

© Springer International Publishing Switzerland 2015
F. Cicalese et al. (Eds.): CPM 2015, LNCS 9133, pp. 65–76, 2015.
DOI: 10.1007/978-3-319-19929-0_6

LCE queries. The input string is given in read-only memory and is not counted in the space complexity. Throughout the paper we use ℓ as a shorthand for $\text{LCE}(i,j)$. The standard trade-offs are as follows: At one extreme we can store a suffix tree combined with an efficient nearest common ancestor (NCA) data structure [7,17]. This solution uses $\mathcal{O}(n)$ space and supports LCE queries in $\mathcal{O}(1)$ time. At the other extreme we do not store any data structure and instead answer queries simply by comparing characters from left-to-right in T. This solution uses $\mathcal{O}(1)$ space and answers an $\text{LCE}(i,j)$ query in $\mathcal{O}(\ell) = \mathcal{O}(n)$ time. Recently, Bille et al. [2] presented a number of results. For a trade-off parameter τ, they gave: (1) a deterministic solution with $\mathcal{O}(\frac{n}{\tau})$ space and $\mathcal{O}(\tau^2)$ query time, (2) a randomized Monte Carlo solution with $\mathcal{O}(\frac{n}{\tau})$ space and $\mathcal{O}(\tau \log(\frac{\ell}{\tau})) = \mathcal{O}(\tau \log(\frac{n}{\tau}))$ query time, where all queries are correct with high probability, and (3) a randomized Las Vegas solution with the same bounds as 2) but where all queries are guaranteed to be correct. Bille et al. [2] also gave a lower bound showing that any data structure for the LCE problem must have a time-space product of $\Omega(n)$ bits.

Our Results. Let τ be a trade-off parameter. We present four new solutions with the following improved bounds. Unless otherwise noted the space bound is the number of words on a standard RAM with logarithmic word size, not including the input string, which is given in read-only memory.

- A deterministic solution with $\mathcal{O}(n/\tau)$ space and $\mathcal{O}(\tau \log^2(n/\tau))$ query time.
- A randomized Monte Carlo solution with $\mathcal{O}(n/\tau)$ space and $\mathcal{O}(\tau)$ query time, such that all queries are correct with high probability.
- A randomized Las Vegas solution with $\mathcal{O}(n/\tau)$ space and $\mathcal{O}(\tau)$ query time.

Table 1. Overview of solutions for the LCE problem. Here $\ell = \text{LCE}(i,j)$, $\varepsilon > 0$ is an arbitrarily small constant and w.h.p. (with high probability) means with probability at least $1 - n^{-c}$ for an arbitrarily large constant c. The data structure is *correct* if it answers all LCE queries correctly.

	Data Structure			Preprocessing		Trade-off range	Reference
	Space	Query	Correct	Space	Time		
Deterministic	1	ℓ	always	1	1	–	Store nothing
	n	1	always	n	n	–	Suffix tree + NCA
	$\frac{n}{\tau}$	τ^2	always	$\frac{n}{\tau}$	$\frac{n^2}{\tau}$	$1 \le \tau \le \sqrt{n}$	[2]
	$\frac{n}{\tau}$	$\tau \log^2 \frac{n}{\tau}$	always	$\frac{n}{\tau}$	n^2	$\frac{1}{\log n} \le \tau \le n$	this paper, Sec. 2
	$\frac{n}{\tau}$	τ	always	$\frac{n}{\tau}$	$n^{2+\varepsilon}$	$1 \le \tau \le n$	this paper, Sec. 4
Monte Carlo	$\frac{n}{\tau}$	$\tau \log \frac{\ell}{\tau}$	w.h.p.	$\frac{n}{\tau}$	n	$1 \le \tau \le n$	[2]
	$\frac{n}{\tau}$	τ	w.h.p.	$\frac{n}{\tau}$	$n \log \frac{n}{\tau}$	$1 \le \tau \le n$	this paper, Sec. 3
Las Vegas	$\frac{n}{\tau}$	$\tau \log \frac{\ell}{\tau}$	always	$\frac{n}{\tau}$	$n(\tau + \log n)$ w.h.p.	$1 \le \tau \le n$	[2]
	$\frac{n}{\tau}$	τ	always	$\frac{n}{\tau}$	$n^{3/2}$ w.h.p.	$1 \le \tau \le n$	this paper, Sec. 3.5

– A derandomized version of the Monte Carlo solution with $\mathcal{O}(n/\tau)$ space and $\mathcal{O}(\tau)$ query time.

Hence, we obtain the first trade-off for the LCE problem with a linear time-space product in the full range from constant to linear space. This almost matches the time-space product lower bound of $\Omega(n)$ *bits*, and improves the best deterministic upper bound by a factor of τ, and the best randomized bound by a factor $\log(\frac{n}{\tau})$. See the columns marked *Data Structure* in Table 1 for a complete overview.

While our main focus is the space and query time complexity, we also provide efficient *preprocessing* algorithms for building the data structures, supporting independent trade-offs between the preprocessing time and preprocessing space. See the columns marked *Preprocessing* in Table 1.

To achieve our results we develop several new techniques and specialized data structures which are likely of independent interest. For instance, in our deterministic solution we develop a novel recursive decomposition of LCE queries and for the randomized solution we develop a new sampling technique for Karp-Rabin fingerprints that allow fast LCE queries. We also give a general technique for efficiently derandomizing algorithms that rely on "few" or "short" Karp-Rabin fingerprints, and apply the technique to derandomize our Monte Carlo algorithm. To the best of our knowledge, this is the first derandomization technique for Karp-Rabin fingerprints.

Preliminaries. We assume an integer alphabet, i.e., T is chosen from some alphabet $\Sigma = \{0, \ldots, n^c\}$ for some constant c, so every character of T fits in $\mathcal{O}(1)$ words. For integers $a \leq b$, $[a, b]$ denotes the range $\{a, a+1, \ldots, b\}$ and we define $[n] = [0, n-1]$. For a string $S = S[1]S[2] \ldots S[|S|]$ and positions $1 \leq i \leq j \leq |S|$, $S[i..j] = S[i]S[i+1]\cdots S[j]$ is a *substring* of length $j - i + 1$, $S[i...] = S[i, |S|]$ is the ith *suffix* of S, and $S[...i] = S[1, i]$ is the ith *prefix* of S.

2   Deterministic Trade-Off

Here we describe a completely deterministic trade-off for the LCE problem with $\mathcal{O}(\frac{n}{\tau} \log \frac{n}{\tau})$ space and $\mathcal{O}(\tau \log \frac{n}{\tau})$ query time for any $\tau \in [1, n]$. Substituting $\hat{\tau} = \tau/\log(n/\tau)$, we obtain the bounds reflected in Table 1 for $\hat{\tau} \in [1/\log n, n]$.

A key component in this solution is the following observation that allows us to reduce an $\text{LCE}(i, j)$ query on T to another query $\text{LCE}(i', j')$ where i' and j' are both indices in either the first or second half of T.

Observation 1. *Let i, j and j' be indices of T, and suppose that $\text{LCE}(j', j) \geq \text{LCE}(i, j)$. Then $\text{LCE}(i, j) = \min(\text{LCE}(i, j'), \text{LCE}(j', j))$.*

We apply Observation 1 recursively to bring the indices of the initial query within distance τ in $\mathcal{O}(\log(n/\tau))$ rounds. We show how to implement each round with a data structure using $\mathcal{O}(n/\tau)$ space and $\mathcal{O}(\tau)$ time. This leads to a solution using $\mathcal{O}(\frac{n}{\tau} \log \frac{n}{\tau})$ space and $\mathcal{O}(\tau \log \frac{n}{\tau})$ query time. Finally in Sect. 2.4, we show how to efficiently solve the LCE problem for indicies within distance τ in $\mathcal{O}(n/\tau)$ space and $\mathcal{O}(\tau)$ time by exploiting periodicity properties of LCEs.

2.1   The Data Structure

We will store several data structures, each responsible for a specific subinterval $I = [a, b] \subseteq [1, n]$ of positions of the input string T. Let $I_{\text{left}} = [a, (a + b)/2]$, $I_{\text{right}} = ((a + b)/2, b]$, and $|I| = b - a + 1$. The task of the data structure for I will be to reduce an $\text{LCE}(i, j)$ query where $i, j \in I$ to one where both indices belong to either I_{left} or I_{right}.

The data structure stores information for $\mathcal{O}(|I|/\tau)$ suffixes of T that start in I_{right}. More specifically, we store information for the suffixes starting at positions $b - k\tau \in I_{\text{right}}$, $k = 0, 1, \ldots, (|I|/2)/\tau$. We call these the *sampled positions* of I_{right}. See Fig. 1 for an illustration.

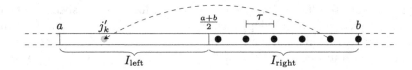

Fig. 1. Illustration of the contents of the data structure for the interval $I = [a, b]$. The black dots are the sampled positions in I_{right}, and each such position has a pointer to an index $j'_k \in I_{\text{left}}$.

For every sampled position $b - k\tau \in I_{\text{right}}$, $k = 0, 1, \ldots, (|I|/2)/\tau$, we store the index j'_k of the suffix starting in I_{left} that achieves the maximum LCE value with the suffix starting at the sampled position, i.e., $T[b - k\tau...]$ (ties broken arbitrarily). Along with j'_k, we also store the value of the LCE between suffix $T[j'_k...]$ and $T[b - k\tau...]$. Formally, j'_k and L_k are defined as follows,

$$j'_k = \operatorname*{argmax}_{h \in I_{\text{left}}} \text{LCE}(h, b - k\tau) \quad \text{and} \quad L_k = \text{LCE}(j'_k, b - k\tau) .$$

Building the Structure. We construct the above data structure for the interval $[1, n]$, and build it recursively for $[1, n/2]$ and $(n/2, n]$, stopping when the length of the interval becomes smaller than τ.

2.2   Answering a Query

We now describe how to reduce a query $\text{LCE}(i, j)$ where $i, j \in I$ to one where both indices are in either I_{left} or I_{right}. Suppose without loss of generality that $i \in I_{\text{left}}$ and $j \in I_{\text{right}}$. We start comparing $\delta < \tau$ pairs of characters of T, starting with $T[i] = T[j]$, until (1) we encounter a mismatch, (2) both positions are in I_{right} or (3) we reach a sampled position in I_{right}. It suffices to describe the last case, in which $T[i, i + \delta] = T[j, j + \delta]$, $i + \delta \in I_{\text{left}}$ and $j + \delta = b - k\tau \in I_{\text{right}}$ for some k. Then by Observation 1, we have that

$$\begin{aligned} \text{LCE}(i, j) &= \delta + \text{LCE}(i + \delta, j + \delta) \\ &= \delta + \min(\text{LCE}(i + \delta, j'_k), \text{LCE}(j'_k, b - k\tau)) \\ &= \delta + \min(\text{LCE}(i + \delta, j'_k), L_k) . \end{aligned}$$

Thus, we have reduced the original query to computing the query $\mathrm{LCE}(i+\delta, j'_k)$ in which both indices are in I_{left}.

2.3 Analysis

Each round takes $\mathcal{O}(\tau)$ time and halves the upper bound for $|i-j|$, which initially is n. Thus, after $\mathcal{O}(\tau \log(n/\tau))$ time, the initial LCE query has been reduced to one where $|i - j| \leq \tau$. At each of the $\mathcal{O}(\log(n/\tau))$ levels, the number of sampled positions is $(n/2)/\tau$, so the total space used is $\mathcal{O}((n/\tau) \log(n/\tau))$.

2.4 Queries with Nearby Indices

We now describe the data structure used to answer a query $\mathrm{LCE}(i,j)$ when $|i - j| \leq \tau$. We first give some necessary definitions and properties of periodic strings. We say that the integer $1 \leq p \leq |S|$ is a *period* of a string S if any two characters that are p positions apart in S match, i.e., $S[i] = S[i + p]$ for all positions i s.t. $1 \leq i < i + p \leq |S|$. The following is a well-known property of periods.

Lemma 1 (Fine and Wilf [5]). *If a string S has periods a and b and $|S| \geq |a| + |b| - \gcd(a, b)$, then $\gcd(a, b)$ is also a period of S.*

The period of S is the smallest period of S and we denote it by $\mathrm{per}(S)$. If $\mathrm{per}(S) \leq |S|/2$, we say S is *periodic*. A periodic string S might have many periods smaller than $|S|/2$, however it follows from the above lemma that

Corollary 1. *All periods smaller than $|S|/2$ are multiples of $\mathrm{per}(S)$.*

The Data Structure. Let $T_k = T[k\tau...(k + 2)\tau - 1]$ denote the substring of length 2τ starting at position $k\tau$ in T, $k = 0, 1, \ldots, n/\tau$. For the strings T_k that are periodic, let $p_k = \mathrm{per}(T_k)$ be the period. For every periodic T_k, the data structure stores the length ℓ_k of the maximum substring starting at position $k\tau$, which has period p_k. Nothing is stored if T_k is aperiodic.

Answering a Query. We may assume without loss of generality that $i = k\tau$, for some integer k. If not, then we check whether $T[i + \delta] = T[j + \delta]$ until $i + \delta = k\tau$. Hence, assume that $i = k\tau$ and $j = i + d$ for some $0 < d \leq \tau$. In $\mathcal{O}(\tau)$ time, we first check whether $T[i + \delta] = T[j + \delta]$ for all $\delta \in [0, 2\tau]$. If we find a mismatch we are done, and otherwise we return $\mathrm{LCE}(i, j) = \ell_k - d$.

Correctness. If a mismatch is found when checking that $T[i + \delta] = T[j + \delta]$ for all $\delta \in [0, 2\tau]$, the answer is clearly correct. Otherwise, we have established that $d \leq \tau$ is a period of T_k, so T_k is periodic and d is a multiple of p_k (by Corollary 1). Consequently, $T[i + \delta] = T[j + \delta]$ for all δ s.t. $d + \delta \leq \ell_k$, and thus $\mathrm{LCE}(i, j) = \ell_k - d$.

2.5   Preprocessing

The preprocessing details appear in the full version of this paper [3].

3   Randomized Trade-Offs

In this section we describe a randomized LCE data structure using $\mathcal{O}(n/\tau)$ space with $\mathcal{O}(\tau + \log \frac{\ell}{\tau})$ query time. In Sect. 3.6 we describe another $\mathcal{O}(n/\tau)$-space LCE data structure that either answers an LCE query in constant time, or provides a certificate that $\ell \leq \tau^2$. Combining the two data structures, shows that the LCE problem can be solved in $\mathcal{O}(n/\tau)$ space and $\mathcal{O}(\tau)$ time.

The randomization comes from our use of Karp-Rabin fingerprints [9] for comparing substrings of T for equality. Before describing the data structure, we start by briefly recapping the most important definitions and properties of Karp-Rabin fingerprints.

3.1   Karp-Rabin Fingerprints

For a prime p and $x \in [p]$ the Karp-Rabin fingerprint [9], denoted $\phi_{p,x}(T[i...j])$, of the substring $T[i...j]$ is defined as

$$\phi_{p,x}(T[i...j]) = \sum_{i \leq k \leq j} T[k] x^{k-i} \mod p .$$

If $T[i...j] = T[i'...j']$ then clearly $\phi_{p,x}(T[i...j]) = \phi_{p,x}(T[i'...j'])$. In the Monte Carlo and the Las Vegas algorithms we present we will choose p such that $p = \Theta(n^{4+c})$ for some constant $c > 0$ and x uniformly from $[p] \setminus \{0\}$. In this case a simple union bound shows that the converse is also true with high probability, i.e., ϕ is *collision-free* on all substring pairs of T with probability at least $1 - n^{-c}$. Storing a fingerprint requires $\mathcal{O}(1)$ space. When p, x are clear from the context we write $\phi = \phi_{p,x}$.

For shorthand we write $f(i) = \phi(T[1, i]), i \in [1, n]$ for the fingerprint of the ith prefix of T. Assuming that we store the exponent $x^i \mod p$ along with the fingerprint $f(i)$, the following two properties of fingerprints are well-known and easy to show.

Lemma 2. *(1) Given $f(i)$, the fingerprint $f(i \pm a)$ for some integer a, can be computed in $\mathcal{O}(a)$ time. (2) Given fingerprints $f(i)$ and $f(j)$, the fingerprint $\phi(T[i..j])$ can be computed in $\mathcal{O}(1)$ time.*

In particular this implies that for a fixed length l, the fingerprint of all substrings of length l of T can be enumerated in $\mathcal{O}(n)$ time using a sliding window.

3.2   Overview

The main idea in our solution is to binary search for the LCE(i,j) value using Karp-Rabin fingerprints. Suppose for instance that $\phi(T[i, i + M]) \neq \phi(T[j, j + M])$ for some integer M, then we know that LCE$(i,j) \leq M$, and thus we can find the true $\mathrm{LCE}(i, j)$ value by comparing $\log(M)$ additional pair of fingerprints. The challenge is to obtain the fingerprints quickly when we are only allowed to use $\mathcal{O}(n/\tau)$ space. We will partition the input string T into n/τ *blocks* each of length τ. Within each block we sample a number of equally spaced positions. The data structure consists of the fingerprints of the prefixes of T that ends at the sampled positions, i.e., we store $f(i)$ for all sampled positions i. In total we sample $\mathcal{O}(n/\tau)$ positions. If we just sampled a single position in each block (similar to the approach in [2]), we could compute the fingerprint of any substring in $\mathcal{O}(\tau)$ time (see Lemma 2), and the above binary search algorithm would take time $\mathcal{O}(\tau \log n)$ time. We present a new sampling technique that only samples an additional $\mathcal{O}(n/\tau)$ positions, while improving the query time to $\mathcal{O}(\tau + \log(\ell/\tau))$.

Preliminary Definitions. We partition the input string T into n/τ blocks of τ positions, and by *block* k we refer to the positions $[k\tau, k\tau + \tau)$, for $k \in [n/\tau]$.

We assume without loss of generality that n and τ are both powers of two. Every position $q \in [1, n]$ can be represented as a bit string of length $\lg n$. Let $q \in [1, n]$ and consider the binary representation of q. We define the leftmost $\lg(n/\tau)$ bits and rightmost $\lg(\tau)$ bits to be the *head*, denoted $h(q)$ and the *tail*, denoted $t(q)$, respectively. A position is *block aligned* if $t(q) = 0$. The *significance* of q, denoted $s(q)$, is the number of trailing zeros in $h(q)$. Note that the τ positions in any fixed block $k \in [n/\tau]$ all have the same head, and thus also the same significance, which we denote by μ_k. See Fig. 2.

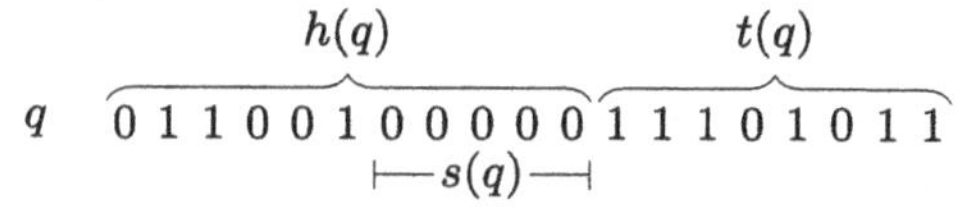

$$\overbrace{\qquad\qquad}^{h(q)} \qquad\qquad \overbrace{\qquad\qquad}^{t(q)}$$
$$q \quad 0\ 1\ 1\ 0\ 0\ 1\ 0\ 0\ 0\ 0\ 0\ 1\ 1\ 1\ 0\ 1\ 0\ 1\ 1$$
$$\underbrace{\qquad\qquad}_{s(q)}$$

Fig. 2. Example of the definitions for the position $q = 205035$ in a string of length $n = 2^{19}$ with block length $\tau = 2^8$. Here $h(q)$ is the first $\lg(n/\tau) = 11$ bits, and $t(q)$ is the last $\lg(\tau) = 8$ bits in the binary representation of q. The significance is $s(q) = 5$.

3.3 The Monte Carlo Data Structure

The data structure consists of the values $f(i)$, $i \in \mathcal{S}$, for a specific set of sampled positions $\mathcal{S} \subseteq [1, n]$, along with the information necessary in order to look up the values in constant time. We now explain how to construct the set $\mathcal{S}$. In block $k \in [n/\tau]$ we will sample $b_k = \min\{2^{\lfloor \mu_k/2 \rfloor}, \tau\}$ evenly spaced positions, where μ_k is the significance of the positions in block k, i.e., $\mu_k = s(k\tau)$. More precisely, in block k we sample the positions $\mathcal{B}_k = \{k\tau + j\tau/b_k \mid j \in [b_k]\}$, and let $\mathcal{S} = \cup_{k \in [n/\tau]} \mathcal{B}_k$. See Fig. 3.

We now bound the size of $\mathcal{S}$. The significance of a block is at most $\lg(n/\tau)$, and there are exactly $2^{\lg(n/\tau)-\mu}$ blocks with significance μ, so

$$|\mathcal{S}| = \sum_{k=0}^{n/\tau-1} b_k \leq \sum_{\mu=0}^{\lg(n/\tau)} 2^{\lg(n/\tau)-\mu} 2^{\lfloor \mu/2 \rfloor} \leq \frac{n}{\tau} \sum_{\mu=0}^{\infty} 2^{-\mu/2} = \left(2 + \sqrt{2}\right) \frac{n}{\tau} = \mathcal{O}\left(\frac{n}{\tau}\right).$$

k	0	1	2	3	4	5	6	7	8	9	10	11	12	13	14	15
μ_k	4	0	1	0	2	0	1	0	3	0	1	0	2	0	1	0
b_k	4	1	1	1	2	1	1	1	2	1	1	1	2	1	1	1

Fig. 3. Illustration of a string T partitioned into 16 blocks each of length τ. The significance μ_k for the positions in each block $k \in [n/\tau]$ is shown, as well as the b_k values. The block dots are the sampled positions $\mathcal{S}$.

3.4 Answering a Query

We now describe how to answer an LCE(i, j) query. We will assume that i is block aligned, i.e., $i = k\tau$ for some $k \in [n/\tau]$. Note that we can always obtain this situation in $\mathcal{O}(\tau)$ time by initially comparing at most $\tau - 1$ pairs of characters of the input string directly.

Algorithm 1 shows the query algorithm. It performs an exponential search to locate the block in which the first mismatch occurs, after which it scans the block directly to locate the mismatch. The search is performed by calls to check(i, j, c), which computes and compares $\phi(T[i...i + c])$ and $\phi(T[j...j + c])$. In other words, assuming that ϕ is collision-free, check(i, j, c) returns true if LCE$(i, j) \geq c$ and false otherwise.

Algorithm 1. Computing the answer to a query LCE(i, j)

1: **procedure** LCE(i, j)
2:     $\hat{\ell} \leftarrow 0$
3:     $\mu \leftarrow 0$
4:     **while** check$(i, j, 2^\mu \tau)$ **do**   ▷ Compute an interval such that $\ell \in [\hat{\ell}, 2\hat{\ell}]$.
5:         $(i, j, \hat{\ell}) \leftarrow (i + 2^\mu \tau, j + 2^\mu \tau, \hat{\ell} + 2^\mu \tau)$
6:         **if** $s(j) > \mu$ **then**
7:             $\mu \leftarrow \mu + 1$
8:     **while** $\mu > 0$ **do**   ▷ Identify the block in which the first mismatch occurs
9:         **if** check$(i, j, 2^{\mu-1} \tau)$ **then**
10:            $(i, j, \hat{\ell}) \leftarrow (i + 2^{\mu-1} \tau, j + 2^{\mu-1} \tau, \hat{\ell} + 2^{\mu-1} \tau)$
11:            $\mu \leftarrow \mu - 1$
12:     **while** $T[i] = T[j]$ **do** ▷ Scan the final block left to right to find the mismatch
13:         $(i, j, \hat{\ell}) \leftarrow (i + 1, j + 1, \hat{\ell} + 1)$
14:     **return** $\hat{\ell}$

Analysis. We now prove that Algorithm 1 correctly computes $\ell = $ LCE(i, j) in $\mathcal{O}(\tau + \log(\ell/\tau))$ time. The algorithm is correct assuming that check$(i, j, 2^\mu \tau)$ always returns the correct answer, which will be the case if ϕ is collision-free.

The following is the key lemma we need to bound the time complexity.

Lemma 3. *Throughout Algorithm 1 it holds that* $\ell \geq (2^\mu - 1)\tau$, $s(j) \geq \mu$, *and* μ *is increased in at least every second iteration of the first **while**-loop.*

Proof. We first prove that $s(j) \geq \mu$. The claim holds initially. In the first loop j is changed to $j + 2^\mu \tau$, and $s(j + 2^\mu \tau) \geq \min\{s(j), s(2^\mu \tau)\} = \min\{s(j), \mu\} = \mu$, where the last equality follows from the induction hypothesis $s(j) \geq \mu$. Moreover, μ is only incremented when $s(j) > \mu$. In the second loop j is changed to $j + 2^{\mu-1} \tau$, which under the assumption that $s(j) \geq \mu$, has significance $s(j + 2^{\mu-1} \tau) = \mu - 1$. Hence the invariant is restored when μ is decremented at line 11.

Now consider an iteration of the first loop where μ is not incremented, i.e., $s(j) = \mu$. Then $\frac{j}{2^\mu \tau}$ is an odd integer, i.e. $\frac{j + 2^\mu \tau}{2^\mu \tau}$ is even, and hence $s(j + 2^\mu \tau) > \mu$, so μ will be incremented in the next iteration of the loop.

In order to prove that $\ell \geq (2^\mu - 1)\tau$ we will prove that $\hat{\ell} \geq (2^\mu - 1)\tau$ in the first loop. This is trivial by induction using the observation that $(2^\mu - 1)\tau + 2^\mu \tau = (2^{\mu+1} - 1)\tau$.   □

Since $\ell \geq (2^\mu - 1)\tau$ and μ is increased at least in every second iteration of the first loop and decreased in every iteration of the second loop, it follows that there are $\mathcal{O}(\log(\ell/\tau))$ iterations of the two first loops. The last loop takes $\mathcal{O}(\tau)$ time. It remains to prove that the time to evaluate the $\mathcal{O}(\log(\ell/\tau))$ calls to check$(i, j, 2^\mu \tau)$ sums to $\mathcal{O}(\tau + \log(\ell/\tau))$.

Evaluating check$(i, j, 2^\mu \tau)$ requires computing $\phi(T[i...i + 2^\mu \tau])$ and $\phi(T[j...j + 2^\mu \tau])$. The first fingerprint can be computed in constant time because i and $i + 2^\mu \tau$ are always block aligned (see Lemma 2). The time to compute the second fingerprint depends on how far j and $j + 2^\mu \tau$ each are from a sampled position, which in turn depends inversely on the significance of the block containing those positions. By Lemma 3, μ is always a lower bound on the significance of j, which implies that μ also lower bounds the significance of $j + 2^\mu \tau$, and thus by the way we sample positions, neither will have distance more than $\tau/2^{\lfloor \mu/2 \rfloor}$ to a sampled position in $\mathcal{S}$. Finally, note that by the way μ is increased and decreased, check$(i, j, 2^\mu \tau)$ is called at most three times for any fixed value of μ. Hence, the total time to compute all necessary fingerprints can be bounded as

$$\mathcal{O}\left(\sum_{\mu=0}^{\lg(\ell/\tau)} 1 + \tau/2^{\lfloor \mu/2 \rfloor}\right) = \mathcal{O}(\tau + \log(\ell/\tau)) .$$

3.5   The Las Vegas Data Structure

We now describe an $\mathcal{O}(n^{3/2})$-time and $\mathcal{O}(n/\tau)$-space algorithm for verifiying that ϕ is *collision-free* on all pairs of substrings of T that the query algorithm compares. If a collision is found we pick a new ϕ and try again. With high probability we can find a collision-free ϕ in a constant number of trials, so we obtain the claimed Las Vegas data structure.

If $\tau \leq \sqrt{n}$ we use the verification algorithm of Bille et al. [2], using $\mathcal{O}(n\tau + n \log n)$ time and $\mathcal{O}(n/\tau)$ space. Otherwise, we use the simple $\mathcal{O}(n^2/\tau)$-time and $\mathcal{O}(n/\tau)$-space algorithm described below.

Recall that all fingerprint comparisions in our algorithm are of the form

$$\phi\big(T[k\tau...k\tau + 2^l\tau - 1]\big) \stackrel{?}{=} \phi\big(T[j...j + 2^l\tau - 1]\big)$$

for some $k \in [n/\tau], j \in [n], l \in [\log(n/\tau)]$. The algorithm checks each $l \in [\log(n/\tau)]$ separately. For a fixed l it stores the fingerprints $\phi(T[k\tau...k\tau + 2^l\tau])$ for all $k \in [n/\tau]$ in a hash table $\mathcal{H}$. This can be done in $\mathcal{O}(n)$ time and $\mathcal{O}(n/\tau)$ space. For every $j \in [n]$ the algorithm then checks whether $\phi\big(T[j...j + 2^l\tau]\big) \in \mathcal{H}$, and if so, it verifies that the underlying two substrings are in fact the same by comparing them character by character in $\mathcal{O}(2^l\tau)$ time. By maintaining the fingerprint inside a sliding window of length $2^l\tau$, the verification time for a fixed l becomes $\mathcal{O}(n2^l\tau)$, i.e., $\mathcal{O}(n^2/\tau)$ time for all $l \in [\log(n/\tau)]$.

3.6   Queries with Long LCEs

In this section we describe an $\mathcal{O}(\frac{n}{\tau})$ space data structure that in constant time either correctly computes $\mathrm{LCE}(i, j)$ or determines that $\mathrm{LCE}(i, j) \leq \tau^2$. The data structure can be constructed in $\mathcal{O}(n \log \frac{n}{\tau})$ time by a Monte Carlo or Las Vegas algorithm.

The Data Structure. Let $\mathcal{S}_\tau \subseteq [1, n]$ called the *sampled positions* of T (to be defined below), and consider the sets A and B of suffixes of T and T^R, respectively.

$$A = \{T[i...] \mid i \in \mathcal{S}_\tau\}, \quad B = \{T[...i]^R \mid i \in \mathcal{S}_\tau\} \, .$$

We store a data structure for A and B, that allows us to perform constant time longest common extension queries on any pair of suffixes in A or any pair in B. This can be achieved by well-known techniques, e.g., storing a sparse suffix tree for A and B, equipped with a nearest common ancestor data structure. To define $\mathcal{S}_\tau$, let $D_\tau = \{0, 1, \dots, \tau\} \cup \{2\tau, \dots, (\tau - 1)\tau\}$, then

$$\mathcal{S}_\tau = \{1 \leq i \leq n \mid i \mod \tau^2 \in D_\tau\} \, . \tag{1}$$

Answering a Query. To answer an LCE query, we need the following definitions. For $i, j \in \mathcal{S}_\tau$ let $\mathrm{LCE}_R(i, j)$ denote the longest common prefix of $T[...i]^R \in B$ and $T[...j]^R \in B$. Moreover, for $i, j \in [n]$, we define the function

$$\delta(i, j) = \big(((i - j) \mod \tau) - i\big) \mod \tau^2 \, . \tag{2}$$

We will write δ instead of $\delta(i, j)$ when i and j are clear from the context.

The following lemma gives the key property that allows us to answer a query.

Lemma 4. *For any $i, j \in [n - \tau^2]$, it holds that $i + \delta, j + \delta \in \mathcal{S}_\tau$.*

Proof. Direct calculation shows that $(i + \delta) \mod \tau^2 \leq \tau$, and that $(j + \delta) \mod \tau = 0$, and thus by definition both $i + \delta$ and $j + \delta$ are in $\mathcal{S}_\tau$.

To answer a query $\mathrm{LCE}(i, j)$, we first verify that $i, j \in [n - \tau^2]$ and that $\mathrm{LCE}_R(i + \delta, j + \delta) \geq \delta$. If this is not the case, we have established that $\mathrm{LCE}(i, j) \leq \delta < \tau^2$, and we stop. Otherwise, we return $\delta + \mathrm{LCE}(i + \delta, j + \delta) - 1$.

Analysis. To prove the correctness, suppose $i, j \in [n - \tau^2]$ (if not clearly $\text{LCE}(i, j) < \tau^2$) then we have that $i + \delta, j + \delta \in \mathcal{S}_\tau$ (Lemma 4). If $\text{LCE}_R(i + \delta, j + \delta) \geq \delta$ it holds that $T[i...i + \delta] = T[j...j + \delta]$ so the algorithm correctly computes $\text{LCE}(i, j)$ as $\delta + 1 + \text{LCE}(i + \delta, j + \delta)$. Conversely, if $\text{LCE}_R(i + \delta, j + \delta) < \delta$, $T[i...i + \delta] \neq T[j...j + \delta]$ it follows that $\text{LCE}(i, j) < \delta < \tau^2$.

Query time is $\mathcal{O}(1)$, since computing δ, $\text{LCE}_R(i + \delta, j + \delta)$ and $\text{LCE}(i + \delta, j + \delta)$ all takes constant time. Storing the data structures for A and B takes space $\mathcal{O}(|A| + |B|) = \mathcal{O}(|\mathcal{S}_\tau|) = \mathcal{O}(\frac{n}{\tau})$. For the preprocessing stage, we can use recent algorithms by I et al. [8] for constructing the sparse suffix tree for A and B in $\mathcal{O}(\frac{n}{\tau})$ space. They provide a Monte Carlo algorithm using $\mathcal{O}(n \log \frac{n}{\tau})$ time (correct w.h.p.), and a Las Vegas algorithm using $\mathcal{O}(\frac{n}{\tau})$ time (w.h.p.).

4   Derandomizing the Monte Carlo Data Structure

Here we give a general technique for derandomizing Karp-Rabin fingerprints, and apply it to our Monte Carlo algorithm. The main result is that for any constant $\varepsilon > 0$, the data structure can be constructed completely deterministically in $\mathcal{O}(n^{2+\varepsilon})$ time using $\mathcal{O}(n/\tau)$ space. Thus, compared to the probabilistic preprocessing of the Las Vegas structure using $\mathcal{O}(n^{3/2})$ time with high probability, it is relatively cheap to derandomize the data structure completely.

Our derandomizing technique is stated in the following lemma.

Lemma 5. *Let* $A, L \subset \{1, 2, \ldots, n\}$ *be a set of positions and lengths respectively such that* $\max(L) = n^{\Omega(1)}$. *For every* $\varepsilon \in (0, 1)$, *there exist a fingerprinting function* ϕ *that can be evaluated in* $\mathcal{O}\left(\frac{1}{\varepsilon}\right)$ *time and has the property that for all* $a \in A, l \in L, i \in \{1, 2, \ldots, n\}$:

$$\phi(T[a...a + (l - 1)]) = \phi(T[i...i + (l - 1)]) \iff T[a...a + (l - 1)] = T[i...i + (l - 1)]$$

We can find such a ϕ *using* $\mathcal{O}\left(\frac{S}{\varepsilon}\right)$ *space and* $\mathcal{O}\left(\frac{n^{1+\varepsilon} \log n}{\varepsilon^2} \frac{|A|}{S} \max(L)\,|L|\right)$ *time, for any value of* $S \in [1, |A|]$.

The proof of the lemma appears in the full version of this paper [3].

Corollary 2. *For any* $\tau \in [1, n]$, *the LCE problem can be solved by a deterministic data structure with* $\mathcal{O}(n/\tau)$ *space usage and* $\mathcal{O}(\tau)$ *query time. The data structure can be constructed in* $\mathcal{O}(n^{2+\varepsilon})$ *time using* $\mathcal{O}(n/\tau)$ *space.*

Proof. We use the lemma with $A = \{k\tau \mid k \in [n/\tau]\}$, $L = \{2^l \tau \mid l \in [\log(n/\tau)]\}$, $S = |A| = n/\tau$ and a suitable small constant $\varepsilon > 0$. $\quad\Box$

References

1. Amir, A., Lewenstein, M., Porat, E.: Faster algorithms for string matching with k mismatches. J. Algorithms **50**(2), 257–275 (2004)
2. Bille, P., Gørtz, I.L., Sach, B., Vildhøj, H.W.: Time-space trade-offs for longest common extensions. J. Discret. Algorithms **25**, 42–50 (2014)

3. Bille, P., Gørtz, I.L., Knudsen, M.B.T., Lewenstein, M., Vildhøj, H.W.: Longest common extensions in sublinear space. arXiv:1504.02671(2015)
4. Cole, R., Hariharan, R.: Approximate string matching: a simpler faster algorithm. SIAM J. Comput. **31**(6), 1761–1782 (2002)
5. Fine, N.J., Wilf, H.S.: Uniqueness theorems for periodic functions. Proc. AMS **16**(1), 109–114 (1965)
6. Gusfield, D., Stoye, J.: Linear time algorithms for finding and representing all the tandem repeats in a string. J. Comput. Syst. Sci. **69**, 525–546 (2004)
7. Harel, D., Tarjan, R.E.: Fast algorithms for finding nearest common ancestors. SIAM J. Comput. **13**(2), 338–355 (1984)
8. Kärkkäinen, T.I J., Kempa, D.: Faster sparse suffix sorting. In: Proceedings of 31st STACS, vol. 25, pp. 386–396, Dagstuhl, Germany (2014)
9. Karp, R.M., Rabin, M.O.: Efficient randomized pattern-matching algorithms. IBM J. Res. Dev. **31**(2), 249–260 (1987)
10. Kolpakov, R., Kucherov, G.: Searching for gapped palindromes. In: Ferragina, P., Landau, G.M. (eds.) CPM 2008. LNCS, vol. 5029, pp. 18–30. Springer, Heidelberg (2008)
11. Landau, G.M., Myers, E.W., Schmidt, J.P.: Incremental string comparison. SIAM J. Comput. **27**(2), 557–582 (1998)
12. Landau, G.M., Schmidt, J.P.: An algorithm for approximate tandem repeats. J. Comput. Biol. **8**(1), 1–18 (2001)
13. Landau, G.M., Vishkin, U.: Fast parallel and serial approximate string matching. J. Algorithms **10**, 157–169 (1989)
14. Main, M.G., Lorentz, R.J.: An O (n log n) algorithm for finding all repetitions in a string. J. Algorithms **5**(3), 422–432 (1984)
15. Manacher, G.: A new linear-time "On-Line" algorithm for finding the smallest initial palindrome of a string. J. ACM **22**(3), 346–351 (1975)
16. Myers, E.W.: An $O(ND)$ difference algorithm and its variations. Algorithmica **1**(2), 251–266 (1986)
17. Weiner, P.: Linear pattern matching algorithms. In: Proceedings of 14th FOCS (SWAT), pp. 1–11 (1973)

Ranked Document Retrieval
with Forbidden Pattern

Sudip Biswas[1], Arnab Ganguly[1]([✉]), Rahul Shah[1], and Sharma V. Thankachan[2]

[1] School of Electrical Engineering and Computer Science,
Louisiana State University, Baton Rouge, USA
{sbiswa7,agangu4}@lsu.edu, rahul@csc.lsu.edu
[2] School of Computational Science and Engineering,
Georgia Institute of Technology, Atlanta, USA
sharma.thankachan@gatech.edu

Abstract. Let $\mathcal{D} = \{\mathsf{T}_1, \mathsf{T}_2, \ldots, \mathsf{T}_D\}$ be a collection of D string documents of n characters in total. The forbidden pattern *document listing* problem asks to report those documents $\mathcal{D}' \subseteq \mathcal{D}$ which contain the pattern P, but not the pattern Q. The top-k forbidden pattern query (P, Q, k) asks to report those k documents in $\mathcal{D}'$ that are most relevant to P. For typical relevance functions (like *document importance, term-frequency, term-proximity*), we present a linear space index with worst case query time of $O(|P| + |Q| + \sqrt{nk})$ for the top-k problem. As a corollary of this result, we obtain a linear space and $O(|P| + |Q| + \sqrt{nt})$ query time solution for the document listing problem, where t is the number of documents reported. We conjecture that any significant improvement over the results in this paper is highly unlikely.

1 Introduction and Related Work

A fundamental problem in information retrieval is indexing a collection of documents for efficient retrieval of the (most relevant) documents corresponding to a query, which may consist of one or more patterns (strings). The traditional approach is to partition the documents into sub-strings (called words), and then build an inverted index over them. In numerous scenarios, such as genome data or texts from some Asian languages, there is no clear demarcation of words, and documents need to be treated as plain strings (see [18] for an excellent survey). In these cases, inverted index approach may either require too much space, or support only limited search functions [22]. Therefore, we require alternative approaches based on full-text indexes such as (compressed) suffix trees/arrays.

In most of the earlier string retrieval problems, the query consists of a single pattern P. Introduced by Matias et al. [16], the most basic problem is *document listing*, which asks to report all unique documents containing P. Later Muthukrishnan [17] gave a linear space and optimal query time solution. The

This research is funded in part by National Science Foundation (NSF) Grants CCF–1017623 and CCF–1218904.

top-k document retrieval, introduced by Hon et al. [14], is an important extension of this problem, and asks to report the k documents that are most relevant to P. Relevance of a document is based on various measures such as document PageRank (which is independent of P), term-frequency (i.e., the number of times a pattern appears in the document), term-proximity (i.e., the distance between the closest appearance of a pattern in the document). See [13,19] for linear space and optimal i.e., $O(|P| + k)$ query time solutions.

For two patterns P and Q, Muthukrishnan [17] showed that by maintaining an $O(n^{3/2} \log^{O(1)} n)$ space index, all t documents containing both P and Q can be reported in time $O(|P| + |Q| + \sqrt{n} + t)$, where n is the total length of all documents. Cohen and Porat [2] observed that the problem can be reduced to the *set intersection* problem, and presented an $O(n \log n)$ space (in words) index with query time $O(|P|+|Q|+\sqrt{nt} \log^{5/2} n)$. Subsequently, Hon et al. [11] improved this to an $O(n)$ space (in words) index with query time $O(|P| + |Q| + \sqrt{nt} \log^{3/2} n)$. Also see [13,14] for a succinct solution, and [7] for a lower bound which states that for $O(|P| + |Q| + \log^{O(1)} n + t)$ query time, $\Omega(n(\log n/ \log \log n)^3)$ bits are required. A recent result [15] on the hardness of this problem states that any improvement other than poly-logarithmic factors is highly unlikely.

The problem of forbidden pattern queries can be seen as a variation of the two-pattern problem described above, and can be defined as follows.

Problem 1 (Document Listing with Forbidden Pattern). *Index a collection $\mathcal{D} = \{T_1, T_2, \ldots, T_D\}$ of D strings (called documents) of n characters in total, such that when two patterns P (called included pattern) and Q (called excluded or forbidden pattern) come as a query, all those d's, where P is a substring of T_d and Q is not a substring of T_d can be reported efficiently.*

Such queries are common in web-search applications, and a typical example is the following [12]: say we are interested in accessing the websites on "jaguar", the big cat. However, searching for "jaguar" will give many web sites related to "jaguar car", which we are not interested in. So, the query can be modeled as a forbidden pattern query with $P = jaguar$ and $Q = car$.

Fischer et al. [7] introduced this problem, and presented an $O(n^{3/2})$-bit solution with query time $O(|P|+|Q|+\sqrt{n}+t)$. Hon et al. [12] presented an $O(n)$-word index with query time $O(|P| + |Q| + \sqrt{nt} \log^{5/2} n)$. Larsen et al. [15] presented a hardness result of this problem via a reduction from boolean matrix multiplication and claimed that any significant (i.e., beyond poly-logarithmic factors) improvement over the existing results are highly unlikely. In this paper, we revisit Problem 1 as well as a more general top-k version of it described below.

Problem 2 (Top-k Document Retrieval with Forbidden Pattern). *Index a collection $\mathcal{D} = \{T_1, T_2, \ldots, T_D\}$ of D strings (called documents) of n characters in total such that when two patterns P and Q, and an integer k come as a query, among all documents containing P, but not Q, those k documents that are the most relevant to P can be reported efficiently.*

Relevance Function. In Problem 2 above, the relevance of a document T_d is determined by a function $\mathsf{score}(P, Q, d)$. In particular, $score(P, Q, d) = -\infty$ if Q occurs in T_d, otherwise it is a function of the set of occurrences of P in T_d. We assume that the relevance function is monotonic increasing i.e., $score(P, Q, d) \leq score(P', Q, d)$, where P' is a prefix of P. Various functions will fall under this category, such as term-frequency, and PageRank. With respect to the patterns P and Q, a document T_d is more relevant than $\mathsf{T}_{d'}$ iff $\mathsf{score}(P, Q, d) > \mathsf{score}(P, Q, d')$. We remark that term-proximity is monotonic decreasing; however, by considering the negation of the proximity function, this can be made to fit the criteria of monotonic increasing; in this case T_d is more relevant than $\mathsf{T}_{d'}$ iff $\mathsf{score}(P, Q, d) < \mathsf{score}(P, Q, d')$. See the bottom-$k$ document retrieval problem in [21] as an example of a relevance function which is not monotonic.

We use the generalized suffix tree GST on the collection of documents $\mathcal{D}$, and use a modified form of the marking scheme introduced by Hon et al. [14]. Specifically, we identify few nodes in the GST as *marked* and *prime* nodes. For each pair of marked and prime node, and properly chosen values of k, we precompute and store the answers in a space-efficient way so that can be retrieved efficiently. Our solution for Problem 2 is summarized in the following theorem.

Theorem 1. *Let $\mathcal{D}$ be a collection of D documents of n characters in total. There exists an $O(n)$ space (in words) and $O(|P| + |Q| + \sqrt{nk})$ query time data structure such that when two patterns P and Q, and an integer k come as a query, among all documents containing P, but not Q, we can report those k documents that are the most relevant to P, where the relevance function is monotonic.*

Using the above result as a black box, we can easily obtain the following solution for Problem 1.

Corollary 1. *There exists an $O(n)$ space (in words) and $O(|P| + |Q| + \sqrt{nt})$ query time data structure for Problem 1.*

Proof. In the query time complexity of Theorem 1, the term $O(|P| + |Q|)$ is due to the time for finding the locus nodes of P and Q in a generalized suffix tree of $\mathcal{D}$. To answer document listing queries using Theorem 1, we perform top-k queries for values of k from $1, 2, 4, 8, \cdots$ up to k', where the number of documents returned by the top-k' query is $< k'$, whereas that of every top-k'' query, $k'' < k'$, is k''. This means the answer to top-k' query is the same as that of a document listing query. Also, $k'/2 \leq t < k'$. Therefore, total time spend over all queries (in addition to the time for initial loci search) can be bounded by $O(\sqrt{n} + \sqrt{2n} + \sqrt{4n} + \cdots + \sqrt{nk'}) = O(\sqrt{nt})$. ∎

A Note on the Tightness of Our Result. In order to show the hardness of Problem 1, let us first define a couple of related problems. Let $\mathcal{S} = \{\mathcal{S}_1, \mathcal{S}_2, \ldots, \mathcal{S}_r\}$ be a collection of sets of total cardinality n. The *set intersection* (resp. *set difference*) problem is to preprocess $\mathcal{S}$ into a data structure so that we can report the elements in $\mathcal{S}_i \cap \mathcal{S}_j$ (resp. $\mathcal{S}_i \setminus \mathcal{S}_j$) efficiently for any query (i, j).

For each element e_x in the collection of sets, create document T_x, where the content of T_x is the sequence of identifiers of all sets containing e_x. Clearly, a forbidden pattern query with $P = i$ and $Q = j$ gives the answer to the set difference problem. We conclude that Problem 1 is at least as hard as the set difference problem. The best known upper bound for the set intersection problem is by Cohen and Porat [2], where the space is $O(n)$ words and time is $O(\sqrt{n|\mathcal{S}_i \cap \mathcal{S}_j|})$. The framework by Cohen and Porat can be used to obtain the same space and time solution for the set difference problem. It is unclear whether a better solution for the set difference problem exists, or not; however, the existence of such a solution seems unlikely.

Organization of the Paper. The rest of the paper is dedicated for proving Theorem 1. In Sect. 2, we discuss standard data-structures, and introduce the terminologies used in this paper. We prove Theorem 1 in Sect. 3, and finally conclude the paper in Sect. 4.

2   Preliminaries

We refer the reader to [10] for standard definitions. Let $\mathcal{D} = \{\mathsf{T}_1, \mathsf{T}_2, \cdots, \mathsf{T}_D\}$ be a collection of documents of total size n characters; each document in $\mathcal{D}$ has a special terminating character that does not appear anywhere in the document.

Generalized Suffix Tree. The generalized suffix tree (denoted by GST) is a compact trie that stores all (non-empty) suffixes of every string in $\mathcal{D}$. The GST consists of n leaves, and at most $n - 1$ internal nodes, and can be stored in space $O(n \log n)$ bits. Let $\mathsf{leaf}(u)$ be the set of leaves in the sub-tree of GST rooted at u, and $\mathsf{leaf}(u \setminus v)$ be the set $\mathsf{leaf}(u) \setminus \mathsf{leaf}(v)$. Leaves in GST are numbered in the lexicographic order of the suffix they represent. We use ℓ_i to denote the ith leftmost leaf of GST, and $\mathsf{doc}(i)$ to denote the index of the document to which the suffix corresponding to ℓ_i belongs. By maintaining an array, referred to as **document array**, of size $O(n \log D)$ bits, in constant time we can find $\mathsf{doc}(i)$ corresponding to leaf ℓ_i. The locus of a pattern P, denoted by $\mathsf{locus}(P)$, is the highest node u such that $\mathsf{path}(u)$ i.e., the concatenation of edge labels on the path from root to u is prefixed by P; we can compute $\mathsf{locus}(P)$ in $O(|P|)$ time, where $|P|$ is the length of P. We refer to $[\mathsf{sp}(p), \mathsf{ep}(p)]$ as the **suffix range** of a pattern P, where $p = \mathsf{locus}(P)$, and $\mathsf{sp}(u)$ (resp. $\mathsf{ep}(u)$) is the index of the leftmost (resp. rightmost) leaf in the sub-tree of u.

Computing the Relevance Function. Note that a pattern P occurs in a document T_d iff $d = \mathsf{doc}(i)$ for some leaf ℓ_i which lies in the suffix range of P. We now prove an important lemma.

Lemma 1. *Given the locus nodes of pattern P and Q in GST, and the identifier d of a document in $\mathcal{D}$, by using $O(n)$ space (in words) data-structures, in constant time we can compute $\mathsf{score}(P, Q, d)$.*

Proof. Let $\mathcal{I}$ be a set of integers drawn from a set $\mathcal{U} = \{0, 1, 2, \cdots, 2^w - 1\}$, where $w \geq \log n$ is the word size. Alstrup et al. [1] presents a data-structure of size $O(|\mathcal{I}|)$ words which for a given $a, b \in \mathcal{U}$, $b \geq a$, can report in $O(1)$ time whether $\mathcal{I} \cap [a, b]$ is empty or not. In case $\mathcal{I} \cap [a, b]$ is not-empty, the data-structure returns any arbitrary value in $\mathcal{I} \cap [a, b]$. We build the above data-structure of Alstrup et al. for the sets $\mathcal{I}_d = \{i \mid \mathsf{doc}(i) = d\}$ for $d = 1, 2, \cdots, D$. The total space can be bounded by $O(n)$ words. Using this, we can answer whether a pattern P' occurs in T_d, or not, by checking if there exists an element in $\mathcal{I}_d \cap [\mathsf{sp}', \mathsf{ep}']$, where $[\mathsf{sp}', \mathsf{ep}']$ is the suffix range of a pattern P'. In case an element exists, we a get a leaf $\ell_i \in \mathsf{GST}$ such that $\mathsf{doc}(i) = d$.

We assign $\mathsf{score}(P, Q, d) = -\infty$ iff T_d contains Q. Otherwise, $\mathsf{score}(P, Q, d)$ equals the relevance of document T_d w.r.t P. Denote by ST_d, the suffix tree of document T_d, and by $\mathsf{path}_d(u)$, the string formed by the concatenation of the edge-labels from root to a node u in ST_d. For every node u in ST_d, we maintain the relevance of the pattern $\mathsf{path}_d(u)$ w.r.t the document T_d. Also, we maintain a pointer from every leaf ℓ_i of GST to that leaf node ℓ_j of $\mathsf{ST}_{\mathsf{doc}(i)}$ for which $\mathsf{path}_{\mathsf{doc}(i)}(\ell_j)$ is same as $\mathsf{path}(\ell_i)$. Figure 1 illustrates this. We now use a more recent result of Gawrychowski et al. [8], which can be summarized as follows: given a suffix tree ST having $|\mathsf{ST}|$ nodes, where every node u has an integer weight $\mathsf{weight}(u) \leq |ST|$, and satisfies the min-heap property, there exists an $O(|\mathsf{ST}|)$ words data structure, such that for any leaf $\ell \in \mathsf{ST}$ and an integer W, in constant time we can find the lowest ancestor v (if any) of ℓ, that satisfies $\mathsf{weight}(v) \leq W$. For every node $u \in \mathsf{ST}_d$, we let $\mathsf{weight}(u) = |\mathsf{path}_d(u)|$. Note that this satisfies the min-heap property, and $|\mathsf{weight}(u)| \leq |\mathsf{ST}_d|$. Using the data-structure of Gawrychowski et al., in constant time we can locate the lowest ancestor v of ℓ_j such that $|\mathsf{path}_d(v)| \leq |P|$. If $|\mathsf{path}_d(v)| = |P|$, then v is the locus node of P in ST_d. Otherwise, one of the children of v is the desired locus, which can be found in constant time by checking the $(|\mathsf{path}_d(v)| + 1)^{th}$ character of P. Therefore, we can compute $\mathsf{score}(P, Q, d)$ in constant time; clearly, the total space required for storing the data-structures is bounded by $O(n)$ words. ∎

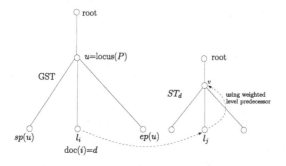

Fig. 1. Illustration of Lemma 1

Marked Nodes and Prime Nodes. We identify certain nodes in GST as marked nodes and prime nodes based on a parameter g called grouping factor [13]. First, starting from the leftmost leaf in GST, we combine every g leaves together to form a group. In particular, the leaves ℓ_1 through ℓ_g forms the first group, ℓ_{g+1} through ℓ_{2g} forms the second, and so on. We mark the lowest common ancestor (LCA) of the first and last leaves of every group. Moreover, for any two marked nodes, we mark their LCA (and continue this recursively). Note that the root node is marked, and the number of marked nodes is at most $2\lceil n/g \rceil$.

Lemma 2 ([11]). *The suffix range* $[\mathsf{sp}, \mathsf{ep}]$ *of any pattern* P, *where* $\mathsf{ep} - \mathsf{sp} + 1 \geq 2g$, *can split into a suffix range* $[\mathsf{sp}', \mathsf{ep}']$ *corresponding to a highest descendant marked node* u^*, *and two other ranges* $[\mathsf{sp}, \mathsf{sp}' - 1]$ *and* $[\mathsf{ep}' + 1, \mathsf{ep}]$ *such that* $\mathsf{sp}' - \mathsf{sp} < g$ *and* $\mathsf{ep} - \mathsf{ep}' < g$.

Note that u^* is essentially the LCA of the leaves ℓ_i and ℓ_j, where ℓ_i (resp. ℓ_j) is the first (resp. last) leaf after (resp. before) sp (resp. ep) such that i (resp. j) is a multiple of g. If u is itself a marked node, then u^* is same as u.

Corresponding to each marked node (except the root), we identify a unique node called the prime node. Specifically, the prime node u' corresponding to a marked node u^* is the node on the path from root to u^*, which is a child of the lowest marked ancestor of u^*; we refer to u' as the lowest prime ancestor of u^*. Since the root node is marked, there is always such a node. If the parent of u^* is marked, then u' is same as u^*. Also, for every prime node u', the corresponding closest marked descendant u^* is unique. Therefore number of prime nodes is one less than the number of marked node. From the definition of marked and prime nodes, we have the following.

Observation 1. *Let* u^* *be a marked node, and* u' *be the lowest prime ancestor of* u^*. *Then* $|\mathsf{leaf}(u' \setminus u^*)| \leq 2g$.

Observation 2. *If a node* u *has no marked descendant in its sub-tree, then* $|\mathsf{leaf}(u)| < 2g$.

We assign each node in the GST its pre-order rank. Thus given a node u (i.e., its pre-order rank), in constant time we can find its highest marked descendant (if any) using Lemma 2. We now present a useful lemma.

Lemma 3. *Let* u *be any node in* GST *such that w.r.t a grouping factor* g, u^* *is its highest marked descendant and* u' *is the lowest prime ancestor of* u^*. *Then by maintaining an* $O(n/g)$ *space array, we can compute the pre-order ranks of* u^* *and* u' *in* $O(n/g)$ *time. We can also compute the number of marked (resp. prime) nodes that comes before* u^* *(resp.* u'*) in pre-order.*

Proof. We maintain two arrays containing all marked and prime nodes (along with their pre-order ranks) of GST. The size of these arrays is bounded by $O(n/g)$. Note that in constant time we can check if a given node is an ancestor/descendant of another by comparing their ranges corresponding to the leaves in its sub-tree. Therefore, by examining all elements in the array one by one, we can identify the nodes corresponding to u^* and u'. We remark that it is possible to achieve constant query time using additional structures; however, we chose to use this simple structure as such an improvement would not affect the overall query complexity of Theorem 1. ∎

3   The Framework

In this section we prove Theorem 1. We start with the following definitions.

Definition 1. Let u and v be any two nodes in GST. Then

- $\mathsf{list}(u, v) = \{\mathsf{doc}(i) \mid \ell_i \in \mathsf{leaf}(u)\} \setminus \{\mathsf{doc}(i) \mid \ell_i \in \mathsf{leaf}(v)\}$ is the set of document identifiers corresponding to the query $(\mathsf{path}(u), \mathsf{path}(v))$.
- $\mathsf{list}_k(u, v)$ is the set of k most relevant document identifiers in $\mathsf{list}(u, v)$.
- $\mathsf{cand}_k(u, v)$, a **candidate set**, is any super set of $\mathsf{list}_k(u, v)$.

Moving forward, we use p and q to denote the loci of the included pattern P and the forbidden pattern Q respectively. Our task is then to report $\mathsf{list}_k(p, q)$.

Lemma 4. *Given a candidate set* $\mathsf{cand}_k(p, q)$, *and a* $\mathsf{score}(P, Q, d)$ *for each* $d \in \mathsf{cand}_k(p, q)$, *we can find* $\mathsf{list}_k(p, q)$ *in time* $O(|\mathsf{cand}_k(p, q)|)$.

Proof. Any document with $\mathsf{score}(P, Q, \cdot) = -\infty$ does not contribute to $\mathsf{list}_k(p, q)$, and can be safely ignored. Let this reduced set of documents be $\mathcal{D}'$. Further, we maintain only the distinct documents in $\mathcal{D}'$ (which can be done easily using an bitmap of size D). Among these distinct documents, we first find the document identifier, say d_k, with the kth largest $\mathsf{score}(P, Q, \cdot)$ value (which can be found in $O(|\mathcal{D}'|)$ time using k^{th} order statistics [3]), and report the identifiers d that satisfy $\mathsf{score}(P, Q, d) \geq \mathsf{score}(P, Q, d_k)$. Clearly, the time required for the entire process can be bound by $O(|\mathsf{cand}_k(u, v)|)$. ∎

Lemma 5. *For any two nodes* u *and* v *in* GST, *let* $u^{\downarrow}$ *be* u *or a descendent of* u *and* $v^{\uparrow}$ *be* v *or an ancestor of* v, *and* $\mathcal{L} = \mathsf{leaf}(u \setminus u^{\downarrow}) \cup \mathsf{leaf}(v^{\uparrow} \setminus v)$. *Then,* $\mathsf{list}_k(u, v) \subseteq \mathsf{list}_k(u^{\downarrow}, v^{\uparrow}) \cup \{\mathsf{doc}(i) \mid \ell_i \in \mathcal{L}\}$.

Proof. Follows from the monotonicity property of the relevance function. ∎

3.1   Index Construction and top-k Query

The data-structure in the following lemma is the most intricate component for retrieving $\mathsf{list}_k(p, q)$.

Lemma 6. *For grouping factor* $g = \sqrt{nk}$, *there exists a data-structure requiring* $O(n)$ *bits of space such that for any marked node* u^* *and prime node* v' *(w.r.t* g), *we can find the documents* d *in* $\mathsf{list}_k(u^*, v')$, *and* $\mathsf{score}(\mathsf{path}(u^*), \mathsf{path}(v'), d)$ *in* $O(\sqrt{nk})$ *time.*

We prove the lemma in the following section. Using the lemma, we describe how to obtain $\mathsf{list}_k(p, q)$ in $O(|P| + |Q| + \sqrt{nk})$ time, thereby proving Theorem 1.

Let $\kappa \geq 1$ be a parameter to be fixed later. For grouping factor $\sqrt{n\kappa}$, construct the data-structure DS_κ in Lemma 6 above. Further, we maintain a data-structure DS'_κ, as described in Lemma 3, such that for any node, we can find its highest marked descendant (if any), and its lowest prime ancestor; this takes $O(\sqrt{n/\kappa} \log n)$ bits. Construct the data-structures DS_κ and DS'_κ for $\kappa = 1, 2, 4, 8, \cdots, D$. Clearly, maintaining the data-structures over all the above values of κ takes $O(n)$ words in total.

For the patterns P and Q, and some integer k for a top-k query, we first locate the loci p and q in $O(|P| + |Q|)$ time. Let $k' = \min\{D, 2^{\lceil \log k \rceil}\}$ and $g' = \sqrt{nk'}$. Note that $k \leq k' < 2k$. Depending on whether p has any marked node below it, we have the following two cases. We show that in either case, $\mathsf{list}_k(p, q)$ can be found in $O(|P| + |Q| + \sqrt{nk})$ time by using data-structures of size $O(n \log n)$ bits in total.

Case 1. Assume that p contains a descendant marked node, and therefore, the highest descendant marked node, say p^*. If p is itself marked, then $p^* = p$. Let q' be the lowest prime ancestor of q. Both p^* and q' are found in $O(n/g') = O(\sqrt{n/k'})$ time (refer to Lemma 3). Now using the data-structure $\mathsf{DS}_{k'}$, we find $\mathsf{list}_{k'}(p^* \setminus q')$ in $O(\sqrt{nk'})$ time. The number of leaves in $\mathcal{L} = \mathsf{leaf}(p \setminus p^*) \cup \mathsf{leaf}(q' \setminus q)$ is bounded by $O(g') = O(\sqrt{nk})$ (see Observation 1). Finally, we compute $\mathsf{list}_k(p, q)$ from $\mathsf{list}_{k'}(p^* \setminus q')$ and $\mathcal{L}$ in $O(\sqrt{n/k'} + \sqrt{nk'}) = O(\sqrt{nk})$ time (refer to Lemmas 4 and 5).

Case 2. If there is no descendant marked node of p, then the size of the suffix range of P is less than $2g'$ (see Observation 2). We find the document identifiers, and their corresponding $\mathsf{score}(P, Q, \cdot)$ corresponding to the leaves in $\mathsf{leaf}(p)$ using the document array in $O(g')$ time (refer to Lemma 1). Denote this set of document identifiers by $\mathcal{D}_p$, where $|\mathcal{D}_p| = O(g')$. Now any document identifier $d \in \mathcal{D}_p$, for which $\mathsf{score}(P, Q, d) = -\infty$, can be safely ignored. The remaining document constitute the candidate set, and as described in Lemma 4, we can find the top-k documents. Clearly, the query time can be bounded by $O(g') = O(\sqrt{nk})$.

3.2   Proof of Lemma 6

A slightly weaker version of the result can be easily obtained as follows: maintain $\mathsf{list}_k(\cdot, \cdot)$ for all pairs of marked and prime nodes explicitly for $g = \sqrt{nk}$. This requires space $O((n/g)^2 k \log D) = O(n \log n)$ bits (off by a factor of $\log n$ from our desired space), but offers $O(k)$ query time (better than the desired time). Note that this saving in time will not have any effect on the time complexity of our final result implying that we can afford to spend any time up to $O(\sqrt{nk})$, but the space cannot be more than $O(n)$ bits. Therefore, we seek to encode these lists in a compressed form, from which any list can be decoded in $O(\sqrt{nk})$ time. The scheme is recursive, and is similar to that used in [4,20]. Before we begin with the proof of Lemma 6, let us first present the following useful result.

Lemma 7 ([5,6,9]). *A set of m integers from $\{1, 2, \cdots, U\}$ can be encoded in $O(m \log (U/m))$ bits, such that they can be decoded back in $O(1)$ time per integer.*

We begin with some notations: let $\log^{(h)} n = \log(\log^{(h-1)} n)$ for $h > 1$, and $\log^{(1)} n = \log n$. Further, let $\log^* n$ be the smallest integer α such that $\log^{(\alpha)} n \leq 2$. Let $g_h = \sqrt{nk} \log^{(h)} n$ (rounded to the next highest power of 2). Note that $g = g_{\log^* n}$. Therefore, our task is to report $\text{list}_k(u^*, v')$, whenever a marked node u^* and a prime node u' comes as a query, where both u^* and v' are based on the grouping factor $g_{\log^* n}$. For $1 \leq h \leq \log^* n$, let N_h^* (resp. N_h') be the set of marked (resp. prime) nodes w.r.t g_h. Also, let u_h^* be highest node in N_h^*, which is either u^* or a descendant of u^*. Likewise, let v_h' be lowest node in N_h', which is either v' or an ancestor of v'. As described in Lemma 3, we maintain the list of all nodes in N_h^* and N_h' for $1 \leq h \leq \log^* n$. Using this we can compute u_h^* and v_h', corresponding to any u^*, v' and h, in $O(n/g_h) = O(\sqrt{n/k}/\log^{(h)} n)$ time. The space required (in words) can be bounded as follows.

$$\sum_{h=1}^{\log^* n} \frac{n}{g_h} = \frac{\sqrt{n}}{\sqrt{k}} \sum_{h=1}^{\log^* n} \frac{1}{\log^{(h)} n} = O(\sqrt{n})$$

We are now ready to present the recursive encoding scheme. Assume there exists a scheme for encoding $\text{list}_k(\cdot, \cdot)$ of all pairs of marked/prime nodes w.r.t. to g_h in S_h bits of space, such that the top-k documents corresponding to any particular pair can be decoded in T_h time. Then,

$$S_{h+1} = S_h + O\left(\frac{n}{\log^{(h+1)} n}\right) \tag{1}$$

$$T_{h+1} = T_h + O\left(\frac{\sqrt{nk}}{\log^{(h)} n}\right) \tag{2}$$

By storing the answers explicitly for $h = 1$, the base case is established: $S_1 = O((n/g_1)^2 k \log n) = O(n/\log n)$ and $T_1 = O(k)$ plus $O(\sqrt{n/k}/\log n)$ time for finding u_1^* and v_1' from u^* and v'. Solving the above recursions leads to space bound $S_{\log^* n}$ (in bits) and time bound $T_{\log^* n}$ as follows:

$$S_{\log^* n} = O\left(\sum_{h=1}^{\log^* n} \frac{n}{\log^{(h)} n}\right) = O(n)$$

$$T_{\log^* n} = O\left(\sum_{h=1}^{\log^* n} \frac{\sqrt{nk}}{\log^{(h)} n}\right) = O(\sqrt{nk})$$

Space Bound. First we show how to arrive at Eq. 1. Specifically, we show how to encode $\text{list}_k(\cdot, \cdot)$ w.r.t g_{h+1} given that $\text{list}_k(\cdot, \cdot)$ w.r.t g_h is already encoded separately. With abuse of notation, let x_h^* (resp. y_h') be any marked (resp. prime) node w.r.t g_h. Further, let x_{h+1}^* (w.r.t g_{h+1}) be the lowest marked node above

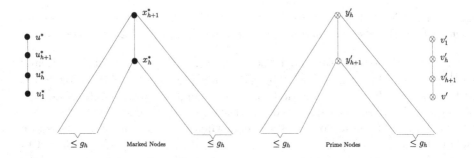

Fig. 2. Recursive encoding scheme.

x_h^*. Likewise, y_{h+1}' is the highest prime node below y_h'. Consider, the set of leaves in $\mathsf{leaf}(x_{h+1}^* \setminus x_h^*)$. The number of such leaves are bounded above by $2g_h$; otherwise, this contradicts the definition of x_{h+1}^*. Likewise, $|\mathsf{leaf}(y_h' \setminus y_{h+1}')| < 2g_h$. Figure 2 illustrates this. Let $f_h = \sqrt{n}/(\sqrt{k}\log^{(h)} n)$ (rounded to the next highest power of 2). We divide the leaves in $\mathsf{leaf}(x_{h+1}^* \setminus x_h^*)$ (resp. $\mathsf{leaf}(y_h' \setminus y_{h+1}'))$ on either side of the suffix range of x_h^* (resp. y_{h+1}') into g_h/f_h chunks. Thus, there are $O(k(\log^{(h)} n)^2)$ chunks, and each chunk has $O(f_h)$ leaves. For x_{h+1}^* and y_{h+1}' pair, we maintain those at most k chunks, leaves of which may contribute to the top-k documents. This can be seen as selecting at most k integers from an universe of size $O(k(\log^{(h)} n)^2)$, and can be encoded in $O(k\log(k(\log^{(h)} n)^2/k)) = O(k\log^{(h+1)} n)$ bits (refer to Lemma 7). Since, the number of marked and prime nodes at the $(h+1)^{th}$ level are both bounded by $O(n/g_{h+1})$, the total size in bits for maintaining the top-k chunks for all pairs of marked and prime nodes w.r.t g_{h+1} can be bounded by $O((n/g_{h+1})^2 k\log^{(h+1)} n) = O(n/\log^{(h+1)} n)$. For g_{h+1}, we also maintain the index w.r.t g_h having size S_h. Thus we have $S_{h+1} = S_h + O(n/\log^{(h+1)} n)$.

Time Bound. Let $\mathcal{L}_h$ be the candidate leaves for u_h^* and v_h'. We show how to obtain $\mathcal{L}_{h+1}$ from $\mathcal{L}_h$, using the index described above, in time T_{h+1} as described in Eq. 2. We first locate u_{h+1}^* and v_{h+1}', which can be achieved in time $O(n/g_{h+1}) = O(\sqrt{n/k}/\log^{(h+1)} n)$ with the aid of the arrays maintaining N_{h+1}^* and N_{h+1}'. By using the index, having size S_{h+1}, at the $(h+1)^{th}$ level of recursion, we decode the top-k chunks for u_{h+1}^* and v_{h+1}', and for each chunk we find the leaves in it. The number of such leaves is bounded by $O(kf_h)$. By Lemma 7, decoding each chunk takes constant time, and therefore finding $\mathcal{L}_{h+1}$, the set of candidate leaves for u_{h+1}^* and v_{h+1}' is achieved in total time $T_{h+1} = T_h + O(kf_h) = T_h + O(\sqrt{nk}/\log^{(h)} n)$.

Retrieving top-k Documents. In the query process described above, at the end of $\log^* n$ levels, we have the candidate leaves for $\mathsf{list}_k(u^*, v')$. The number of such leaves is bounded by $O(\sqrt{nk})$. By using these leaves and document array, as described in Lemma 1, we find the top-k (unique) document identifiers d, and corresponding $\mathsf{score}(\mathsf{path}(u^*), \mathsf{path}(v'), d)$ in time $O(\sqrt{nk})$. This completes the proof of Lemma 6.

4   Conclusion

In this paper, we revisit the problem of reporting the documents which contains a pattern P, and does not contain a pattern Q. We present a linear space index which reports such documents in time $O(|P| + |Q| + \sqrt{nt})$, where t is the number of documents reported. This improves the previously best known query time of $O(|P| + |Q| + \sqrt{nt} \log^{5/2} n)$ using linear space data-structures. For solving the problem, we actually solve a general version, in which we report the top-k documents in time $O(|P| + |Q| + \sqrt{nk})$, where document relevance function is monotonic increasing. We remark that our framework can be used to obtain same space and time solution for the document retrieval problem (as well as the top-k version of it) with two (included) patterns. In future, we will like to present solutions to the top-k version of the problem with excluded patterns, where the relevance metric is not limited to monotonic functions.

References

1. Alstrup, S., Brodal, G.S., Rauhe, T.: Optimal static range reporting in one dimension. In: Proceedings on 33rd Annual ACM Symposium on Theory of Computing, Heraklion, Crete, Greece, pp. 476–482, 6–8 July 2001

2. Cohen, H., Porat, E.: Fast set intersection and two-patterns matching. In: López-Ortiz, A. (ed.) LATIN 2010. LNCS, vol. 6034, pp. 234–242. Springer, Heidelberg (2010)

3. Cormen, T.H., Stein, C., Rivest, R.L., Leiserson, C.E.: Introduction to Algorithms, 2nd edn. McGraw-Hill Higher Education, New York (2001)

4. Durocher, S., Shah, R., Skala, M., Thankachan, S.V.: Linear-space data structures for range frequency queries on arrays and trees. In: Chatterjee, K., Sgall, J. (eds.) MFCS 2013. LNCS, vol. 8087, pp. 325–336. Springer, Heidelberg (2013)

5. Elias, P.: Efficient storage and retrieval by content and address of static files. J. ACM **21**(2), 246–260 (1974)

6. Fano, R.M.: On the number of bits required to implement an associative memory. Massachusetts Institute of Technology, Project MAC, Cambridge (1971)

7. Fischer, J., Gagie, T., Kopelowitz, T., Lewenstein, M., Mäkinen, V., Salmela, L., Välimäki, N.: Forbidden patterns. In: Fernández-Baca, D. (ed.) LATIN 2012. LNCS, vol. 7256, pp. 327–337. Springer, Heidelberg (2012)

8. Gawrychowski, P., Lewenstein, M., Nicholson, P.K.: Weighted ancestors in suffix trees. In: Schulz, A.S., Wagner, D. (eds.) ESA 2014. LNCS, vol. 8737, pp. 455–466. Springer, Heidelberg (2014)

9. Grossi, R., Vitter, J.S.: Compressed suffix arrays and suffix trees with applications to text indexing and string matching (extended abstract). In: Proceedings of the Thirty-Second Annual ACM Symposium on Theory of Computing, Portland, OR, USA, pp. 397–406, 21–23 May 2000

10. Gusfield, D.: Algorithms on Strings, Trees, and Sequences - Computer Science and Computational Biology. Cambridge University Press, New York (1997)

11. Hon, W.-K., Shah, R., Thankachan, S.V., Vitter, J.S.: String retrieval for multi-pattern queries. In: Chavez, E., Lonardi, S. (eds.) SPIRE 2010. LNCS, vol. 6393, pp. 55–66. Springer, Heidelberg (2010)

12. Hon, W.-K., Shah, R., Thankachan, S.V., Vitter, J.S.: Document listing for queries with excluded pattern. In: Kärkkäinen, J., Stoye, J. (eds.) CPM 2012. LNCS, vol. 7354, pp. 185–195. Springer, Heidelberg (2012)

13. Hon, W., Shah, R., Thankachan, S.V., Vitter, J.S.: Space-efficient frameworks for top-k string retrieval. J. ACM **61**(2), 9 (2014)

14. Hon, W., Shah, R., Vitter, J.S.: Space-efficient framework for top-k string retrieval problems. In: 50th Annual IEEE Symposium on Foundations of Computer Science, FOCS 2009, Atlanta, Georgia, USA, pp. 713–722, 25–27 October 2009

15. Larsen, K.G., Munro, J.I., Nielsen, J.S., Thankachan, S.V.: On hardness of several string indexing problems. In: Kulikov, A.S., Kuznetsov, S.O., Pevzner, P. (eds.) CPM 2014. LNCS, vol. 8486, pp. 242–251. Springer, Heidelberg (2014)

16. Matias, Y., Muthukrishnan, S.M., Şahinalp, S.C., Ziv, J.: Augmenting suffix trees, with applications. In: Bilardi, G., Pietracaprina, A., Italiano, G.F., Pucci, G. (eds.) ESA 1998. LNCS, vol. 1461, p. 67. Springer, Heidelberg (1998)

17. Muthukrishnan, S.: Efficient algorithms for document retrieval problems. In: Proceedings of the Thirteenth Annual ACM-SIAM Symposium on Discrete Algorithms, San Francisco, CA, USA, pp. 657–666, 6–8 January 2002

18. Navarro, G.: Spaces, trees, and colors: the algorithmic landscape of document retrieval on sequences. ACM Comput. Surv. **46**(4), 52 (2013)

19. Navarro, G., Nekrich, Y.: Top-k document retrieval in optimal time and linear space. In: Proceedings of the Twenty-Third Annual ACM-SIAM Symposium on Discrete Algorithms, SODA 2012, Kyoto, Japan, pp. 1066–1077, 17–19 January 2012

20. Navarro, G., Thankachan, S.V.: New space/time tradeoffs for top-k document retrieval on sequences. Theor. Comput. Sci. **542**, 83–97 (2014)

21. Navarro, G., Thankachan, S.V.: Bottom-k document retrieval. J. Discret. Algorithms **32**, 69–74 (2015). StringMasters 2012; 2013 Special Issue (Volume 2)

22. Patil, M., Thankachan, S.V., Shah, R., Hon, W., Vitter, J.S., Chandrasekaran, S.: Inverted indexes for phrases and strings. In: Proceeding of the 34th International ACM SIGIR Conference on Research and Development in Information Retrieval, SIGIR 2011, Beijing, China, pp. 555–564, 25–29 July 2011

Parameterized Complexity of Superstring Problems

Ivan Bliznets[2], Fedor V. Fomin[1,2], Petr A. Golovach[1,2(✉)], Nikolay Karpov[2], Alexander S. Kulikov[2], and Saket Saurabh[1,3]

[1] Department of Informatics, University of Bergen, Bergen, Norway
[2] Steklov Institute of Mathematics at St. Petersburg,
Russian Academy of Sciences, St. Petersburg, Russia
petr.golovach@ii.uib.no
[3] Institute of Mathematical Sciences, Chennai, India

Abstract. In the SHORTEST SUPERSTRING problem we are given a set of strings $S = \{s_1, \ldots, s_n\}$ and an integer ℓ and the question is to decide whether there is a superstring s of length at most ℓ containing all strings of S as substrings. We obtain several parameterized algorithms and complexity results for this problem.

In particular, we give an algorithm which in time $2^{O(k)} \operatorname{poly}(n)$ finds a superstring of length at most ℓ containing at least k strings of S. We complement this by the lower bound showing that such a parameterization does not admit a polynomial kernel up to some complexity assumption. We also obtain several results about "below guaranteed values" parameterization of the problem. We show that parameterization by compression admits a polynomial kernel while parameterization "below matching" is hard.

1 Introduction

We consider the SHORTEST SUPERSTRING problem defined as follows:

SHORTEST SUPERSTRING
Input: A set of n strings $S = \{s_1, \ldots, s_n\}$ over an alphabet Σ and a non-negative integer ℓ.
Question: Is there a string s of length at most ℓ containing all strings from S as substrings?

This is a well-known NP-complete problem [10] with a range of practical applications from DNA assembly [7] to data compression [9]. Due to this fact approximation algorithms for it are widely studied. The currently best known approximation guarantee $2\frac{11}{23}$ is due to Mucha [17]. At the same time the best known exact algorithms run in roughly 2^n steps and are known for more than

---

The research leading to these results has received funding from the Government of the Russian Federation (grant 14.Z50.31.0030) and Grant of the President of Russian Federation (MK-6550.2015.1).

F. Cicalese et al. (Eds.): CPM 2015, LNCS 9133, pp. 89–99, 2015.
DOI: 10.1007/978-3-319-19929-0_8

50 years already. More precisely, using known algorithms for the TRAVELING SALESMAN problem, SHORTEST SUPERSTRING can be solved either in time $O^*(2^n)$ and the same space by dynamic programming over subsets [3,13] or in time $O^*(2^n)$ and only polynomial space by inclusion-exclusion [14,16] (here, $O^*(\cdot)$ hides factors that are polynomial in the input length, i.e., $\sum_{i=1}^{n} |s_i|$). Such algorithms can only be used in practice to solve instances of very moderate size. Stronger upper bounds are known for a special case when input strings have bounded length [11,12]. There are heuristic methods for solving TRAVELING SALESMAN, and hence also SHORTEST SUPERSTRING, they are efficient in practice, however have no efficient provable guarantee on the running time (see, e.g., [1]).

In this paper, we study the SHORTEST SUPERSTRING problem from the parameterized complexity point of view. This field studies the complexity of computational problems with respect not only to input size, but also to some additional parameters and tries to identify parameters of input instances that make the problem tractable. Interestingly, prior to our work, except observations following from the known reductions to TRAVELING SALESMAN, not much about the parameterized complexity of SHORTEST SUPERSTRING was known. We refer to the survey of Bulteau et al. [4] for a nice overview of known results on parameterized algorithms and complexity of strings problems. Thus our work can be seen as the first non-trivial step towards the study of this interesting and important problem from the perspective of parameterized complexity.

Our Results. In this paper we study two types of parameterization for SHORTEST SUPERSTRING and present two kind of results. The first set of results concerns "natural" parameterization of the problem. We consider the following generalization of SHORTEST SUPERSTRING:

PARTIAL SUPERSTRING
Input: A collection (multiset) of strings S over an alphabet Σ, and non-negative integers k, ℓ.
Question: Is there a string s of length at most ℓ such that s is a superstring of a collection of at least k strings $S' \subseteq S$?

If $k = |S|$, then this is SHORTEST SUPERSTRING. Notice that S can contain copies of the same string and a string of S can be a substring of another string of the collection. For SHORTEST SUPERSTRING, such cases could be easily avoided, but for PARTIAL SUPERSTRING it is natural to assume that we have such possibilities.

Here we show that PARTIAL SUPERSTRING is fixed parameter tractable (FPT) when parameterized by k or ℓ. We complement this result by showing that it is unlikely that the problem admits a polynomial kernel with respect to these parameters.

The second set of results concerns "below guaranteed value" parameterization. Note that an obvious (non-optimal) superstring of $S = \{s_1, \ldots, s_n\}$ is a string of length $\sum_{i=1}^{n} |s_i|$ formed by concatenating all strings from S. For a superstring s of S the value $\sum_{i=1}^{n} |s_i| - |s|$ is called by *compression of s with respect to S*. Then finding a shortest superstring is equivalent to finding an order of $s_1, \ldots, s_n$ such that the consecutive strings have the largest possible total overlap. We first show that it is FPT with respect to r to check whether one can achieve a compression at least r by construction a kernel of size $O(r^4)$. We complement this result by a hardness result about "stronger" parameterization. Let us partition n input strings into $n/2$ pairs such that the sum of the $n/2$ resulting overlaps is maximized. Such a partition can be found in polynomial time by constructing a maximum weight matching in an auxiliary graph. Then this total overlap provides a lower bound on the maximum compression (or, equivalently, an upper bound on the length of a shortest superstring). We show that already deciding whether at least one additional symbol can be saved beyond the maximum weight matching value is already NP-complete.

2   Basic Definitions and Preliminaries

Strings. Let s be a string. By $|s|$ we denote the *length* of s. By $s[i]$, where $1 \leq i \leq |s|$, we denote the i-th symbol of s, and $s[i, j] = s[i] \ldots s[j]$ for $1 \leq i \leq j \leq |s|$. We assume that $s[i, j]$ is the empty string if $i > j$. We denote $\mathrm{prefix}_i(s) = s[1, i]$ and $\mathrm{suffix}_i(s) = s[|s| - i + 1, |s|]$ the *i-th prefix* and *i-th suffix* of s respectively for $i \in \{1, \ldots, |s|\}$; $\mathrm{prefix}_0(s) = \mathrm{suffix}_0(s)$ is the empty string. Let s, s' be strings. We write $s \subseteq s'$ to denote that s is a *substring* of s'. If $s \subseteq s'$, then s' is a *superstring* of s. We write $s \subset s'$ and $s \supset s'$ to denote proper sub and superstrings. For a collection of strings S, a string s is a superstring of S if s is a superstring of each string in S. The *compression measure* of a superstring s of a collection of strings S is $\sum_{x \in S} |x| - |s|$. If $s \subseteq s'$, then $\mathrm{overlap}(s, s') = \mathrm{overlap}(s', s) = s$; otherwise, if $s \not\subseteq s'$ and $s' \not\subseteq s$, then $\mathrm{overlap}(s, s') = \mathrm{suffix}_r(s) = \mathrm{prefix}_r(s')$, where $r = \max\{i \mid 0 \leq i \leq \min\{|s|, |s'|\}, \mathrm{suffix}_i(s) = \mathrm{prefix}_i(s')\}$. We denote by ss' the *concatenation* of s and s'. For strings s, s', we define the *concatenation with overlap* $s \circ s'$ as follows. If $s \subseteq s'$, then $s \circ s' = s' \circ s = s'$. If $s \not\subseteq s'$ and $s' \not\subseteq s$, then $s \circ s' = \mathrm{prefix}_p(s)\mathrm{overlap}(s, s')\mathrm{suffix}_q(s')$, where $p = |s| - |\mathrm{overlap}(s, s')|$ and $q = |s'| - |\mathrm{overlap}(s, s')|$.

We need the following folklore property of superstrings.

Lemma 1. *Let s be a superstring of a collection S of strings. Let $S' = \{s_1, \ldots, s_n\}$ be a set of inclusion maximal pairwise distinct strings of S such that each string of S is a substring of a string from S'. Let indices $p_i, q_i \in \{1, \ldots, |s|\}$ be such that $s_i = s[p_i, q_i]$ for $i \in \{1, \ldots, n\}$ and assume that $p_1 < \cdots < p_n$. Then $s' = s_1 \circ \cdots \circ s_n$ is a superstring of S of length at most $|s|$.*

Graphs. We consider finite directed and undirected graphs without loops or multiple edges. The vertex set of a (directed) graph G is denoted by $V(G)$, the edge set of an undirected graph and the arc set of a directed graph G is denoted by $E(G)$. To distinguish edges and arcs, the edge with two end-vertices u, v is denoted by $\{u, v\}$, and we write (u, v) for the corresponding arc. For an arc $e = (u, v)$, v is the *head* of e and u is the tail. Let G be a directed graph. For a vertex $v \in V(G)$, we say that u is an *in-neighbor* of v if $(u, v) \in E(G)$. The set of all in-neighbors of v is denoted by $N_G^-(v)$. The *in-degree* $d_G^-(v) = |N_G^-(v)|$. Respectively, u is an *out-neighbor* of v if $(v, u) \in E(G)$, the set of all out-neighbors of v is denoted by $N_G^+(v)$, and the *out-degree* $d_G^+(v) = |N_G^+(v)|$. For a directed graph G, a (directed) *trail* of *length* k is a sequence $v_0, e_1, v_1, e_2, \ldots, e_k, v_k$ of vertices and arcs of G such that $v_0, \ldots, v_k \in V(G)$, $e_1, \ldots, e_k \in E(G)$, the arcs $e_1, \ldots, e_k$ are pairwise distinct, and for $i \in \{1, \ldots, k\}$, $e_i = (v_{i-1}, v_i)$. We omit the word "directed" if it does not create a confusion. Slightly abusing notations we often write a trail as a sequence of its vertices $v_0, \ldots, v_k$ or arcs $e_1, \ldots, e_k$. If $v_0, \ldots, v_k$ are pairwise distinct, then $v_0, \ldots, v_k$ is a (directed) path. Recall that a path of length $|V(G)| - 1$ is a *Hamiltonian* path. For an undirected graph G, a set $U \subseteq V(G)$ is a *vertex cover* of G if for any edge $\{u, v\}$ of G, $u \in U$ or $v \in U$. A set of edges M with pairwise distinct end-vertices is a *matching*.

Parameterized Complexity. Parameterized complexity is a two dimensional framework for studying the computational complexity of a problem. One dimension is the input size and another one is a parameter. We refer to the books of Downey and Fellows [5], Flum and Grohe [8], and Niedermeier [19] for detailed introductions to parameterized complexity.

Formally, a parameterized problem $\mathcal{P} \subseteq \Sigma^* \times \mathbb{N}$, where Σ is a finite alphabet, i.e., an instance of $\mathcal{P}$ is a pair (I, k) for $I \in \Sigma^*$ and $k \in \mathbb{N}$, where I is an input and k is a parameter. It is said that a problem is *fixed parameter tractable* (or FPT), if it can be solved in time $f(k) \cdot |I|^{O(1)}$ for some function f. A *kernelization* for a parameterized problem is a polynomial algorithm that maps each instance (I, k) to an instance (I', k') such that

(i) (I, k) is a yes-instance if and only if (I', k') is a yes-instance of the problem, and
(ii) the size of I' and k' are bounded by $f(k)$ for a computable function f.

The output (I', k') is called a *kernel*. The function f is said to be a *size* of a kernel. Respectively, a kernel is *polynomial* if f is polynomial. While a parameterized problem is FPT if and only if it has a kernel, it is widely believed that not all FPT problems have polynomial kernels.

We use randomized algorithms for our problems. Recall that a *Monte Carlo algorithm* is a randomized algorithm whose running time is deterministic, but whose output may be incorrect with a certain (typically small) probability. A Monte-Carlo algorithm is *true-biased* (*false-biased* respectively) if it always returns a correct answer when it returns a yes-answer (a no-answer respectively).

3   FPT-Algorithms for Partial Superstring

In this section we show that PARTIAL SUPERSTRING is FPT, when parameterized by k or ℓ. For technical reasons, we consider the following variant of the problem with weights:

PARTIAL WEIGHTED SUPERSTRING
Input: A collection of strings S over an alphabet Σ with a weight function $w \colon S \to \mathbb{N}_0$, and non-negative integers k, ℓ and W.
Question: Is there a string s of length at most ℓ such that s is a superstring of a collection of k strings $S' \subseteq S$ with $w(S') \geq W$?

Clearly, if $w \equiv 1$ and $W = k$, then we have the PARTIAL SUPERSTRING problem.

Theorem 1. PARTIAL WEIGHTED SUPERSTRING *can be solved in time* $O((2e)^k \cdot kn^2 m \log W)$ *by a true-biased Monte-Carlo algorithm and in time* $(2e)^k k^{O(\log k)} \cdot n^2 \log n \cdot m \log W$ *by a deterministic algorithm for a collection of n strings of length at most m.*

Proof. First, we describe the randomized algorithm and then explain how it can be derandomized. The algorithm uses the color coding technique proposed by Alon, Yuster and Zwick [2].

If $\ell \geq km$, then the problem is trivial, as the concatenation of any k strings of S has length at most ℓ and we can greedily choose k strings of maximum weight. Assume that $\ell < km$.

We color the strings of S by k colors $1, \ldots, k$ uniformly at random independently from each other. Now we are looking for a string s that is a superstring of k strings of maximum total weight that have pairwise distinct colors.

To do it, we apply the dynamic programming across subsets. For simplicity, we explain only how to solve the decision problem, but our algorithm can be modified to find a colorful superstring as well. For $X \subseteq \{1, \ldots, k\}$, a string $x \in S$ and a positive integer $h \in \{1, \ldots, \ell\}$, the algorithm computes the maximum weight $W(X, x, h)$ of a string s of length at most h such that

(i) s is a superstring of a collection of $k' = |X|$ strings $S' \subseteq S$ of pairwise distinct colors from X,
(ii) x is inclusion maximal string of S' and $x = \text{suffix}_{|x|}(s)$.

If such a string s does not exist, then $W(X, x, h) = -\infty$.

We compute the table of values of $W(X, x, h)$ consecutively for $|X| = 1, \ldots, k$. To simplify computations, we assume that $W(X, x, h) = -\infty$ for $h < 0$. If $|X| = 1$, then for each string $x \in S$, we set $W(X, x, h) = w(x)$ if x is colored by the unique color of X and $|x| \leq h$. In all other cases $W(X, x, h) = -\infty$. Assume that $|X| = k' \geq 2$ and the values of $W(X', x, h)$ are already computed for $|X'| < k'$. Let

$$W' = \max\{W(X \setminus \{c\}, x, h) + w(y) \mid y \subseteq x \text{ has color } c \in X\},$$

and

$$W'' = \max\{W(X \setminus \{c\}, y, h - |x| + |\text{overlap}(y, x)|) + w(x) \mid x \not\subseteq y, y \not\subseteq x\},$$

where c is the color of x; we assume that $W' = -\infty$ if there is no substring y of x of color $c \in X$, and $W'' = -\infty$ if every string y is a sub or superstring of x. We set $W(X, x, h) = \max\{W', W''\}$.

We show that $\max\{W(\{1, \ldots, k\}, x, \ell) \mid x \in S\}$ is the maximum weight of k strings of S colored by distinct colors that have a superstring of length at most ℓ; if this value equals $-\infty$, then there is no string of length at most ℓ that is a superstring of k string of S of distinct colors.

To prove this, it is sufficient to show that the values $W(X, x, h)$ computed by the algorithms are the maximum weights of strings of length at most h that satisfy (i) and (ii). The proof is by induction on the size of $|X|$. It is straightforward to verify that it holds if $|X| = 1$. Assume that $|X| > 1$ and the claim holds for sets of lesser size. Denote by $W^*(X, x, h)$ the maximum weight of a string s of length at most h that satisfies (i) and (ii). By the description of the algorithm, $W^*(X, x, h) \geq W(X, x, h)$. We show that $W^*(X, x, h) \leq W(X, x, h)$.

Let S' be a collection of k' strings of pairwise distinct colors from X that have s as a superstring. Denote by S'' a set of inclusion maximal distinct strings of S' that contains x such that every string of S' is a substring of a string of S''. Assume that $S'' = \{x_1, \ldots, x_r\}$ and $x_i = s[p_i, q_i]$ for $i \in \{1, \ldots, r\}$. Clearly, $x = x_r$.

Suppose that there is $y \in S' \setminus \{x\}$ such that $y \subseteq x$. Let $c \in X$ be a color of y. Then s is a superstring of $S' \setminus \{y\}$ and the total weight of these string is $W^*(X, x, h) - w(y)$. By induction, $W^*(X, x, h) - w(y) \leq W(X \setminus \{c\}, x, h)$ and we have that $W^*(X, x, h) \leq W(X \setminus \{c\}, x, h) + w(y) \leq W' \leq W(X, x, h)$.

Suppose now that $S' \setminus \{x\}$ does not contain substrings of x. Then $r \geq 2$. Let $y = s_{r-1}$ and $s' = s[1, q_{i-1}]$. Observe that $y = \text{suffix}_{|y|}(s')$. Notice that s' is a superstring of $S'' \setminus x$. Because $S' \setminus \{x\}$ has no substrings of x, every string in $S' \setminus \{x\}$ is a substring of any superstring of $S'' \setminus \{x\}$ and, therefore, s' is a superstring of $S' \setminus \{x\}$ of length at most $|s| - |x| + |\text{overlap}(y, x)| \leq h - |x| + |\text{overlap}(y, x)|$. The weight of $S' \setminus \{x\}$ is $W^*(X, x, h) - w(x)$. By induction, $W^*(X, x, h) - w(x) \leq W(X \setminus \{c\}, y, h - |x| + |\text{overlap}(y, x)|)$. Hence $W^*(X, x, h) \leq W(X \setminus \{c\}, y, h - |x| + |\text{overlap}(y, x)|) + w(x) \leq W'' \leq W(X, x, h)$.

To evaluate the running time of the dynamic programming algorithm, observe that we can check whether y is a substring of x or find $\text{overlap}(y, x)$ in time $O(m)$ using, e.g., the algorithm of Knuth, Morris, and Pratt [15], and we can construct the table of the overlaps and their sizes in time $O(n^2 m)$. Hence, for each X, the values $W(X, x, h)$ can be computed in time $O(n^2 km \log W)$, as $h \leq \ell < km$. Therefore, the running time is $O(2^k \cdot n^2 km \log W)$.

We proved that an optimal colorful solution can be found in time $O(2^k \cdot n^2 km \log W)$. Using the standard color coding arguments (see [2]), we obtain that it is sufficient to consider $N = e^k$ random colorings of S to claim that with probability $\alpha > 0$, where α is a constant that does not depend on the input size and the parameter, we get a coloring for which k string of S that have a superstring of length at most ℓ and the total weight at least W are colored by distinct colors if such a string exists. It implies that PARTIAL WEIGHTED SUPERSTRING can be solved in time $O((2e)^k \cdot kn^2 m \log W)$ by our randomized algorithm.

To derandomize the algorithm, we apply the technique proposed by Alon, Yuster and Zwick [2] using the k-perfect hash functions constructed by Naor, Schulman and Srinivasan [18]. The random colorings are replaced by the family of at most $e^k k^{\log k} \log n$ hash functions $c: S \to \{1, \ldots, k\}$ that have the following property: there is a hash function c that colors k string of S that have a superstring of length at most ℓ and the total weight at least W by distinct colors if such a string exists. It implies that PARTIAL WEIGHTED SUPERSTRING can be solved in time $(2e)^k k^{O(\log k)} \cdot n^2 \log n \cdot m \log W$ deterministically.   □

Because PARTIAL SUPERSTRING is a special case of PARTIAL WEIGHTED SUPER-STRING, Theorem 1 implies that this problem is FPT when parameterized by k. We show that the same holds if we parameterize the problem by ℓ.

Corollary 1. PARTIAL SUPERSTRING *is* FPT *when parameterized by* ℓ.

Proof. Consider an instance (S, k, ℓ) of PARTIAL SUPERSTRING. Recall that S can contain several copies of the same string. We construct a set of weighted strings S' by replacing a string s that occurs r times in S by the single copy of s of weight $w(s) = r$. Let $W = k$. Observe that there is a string s of length at most ℓ such that s is a superstring of a collection of at least k strings of S if and only if there a string s of length at most ℓ such that s is a superstring of a set of strings of S' of total weight at least W. A string of length at most ℓ has at most $\ell(\ell-1)/2$ distinct substrings. We consider the instances (S', w, k', ℓ, W) of PARTIAL WEIGHTED SUPERSTRING for $k' \in \{1, \ldots, \ell(\ell-1)/2\}$. For each of these instances, we solve the problem using Theorem 1. It remains to observe that there is a string s of length at most ℓ such that s is a superstring of a set of strings of S' of total weight at least W if and only if one of the instances (S', w, k', ℓ, W) is a yes-instance of PARTIAL WEIGHTED SUPERSTRING.   □

We complement the above algorithmic results by showing that we hardly can expect that PARTIAL SUPERSTRING has a polynomial kernel when parameterized by k or ℓ.

Theorem 2. PARTIAL SUPERSTRING *does not admit a polynomial kernel when parameterized by* $k+m$ *or* $\ell+m$ *for strings of length at most* m *over the alphabet* $\Sigma = \{0, 1\}$ *unless* NP $\subseteq$ coNP /poly.

4   Shortest Superstring Below Guaranteed Values

In this section we discuss SHORTEST SUPERSTRING parameterized by the difference between upper bounds for the length of a shortest superstring and the length of a solution superstring. For a collection of strings S, the length of the shortest superstring is trivially upper bounded by $\sum_{x \in S} |x|$. We show that SHORTEST SUPERSTRING admits a polynomial kernel when parameterized by the compression measure of a solution.

Theorem 3. SHORTEST SUPERSTRING *admits a kernel of size* $O(r^4)$ *when parameterized by* $r = \sum_{x \in S} |x| - \ell$.

Proof. Let (S, ℓ) be an instance of SHORTEST SUPERSTRING, $r = \sum_{x \in S} |x| - \ell$. First, we apply the following reduction rules for the instance.

Rule 1. If there are distinct elements x and y of S such that $x \sqsubseteq y$, then delete x and set $r = r - |x|$. If $r \leq 0$, then return a yes-answer and stop.

Rule 2. If there is $x \in S$ such that for any $y \in S \setminus \{x\}$, $|\text{overlap}(x,y)| = |\text{overlap}(y,x)| = 0$, then delete x and set $\ell = \ell - |x|$. If $S = \emptyset$ and $\ell \geq 0$, then return a yes-answer and stop. If $\ell < 0$, then return a no-answer and stop.

Rule 3. If there are distinct elements x and y of S such that $|\text{overlap}(x,y)| \geq r$, then return a yes-answer and stop.

It is straightforward to verify that these rules are *safe*, i.e., by the application of a rule we either solve the problem or obtain an equivalent instance. We exhaustively apply Rules 1–3. To simplify notations, we assume that S is the obtained set of strings and ℓ and r are the obtained values of the parameters. Notice that all strings in S are distinct and no string is a substring of another. Our next aim is to bound the lengths of considered strings.

Rule 4. If there is $x \in S$ with $|x| > 2r$, then set $\ell = \ell - |x| + 2r$ and $x = \text{prefix}_r(x)\text{suffix}_r(x)$. If $\ell < 0$, then return a no-answer and stop.

To see that the rule is safe, recall that x is not a sub or superstring of any other string of S, and $|\text{overlap}(x,y)| < r$ and $|\text{overlap}(y,x)| < r$ for any $y \in S$ distinct from x after the applications of Rule 3. As before, we apply Rule 4 exhaustively.

Now we construct an auxiliary graph G with the vertex set S such that two distinct $x, y \in S$ are adjacent in G if and only if $|\text{overlap}(x,y)| > 0$ or $|\text{overlap}(y,x)| > 0$. We greedily select a maximal matching M in G and apply the following rule.

Rule 5. If $|M| \geq r$, then return a yes-answer and stop.

To show that the rule is safe, it is sufficient to observe that if $M = \{x_1, x'_1\}, \ldots, \{x_h, x'_h\}$, $|\text{overlap}(x_i, x'_i)| > 0$ for $i \in \{1, \ldots, h\}$ and $h \geq r$, then the string s obtained by the consecutive concatenations with overlaps of $x_1, x'_1, \ldots, x_h, x'_h$ and then all the other strings of S in arbitrary order, then the compression measure of s is at least r.

Assume from now that we do not stop here, i.e., $|M| \leq r - 1$. Let $X \subseteq S$ be the set of end-vertices of the edges of M and $Y = S \setminus X$. Let $X = \{x_1, \ldots, x_h\}$. Clearly, $h \leq 2(r - 1)$. Observe that X is a vertex cover of G and Y is an independent set of G.

For each ordered pair (i, j) of distinct $i, j \in \{1, \ldots, h\}$, find an ordering $y_1, \ldots, y_t$ of the elements of Y sorted by the decrease of $|\text{overlap}(x_i, y_p)| + |\text{overlap}(y_p, x_j)|$ for $p \in \{1, \ldots, t\}$. We construct the set $R_{(i,j)}$ that contains the first $\min\{2h, t\}$ elements of the sequence.

For each $i \in \{1, \ldots, h\}$, find an ordering $y_1, \ldots, y_t$ of the elements of Y sorted by the decrease of $|\text{overlap}(y_p, x_i)|$ for $p \in \{1, \ldots, t\}$. We construct the set S_i that contains the first $\min\{2h, t\}$ elements of the sequence.

For each $i \in \{1, \ldots, h\}$, find an ordering $y_1, \ldots, y_t$ of the elements of Y sorted by the decrease of $|\text{overlap}(x_i, y_p)|$ for $p \in \{1, \ldots, t\}$. We construct the set T_i that contains the first $\min\{2h, t\}$ elements of the sequence.

Let

$$S' = X \cup \left(\bigcup_{(i,j),\ i,j \in \{1,\ldots,h\}, i \neq j} R_{(i,j)} \right) \cup \left(\bigcup_{i \in \{1,\ldots,h\}} S_i \right) \cup \left(\bigcup_{i \in \{1,\ldots,h\}} T_i \right).$$

Claim (∗). *There is a superstring s of S with the compression measure at least r if and only if there is a superstring s' of S' with the compression measure at least r.*

Proof (of Claim (∗)). If s' is a superstring of S' with the compression measure at least r, then the string s obtained from s' by the concatenation of s' and the strings of $S \setminus S'$ (in any order) is a superstring of S with the same compression measure as s'.

Suppose that s is a shortest superstring of S and the compression measure at least r. By Lemma 1, $s = s_1 \circ \ldots \circ s_n$, where $S = \{s_1, \ldots, s_n\}$. Let

$$Z = \{s_i \mid s_i \in Y, |\mathrm{overlap}(s_{i-1}, s_i)| > 0 \text{ or } |\mathrm{overlap}(s_i, s_{i+1})| > 0, 1 \leq i \leq n\};$$

we assume that s_0, s_{n+1} are empty strings.

We show that $|Z| \leq 2h$. Suppose that $s_i \in Z$. If $|\mathrm{overlap}(s_{i-1}, s_i)| > 0$, then $s_{i-1} \in X$, because $s_i \in Y$ and any two strings of Y have the empty overlap. By the same arguments, if $|\mathrm{overlap}(s_i, s_{i+1})| > 0$, then $s_{i+1} \in X$. Because $|X| = h$, we have that $|Z| \leq 2h$.

Suppose that the shortest superstring s is chosen in such a way that $|Z \setminus S'|$ is minimum. We prove that $Z \subseteq S'$ in this case. To obtain a contradiction, assume that there is $s_i \in Z \setminus S'$. We consider three cases.

Case 1. $|\mathrm{overlap}(s_{i-1}, s_i)| > 0$ and $|\mathrm{overlap}(s_i, s_{i+1})| > 0$. Recall that $s_{i-1}, s_{i+1} \in X$ in this case. Since $s_i \notin S'$, $s_i \notin R_{(p,q)}$ for $x_p = s_{i-1}$ and $x_q = s_{i+1}$. In particular, it means that $|R_{(p,q)}| = 2h$. As $|Z| \leq 2h$ and $|R_{(p,q)}| = 2h$, there is $s_j \in R_{(p,q)}$ such that $s_j \notin Z$, i.e., $|\mathrm{overlap}(s_{j-1}, s_j)| = |\mathrm{overlap}(s_j, s_{j+1})| = 0$. By the definition of $R_{(p,q)}$, $|\mathrm{overlap}(s_{i-1}, s_j)| + |\mathrm{overlap}(s_j, s_{i+1})| \geq |\mathrm{overlap}(s_{i-1}, s_i)| + |\mathrm{overlap}(s_i, s_{i+1})|$. Consider $s^* = s_1 \circ \ldots \circ s_{i-1} \circ s_j \circ s_{i+1} \ldots \circ s_{j-1} \circ s_i \circ s_j \circ \ldots \circ s_n$ assuming that $i < j$ (the other case is similar). Because $|\mathrm{overlap}(s_{i-1}, s_j)| + |\mathrm{overlap}(s_j, s_{i+1})| \geq |\mathrm{overlap}(s_{i-1}, s_i)| + |\mathrm{overlap}(s_i, s_{i+1})|$, $|s^*| \leq |s|$. Moreover, since s is a shortest superstring of S, $|s| \geq |s^*|$ and, therefore, $|\mathrm{overlap}(s_{j-1}, s_i)| = |\mathrm{overlap}(s_i, s_{j+1})| = 0$. But then for the set Z^* constructed for s^* in the same way as the set Z for s, we obtain that $|Z^* \setminus S'| < |Z \setminus S'|$; a contradiction.

Case 2. $|\mathrm{overlap}(s_{i-1}, s_i)| = 0$ and $|\mathrm{overlap}(s_i, s_{i+1})| > 0$. Then $s_{i+1} \in X$. Since $s_i \notin S'$, $s_i \notin S_p$ for $x_p = s_{i+1}$ and $|S_p| = 2h$. As $|Z| \leq 2h$ and $|S_p| = 2h$, there is $s_j \in S_p$ such that $s_j \notin Z$, i.e., $|\mathrm{overlap}(s_{j-1}, s_j)| = |\mathrm{overlap}(s_j, s_{j+1})| = 0$. By the definition of S_p, $|\mathrm{overlap}(s_j, s_{i+1})| \geq |\mathrm{overlap}(s_i, s_{i+1})|$. As in Case 1, consider s^* obtained by the exchange of s_i and s_j in the sequence of strings that is used for the concatenations with overlaps. In the same way, we obtain a contradiction with the choice of Z, because for the set Z^* constructed for s^* in the same way as the set Z for s, we obtain that $|Z^* \setminus S'| < |Z \setminus S'|$.

Case 3. $|\mathrm{overlap}(s_{i-1}, s_i)| > 0$ and $|\mathrm{overlap}(s_i, s_{i+1})| = 0$. To obtain contradiction in this case, we use the same arguments as in Case 2 using symmetry. Notice that we should consider T_p instead of S_p.

Now let $s' = s_{i_1} \circ \ldots \circ s_{i_p}$, where $s_{i_1}, \ldots, s_{i_p}$ is the sequence of string of S' obtained from $s_1, \ldots, s_n$ by the deletion of the strings of $S \setminus S'$. Because we have that $Z \subseteq S'$, the overlap of each deleted string with its neighbors is empty and, therefore, s' has the same compression measure as s

To finish the construction of the kernel, we define $\ell' = \ell - \sum_{x \in S \setminus S'} |x|$ and apply the following rule that is safe by Claim $(*)$.

Rule 6. If $\ell' < 0$, then return a no-answer and stop. Otherwise, return the instance (S', ℓ') and stop.

Since $|X| = h \leq 2(r-1)$, $|S'| \leq h + h^2 \cdot 2h + h \cdot 2h + h \cdot 2h = 2h^3 + 4h^2 + h = O(h^3) = O(r^3)$. Because each string of S' has length at most $2r$, the kernel has size $O(r^4)$.

It is easy to see that Rules 1-3 can be applied in polynomial time. Then graph G and M can be constructed in polynomial time and, trivially, Rule 5 demands $O(1)$ time. The sets X, Y, $R_{(i,j)}$, S_i and T_i can be constructed in polynomial time. Hence, S' and ℓ' can be constructed in polynomial time. Because Rule 6 can be applied in time $O(1)$, we conclude that the kernel is constructed in polynomial time. $\square$

Now we consider another upper bound for the length of the shortest superstring. Let S be a collection of strings. We construct an auxiliary weighted graph $G(S)$ with the vertex set S by assigning the weight $w(\{x,y\}) = \max\{|\text{overlap}(x,y)|, |\text{overlap}(y,x)|\}$ for any two distinct $x, y \in S$. Let $\mu(S)$ be the size of a maximum weighted matching in G. Clearly, G can be constructed in polynomial time and the computation of $\mu(G)$ is well known to be polynomial [6]. If $M = \{x_1, y_1\}, \ldots, \{x_h, y_h\}$ and $|\text{overlap}(x_i, y_i)| = w(\{x_i, y_i\})$ for $i \in \{1, \ldots, h\}$, then the string s obtained by the consecutive concatenations with overlaps of $x_1, y_1, \ldots, x_h, y_h$ and then (possibly) the remaining string of S has the compression measure at least $\mu(G)$. Hence, $\sum_{x \in S} |x| - \mu(G)$ is the upper bound for the length of the shortest superstring of G. We show that it is NP-hard to find a superstring that is shorter than this bound.

Theorem 4. SHORTEST SUPERSTRING *is NP-complete for* $\ell = \sum_{x \in S} |x| - \mu(S) - 1$ *even if restricted to the alphabet* $\Sigma = \{0, 1\}$.

References

1. Concorde TSP Solver. http://www.math.uwaterloo.ca/tsp/concorde.html
2. Alon, N., Yuster, R., Zwick, U.: Color-coding. J. ACM **42**(4), 844–856 (1995)
3. Bellman, R.: Dynamic programming treatment of the travelling salesman problem. J. ACM (JACM) **9**(1), 61–63 (1962)
4. Bulteau, L., Hüffner, F., Komusiewicz, C., Niedermeier, R.: Multivariate algorithmics for NP-hard string problems. Bull. EATCS **114**, 31–73 (2014)
5. Downey, R.G., Fellows, M.R.: Fundamentals of Parameterized Complexity. Texts in Computer Science. Springer, Berlin (2013)
6. Edmonds, J.: Maximum matching and a polyhedron with 0, 1-vertices. J. Res. Nat. Bur. Standards Sect. B **69B**, 125–130 (1965)
7. Evans, P.A., Wareham, T.: Efficient restricted-case algorithms for problems in computational biology. In: Zomaya, A.Y., Elloumi, M. (eds.) Algorithms in Computational Molecular Biology: Techniques, Approaches and Applications. Wiley Series in Bioinformatics, pp. 27–49. Wiley, Chichester (2011)

8. Flum, J., Grohe, M.: Parameterized Complexity Theory. Texts in Theoretical Computer Science. An EATCS Series. Springer-Verlag, Berlin (2006)
9. Gallant, J., Maier, D., Storer, J.A.: On finding minimal length superstrings. J. Comput. Syst. Sci. **20**(1), 50–58 (1980)
10. Garey, M.R., Johnson, D.S.: Computers and Intractability: A Guide to the Theory of NP-Completeness. W. H. Freeman, New York (1979)
11. Golovnev, A., Kulikov, A.S., Mihajlin, I.: Solving SCS for bounded length strings in fewer than 2^n steps. Inf. Process. Lett. **114**(8), 421–425 (2014)
12. Golovnev, A., Kulikov, A.S., Mihajlin, I.: Solving 3-superstring in $3^{n/3}$ time. In: Chatterjee, K., Sgall, J. (eds.) MFCS 2013. LNCS, vol. 8087, pp. 480–491. Springer, Heidelberg (2013)
13. Held, M., Karp, R.M.: A dynamic programming approach to sequencing problems. J. Soc. Ind. Applied Math. **10**(1), 196–210 (1962)
14. Karp, R.M.: Dynamic programming meets the principle of inclusion and exclusion. Oper. Res. Lett **1**(2), 49–51 (1982)
15. Knuth, D.E., Morris, J.H.J., Pratt, V.R.: Fast pattern matching in strings. SIAM J. Comput. **6**(2), 323–350 (1977)
16. Kohn, S., Gottlieb, A., Kohn, M.: A generating function approach to the traveling salesman problem. In: Proceedings of the 1977 Annual Conference, pp. 294–300. ACM (1977)
17. Mucha, M.: Lyndon words and short superstrings. In: Proceedings of the Twenty-Fourth Annual ACM-SIAM Symposium on Discrete Algorithms, pp. 958–972. SIAM (2013)
18. Naor, M., Schulman, L.J., Srinivasan, A.: Splitters and near-optimal derandomization. In: FOCS, pp. 182–191. IEEE Computer Society (1995)
19. Niedermeier, R.: Invitation to Fixed-Parameter Algorithms. Oxford Lecture Series in Mathematics and its Applications, vol. 31. Oxford University Press, Oxford (2006)

On the Fixed Parameter Tractability and Approximability of the Minimum Error Correction Problem

Paola Bonizzoni[1], Riccardo Dondi[2], Gunnar W. Klau[3,5], Yuri Pirola[1], Nadia Pisanti[4,5], and Simone Zaccaria[1 (✉)]

[1] DISCo, Univ. degli Studi di Milano-Bicocca, Milan, Italy
{simone.zaccaria,bonizzoni,pirola}@disco.unimib.it
[2] Dip. di Scienze Umane e Sociali, Univ. degli Studi di Bergamo, Bergamo, Italy
riccardo.dondi@unibg.it
[3] Life Sciences, Centrum Wiskunde & Informatica (CWI),
Amsterdam, The Netherlands
gunnar.klau@cwi.nl
[4] Dipartimento di Informatica, Univ. degli Studi di Pisa, Pisa, Italy
pisanti@di.unipi.it
[5] Erable Team, INRIA, Lyon, France

Abstract. Haplotype assembly is the computational problem of reconstructing the two parental copies, called *haplotypes*, of each chromosome starting from sequencing reads, called *fragments*, possibly affected by sequencing errors. *Minimum Error Correction* (MEC) is a prominent computational problem for haplotype assembly and, given a set of fragments, aims at reconstructing the two haplotypes by applying the minimum number of base corrections.

By using novel combinatorial properties of MEC instances, we are able to provide new results on the fixed-parameter tractability and approximability of MEC. In particular, we show that MEC is in FPT when parameterized by the number of corrections, and, on "gapless" instances, it is in FPT also when parameterized by the length of the fragments, whereas the result known in literature forces the reconstruction of complementary haplotypes. Then, we show that MEC cannot be approximated within any constant factor while it is approximable within factor $O(\log nm)$ where nm is the size of the input. Finally, we provide a practical 2-approximation algorithm for the Binary MEC, a variant of MEC that has been applied in the framework of clustering binary data.

1 Introduction

The genome of diploid organisms, as humans, is composed of two parental copies, called *haplotypes*, for each chromosome. The most frequent form of genetic variations between the two haplotypes of the same chromosome are the *Single Nucleotide Polymorphisms (SNPs)*. Haplotype analysis is of fundamental importance for a variety of applications including agricultural research, medical diagnostic, and drug design [3, 4, 22].

© Springer International Publishing Switzerland 2015
F. Cicalese et al. (Eds.): CPM 2015, LNCS 9133, pp. 100–113, 2015.
DOI: 10.1007/978-3-319-19929-0_9

The task of the *haplotyping problem* is the reconstruction of each pair of haplotypes. However, large scale direct experimental reconstruction from the collected samples is not yet cost-effective. One of the computational approaches that have been proposed, *haplotype assembly*, considers the high-throughput sequencing reads (also called *fragments*) that have to be bipartitioned in order to reconstruct the two haplotypes. Since for most of the SNP positions only two nucleotides are seen, the haplotypes can be represented as binary vectors. The fragments obtained from sequencing may not cover some positions of the haplotypes. These uncovered positions are called *holes*, whereas a sequence of holes within a fragment is called *gap*. However, the presence of sequencing and (possible) mapping errors makes the haplotype assembly problem a challenging task. In literature, different combinatorial formulations of the problem have been proposed [1,7,17,18]. Among them, *Minimum Error Correction* (MEC) [18] has been proved particularly successful in the reconstruction of accurate haplotypes [5,13,20]. However, MEC is a computationally hard problem. Indeed, MEC is APX-hard even if the fragments have at least one gap [6] and remains NP-hard even if the fragments do not contain gaps (*Gapless MEC*) [6]. Instead, the computational complexity of MEC on instances without holes – called *Binary MEC* – is still unknown. Many successful approaches for coping with the computational intractability of MEC are based on the parameterized complexity framework. In particular, MEC is in FPT when parameterized by the "coverage" [20], that is the maximum number of fragments with non-hole values on a SNP position. Moreover, MEC is in FPT also when parameterized by the length of the fragments [13], but only under the *all-heterozygous assumption*, that forces to reconstruct complementary haplotypes. In fact, this assumption allows the dynamic programming algorithm of [13] to focus on the reconstruction of a single haplotype and, hence, to limit the possible combinations for each SNP position.

Despite the significant amount of work present in the literature, some important questions related to the fixed-parameter tractability and approximability of MEC are still open. Two significant open problems are whether there exists a constant approximation algorithm for MEC and whether MEC is in FPT when parameterized by parameters of classical or practical interest, such as the total number of corrections or the length of the fragments. Indeed, removing the dependency on the all-heterozygous assumption from [13] does not appear straightforward and, hence, fixed-parameter tractability of MEC when parameterized by the fragment length is still an open problem.

The binary restriction of MEC where the fragments do not contain holes is particularly interesting from a mathematical point of view, and is the variant of the well-known *Hamming k-Median Clustering Problem* [6,16], when $k = 2$. This clustering problem asks for k representative "consensus" (also called "median") strings with the goal of minimizing the distance between each input string and its closest consensus string. Hamming 2-Median Clustering is well studied from the approximation viewpoint, and a Polynomial Time Approximation

Scheme (PTAS) has been proposed, both in a randomized [19] and deterministic form [14].

In this work, we present advances in the characterization of the fixed-parameter tractability and the approximability of MEC problem in the general, gapless, and binary cases. We first show that MEC is not in APX, *i.e.*, it is not approximable within constant factor. However, we also show that a reduction previously known [8] can be adapted to prove that MEC is approximable within factor $O(\log nm)$ (where n is the number of fragments and m is the number of SNPs) and that MEC is in FPT when parameterized by the total number of corrections.

Furthermore, by inspecting novel combinatorial properties of gapless instances, we show that Gapless MEC is in FPT when parameterized by the length of the fragments and that Binary MEC can be approximated within factor 2. Although Binary MEC is known to admit a PTAS, the 2-approximation algorithm we give is more practical and intuitive than the previous approximation results.

2   Preliminary Definitions

In this section, we introduce some basic notions and the formal definition of the MEC problem. In the rest of the work, we indicate, as usual, the value of a vector s at position t as $s[t]$.

A *fragment matrix* is a matrix $\mathcal{M}$ composed of n rows and m columns such that each entry contains a value in $\{0, 1, -\}$. Each row of $\mathcal{M}$ represents a *fragment* and, formally, is a vector belonging to $\{0, 1, -\}^m$. Symmetrically, each column of $\mathcal{M}$ corresponds to an SNP position and is a vector belonging to $\{0, 1, -\}^n$. We denote by f_i the i-th row of $\mathcal{M}$ and by p_j the j-th column of $\mathcal{M}$. As a consequence, the entry of $\mathcal{M}$ at the i-th row and j-th column is denoted by $f_i[j]$ or $p_j[i]$. The *length* ℓ_i of a fragment f_i is defined as the number of elements in f_i between the rightmost and the leftmost non-hole elements (included) and we denote by ℓ the maximum length over all the fragments in $\mathcal{M}$. Moreover, we say that a column p_j *covers* a row f_i if $p_j[i] \in \{0, 1\}$ and we define the *active fragments* of p_j as the set $active(p_j)$ of all the covered rows, that is $active(p_j) = \{f_i \mid p_j[i] \in \{0, 1\}\}$ (Notice that we denote by $active(p_{j_1}, p_{j_2})$ the intersection $active(p_{j_1}) \cap active(p_{j_2})$ for two columns p_{j_1} and p_{j_2}). A column p_j is *heterozygous* if it contains both 0's and 1's, otherwise is *homozygous*. A *hole* is an entry $f_i[j]$ of $\mathcal{M}$ equal to the symbol $-$. A *gap* in a fragment f_i is a maximal subvector of holes in f_i surrounded by non-hole entries (that is, there exist two positions j_1 and j_2 with $j_1 + 1 < j_2$ such that $f_i[j_1], f_i[j_2] \neq -$ and $f_i[t] = -$ for all t with $j_1 < t < j_2$). A fragment matrix is *gapless* if no fragment contains a gap.

Two rows f_{i_1} and f_{i_2} are in *conflict* when there exists a position j, with $1 \leq j \leq m$, such that $f_{i_1}[j] \neq f_{i_2}[j]$, and $f_{i_1}[j], f_{i_2}[j] \neq -$. Otherwise, we say that f_{i_1} and f_{i_2} are in *agreement*. A collection $\mathcal{F}$ of fragments is in *agreement* if any pair of fragments f_1, f_2 in $\mathcal{F}$ are in agreement. A fragment matrix $\mathcal{M}$ is *conflict free* if there exists a bipartition $(\mathcal{F}_1, \mathcal{F}_2)$ of its fragments such that both $\mathcal{F}_1$ and $\mathcal{F}_2$ are in agreement.

When a fragment matrix $\mathcal{M}$ is conflict free, all the fragments in each part of the bipartition can be merged in order to reconstruct a haplotype, intended as a fragment without holes. Unfortunately, a fragment matrix $\mathcal{M}$ is not always conflict free. The Minimum Error Correction problem deals precisely with this issue by asking for a minimum set of *corrections* that make a fragment matrix conflict free, where a correction of a given fragment f_i at position j, with $f_i[j] \neq -$, is the flip of the value $f_i[j]$, replacing a 0 with a 1, or a 1 with a 0.

Problem 1 (Minimum Error Correction (MEC) problem).
Input: a fragment matrix $\mathcal{M}$ of n rows and m columns.
Output: a conflict free matrix $\mathcal{M}'$ obtained from $\mathcal{M}$ with the minimum number of corrections.

Gapless MEC is the restriction of MEC where the input fragment matrix $\mathcal{M}$ is gapless, while *Binary MEC* is the restriction of (Gapless) MEC where the matrix $\mathcal{M}$ does not contain holes (that is, when $\mathcal{M}$ is a binary matrix).

Given a conflict free fragment matrix $\mathcal{M}$, any heterozygous column p_j encodes a bipartition of the fragments covered by p_j indicating which one belongs to one haplotype and which one belongs to other. Instead, any homozygous column p_j gives no information on how the covered fragments have to be partitioned, and it is "in accordance" with any other bipartition or heterozygous column. More formally, we say that two columns p_{j_1}, p_{j_2} of a fragment matrix are in *accordance* if (1) at least one of p_{j_1}, p_{j_2} is homozygous, or (2) p_{j_1}, p_{j_2} are both heterozygous and are identical or complementary on the fragments covered by both.

As stated in the following lemma, pairwise column accordance on gapless matrices is a necessary and sufficient condition for being conflict free.

Lemma 2. *Let $\mathcal{M}$ be a gapless fragment matrix. Then, $\mathcal{M}$ is conflict free if and only if each pair of columns is in accordance.*

Proof. By definition, if $\mathcal{M}$ is conflict free, each pair of columns is in accordance. For this reason, we just prove by induction on the number m of columns in $\mathcal{M}$ that if each pair of columns is in accordance, then $\mathcal{M}$ is conflict free.

If $h = 1$, the lemma obviously holds.

Assume by induction that the lemma holds for the first h columns in $\mathcal{M}$, we need to prove that the lemma still holds for the first $h + 1$ columns. The submatrix on the first h columns is conflict free by induction and, for this reason, a bipartition (P_1, P_2) of the corresponding fragments exists. By assumption, p_{h+1} and p_h are in accordance. Hence, p_{h+1} and p_h define the same bipartition on the fragments in $active(p_h, p_{h+1})$. Since $\mathcal{M}$ is gapless, there is no column p_y in $\{p_1, \ldots, p_{h-1}\}$ such that $active(p_y, p_{h+1}) \setminus active(p_h) \neq \emptyset$, hence $active(p_{h+1}) \setminus active(p_h) \not\subseteq active(p_y)$ for $1 \leq y \leq h-1$. It follows that there exists a bipartition $(P_1 \cup P_1', P_2 \cup P_2')$ for all the fragments active on the first $h + 1$ columns, where (P_1', P_2') is the bipartition induced by p_{h+1} on the fragments in $active(p_{h+1}) \setminus active(p_h)$. As a consequence the submatrix on the first $h + 1$ columns is conflict free. $\square$

Such a property is particularly important when designing exact algorithms for Gapless MEC, as it allows to test only for pairwise column accordance in order to ensure that the matrix is conflict free. In fact, the fixed-parameter algorithm for Gapless MEC that we present in Sect. 4 is based on this property. Furthermore, notice that if we relax the requirement that $\mathcal{M}$ is gapless, then the property does not hold. Consider, for example, the fragment matrix $\mathcal{M}$ composed of three fragments $f_1 = 01-$, $f_2 = -01$, and $f_3 = 1 - 0$. The three columns are pairwise in accordance, but the matrix is not conflict free (and, in fact, f_3 contains a gap).

Given two columns p_{j_1}, p_{j_2} of a fragment matrix $\mathcal{M}$, we define their (generalized) Hamming distance $d_H(p_{j_1}, p_{j_2})$ as $|\{i \mid \{p_{j_1}[i], p_{j_2}[i]\} = \{0, 1\}\}|$ while their *correction distance* $d(p_{j_1}, p_{j_2})$ as the minimum between $d_H(p_{j_1}, p_{j_2})$ and $d_H(\overline{p_{j_1}}, p_{j_2})$ (where $\overline{p}$ is the complement of p on non-hole entries). Notice that the correction distance is non-negative and symmetric, but does not satisfy the triangle inequality, hence, despite the name, is not a metric. We also define the *homozygous distance* $H(p_j)$ as the minimum between the number of 0's and 1's contained in p_j. Intuitively, the correction distance is the cost of making a column equal or complementary to another column, while the homozygous distance is the cost of making a column homozygous.

A solution of MEC over a fragment matrix $\mathcal{M}$ is a bipartition of its fragments, that can be encoded as a binary vector O. It is easy to see that the cost of that solution is: $cost_\mathcal{M}(O) = \sum_{j=1}^{m} \min(d(O, p_j), H(p_j))$.

3   Inapproximability of MEC

In this section, we show that MEC is not in APX, that is MEC cannot be approximated within constant factor. We achieve this result by introducing an L-reduction from the Edge Bipartization problem to MEC.

The Edge Bipartization problem is defined as follows.

Problem 3 (Edge Bipartization (EB) problem [9]).
Input: an undirected graph $G = (V, E)$.
Output: $E' \subseteq E$ of minimum size such that $G' = (V, E \setminus E')$ is bipartite.

Now, we present the details of the reduction. Given an undirected graph $G = (V, E)$, we build the associated fragment matrix $\mathcal{M}(G)$ (with $|V|$ rows and $|E|$ columns) by setting, at each column p_j associated with edge $e_j = \{u, v\} \in E$, $f_u[j] = 0$, $f_v[j] = 1$, and $f_z[j] = -$ for $z \neq u, v$. Notice that, by construction, there exists a conflict in $\mathcal{M}(G)$ between fragments f_u and f_v if and only if $\{u, v\} \in E$.

Lemma 4. *Let $G = (V, E)$ be an undirected graph and $\mathcal{M}(G)$ be the associated fragment matrix. Given a solution E' of EB over G, we can compute in polynomial time a solution of MEC over $\mathcal{M}(G)$ with $|E'|$ corrections. Symmetrically, given a solution of MEC over $\mathcal{M}(G)$ with h corrections, we can compute in polynomial time a solution E' of EB over G of size at most h.*

Proof. ($\Rightarrow$) Let E' be a set of edges such that $(V_1 \uplus V_2, E \setminus E')$ is bipartite, where V_1 and V_2 are the parts of the bipartition. Build a matrix $\mathcal{M}'(G)$ from $\mathcal{M}(G)$ by flipping, for each $e_j = \{u, v\} \in E'$, the entry $f_u[j]$. Clearly, $\mathcal{M}'(G)$ is obtained from $\mathcal{M}(G)$ with $|E'|$ corrections and it does not contain conflicts induced by edges in E'. Let $(\mathcal{F}_1, \mathcal{F}_2)$ be the bipartition of fragments of $\mathcal{M}'(G)$ such that $\mathcal{F}_i := \{f_u \mid v_u \in V_i\}$ (for $i \in \{1, 2\}$). Each $\mathcal{F}_i$ is in agreement because it does not contain a pair of fragments associated with the endpoints of an edge of $E \setminus E'$. Hence, $\mathcal{M}'(G)$ is conflict free.

($\Leftarrow$) Let $\mathcal{M}'(G)$ be a conflict free matrix obtained from $\mathcal{M}(G)$ with h corrections and let C' be the subset of columns of $\mathcal{M}'(G)$ that contain exactly one correction. Consider the set $E' := \{e_j \in E \mid p_j \in C'\}$. Clearly, $|E'| \leq h$. Since $\mathcal{M}'(G)$ is conflict free, there exists a bipartition $(\mathcal{F}_1, \mathcal{F}_2)$ of the fragments such that both $\mathcal{F}_1, \mathcal{F}_2$ are in agreement. Build sets V_1, V_2 such that $V_i := \{v_u \mid f_u \in \mathcal{F}_i\}$ (with $i \in \{1, 2\}$). We claim that $(V_1 \uplus V_2, E \setminus E')$ is bipartite. Suppose to the contrary that there exists an edge $e_j = \{u, v\} \in E \setminus E'$ such that $u, v \in V_i$, $i \in \{1, 2\}$. Since $f_u[j] = f_v[j]$ in $\mathcal{M}'(G)$, this implies that exactly one of $f_u[j]$ and $f_v[j]$ has been corrected (since $f_u[j] \neq f_v[j]$ in $\mathcal{M}(G)$). As a consequence, we have that $e_j \in E'$, contradicting the assumption. $\square$

Khot [15] proved that, under the Unique Games Conjecture, EB is not in APX. Since Lemma 4 proves that MEC is L-reducible to EB, we have the following result.

Theorem 5. *Under the Unique Games Conjecture [15], MEC is not in* APX.

The inapproximability result given in Theorem 5 nicely complements an approximation (and fixed-parameter tractable) result that can be easily inferred by a reduction presented in [8]. In [8], MEC is reduced to the Maximum Bipartite Induced Subgraph problem (MBIS). Given a vertex-weighted graph G, MBIS asks for a maximum weight subset of vertices of G that induces a bipartite graph. The reduction defines a graph, called *fragment graph*, whose set of nodes is the union of two sets: a set of nodes, called *fragment nodes*, one for each fragment, and a set of nodes, called *entry nodes*, one for each entry of the matrix. In order to avoid the removal of fragments nodes, they are assigned a sufficiently large weight.

The reduction can be easily reworked in order to prove approximation and fixed-parameter tractability results for MEC. More precisely, MEC is now reduced to the *Graph Bipartization* (GB) problem, a problem related to MBIS. Given an unweighted graph G, GB asks for the minimum number of vertex removals so that the resulting graph is bipartite. The reduction given in [8] can be modified by defining a new version of the fragment graph (see Fig. 1), where each (weighted) fragment node is substituted with a sufficiently large set of fragment nodes. From the construction of the fragment graph, it follows that a fragment matrix $\mathcal{M}$ is conflict free if and only if the corresponding fragment graph is bipartite and that a solution of MEC with k corrections corresponds to a solution of GB that removes k vertices.

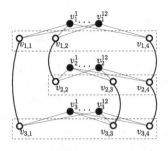

	p_1	p_2	p_3	p_4
f_1	1	0	-	1
f_2	-	1	0	0
f_3	0	-	1	1

Fig. 1. A 3×4 fragment matrix (left) and the associated *fragment graph* (right). Fragment-nodes are in black, while entry-nodes are in white.

Since GB can be approximated within factor $O(\log |V|)$ [10] and is in FPT when parameterized by the number of removed vertices [11,23], we have that:

Theorem 6.

(1) *MEC can be approximated in polynomial time within factor $O(\log nm)$ where n is the number of fragments and m is the number of SNP positions.*
(2) *MEC is in FPT when parameterized by the total number of corrections.*

4   Gapless MEC Is in FPT When Parameterized by the Fragment Length

In this section, we introduce a fixed-parameter tractable algorithm for Gapless MEC when parameterized by the maximum length ℓ of the fragments. The algorithm is based on a dynamic programming approach and aims at finding a specific tripartition for the columns of a gapless fragment matrix $\mathcal{M}$. In this section, we assume w.l.o.g. that $\mathcal{M}$ is a gapless fragment matrix and the fragments of $\mathcal{M}$ are sorted by starting position.

Lemma 2 provides a sufficient and necessary condition for the reconstruction of a solution for Gapless MEC, that is a conflict free fragment matrix. For this reason, the gapless condition is required by this algorithm. In fact, if the fragment matrix contains gaps, the accordance of the columns is not sufficient to ensure that there are no conflicts. Therefore, we firstly show a result that directly derives from Lemma 2. The following proposition stresses the relationship between a bipartition of the fragments and a tripartition of the columns in a gapless fragment matrix $\mathcal{M}$ that is conflict free.

Proposition 7. *Given a gapless fragment matrix $\mathcal{M}$, the following assertions are equivalent:*

1. *$\mathcal{M}$ is conflict free.*
2. *There exists a bipartition $(\mathcal{F}_1, \mathcal{F}_2)$ of the fragments, where both $\mathcal{F}_1$ and $\mathcal{F}_2$ are in agreement.*

3. *There exists a tripartition $T = (L, H, R)$ of the columns such that each column in H is homozygous, each column in $L \cup R$ is heterozygous, $d_H(p_{j_1}, p_{j_2}) = 0$ for all the columns $p_{j_1}, p_{j_2} \in L$ ($p_{j_1}, p_{j_2} \in R$, resp.) and $d_H(\overline{p_{j_1}}, p_{j_2}) = 0$ for each column $p_{j_1} \in L$ and each column $p_{j_2} \in R$.*

Based on Proposition 7, we introduce an algorithm for Gapless MEC that builds a tripartition of the columns of $\mathcal{M}$ in order to find a conflict free matrix $\mathcal{M}'$ obtained from $\mathcal{M}$ with the minimum number of corrections. Notice that in the rest of this section we implicitly refer only to tripartitions built as reported in the third assertion of Proposition 7.

The algorithm iteratively proceeds row-wise and, at each step, computes a tripartition for the columns considered so far. In particular, the key observation that allows to bound the exponential complexity of the algorithm to the parameter ℓ is that we can build any tripartition for all the columns in $\mathcal{M}$ by adding only a subset of columns, called *active columns*, for each row. This subset contains the columns covering the current fragment and the columns covering both previous and successive fragments. Indeed, we need to remember the tripartition established by previous fragments for columns that are covered by successive fragments. More formally, we define the set *active columns* for a fragment f_i as:

$$\mathcal{A}(i) = \{p_j \mid (p_j[i] \neq -) \vee (\exists x, y \text{ with } x < i < y \mid p_j[x], p_j[y] \neq -)\}$$

Figure 2 represents the active columns $\mathcal{A}(i)$ of a fragment f_i. The cardinality of $\mathcal{A}(i)$ is bounded by ℓ. In fact, considering a row f_i, notice that $\ell_i \leq \ell$ and no column p_k, to the left of f_i, is in $\mathcal{A}(i)$. Recall that fragments are sorted by starting position and assume that r is the number of columns p_j to the right of f_i, such that there are f_b, f_q with $b < i < q$ and $p_j[b], p_j[q] \neq -$. Since the r columns must be contained in $\mathcal{A}(b)$ for a fragment f_b with a starting position preceding the one of f_i, it holds that $\ell_i + r \leq \ell_b \leq \ell$. It clearly follows that $|\mathcal{A}(i)| = \ell_i + r \leq \ell$.

Considering two rows f_{i_1} and f_{i_2}, with $i_1 < i_2$, a tripartition for all the columns in $\mathcal{A}(i_1) \cup \mathcal{A}(i_2)$ can be computed by combining a tripartition T_1 for $\mathcal{A}(i_1)$ and a tripartition T_2 for $\mathcal{A}(i_2)$, only if T_1 and T_2 are "in accordance", that is, they are partitioning the shared columns in the same way. For this reason, we say that a tripartition $T_2 = (L_2, H_2, R_2)$ for $\mathcal{A}(i_2)$ *extends* another tripartition

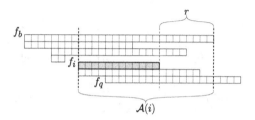

Fig. 2. The set $\mathcal{A}(i)$ of active columns for a fragment f_i.

$T_1 = (L_1, H_1, R_1)$ for $\mathcal{A}(i_1)$ if and only if $L_1 \cap \mathcal{A}(i_2) \subseteq L_2$, $H_1 \cap \mathcal{A}(i_2) \subseteq H_2$, and $R_1 \cap \mathcal{A}(i_2) \subseteq R_2$.

At each step i, the algorithm computes a tripartition T for $\mathcal{A}(i)$ extending a tripartition T' for $\mathcal{A}(i-1)$. Since $\mathcal{A}(i-1)$ also contains all the columns p_j with $p_j[i-1] = -$ such that there exists $y < i - 1$ with $p_j[y] \neq -$ and $p_j[i] \neq -$, it follows that T even extends any tripartition computed at the previous steps extended by T'. As a consequence, we prove the following implication.

Lemma 8. *If there exists a conflict free matrix $\mathcal{M}''$ obtained from $\mathcal{M}$ on the first $i - 1$ rows that induces a tripartition T' for the columns in $\mathcal{A}(i-1)$, and if T is a tripartition for the columns in $\mathcal{A}(i)$ extending T', then there exists a conflict free matrix $\mathcal{M}'$ obtained from $\mathcal{M}$ on the first i rows that induces the tripartition T for the columns in $\mathcal{A}(i)$.*

Proof. By definition, $p_j[i] \neq -$ and $p_j[y] = -$ for each column $p_j \in \mathcal{A}(i) \backslash \mathcal{A}(i-1)$ and for each $y < i$. By assumption T extends T', hence build $\mathcal{M}'$ such that the columns covered by the first $i - 1$ rows are tripartitioned as in $\mathcal{M}''$ and the remaining columns only covered by f_i are tripartitioned according to T. By construction, $\mathcal{M}'$ induces the tripartition T for $\mathcal{A}(i)$. Since $\mathcal{M}''$ is conflict free, it follows that $\mathcal{M}'$ is conflict free by Proposition 7. $\qquad\square$

At each step i and for each tripartition $T = (L, H, R)$ for $\mathcal{A}(i)$, the algorithm chooses the tripartition T' extended by T for $\mathcal{A}(i-1)$ that induces the minimum cost (*recursive step*) and computes the minimum number of corrections to add on the current fragment f_i in order to tripartition all the columns in $\mathcal{A}(i)$ according to T (*local contribution*). In particular, the algorithm considers the minimum number of corrections on f_i such that $p_j[i] = 1$ or $p_j[i] = 0$ for all p_j in L and, on the contrary, $p_j[i] = 0$ or $p_j[i] = 1$ for all p_j in R. At the same time, the minimum number of corrections on the fragment f_i is computed for each column p_j in H such that p_j on the first i rows can be optimally transformed into a homozygous column. Therefore, we define $D[i, T]$ as the minimum number of corrections to obtain a conflict free matrix $\mathcal{M}'$ from $\mathcal{M}$ on the first i rows that induces a tripartition T for $\mathcal{A}(i)$. The algorithm proceeds row-wise computing the value $D[i, T]$ for each fragment f_i and for each tripartition T for $\mathcal{A}(i)$ by the following recursive equation:

$$D[i, T] = \Delta(i, T) + \min_{T' \text{ extended by } T} D[i-1, T'] \qquad (1)$$

where T' is a tripartition for $\mathcal{A}(i-1)$. In the recursion, we consider only the tripartitions T' extended by T, since the shared columns have to be partitioned in the same way. In conclusion, the local contribution is defined as:

$$\Delta(i, T) = O(i, H) + \min \begin{cases} E^0(i, L) + E^1(i, R) \\ E^1(i, L) + E^0(i, R) \end{cases} \qquad \text{where } T = (L, H, R) \quad (2)$$

such that $E^x(i, F)$ is the cost of correcting the columns in F for fragment f_i to value x, that is $E^x(i, F) = |\{j \mid j \in F \wedge p_j[i] \notin \{x, -\}\}|$, and $O(i, H)$

is the minimum number of corrections to apply on fragment f_i such that the columns in H, considered on the first i rows, can be turned into homozygous columns with minimum cost. Denote by $\#_{i,j}^x$ the number of values equal to x in $\{p_j[1], \ldots, p_j[i]\}$. The minimum between $\#_{i,j}^0$ and $\#_{i,j}^1$ states the minimum number of corrections necessary to turn a column p_j on the first i rows into a homozygous column. Since $O(i, H)$ refers only to the corrections on fragment f_i, we can compute $O(i, H)$ as:

$$O(i, H) = \sum_{j \in H} \begin{cases} 1 & p_j[i] = 0 \text{ and } \#_{i,j}^0 \leq \#_{i,j}^1 \\ 1 & p_j[i] = 1 \text{ and } \#_{i,j}^1 \leq \#_{i,j}^0 \\ 0 & \text{otherwise} \end{cases} \tag{3}$$

Given a set of columns F, it is easy to see that $\sum_{i \in \{1,\ldots,n\}} O(i, F) = \sum_{p_j \in F} H(p_j)$.

The base case of the recurrence is $D[1, T] = \Delta(1, T)$ for each tripartition T for $\mathcal{A}(1)$. The algorithm returns the optimum corresponding to $\min_T D[n, T]$ where T is a tripartition for $\mathcal{A}(n)$. Furthermore, an optimal tripartition for all the columns can be computed by backtracking.

The algorithm computes all the values $D[i, T]$ for each tripartition T of the columns in $\mathcal{A}(i)$ and for each i in $\{1, \ldots, n\}$. It follows that there are $O(3^\ell \cdot n)$ entries and, therefore, the space complexity is equal to $O(3^\ell \cdot n)$. Given a tripartition T, we need $O(3^\ell)$ time to enumerate all the tripartitions T' extended by T because we have to tripartition all the columns in $|\mathcal{A}(i-1) \setminus \mathcal{A}(i)|$ with $\mathcal{A}(i-1) \leq \ell$ and, consequently, $|\mathcal{A}(i-1) \setminus \mathcal{A}(i)| \leq \ell$. Since $\Delta(i, T)$ can be computed in $O(\ell)$ time, each entry $D[i, T]$ can be computed in $O(3^\ell \cdot \ell)$. It follows that the total running time of the algorithm is $O(3^{2\ell} \cdot \ell \cdot n)$. Notice that storing partial information during the computation (using an approach similar to the one presented in [20]) we can decrease the complexity to $O(3^\ell \cdot \ell \cdot n)$.

We now show the correctness of the algorithm.

Lemma 9. *Consider a gapless fragment matrix $\mathcal{M}$.*

1. *If $D[i, T] = h$, then there exists a conflict free matrix $\mathcal{M}'$ obtained from $\mathcal{M}$ on the first i rows with h corrections that induces a tripartition T for the columns in $\mathcal{A}(i)$.*
2. *If $\mathcal{M}'$ is a conflict free matrix obtained from $\mathcal{M}$ on the first i rows with h corrections that induces a tripartition T for the columns in $\mathcal{A}(i)$, $D[i, T] \leq h$.*

Proof. We prove the lemma by induction on the number n of rows of $\mathcal{M}$. Both the statements obviously hold for $i = 1$. Assume that lemma holds for $i - 1$, we show that both the statements hold for i.

(1) By Eq. (1), there exists a tripartition T' for $\mathcal{A}(i-1)$ such that T extends T' and $D[i, T] = h = \Delta(i, T) + D[i-1, T']$. Assuming $D[i-1, T'] = h'$, by induction there exists a conflict free matrix $\mathcal{M}''$ obtained from $\mathcal{M}$ on the first $i-1$ rows with h' corrections that induces a tripartition T' for $\mathcal{A}(i-1)$. By Proposition 8, there exists a conflict free matrix $\mathcal{M}'$ obtained from $\mathcal{M}$ on the first i rows that

induces a tripartition T for $\mathcal{A}(i)$. Since T extends T', by construction we can add $\Delta(i, T)$ corrections on fragment f_i in order to build $\mathcal{M}'$ starting from $\mathcal{M}''$. It follows that $\mathcal{M}'$ is obtained from $\mathcal{M}$ with $\Delta(i, T) + h' = h$ corrections.

(2) Assume that $\mathcal{M}''$ is the submatrix of $\mathcal{M}'$ obtained from $\mathcal{M}$ on the first $i - 1$ rows with h' corrections that induces a tripartition T' for $\mathcal{A}(i-1)$. Clearly, T' is extended by T due to the fact that $\mathcal{M}''$ is equal to $\mathcal{M}'$ on the first $i - 1$ rows. Since $\mathcal{M}'$ contains $\Delta(i, T)$ corrections on the row f_i by construction, it follows that $h = \Delta(i, T) + h'$. Moreover, we know that $D[i - 1, T'] \leq h'$ by induction and by Eq. (1) that $D[i, T] = \Delta(i, T) + \min_{T'' \text{ extended by } T} D[i-1, T'']$. Hence, since $\min_{T'' \text{ extended by } T} D[i - 1, T''] \leq D[i - 1, T']$, we conclude that $D[i, T] \leq \Delta(i, T) + h'$ and, consequently, $D[i, T] \leq h$.   □

From the correctness of the algorithm, it directly follows that:

Theorem 10. *Gapless MEC (without the all-heterozygous assumption) is in FPT when parameterized by the length of the fragments and it can be solved in $O(3^\ell \cdot \ell \cdot n)$ time.*

5   A 2-Approximation Algorithm for Binary MEC

In this section we present a 2-approximation algorithm for Binary MEC, that is the restriction of MEC where the fragment matrix does not contain holes. The approximation algorithm is based on the observation that heterozygous columns in binary matrices naturally encode bipartitions of the fragments and that, by Lemma 2, if the columns of a gapless fragment matrix are pairwise in accordance then the matrix is conflict free. In particular, Algorithm 1 builds a feasible solution $\mathtt{SOL}[t]$ for each t in $\{1, \ldots, m\}$ assuming that p_t is the closest column to an (unknown) optimal bipartition O of the fragments. Each solution $\mathtt{SOL}[t]$ corrects columns $p_{j'}$ with cost $H(p_{j'}) \leq d(p_t, p_{j'})$ into homozygous columns (equal to $\underline{1}$ or $\underline{0}$ depending on best choice), whereas it corrects the remaining columns $p_{j''}$ with cost $d(p_t, p_{j''}) < H(p_{j''})$ into heterozygous columns equal (or complementary, depending on the best choice) to p_t. It is easy to see that $\mathtt{SOL}[t]$ for each t in $\{1, \ldots, m\}$ is a feasible solution (by Lemma 2) and that its cost is exactly $cost_{\mathcal{M}}(p_t)$.

Algorithm 1. A 2-approximation algorithm for Binary MEC

Require: A $n \times m$ binary matrix $\mathcal{M}$
    for $t = 1$ **to** m **do**   ▷ Assume that p_t is the column "closest" to O
        for $j = 1$ **to** m **do**
            if $H(p_j) \leq d(p_t, p_j)$ **then**
                Set p_j homozygous in $\mathtt{SOL}[t]$
            else
                Set p_j equal/complementary to p_t in $\mathtt{SOL}[t]$
        return $\arg\min_{\mathtt{SOL}[t]} cost_{\mathcal{M}}(p_t)$

Algorithm 1 is a 2-approximation algorithm for Binary MEC.

Lemma 11. *Given a fragment matrix $\mathcal{M}$ without holes, if OPT is the optimum for Binary MEC on input $\mathcal{M}$, then Algorithm 1 returns in $O(m^2 n)$ time a feasible solution with cost OPT' such that $OPT' \leq 2 \cdot OPT$.*

Proof. Assume that p_O is the column of $\mathcal{M}$ closest to an optimal bipartition O, that is $d(O, p_O) \leq d(O, p_j)$ for each j in $\{1, \ldots, m\}$ and assume that $d_H(O, p_O) \leq d_H(\overline{O}, p_O)$ (if $d_H(\overline{O}, p_O) < d_H(O, p_O)$ we can substitute O with $\overline{O}$ since they encode the same bipartition). Clearly, one such a column exists and $d_H(O, p_O) \leq d(O, p_j)$ for each j in $\{1, \ldots, m\}$. We show that, under this assumption, $d(p_O, p_j) \leq 2d(O, p_j)$. By the triangle inequality, $d_H(p_O, p_j) \leq d_H(p_O, O) + d_H(O, p_j)$. Hence, since $d_H(p_O, O) \leq d(O, p_j) \leq d_H(O, p_j)$, we have $d_H(p_O, p_j) \leq 2d_H(O, p_j)$. Similarly, we can prove that $d_H(p_O, \overline{p_j}) \leq 2d_H(O, \overline{p_j})$. As a consequence we have that $d(p_O, p_j) \leq 2d_H(O, p_j)$ and that $d(p_O, p_j) \leq 2d_H(O, \overline{p_j})$, which then imply $d(p_O, p_j) \leq 2d(O, p_j)$. Clearly, since $d(p_O, p_j) \leq 2d(O, p_j)$, we also have that $\min(d(p_O, p_j), H(p_j)) \leq 2\min(d(O, p_j), H(p_j))$.

Since Algorithm 1 iteratively assumes that each column p_j is the closest column to the unknown optimal bipartition O, we have that the cost of the returned solution is $OPT' \leq cost_{\mathcal{M}}(p_O) \leq 2\sum_{j=1}^{m} \min(d(O, p_j), H(p_j)) = 2OPT$. Since each iteration t of the algorithm computes $\texttt{SOL}[t]$ in $O(mn)$ time, the total running time is clearly equal to $O(m^2 n)$. □

Algorithm 1 runs in $O(m^2 n)$ time and, due to its simplicity, it is a more direct and practical approach than the PTAS algorithms known in literature [14, 19].

6   Conclusions

Minimum Error Correction is a prominent combinatorial problem for haplotype assembly. Investigating the approximation complexity and the fixed-parameter tractability of MEC has proven useful to develop practical haplotype assembly tools [2, 13, 20]. Despite in this paper we addressed some issues that were left open, some other theoretical questions still need an answer.

In this work, we showed that, under the Unique Games Conjecture, MEC is not approximable within any constant factor. However, the approximation complexity of Gapless MEC and the computational complexity of Binary MEC are still unknown. It would be interesting to explore whether Lemma 2, that we used in this paper for achieving a direct 2-approximation algorithm for Binary MEC and an FPT algorithm for Gapless MEC, is also useful for answering to these open questions. Similarly, the design of practical FPT algorithms for the general MEC parameterized by the fragment length is an interesting research direction.

Recent advances in sequencing technologies are radically changing the characteristics of the produced data. For example, long gapless reads with sequencing errors uniformly distributed will likely be common in the near future. The design of FPT algorithms that exploit these characteristics is another important research direction. Furthermore, the drop in sequencing costs allows large-scale studies of rare diseases. In fact, they are usually caused by rare mutations that can only be reliably discovered by sequencing many related individuals. Hence, we expect an increasing interest in the study of new formulations extending MEC on structured populations (where additional constraints induced by the Mendelian laws of inheritance improve the accuracy of the reconstructed haplotypes [21]), as initially done in [12].

Acknowledgements. This work has been stimulated by discussions between PB, GK, and NP during the No.045 NII Shonan workshop on Exact Algorithms for Bioinformatics Research, March 2014, Japan.

The authors acknowledge the support of the MIUR PRIN 2010-2011 grant 2010LYA9RH (Automi e Linguaggi Formali: Aspetti Matematici e Applicativi), of the Cariplo Foundation grant 2013-0955 (Modulation of anti cancer immune response by regulatory non-coding RNAs), of the FA 2013 grant (Metodi algoritmici e modelli: aspetti teorici e applicazioni in bioinformatica).

References

1. Aguiar, D., Istrail, S.: HapCompass: a fast cycle basis algorithm for accurate haplotype assembly of sequence data. J. Comput. Biol. **19**(6), 577–590 (2012)
2. Bansal, V., Bafna, V.: HapCUT: an efficient and accurate algorithm for the haplotype assembly problem. Bioinformatics **24**(16), i153–i159 (2008)
3. Bonizzoni, P., Della Vedova, G., Dondi, R., Li, J.: The haplotyping problem: an overview of computational models and solutions. J. Comput. Sci. Techol. **18**(6), 675–688 (2003)
4. Browning, B., Browning, S.: Haplotypic analysis of Wellcome Trust case control consortium data. Hum. Genet. **123**(3), 273–280 (2008)
5. Chen, Z.Z., Deng, F., Wang, L.: Exact algorithms for haplotype assembly from whole-genome sequence data. Bioinformatics **29**(16), 1938–45 (2013)
6. Cilibrasi, R., Van Iersel, L., Kelk, S., Tromp, J.: The complexity of the single individual SNP haplotyping problem. Algorithmica **49**(1), 13–36 (2007)
7. Dondi, R.: New results for the Longest Haplotype Reconstruction problem. Discrete Appl. Math. **160**(9), 1299–1310 (2012)
8. Fouilhoux, P., Mahjoub, A.: Solving VLSI design and DNA sequencing problems using bipartization of graphs. Comput. Optim. Appl. **51**(2), 749–781 (2012)
9. Garey, M.R., Johnson, D.S.: Computers and Intractability: A Guide to the Theory of NP-Completeness. W. H. Freeman, New York (1979)
10. Garg, N., Vazirani, V.V., Yannakakis, M.: Approximate max-flow min-(multi) cut theorems and their applications. SIAM J. Comput. **25**(2), 235–251 (1996)
11. Guo, J., et al.: Compression-based fixed-parameter algorithms for feedback vertex set and edge bipartization. J. Comput. Syst. Sci. **72**(8), 1386–1396 (2006)
12. Halldórsson, B.V., Aguiar, D., Istrail, S.: Haplotype phasing by multi-assembly of shared haplotypes: phase-dependent interactions between rare variants. In: PSB, pp. 88–99. World Scientific Publishing (2011)
13. He, D., et al.: Optimal algorithms for haplotype assembly from whole-genome sequence data. Bioinformatics **26**(12), i183–i190 (2010)
14. Jiao, Y., Xu, J., Li, M.: On the k-closest substring and k-consensus pattern problems. In: Sahinalp, S.C., Muthukrishnan, S.M., Dogrusoz, U. (eds.) CPM 2004. LNCS, vol. 3109, pp. 130–144. Springer, Heidelberg (2004)

15. Khot, S.: On the power of unique 2-prover 1-round games. In: STOC, pp. 767–775. ACM (2002)
16. Kleinberg, J., Papadimitriou, C., Raghavan, P.: Segmentation problems. In: STOC, pp. 473–482. ACM (1998)
17. Lancia, G., Bafna, V., Istrail, S., Lippert, R., Schwartz, R.: SNPs problems, complexity, and algorithms. In: Meyer auf der Heide, F. (ed.) ESA 2001. LNCS, vol. 2161, pp. 182–193. Springer, Heidelberg (2001)
18. Lippert, R., Schwartz, R., Lancia, G., Istrail, S.: Algorithmic strategies for the single nucleotide polymorphism haplotype assembly problem. Brief. Bioinform. 3(1), 23–31 (2002)
19. Ostrovsky, R., Rabani, Y.: Polynomial-time approximation schemes for geometric min-sum median clustering. J. ACM 49(2), 139–156 (2002)
20. Patterson, M., Marschall, T., Pisanti, N., van Iersel, L., Stougie, L., Klau, G.W., Schönhuth, A.: WHATSHAP: haplotype assembly for future-generation sequencing reads. In: Sharan, R. (ed.) RECOMB 2014. LNCS (LNBI), vol. 8394, pp. 237–249. Springer, Heidelberg (2014)
21. Pirola, Y., Bonizzoni, P., Jiang, T.: An efficient algorithm for haplotype inference on pedigrees with recombinations and mutations. IEEE/ACM Trans. Comput. Biol. Bioinform. 9(1), 12–25 (2012)
22. Pirola, Y., et al.: Haplotype-based prediction of gene alleles using pedigrees and SNP genotypes. In: BCB, pp. 33–41. ACM (2013)
23. Reed, B., Smith, K., Vetta, A.: Finding odd cycle transversals. Oper. Res. Lett. 32(4), 299–301 (2004)

Fast String Dictionary Lookup with One Error

Timothy Chan[1] and Moshe Lewenstein[2]([⊠])

[1] Cheriton School of Computer Science, University of Waterloo,
Waterloo, Canada
`tmchan@uwaterloo.ca`
[2] Department of Computer Science, Bar-Ilan University,
Ramat Gan, Israel
`moshe@cs.biu.ac.il`

Abstract. A set of strings, called a *string dictionary*, is a basic string data structure. The most primitive query, where one seeks the existence of a pattern in the dictionary, is called a *lookup query*. Approximate lookup queries, i.e., to lookup the existence of a pattern with a bounded number of errors, is a fundamental string problem. Several data structures have been proposed to do so efficiently. Almost all solutions consider a single error, as will this result. Lately, Belazzougui and Venturini (CPM 2013) raised the question whether one can construct efficient indexes that support lookup queries with one error in optimal query time, that is, $O(|p|/\omega + occ)$, where p is the query, ω the machine word-size, and occ the number of occurrences.

Specifically, for the problem of one mismatch and constant alphabet size, we obtain optimal query time. For a dictionary of d strings our proposed index uses $O(\omega d \log^{1+\epsilon} d)$ additional bit space (beyond the space required to access the dictionary data, which can be maintained in compressed form). Our results are parameterized for a space-time tradeoff.

We propose more results for the case of lookup queries with one insertion/deletion on dictionaries over a constant sized alphabet. These results are especially effective for large patterns.

1 Introduction

Data mining, information retrieval, web search and database tasks are often variants of string processing. Many of these tasks involve storing a set of strings, also known as a *string dictionary*. These dictionaries may be very large, for example such is the case for search engines, applications in bioinformatics, RDF graphs, and meteorological data. Hence, it is desired to maintain the dictionaries in some succinct format while still allowing quick access to the data at hand. One basic primitive operation necessary on a dictionary is a lookup query. A *lookup* query on a string dictionary is a string for which the answer is yes if it exists in the dictionary, or more generally returns a pointer to the satellite data of that

T. Chan—The research is supported by an NSERC grant.

M. Lewenstein—This research is supported by a BSF grant 2010437 and a GIF grant 1147/2011.

© Springer International Publishing Switzerland 2015
F. Cicalese et al. (Eds.): CPM 2015, LNCS 9133, pp. 114–123, 2015.
DOI: 10.1007/978-3-319-19929-0_10

string in the dictionary. Maintaining the dictionaries in a compressed form while allowing lookups has garnered much interest in the last decade.

While exact lookups are interesting, often one desires *approximate lookups*. For example, if one queries a search engine and there is one or two typing errors in the query, it is advantageous to find the correct answer nevertheless. In this case, one actually needs to propose all answers that are within the criteria of the number of errors. Clearly, in many applications one desires to find approximate matches and not only matches.

Errors come in different forms. Three of the most common errors are substitutions, insertions, and deletions of characters. Two widely considered distances between strings are based on these errors. The former is *Hamming distance*, which is the minimal number of substitutions necessary to transform one string into another. The latter is *edit distance* [18], which is the minimal number of any combination of the three operations needed to transform one string into another.

Approximate lookups for one error have received a lot of attention, e.g. [3,4, 7,8,24] and along this similar line also text indexing with one error [1], dictionary matching with one error [1,14,16] and both with one wildcard [2,5,20]. Extensions to k errors, even to 2 errors, is much more difficult. See [9,10,12,17,21,23] for results of this form.

In numerous data structure papers over the last decade there has been a separation between the *encoding* model and the *indexing* model. In the encoding model we preprocess our input I (in our case the dictionary of strings) to create a data structure *enc* and queries have to be answered using *enc* only, *without* access to I. In the indexing model, we create an index *idx* and are able to refer to I when answering queries. In the indexing model we measure the *additional space* required. This model difference was already noted in [13]. For more discussion on the modeling differences see [6].

Interestingly, Belazzougui and Venturini [4] proposed a data structure for lookups with edit distance at most one that answers queries in $O(|p| + occ)$ time, where p is the query string and occ is the number of answers. The space required is $2nH_k + o(n) + 2d \log d$, where d is the number of strings in the dictionary, n is the total length of the dictionary and H_k is the k-th order entropy of the concatenated strings of the dictionary. While the model is set as an encoding model result, it actually is an indexing result with $nH_k + o(n)$ dedicated to the compressed dictionary and $nH_k + o(n) + 2d \log d$ additional bits necessary for the data structure.

They raised the question in [4] whether one can answer queries in optimal $O(|p|/\omega + occ)$ time while maintaining succinct space. We answer this question affirmatively. For the case of Hamming distance we propose a data structure that requires $O(\omega d \log^{1+\epsilon} d)$ additional bit space (beyond the dictionary which can be maintained in compressed form). For the case of edit distance we can obtain $\delta |p| d$, for arbitrarily small constant $\delta > 0$, additional bit space ($|p|d$ is the size of the open dictionary) and $O((|p|/\log |p|) \log^\epsilon d + occ)$ query time. This is an improvement over the times in [4] for $|p| \gg \log d$. However, we do note that the alphabet size in [4] is general, whereas the alphabet size here is constant. Our solution can be generalized to an alphabet of size σ at a cost of a σ factor in time and space.

2   Previous work

Yao and Yao [24] were the first to consider string dictionaries that support lookup queries with one mismatch. The dictionary they suggested had, wlog, all strings of equal size m. The alphabet was binary. They suggested an algorithm in the bit probe model that uses $O(md \log m)$ bits and answers queries in $O(m \log \log d)$ bit probes. Brodal and Gasieniec [7] considered the standard unit-cost RAM model in the same setting, i.e., one mismatch, all strings of length m and a binary alphabet. They proposed a different solution using a trie for the lexicographically ordered strings of $\mathcal{D}$ and a trie for the lexicographically ordered reversed strings of $\mathcal{D}$. The space they used was $O(md)$ words. The query time was $O(m)$. Later, Brodal and Venkatesh [8] considered a perfect-hash solution in the cell-probe model with word-size ω **and** string size $m = \omega$. They proposed a data structure that uses space $O(d \log \omega)$. We elaborate on and generalize their solution in Sect. 5.

Belazzougui [3] proposed the first $O(|p| + occ)$ time algorithm. The space of the solution is $O(n)$ bits, where n is the total dictionary size. The solution used Karp–Rabin fingerprinting and, hence, runs with high probability. As formerly mentioned, in [4] a result was obtained for dictionary lookups with edit distance one that answers queries in $O(|p| + occ)$ time, and with $2nH_k + o(n) + 2d \log d$, where n is the total length of the dictionary and H_k is the k-th order entropy of the concatenated strings of the dictionary.

3   Outline of Our Results

Our goal is to solve the dictionary matching with one error where the query time is optimal $O(|p|/\omega + occ)$ and the space is succinct. Our method is based on succinct bidirectional indexing structures and range searching data structures, see [19]. The method of this search has been used numerous times before, and was first used to solve a one-error problem in [1]. However, the unique feature in this paper is a succinct code for each string which allows optimal query time while maintaining very efficient space. The encoding of the string is a novel folding of the string which turns out to do the trick. The idea is to take a string s, partition it into equal length substrings, say of length b. Then we do a bitwise exclusive-or among the substrings. This folding of strings reduces the space of the string down to a small size and allows to obtain succinct representations of the strings. The encoding, assisted by the range searching techniques, remains powerful enough to deduce the answers required.

4   Preliminaries

Given a string S, $|S|$ is the length of S. An integer i is a *location* or a *position* in S if $i = 1, \ldots, |S|$. The substring $S[i \ldots j]$ of S, for any two positions $i \leq j$, is the substring of S that begins at index i and ends at index j.

A set of strings is called a *dictionary* and is denoted with $\mathcal{D} = \{S_1, \ldots, S_d\}$. That is the number of strings is d and we denote the total size $n = \sum_{i=1}^{d} |S_i|$. We may safely assume that all strings in the dictionary are of the same size. If this is not the case then $\mathcal{D}$ can be partitioned into $\mathcal{D}_l = \{s \in \mathcal{D} \mid |s| = l\}$. For one substitution one accesses $\mathcal{D}_{|p|}$ and for insertion/deletion one accesses $\mathcal{D}_{|p|+1}$ and $\mathcal{D}_{|p|-1}$.

Let $\mathcal{D} = \{S_1, S_2, \ldots, S_d\}$ be a dictionary of strings. The problem of *String Dictionary with Distance One* is the problem of indexing $\mathcal{D}$ to support *lookup queries at distance* 1, that is, for a lookup query p find all strings in the dictionary within distance 1 of p. The desired distance will be either Hamming distance or edit distance, depending on the problem at hand. We will consider both. The desire will be to maintain the dictionary in some compressed form and to answer the lookup queries of distance 1 quickly.

5   The Brodal–Venkatesh Algorithm

The Brodal and Venkatesh [8] algorithm is defined on a dictionary in which all strings are binary and have length exactly ω. However, this can be generalized. We describe this now. We still assume, wlog, that all strings are of the same length.

The proposed scheme is a straightforward solution for the problem based on hashing. Let $Ham(s, x)$ denote the Hamming distance between two equal length strings s and x. Let $H(s) = \{x \in \{0, 1\}^{|s|} \mid Ham(s, x) = 1\}$ and let $H = \cup_{s \in \mathcal{D}} H(s)$, i.e., all strings at Hamming distance 1 from a string $s \in \mathcal{D}$. Generate a perfect hash function for $H \cup D$. Queries p, also of the same length as the strings of the dictionary, are read. Applying the hash function on p yields the answers. Recall that the strings are over a binary alphabet. So, by reading ω bits at a time, that is, treating each ω bits as one ω-bit character, the hashing can be implemented on strings with query time of $O(|p|/\omega + occ)$. The space required is the size of the hash table, which is $O(d|p| \log d)$ bit-space[1].

6   Algorithm for Dictionary Lookup with One Mismatch

We are interested in solving the one mismatch case with the same $O(|p|/\omega + occ)$ time. We still consider a binary alphabet, but point out that a general alphabet of size σ is reducible to the binary alphabet with σ blowup. Note that the size of the dictionary is $O(|p|d)$ bits. Hence, the Brodal and Venkatesh [8] algorithm is unsatisfactory as it uses $O(d|p| \log d)$ bits. The dictionary is not even included in this space, but it is not really necessary for their result.

We desire to obtain a result where the additional bit space is strictly sublinear in the size of the dictionary. Our method will use range queries on strings.

---

[1] The space attributed to this algorithm in [4] is $O(d|p|^2 \log d)$ bits. However, this is probably because it was assumed that the generated strings, which are of size $O(d|p|^2 \log d)$ bits, need to be maintained. However, this is not the case. It is sufficient to maintain the hash function and not the fully generated strings.

However, the encoding of the strings to maintain a small data structure is the essence of our algorithm. We now describe the details of the solution.

Define for string $s \in \mathcal{D}$ a point $(x(s), y(s))$ on a 2D $d \times d$ geometric grid. Let $x(s)$ = the rank of s in the lexicographical sort of $\mathcal{D}$ and $y(s)$ = the rank of s^R, s reversed, in the lexicographical sort of $\mathcal{D}$ after reversing all strings.

String Encodings: Fix a parameter b (think of b as polylogarithmic and assume, wlog, that b divides s). Divide each of the strings s into b-bit words $s_1, \ldots, s_{|s|/b}$ and let $c(s) = \oplus_{i=1}^{|s|/b} s_i$, where $\oplus$ denotes the bitwise exclusive-or of the s_i's. Note that $c(s)$ itself is a b-bit word. Let $C(s)$ be all b-bit words that have Hamming distance 1 from $c(s)$. Note that $|C(s)| = b$. We think of s as a point $(x(s), y(s))$ in 2D, assigned multiple colors, one from each member of $C(s)$. See [22] for a string encoding along the same lines.

We are now ready to construct the data structure.

The Data Structure: We build an orthogonal range reporting structure for each non-empty color class. There are 2^b color classes, but each point is in b color classes. Specifically, consider string $s \in \mathcal{D}$ and $c(s)$ that is associated with it. There are exactly b strings with one bit of $c(s)$ flipped, which is the set $C(s)$. Now, visualize a 3D grid of $d \times d \times 2^b$ where for every string $s \in \mathcal{D}$ we generate grid points $(x(s), y(s), c)$, where $c \in C(s)$. However, we maintain separate orthogonal range reporting structure for each possible c.

Overall there are db points, hence the number of non-empty color classes is bounded by db, but will likely be a lot less. We will use a perfect hash function on these non-empty color classes $\subseteq [2^b]$ so that we can access the, at most, db orthogonal range reporting structures that exist in constant time.

The data structure supports dictionary lookup queries with one error as follows.

Query: Given a pattern p, divide it into b-bit words $p_1, \ldots, p_{|p|/b}$ and let $c(p) = \oplus_{i=1}^{|p|/b} p_i$. For each i, we want to search for all s in $\mathcal{D}$ such that:

1. s has prefix $p_1 \ldots p_{i-1}$ and
2. s has suffix $p_{i+1} \ldots p_{|p|/b}$ and
3. s_i and p_i have Hamming distance 1.

Property (1) is equivalent to having $x(s)$ lie in the interval of the lexicographical sort of $\mathcal{D}$ that is associated with the prefix $p_1 \ldots p_{i-1}$, and (2) is equivalent to having $y(s)$ lie in the interval of the lexicographical sort of the reversed strings of $\mathcal{D}$ that is associated with the suffix $p_{i+1} \ldots p_{|p|/b}$.

To implement (1) and (2) one needs to find the above-described intervals. This can be done using bidirectional tries, as has been done in some of the previous results.

Specifically, divide s into ω-bit words $s_1, \ldots, s_{|s|/\omega}$. The b discussed previously will be a multiple of ω. The current partition of words into $|s|/\omega$ is our choice for the compacted tries construction, whereas the partition of words into $|s|/b$ words will be for the range searching structure.

Construct a compacted trie T of all lexicographically sorted dictionary strings, treating each as a string over alphabet $= [2^\omega]$. To allow constant time traversal from each node in the trie we generate a hash on all first (ω-length) characters emanating from a node. That is if edge e is labeled in the compacted trie with $l(e)$ then for the set $\{(v, a, u) \mid e = (v, u) \in T, l(e) = ax\}$ we generate a perfect hash function h where, in constant time, we can access u from $h(v, a)$. When, traversing with the pattern p, which we also partition into $p_1, \ldots, p_{|p|/\omega}$, we only evaluate the appropriate (ω-length) character of p with the first (ω-length) character on the edge. Once we reach a leaf, or cannot traverse further in the trie - in which case we choose an arbitrary descendant leaf, we use the dictionary string s represented by the leaf to evaluate how far p matches in the trie, by comparing p and s, in comparisons of ω-length characters using the compressed text. We construct a symmetric compacted trie T^R over the reversed strings of the dictionary.

Once we know the path know where p matches in T we traverse this path to compute the boundaries of the range searches described above. That is, after every b binary characters or, in other words, after every b/ω ω-length characters, we need the range of the array of lexicographically ordered strings described by this node. At each such node, we maintain two indices to describe the range. This is the information needed for the range queries. In T we traverse the path from top to bottom and in T^R we traverse from bottom to top.

Now, assuming that (1) and (2) are true, we can show an appropriate condition for (3) to hold.

Lemma 1. *Assume that s has prefix $p_1 \ldots p_{i-1}$ and s has suffix $p_{i+1} \ldots p_{|p|/b}$. Then $c(s)$ and $c(p)$ have Hamming distance 1, i.e., $c(p) \in C(s)$, iff s_i and p_i have Hamming distance 1.*

Proof. Since s has prefix $p_1 \ldots p_{i-1}$ and suffix $p_{i+1} \ldots p_{d/b}$ it follows that $\forall j \neq i$ and $\forall l : s[jb + l] = p[jb + l]$. Hence, $s[ib + l] = p[ib + l]$ iff for the l-th bit $c(s)_l = c(p)_l$. So we can conclude that $c(s)$ and $c(p)$ have Hamming distance 1 iff s_i and p_i have Hamming distance 1.   □

It follows from the lemma that it is sufficient to verify whether $c(p) \in C(s)$ for all dictionary strings s that have prefix $p_1 \ldots p_{i-1}$ and suffix $p_{i+1} \ldots p_{d/b}$. This translates into a single orthogonal 4-sided range reporting query with the 2 ranges defined by prefix $p_1 \ldots p_{i-1}$ and suffix $p_{i+1} \ldots p_{d/b}$ found using the bidirectional tries. The query is asked in the orthogonal range searching structure associated with the color class of $c(p)$. We access this range searching structure, in constant time, with the above-described hash function. Once we have accessed the correct data structure it is a straightforward range query.

6.1   Time and Space

We note that the time and space have interdependencies which we will shortly address. These are affected by the way the dictionary text is saved, by the implementation of the range searching data structures and by the choice of our parameter b. We first explain the space and time and then offer a couple of possible choices of parameters.

Space: Since we only save the skeleton of T, and T^R, their size, including the data saved on the edges and nodes, is $O(d)$ words, or $O(d \log d)$ bits. The perfect hash function table for the data on the edges is also of size $O(d \log d)$ bits.

Hence, the two main space factors are the dictionary size and the range searching data structures. The dictionary itself is only accessed to read substrings. Hence, it can be saved either in open format, in which case the space used will be $|p|d$ bits of space for the dictionary or it can be saved in accessible compressed format. This allows one to analyze the results in either the encoding or indexing model. We give a parameter allowing the user to insert the data structure of their choice. One possible data structure is the following:

Lemma 2. [15] *Given a text T of length t over constant-sized alphabet there exists a compressed data structure that supports the access in constant time of any substring of T of length $O(\log t)$ bits requiring $tH_k(T) + o(t)$, where $H_k(T)$ denotes the kth empirical entropy of T and $k = o(\log t)$.*

Finally the space required by the range searching data structures is dependent on the implementation used which affects the query time as well. To summarize the space: we maintain the dictionary itself in $|DS(\mathcal{D})|$ bit space, where $DS(T)$ is the data structure of choice for maintaining T. The additional space required is $O(d \log d + bS_{\mathrm{rs}}(d))$ bits of space, where $S_{\mathrm{rs}}(d)$ is the number of bits required by the implementation of the range searching data structure on d values.

Query Time: First we read the pattern in $O(|p|/\omega)$ time. We also generate the encoding $c(p)$ in this time. Finally, we find all the ranges within the bidirectional tries within $O(|p|/\omega)$ time. This is true because we make one pass on the trie for each pattern using the hash functions on the nodes and access the dictionary text in parallel which we assume can be done in $O(1)$ for each machine word. While walking down in the tree ω bits at a time, we stop on the nodes which are at depths of multiples of b (bits), that is, after every b/ω characters (of ω-bits each). There we learn the ranges by the data stored in the trie. Hence, all the above is done in $O(|p|/\omega)$ time.

The next phase is the query on the orthogonal data structure. We need to access the orthogonal range data structure maintaining the data for color class $c(p)$. This is accessed by hashing $c(p)$. This is done in $O(|p|/\omega)$ time whilst generating $c(p)$. Finally the query itself is a tradeoff based on the implementation of the range searching data structure in use. Let us denote with $Q_{\mathrm{rs}}(d) + occ\, Q'_{\mathrm{rs}}(d)$ the query cost of the $2D$ orthogonal range reporting of our choice. Hence, the overall query time is:

$$O(|p|/\omega + (|p|/b)Q_{\mathrm{rs}}(d) + occ\, Q'_{\mathrm{rs}}(d)).$$

The current best results on 2D orthogonal range reporting in the word RAM model are due to Chan, Larsen, and Pătraşcu [11]. One possible choice is $S_{\mathrm{rs}}(d) = O(d)$, $Q_{\mathrm{rs}}(d) = O(\log^\epsilon d)$, and $Q'_{\mathrm{rs}}(d) = O(1)$. In this case we have a solution for lookups with 1 mismatch in the binary alphabet setting which requires $O(bd \log d)$ additional bits of space and $O(|p|/w + (|p|/b) \log^\epsilon d + occ)$ query time. Set $b = w \log^\epsilon d$, and we have optimal query time with $O(dw \log^{1+\epsilon} d)$-bit space. We note that if the string lengths $= |p|$ are $\leq w \log^\epsilon d$ then one can use the data structure of [8].

Another option is as follows. Assume that the additional bits of space is bounded by $c(bd \log d)$, for some constant c. We can set $b = \delta|p|/c \log d$ for arbitrary small constant δ and get $\delta|p|d$ bits and $O((1/\delta) \log^{1+\epsilon} d + |p|/w + occ)$ query time. This answers the open question of Belazzougui and Venturini [4] in the uncompressed setting for sufficiently large $|p| \gg w \log^{1+\epsilon} d$.

Another range searching alternative [11] has $S_{rs}(d) = O(d \log \log d)$ and $Q_{rs}(d) = Q'_{rs}(d) = O(\log \log d)$. This gives $O(|p|d)$ bits and $O(\log d \log^2 \log d + |p|/w + occ \log \log d)$ time.

7   Dictionary Lookup with Edit Distance One

We would like to extend the previous idea of using the xor function to the case of one insertion or deletion. However, an insertion or deletion can skew the entire b-bit encoding and make the dictionary string encodings and the pattern encodings incompatible. Hence, we do something slightly different. We still maintain the idea of the xor encoding, and we generate range queries for them, but we do it differently.

For every $u \in \{0,1\}^{b+1}$ and $v \in \{0,1\}^b$, define the subset:

$\mathcal{D}(u,v) = \{s \in \mathcal{D} : c(s) \oplus v$ can be obtained by deleting 1 character from $u\}$.

Now build a 2D orthogonal range searching structure for $\{(x(s), y(s)) \mid s \in \mathcal{D}(u,v)\}$, where $x(s)$ and $y(s)$ are defined as before.

Note that each $s \in \mathcal{D}$ belongs to $O(b2^b)$ $\mathcal{D}(u,v)$'s (because there are 2^b choices for v and $O(b)$ ways to insert 1 character to $c(s) \oplus v$). So, space blows up by a factor $O(b2^b)$.

7.1   Query Algorithm for One Character Deletion from p

Once again we use a trie for the lexicographically sorted strings of $\mathcal{D}$ and a trie for the sorted reversed strings of $\mathcal{D}$. We traverse both similarly to the one mismatch case.

At the i-th iteration, write p as $\alpha_i p_i \beta_i$ with $|\alpha_i| = bi, |p_i| = b + 1, |\beta_i| = b(|s|/b - i - 1)$. We are looking for all $s \in \mathcal{D}$ such that

1. s has prefix α_i and
2. s has suffix β_i and
3. s_i can be obtained by deleting 1 character from p_i.

We make a claim here that is appropriate for the case of one deletion from the pattern.

Lemma 3. *Given that s has prefix α_i and suffix β_i, s_i can be obtained by deleting 1 character from $p_i \iff s \in \mathcal{D}(p_i, c(\alpha_i) \oplus c(\beta_i))$.*

Proof. Given that s has prefix α_i and suffix β_i, we have $c(s) = s_i \oplus c(\alpha_i) \oplus c(\beta_i)$ which directly implies that $s_i = c(s) \oplus c(\alpha_i) \oplus c(\beta_i)$. Set $u = p_i$ and $v = c(\alpha_i) \oplus c(\beta_i)$. Hence, $s_i(= c(s) \oplus v)$ can be obtained by deleting 1 character from $p_i(= u)$ is equivalent by definition to $s \in \mathcal{D}(u, v) = \mathcal{D}(p_i, c(\alpha_i) \oplus c(\beta_i))$. $\square$

We use the same technique on the trie and reverse trie as for the one mismatch case. That is, we have a range in the trie of the dictionary that is appropriate to α_i and a range defined by β_i in the trie of reversed strings. During the traversal of the query we compute $c(\alpha_i) \oplus c(\beta_i)$ at every stage. Now we need to access the orthogonal range reporting structure that is $\mathcal{D}(p_i, c(\alpha_i) \oplus c(\beta_i))$ which is accessible in constant time by a hash based on u and v to $\mathcal{D}(u, v)$, which in our case is p_i and $c(\alpha_i) \oplus c(\beta_i)$ to $\mathcal{D}(p_i, c(\alpha_i) \oplus c(\beta_i))$. Once in the right orthogonal range reporting data structure we ask a 4-sided query based on the ranges we found.

Time and Space Analysis. By following an analysis similar to the mismatch case, we can conclude the following.

The space and time analysis is: $O(b2^b d \log d)$ additional bits of space (over the compressed or uncompressed dictionary) and $O(|p|/\omega + (|p|/b) \log^\epsilon d + occ)$ query time.

We can set $b = \log(\delta|p|/\log d \log(\delta|p|))$ for an arbitrary constant $\delta > 0$ and get $\delta|p|d$ bits and $O(((\delta|p|)/\log|p|) \log^\epsilon d + occ)$ query time for $|p| \gg \log d$, which is a speedup when $|p|$ is large.

Alternatively, we can get $O((|p|/\log|p|) \log \log d + occ \log \log d)$.

7.2   Query Algorithm for Inserting One Character to p

The case for insertion to p is symmetrical to the deletion case. Hence, we only give the changed definition of $\mathcal{D}(u, v)$.

For every $u \in \{0, 1\}^{b-1}$ and $v \in \{0, 1\}^b$, redefine the subset $\mathcal{D}(u, v) = \{s \in \mathcal{D} \mid c(s) \oplus v$ can be obtained by inserting 1 character to $u\}$.

References

1. Amir, A., Keselman, D., Landau, G.M., Lewenstein, M., Lewenstein, N., Rodeh, M.: Text indexing and dictionary matching with one error. J. Algorithms **37**(2), 309–325 (2000)
2. Amir, A., Levy, A., Porat, E., Shalom, B.R.: Dictionary matching with one gap. In: Kulikov, A.S., Kuznetsov, S.O., Pevzner, P. (eds.) CPM 2014. LNCS, vol. 8486, pp. 11–20. Springer, Heidelberg (2014)
3. Belazzougui, D.: Faster and space-optimal edit distance "1" dictionary. In: Kucherov, G., Ukkonen, E. (eds.) CPM 2009 Lille. LNCS, vol. 5577, pp. 154–167. Springer, Heidelberg (2009)
4. Belazzougui, D., Venturini, R.: Compressed string dictionary look-up with edit distance one. In: Kärkkäinen, J., Stoye, J. (eds.) CPM 2012. LNCS, vol. 7354, pp. 280–292. Springer, Heidelberg (2012)

5. Bille, P., Gørtz, I.L., Vildhøj, H.W., Vind, S.: String indexing for patterns with wildcards. Theory Comput. Syst. **55**(1), 41–60 (2014)
6. Brodal, G.S., Davoodi, P., Rao, S.S.: On space efficient two dimensional range minimum data structures. Algorithmica **63**(4), 815–830 (2012)
7. Brodal, G.S., Gasieniec, L.: Approximate dictionary queries. In: Hirschberg, D.S., Meyers, G. (eds.) CPM 1996. LNCS, vol. 1075, pp. 65–74. Springer, Heidelberg (1996)
8. Brodal, G.S., Venkatesh, S.: Improved bounds for dictionary look-up with one error. Inf. Process. Lett. **75**(1–2), 57–59 (2000)
9. Chan, H., Lam, T.W., Sung, W., Tam, S., Wong, S.: Compressed indexes for approximate string matching. Algorithmica **58**(2), 263–281 (2010)
10. Chan, H.-L., Lam, T.-W., Sung, W.-K., Tam, S.-L., Wong, S.-S.: A linear size index for approximate pattern matching. J. Discrete Algorithms **9**(4), 358–364 (2011)
11. Chan, T.M., Larsen, K.G., Pătraşcu, M.: Orthogonal range searching on the RAM, revisited. In: Proceedings of the 27th ACM Symposium on Computational Geometry, Paris, France, June 13–15, 2011, pp. 1–10 (2011)
12. Cole, R., Gottlieb, L.-A., Lewenstein, M.: Dictionary matching and indexing with errors and don't cares. In: Proceedings of Symposium on Theory of Computing (STOC), pp. 91–100 (2004)
13. Demaine, E.D., López-Ortiz, A.: A linear lower bound on index size for text retrieval. J. Algorithms **48**(1), 2–15 (2003)
14. Ferragina, P., Muthukrishnan, S., de Berg, M.: Multi-method dispatching: a geometric approach with applications to string matching problems. In: Proceedings of Symposium on Theory of Computing (STOC), pp. 483–491 (1999)
15. Ferragina, P., Venturini, R.: A simple storage scheme for strings achieving entropy bounds. Theor. Comput. Sci. **372**(1), 115–121 (2007)
16. Hon, W.-K., Ku, T.-H., Shah, R., Thankachan, S.V., Vitter, J.S.: Compressed dictionary matching with one error. In: Data Compression Conference (DCC), pp.13–122 (2011)
17. Lam, T.-W., Sung, W.-K., Wong, S.-S.: Improved approximate string matching using compressed suffix data structures. Algorithmica **51**(3), 298–314 (2008)
18. Levenshtein, V.I.: Binary codes capable of correcting deletions, insertions, and reversals. Sov. Phys. Doklady **10**, 707–710 (1966)
19. Lewenstein, M.: Orthogonal range searching for text indexing. In: Brodnik, A., López-Ortiz, A., Raman, V., Viola, A. (eds.) Ianfest-66. LNCS, vol. 8066, pp. 267–302. Springer, Heidelberg (2013)
20. Lewenstein, M., Munro, J.I., Nekrich, Y., Thankachan, S.V.: Document retrieval with one wildcard. In: Csuhaj-Varjú, E., Dietzfelbinger, M., Ésik, Z. (eds.) MFCS 2014, Part II. LNCS, vol. 8635, pp. 529–540. Springer, Heidelberg (2014)
21. Lewenstein, M., Nekrich, Y., Vitter, J.S.: Space-efficient string indexing for wildcard pattern matching. In: 31st International Symposium on Theoretical Aspects of Computer Science (STACS 2014), pp. 506–517 (2014)
22. Policriti, A., Prezza, N.: Hashing and indexing: succinct data structures and smoothed analysis. In: Ahn, H.-K., Shin, C.-S. (eds.) ISAAC 2014. LNCS, vol. 8889, pp. 157–168. Springer, Heidelberg (2014)
23. Tsur, D.: Fast index for approximate string matching. J. Discrete Algorithms **8**(4), 339–345 (2010)
24. Yao, A.C., Yao, F.F.: Dictionary look-up with one error. J. Algorithms **25**(1), 194–202 (1997)

On the Readability of Overlap Digraphs

Rayan Chikhi[1,2], Paul Medvedev[2(✉)], Martin Milanič[3],
and Sofya Raskhodnikova[2]

[1] CNRS, UMR 9189, Lille, France
[2] The Pennsylvania State University, State College, USA
paul.medvedev@psu.edu
[3] University of Primorska, Koper, Slovenia

Abstract. We introduce the graph parameter *readability* and study it
as a function of the number of vertices in a graph. Given a digraph D,
an injective overlap labeling assigns a unique string to each vertex such
that there is an arc from x to y if and only if x properly overlaps y. The
readability of D is the minimum string length for which an injective overlap labeling exists. In applications that utilize overlap digraphs (e.g., in
bioinformatics), readability reflects the length of the strings from which
the overlap digraph is constructed. We study the asymptotic behaviour
of readability by casting it in purely graph theoretic terms (without any
reference to strings). We prove upper and lower bounds on readability
for certain graph families and general graphs.

1 Introduction

In this paper, we introduce and study a graph parameter called readability,
motivated by applications of overlap graphs in bioinformatics. A string x *overlaps*
a string y if there is a suffix of x that is equal to a prefix of y. They overlap
properly if, in addition, the suffix and prefix are both proper. The *overlap digraph*
of a set of strings S is a digraph where each string is a vertex and there is an arc
from x to y (possibly with $x = y$) if and only if x properly overlaps y. Walks in
the overlap digraph of S represent strings that can be spelled by stitching strings
of S together, using the overlaps between them. Overlap digraphs have various
applications, e.g., they are used by approximation algorithms for the Shortest
Superstring Problem [Swe00]. Their most impactful application, however, has
been in bioinformatics. Their variants, such as de Bruijn graphs [IW95] and
string graphs [Mye05], have formed the basis of nearly all genome assemblers
used today (see [MKS10,NP13] for a survey), successful despite results showing
that assembly is a hard problem in theory [BBT13,NP09,MGMB07]. In this
context, the strings of S represent known fragments of the genome (called *reads*),
and the genome is represented by walks in the overlap digraph of S. However,
do the overlap digraphs generated in this way capture all possible digraphs, or
do they have any properties or structure that can be exploited?

Braga and Meidanis [BM02] showed that overlap digraphs capture all possible
digraphs, i.e., for every digraph D, there exists a set of strings S such that their

F. Cicalese et al. (Eds.): CPM 2015, LNCS 9133, pp. 124–137, 2015.
DOI: 10.1007/978-3-319-19929-0_11

overlap digraph is D. Their proof takes an arbitrary digraph and shows how to construct an *injective overlap labeling*, that is, a function assigning a unique string to each vertex, such that (x, y) is an arc if and only if the string assigned to x properly overlaps the string assigned to y. However, the *length* of strings produced by their method can be exponential in the number of vertices. In the bioinformatics context, this is unrealistic, as the read size is typically much smaller than the number of reads.

To investigate the relationship between the string length and the number of vertices, we introduce a graph parameter called *readability*. The readability of a digraph D, denoted $r(D)$, is the smallest nonnegative integer r such that there exists an injective overlap labeling of D with strings of length r. The result by [BM02] shows that readability is well defined and is at most $2^{\Delta+1} - 1$, where Δ is the maximum of the in- and out-degrees of vertices in D. However, nothing else is known about the parameter, though there are papers that look at related notions [BFK+02, BFKK02, BHKdW99, GP14, LZ07, LZ10, PSW03, TU88].

In this paper, we study the asymptotic behaviour of readability as a function of the number of vertices in a graph. We define readability for undirected bipartite graphs and show that the two definitions of readability are asymptotically equivalent. We capture readability using purely graph theoretic parameters (i.e., without any reference to strings). For trees, we give a parameter that characterizes readability exactly. For the larger family of bipartite C_4-free graphs, we give a parameter that approximates readability to within a factor of 2. Finally, for general bipartite graphs, we give a parameter that is bounded on the same sets of graphs as readability.

We apply our purely graph theoretic interpretation to prove readability upper and lower bounds on several graph families. We show, using a counting argument, that almost all digraphs and bipartite graphs have readability of at least $\Omega(n/\log n)$. Next, we construct a graph family inspired by Hadamard codes and prove that it has readability $\Omega(n)$. Finally, we show that the readability of trees is bounded from above by their radius, and there exist trees of arbitrary readability that achieve this bound.

2   Preliminaries

General Definitions and Notation. Let x be a string. We denote the length of x by $|x|$. We use $x[i]$ to refer to the ith character of x, and denote by $x[i..j]$ the substring of x from the ith to the jth character, inclusive. We let $\text{pre}_i(x)$ denote the prefix $x[1..i]$ of x, and we let $\text{suf}_i(x)$ denote the suffix $x[|x| - i + 1..|x|]$. Let y be another string. We denote by $x \cdot y$ the concatenation of x and y. We say that x *overlaps* y if there exists an i with $1 \leq i \leq \min\{|x|, |y|\}$ such that $\text{suf}_i(x) = \text{pre}_i(y)$. In this case, we say that x overlaps y by i. If $i < \min\{|x|, |y|\}$, then we call the overlap *proper*. Define $\text{ov}(x, y)$ as the minimum i such that x overlaps y by i, or 0 if x does not overlap y. For a positive integer n, we denote by $[n]$ the set $\{1, \ldots, n\}$.

We refer to finite simple undirected graphs simply as graphs and to finite directed graphs without parallel arcs in the same direction as digraphs. For a vertex v in a graph, we denote the set of neighbors of v by $N(v)$. A *biclique* is a complete bipartite graph. Note that the one-vertex graph is a biclique (with one of the parts of its bipartition being empty). Two vertices u, v in a graph are called *twins* if they have the same neighbors, i.e., if $N(u) = N(v)$. If, in addition, $N(u) = N(v) \neq \emptyset$, vertices u, v are called *non-isolated twins*. A *matching* is a graph of maximum degree at most 1, though we will sometimes slightly abuse the terminology and not distinguish between matchings and their edge sets. A cycle (respectively, path) on i vertices is denoted by C_i (respectively, P_i). For graph terms not defined here, see, e.g., [BM08].

Readability of Digraphs. A *labeling* ℓ of a graph or digraph is a function assigning a string to each vertex such that all strings have the same length, denoted by $len(\ell)$. We define $\text{ov}_\ell(u, v) = \text{ov}(\ell(u), \ell(v))$. An *overlap labeling* of a digraph $D = (V, A)$ is a labeling ℓ such that $(u, v) \in A$ if and only if $0 < \text{ov}_\ell(u, v)) < len(\ell)$. An overlap labeling is said to be *injective* if it does not generate duplicate strings. Recall that the readability of a digraph D, denoted $r(D)$, is the smallest nonnegative integer r such that there exists an injective overlap labeling of D of length r. We note that in our definition of readability we do not place any restrictions on the alphabet size. Braga and Meidanis [BM02] gave a reduction from an overlap labeling of length ℓ over an arbitrary alphabet Σ to an overlap labeling of length $\ell \log |\Sigma|$ over the binary alphabet.

Readability of Bipartite Graphs. We also define a modified notion of readability that applies to balanced bipartite graphs as opposed to digraphs. We found that readability on balanced bipartite graphs is simpler to study but is asymptotically equivalent to readability on digraphs. Let $G = (V, E)$ be a bipartite graph with a given bipartition of its vertex set $V(G) = V_s \cup V_p$. (We also use the notation $G = (V_s, V_p, E)$). We say that G is *balanced* if $|V_s| = |V_p|$. An *overlap labeling of G* is a labeling ℓ of G such that for all $u \in V_s$ and $v \in V_p$, $(u, v) \in E$ if and only if $\text{ov}_\ell(u, v) > 0$. In other words, overlaps are exclusively between the suffix of a string assigned to a vertex in V_s and the prefix of a string assigned to a vertex in V_p. The *readability* of G is the smallest nonnegative integer r such that there exists an overlap labeling of G of length r. Note that we do not require injectivity of the labeling, nor do we require the overlaps to be proper. As before, we use $r(G)$ to denote the readability of G.

We note that in our definition of readability we do not place any restrictions on the alphabet size. Braga and Meidanis [BM02] gave a reduction from an overlap labeling of length ℓ over an arbitrary alphabet Σ to an overlap labeling of length $\ell \log |\Sigma|$ over the binary alphabet.

For a labeling ℓ, we define $inner_i(\ell(v)) = \text{suf}_i(\ell(v))$ if $v \in V_s$ and $inner_i(\ell(v)) = \text{pre}_i(\ell(v))$ if $v \in V_p$. Similarly, we define $outer_i(\ell(v)) = \text{pre}_i(\ell(v))$ if $v \in V_s$ and $outer_i(\ell(v)) = \text{suf}_i(\ell(v))$ if $v \in V_p$.

Let $\mathcal{B}_{n \times n}$ be the set of balanced bipartite graphs with nodes $[n]$ in each part, and let $\mathcal{D}_n$ be the set of all digraphs with nodes $[n]$. The readabilities of digraphs and of bipartite graphs are connected by the following theorem, which implies that they are asymptotically equivalent.

Theorem 1. *There exists a bijection* $\psi : \mathcal{B}_{n \times n} \to \mathcal{D}_n$ *with the property that for any* $G \in \mathcal{B}_{n \times n}$ *and* $D \in \mathcal{D}_n$, *such that* $D = \psi(G)$, *we have that* $r(G) < r(D) \leq 2 \cdot r(G) + 1$.

As a result, we can study readability of balanced bipartite graphs, without asymptotically affecting our bounds. For example, we show in Sect. 4.2 (in Theorem 6) that there exists a family of balanced bipartite graphs with readability $\Omega(n)$, which leads to the existence of digraphs with readability $\Omega(n)$.

3   Graph Theoretic Characterizations

In this section, we relate readability of balanced bipartite graphs to several purely graph theoretic parameters, without reference to strings.

3.1   Trees and C_4-free Graphs

For trees, we give an exact characterization of readability, while for C_4-free graphs, we give a parameter that is a 2-approximation to readability. A *decomposition of size k* of a bipartite graph $G = (V_s, V_p, E)$ is a function on the edges of the form $w : E \to [k]$. Note that a labeling ℓ of G implies a decomposition of G, defined by $w(e) = \text{ov}_\ell(e)$ for all $e \in E$. We call this the ℓ-decomposition. We say that a labeling ℓ of G *achieves w* if it is an overlap labeling and w is the ℓ-decomposition. Note that we can express readability as

$$r(G) = \min\{k \mid w \text{ is a decomposition of size } k , \exists \text{ a labeling } \ell \text{ that achieves } w\}.$$

Our goal is to characterize in graph theoretic terms the properties of w which are satisfied if and only if w is the ℓ-decomposition, for some ℓ. While this proves challenging in general, we can achieve this for trees using a condition which we call the P_4-rule. We say that w satisfies the P_4-rule if for every induced four-vertex path $P = (e_1, e_2, e_3)$ in G, the following condition holds: if $w(e_2) = \max\{w(e_1), w(e_2), w(e_3)\}$, then $w(e_2) \geq w(e_1) + w(e_3)$. We will prove:

Theorem 2. *Let T be a tree. Then $r(T) = \min\{k \mid w$ is a decomposition of size k that satisfies the P_4-rule$\}$.*

Note that for cycles, the equality does not hold. For example, consider the decomposition w of C_6 given by the weights $2, 4, 2, 2, 3, 1$. This decomposition satisfies the P_4 rule but it can be shown using case analysis that there does not exist a labeling ℓ achieving w.

However, we can give a characterization of readability for C_4-free graphs in terms of a parameter that is asymptotically equivalent to readability, using a condition which we call the strict P_4-rule. The strict P_4-rule is identical to the P_4-rule accept that the inequality becomes strict. That is, w satisfies the *strict P_4-rule* if for every induced four-vertex path $P = (e_1, e_2, e_3)$, if $w(e_2) = \max\{w(e_1), w(e_2), w(e_3)\}$, then $w(e_2) > w(e_1) + w(e_3)$. Note that a decomposition that satisfies the strict P_4-rule automatically satisfies the P_4-rule, but not vice-versa. We will prove:

Theorem 3. *Let G be a C_4-free bipartite graph. Let $t = \min\{k \mid w$ is a decomposition of size k that satisfies the strict P_4-rule$\}$. Then $t/2 < r(G) \leq t$.*

We note that this characterization cannot be extended to graphs with a C_4. The example in Fig. 1a shows a graph with a decomposition which satisfies the strict P_4-rule but it can be shown using case analysis that there does not exists a labeling ℓ achieving this decomposition.

In the remainder of this section, we will prove these two theorems. We first show that an ℓ-decomposition satisfies the P_4-rule (proof in the full version).

Lemma 1. *Let ℓ be an overlap labeling of a bipartite graph G. Then the ℓ-decomposition satisfies the P_4-rule.*

Now, consider a C_4-free bipartite graph $G = (V_s, V_p, E)$ and let w be a decomposition satisfying the P_4-rule. We will prove both Theorems 2 and 3 by constructing the following labeling. Let us order the edges $e_1, \ldots, e_{|E|}$ in order of non-decreasing weight. For $0 \leq j \leq |E|$, we define the graph $G^j = (V_s, V_p, \{e_i \in E \mid i \leq j\})$. For a vertex u, define $len_j(u) = \max\{w(e_i) \mid i \leq j, e_i$ is incident with $u\}$, if the degree of u in G^j is positive, and 0 otherwise. We will recursively define a labeling ℓ_j of G^j such that $|\ell_j(u)| = len_j(u)$ for all u. The initial labeling ℓ_0 assigns ϵ to every vertex. Suppose we have a labeling ℓ_j for G^j, and $e_{j+1} = (u, v)$. Recall that because w satisfies the P_4-rule and G is C_4-free, $w(u, v) \geq len_j(u) + len_j(v) = |\ell_j(u)| + |\ell_j(v)|$. (Note that the inequality holds also in the case when one of the two summands is 0.) Let A be a (possibly empty) string of length $w(u, v) - |\ell_j(u)| - |\ell_j(v)|$ composed of non-repeating characters that do not exist in ℓ_j. Define ℓ_{j+1} as $\ell_{j+1}(x) = \ell_j(x)$ for all $x \notin \{u, v\}$, and $\ell_{j+1}(u) = \ell_{j+1}(v) = \ell_j(v) \cdot A \cdot \ell_j(u)$. We denote the labeling of G as $\ell = \ell_{|E|}$. We will slightly abuse notation in this section, ignoring the fact that a labeling must have labels of the same length. This is inconsequential, because strings can always be padded from the beginning or end with distinct characters without affecting any overlaps.

First, we state a useful Lemma, that two vertices share a character in the labeling only if they are connected by a path (proof in the full version).

Lemma 2. *Let c be a character that is contained in $\ell_j(u)$ and in $\ell_j(v)$, for some pair of distinct vertices. Then there exists a path between u and v in G^j.*

We are now ready to show that ℓ achieves w for trees, and, if w also satisfies the strict P_4-rule, for C_4-free graphs.

Lemma 3. *Let G be a C_4-free bipartite graph and let w be a decomposition that satisfies the P_4-rule. Then the above defined labeling ℓ achieves w if w satisfies the strict P_4-rule or if G is acyclic.*

Proof. We prove by induction on j that ℓ_j achieves w on G^j. Suppose that the Lemma holds for ℓ_j and consider the effect of adding $e_{j+1} = (u, v)$. Notice that to obtain ℓ_{j+1} we only change labels by adding outer characters, hence, any two vertices that overlap by i in ℓ_j will also overlap by i in ℓ_{j+1}. Moreover, only the labels of u and v are changed, and an overlap between u and v of length $w(u, v)$ is created. It remains to show that no shorter overlap is created between u and v and that no new overlap is created involving u or v, except the one between u and v.

First, consider the case when $w(u, v) > |\ell_j(u)| + |\ell_j(v)|$ and so the middle string (A) of the new labels is non-empty. Because the characters of A do not appear in ℓ_j, we do not create any new overlaps except besides the one between u and v and the only overlap between u and v must be of length $w(u, v)$ since the characters of A must align. Thus ℓ_{j+1} achieves w on G^{j+1}.

Next, consider the case when $w(u, v) = |\ell_j(v)|$ (the case when $w(u, v) = |\ell_j(u)|$ is symmetric). In this case, $A = \epsilon$, $\ell_j(u) = \epsilon$, and $|\ell_j(v)| > 0$ (since $w(u, v) > 0$). Suppose for the sake of contradiction that there exists a vertex $v' \neq v$ such that (u, v') is not an edge but $inner_k(\ell_{j+1}(u)) = inner_k(\ell_{j+1}(v'))$, for some $0 < k \leq w(u, v)$. We know, from the construction of ℓ_j, that there exists a vertex u' such that $w(u', v) = |\ell_j(v)|$. We then have $inner_k(\ell_j(u')) = outer_k(\ell_j(v)) = inner_k(\ell_{j+1}(u)) = inner_k(\ell_{j+1}(v')) = inner_k(\ell_j(v'))$. By the induction hypothesis, there is an edge (u', v') and $w(u', v') \leq k$. The edges $(u, v), (v, u'), (u', v')$ form a P_4, which is also induced because G is C_4-free. Because $w(u, v) = w(u', v) \geq w(u', v') > 0$, the P_4-rule is violated, a contradiction. Therefore no new overlaps are created involving u. To show that there are no overlaps from u to v smaller than $w(u, v)$, observe that any such overlap would also be an overlap between u' and v that is smaller than $w(u', v)$, contradicting the induction hypothesis. Therefore, ℓ_{j+1} achieves w on G^{j+1}.

It remains to consider the case when $w(u, v) = |\ell_j(u)| + |\ell_j(v)|$ and $\ell_j(u) \neq \epsilon \neq \ell_j(v)$. We first show that this case cannot arise if w satisfies the strict P_4-rule. There must exist edges in G^j of weights $|\ell_j(u)|$ and $|\ell_j(v)|$ incident with u and v, respectively. These edges, together with (u, v) in the middle, form a P_4, which must be induced since G does not contain a C_4. Furthermore, (u, v) achieves the maximum weight. The strict P_4-rule implies $w(u, v) > |\ell_j(u)| + |\ell_j(v)|$, a contradiction.

Now, assume that G is acyclic, and suppose for the sake of contradiction that the new labeling creates an overlap between v and a vertex $u' \neq u$ (the case of an overlap between u and $v' \neq v$ is symmetric). Consider the character c at position $|\ell_j(v)| + 1$ of $\ell_{j+1}(v)$. The length of the overlap between $\ell_{j+1}(v)$ and $\ell_{j+1}(u') = \ell_j(u')$ must be greater than $|\ell_j(v)|$, otherwise it would have been an overlap in ℓ_j. Thus, $\ell_j(u')$ must contain c. By construction of v's new label, $\ell_j(u)$ must also contain c. Applying Lemma 2, there must be a path between u' and u in G^j. On the other hand, the overlap between v and u' spans $(\ell_j(v))[1]$, and hence $\ell_j(v)$ and $\ell_j(u')$ must share a character. Applying Lemma 2, there must exist a path between u' and v in G^j. Consequently, there exists a path from u to v in G^j. Combining this path with $e_{j+1} = (u, v)$, we get a cycle in G^{j+1}, which is a contradiction.

Finally suppose, for the sake of contradiction, that $\ell_{j+1}(u)$ overlaps $\ell_{j+1}(v)$ by some $k < w(u,v)$. By the induction hypothesis, $k > |\ell_j(v)|$. Consider the last character c of $\ell_j(v)$. It must also appear as the inner position $i = k - |\ell_j(v)| + 1$ in $\ell_{j+1}(u)$. Since $k \le w(u,v) - 1$, we have $i \le w(u,v) - |\ell_j(v)| = |\ell_j(u)|$, and the ith inner position in $\ell_{j+1}(u)$ is also the ith inner position in $\ell_j(u)$. Applying Lemma 2 to c in $\ell_j(v)$ and $\ell_j(u)$, there must exist a path between u and v in G^j. Combining this path with $e_{j+1} = (u,v)$, we get a cycle in G^{j+1}, which is a contradiction. □

We can now prove Theorems 2 and 3.

Proof of Theorem 2. Let $t = \min\{k \mid w$ is a decomposition of size k that satisfies the P_4-rule$\}$. First, let w be a decomposition of size t satisfying the P_4-rule. Lemma 3 states that the above defined labeling ℓ achieves w and so $r(T) \le \max_e(w_e) = t$. For the other direction, consider an overlap labeling b of T of minimum length. By Lemma 1, the b-decomposition satisfies the P_4-rule. Hence, $r(T) = len(b) \ge t$. □

Proof of Theorem 3. Let w be a decomposition of size t satisfying the strict P_4-rule. By Lemma 3, the above defined labeling ℓ achieves w and so $r(G) \le \max_e(w_e) = t$. On the other hand, let b be an overlap labeling of length $r(G)$. Define $w(e) = 2ov_b(e) - 1$, for all $e \in E(G)$. We claim that w satisfies the strict P_4-rule, which will imply that $t \le \max_e w(e) = 2r(G) - 1$. To see this, let e_1, e_2, e_3 be the edges of an arbitrary induced P_4. Observe that $w(e_2) = \max\{w(e_1), w(e_2), w(e_3)\}$ if and only if $ov_b(e_2) = \max\{ov_b(e_1), ov_b(e_2), ov_b(e_3)\}$. Furthermore, it can be algebraicly verified that if $ov_b(e_2) \ge ov_b(e_1) + ov_b(e_3)$ then $w(e_2) > w(e_1) + w(e_3)$. By Lemma 1, the b-decomposition satisfies the P_4-rule and, therefore, w satisfies the strict P_4-rule. □

3.2   General Graphs

In the previous subsection, we derived graph theoretic characterizations of readability that are exact for trees and approximate for C_4-free bipartite graphs. Unfortunately, for a general graph, it is not clear how to construct an overlap labeling from a decomposition satisfying the P_4-rule (as we did in Lemma 3). In this subsection, we will consider an alternate rule (HUB-rule), which we then use to construct an overlap labeling.

Given $G = (V_s, V_p, E)$ and a decomposition w of size k, we define G_i^w, for $i \in [k]$, as a graph with the same vertices as G and edges given by $E(G_i^w) = \{e \in E \mid w(e) = i\}$. When w is obvious from the context, we will write G_i instead of G_i^w. Observe that the edge sets of $G_1^w, \dots, G_k^w$ form a partition of E. We say that w satisfies the *hierarchical-union-of-bicliques* rule, abbreviated as the *HUB-rule*, if the following conditions hold: (i) for all $i \in [k]$, G_i^w is a disjoint union of bicliques, and (ii) if two distinct vertices u and v are non-isolated twins in G_i^w for some $i \in \{2, \dots, k\}$ then, for all $j \in [i-1]$, u and v are (possibly isolated) twins in G_j^w. An example of a decomposition satisfying the HUB-rule is any $w : E \to [k]$ such that G_1^w is an (arbitrary) disjoint union of bicliques and

$G_2^w, \ldots, G_k^w$ are matchings. We can show that the decomposition implied by any overlap labeling must satisfy the HUB-rule (proof in the full version).

Lemma 4. *Let ℓ be an overlap labeling of a bipartite graph G. Then the ℓ-decomposition satisfies the HUB-rule.*

We define *the HUB number of G* as the minimum size of a decomposition of G that satisfies the HUB-rule, and denote it by $hub(G)$. Observe that a decomposition of a graph into matchings (i.e. each G_i^w is a matching) satisfies the HUB-rule. By König's Line Coloring Theorem, any bipartite graph G can be decomposed into $\Delta(G)$ matchings, where $\Delta(G)$ is the maximum degree of G. Thus, $hub(G) \in [\Delta(G)]$. Clearly, a graph G has $hub(G) = 1$ if and only if G is a disjoint union of bicliques. The HUB number captures readability in the sense that the readability of a graph family is bounded (by a uniform constant independent of the number of vertices) if and only if its HUB number is bounded. This is captured by the following theorem:

Theorem 4. *Let G be a bipartite graph. Then $hub(G) \leq r(G) \leq 2^{hub(G)} - 1$.*

In the remainder of this section, we will prove this theorem. The first inequality directly follows from Lemma 4 because, by definition of readability, there exists an overlap labeling ℓ of length $r(G)$. Then the ℓ-decomposition of G is of size $r(G)$ and satisfies the HUB-rule, implying $hub(G) \leq r(G)$. To prove the second inequality, we will need to show:

Lemma 5. *Let w be a decomposition of size k satisfying the HUB-rule of a bipartite graph G. Then there is an overlap labeling of G of length $2^k - 1$.*

The second inequality of Theorem 4 follows directly by choosing a minimum decomposition satisfying the HUB-rule, in which case $k = hub(G)$. Thus, it only remains to prove Lemma 5.

We now define the labeling t that is used to prove Lemma 5. Our construction of the labeling applies the following operation due to Braga and Meidanis [BM02]. Given two vertices $u \in V_s$ and $v \in V_p$, a labeling t, and a filler character a not used by t, the *BM operation* transforms t by relabeling both u and v with $t(v) \cdot a \cdot t(u)$.

We start by labeling G_1 as follows: each biclique B in G_1 gets assigned a unique character a_B, and each node v in a biclique B gets label $t(v) = a_B$. Next, for $i \in [k-1]$, we iteratively construct a labeling of $G_1 \cup \cdots \cup G_{i+1}$ from a labeling t of $G_1 \cup \cdots \cup G_i$. We show by induction that the constructed labeling has an additional property that all twins in $G_1 \cup \cdots \cup G_{i+1}$ have the same labels and that the length of the labeling is $2^{i+1} - 1$. Observe that the labeling of G_1 satisfies this property.

We choose a unique (not previously used) character a_B for each biclique B of G_{i+1}. If B consists of a single vertex v, then we assign to v the label $a_B \cdot t(v)$ if $v \in V_s$, and $t(v) \cdot a_B$ if $v \in V_p$. Otherwise, since w satisfied the HUB-rule, all vertices in $B \cap V_s$ are twins in $G_1 \cup \cdots \cup G_i$ and, by the induction hypothesis, are assigned the same labels in t. Analogously, t will assign the same labels to all nodes in $B \cap V_p$. Consider an arbitrary edge (u, v) in B. We apply the BM operation with character a_B to (u, v) and assign the resulting label $t(v) \cdot a_B \cdot t(u)$ to all nodes in B. This completes the construction of labeling of $G_1 \cup \cdots \cup G_{i+1}$. Observe that it assigns the same labels to all twins in $G_1 \cup \cdots \cup G_{i+1}$, and that the length is $2^{i+1} - 1$. To complete the proof of Theorem 4, we show in the full version that the final labeling is an overlap labeling of G.

Note that if w is a decomposition into matchings, then our labeling algorithm behaves identically to the Braga-Meidanis (BM) algorithm [BM02]. However, in the case that w is of size $o(\Delta(G))$, our labeling algorithm gives a better bound than BM. For example, for the $n \times n$ biclique, our algorithm gives a labeling of length 1, while BM gives a labeling of length $2^n - 1$.

4   Lower and Upper Bounds on Readability

In this section, we prove several lower and upper bounds on readability, making use of the characterizations of the previous section.

4.1   Almost All Graphs Have Readability $\Omega(n/\log n)$

In this subsection, we show that, in both the bipartite and directed graph models, there exist graphs with readability at least $\Omega(n/\log n)$, and that in fact almost all graphs have at least this readability.

Theorem 5. *Almost all graphs in $\mathcal{B}_{n \times n}$ (and, respectively, $\mathcal{D}_n$) have readability $\Omega(n/\log n)$. When restricted to a constant sized alphabet, almost all graphs in $\mathcal{B}_{n \times n}$ (and, respectively, $\mathcal{D}_n$) have readability $\Omega(n)$.*

Proof (constant sized alphabet case). We prove the lemma by a counting argument. Since there are n^2 pairs of nodes in $[n]^2$ that can form edges in a graph in $\mathcal{B}_{n \times n}$, the size of $\mathcal{B}_{n \times n}$ is 2^{n^2}. Let a be the size of the alphabet. The number of labelings of $2n$ nodes with strings of length s is at most a^{2ns}. In particular, labelings of length $s = n/(3 \log a)$ can generate no more than $a^{2n^2/(3 \log a)} = 2^{2n^2/3}$ bipartite graphs, which is in $o(2^{n^2})$. Consequently, almost all graphs in $\mathcal{B}_{n \times n}$ have readability $\Omega(s) = \Omega(n/\log a) = \Omega(n)$. The proof for $\mathcal{D}_n$ is analogous and is omitted. The proof for variable sized alphabets is given in the full version. $\square$

4.2   Distinctness and a Graph Family with Readability $\Omega(n)$

In this subsection, we will give a technique for proving lower bounds and use it to show a family of graphs with readability $\Omega(n)$. For any two vertices u and v, the *distinctness* of u and v is defined as $DT(u, v) = \max\{|N(u) \setminus N(v)|, |N(v) \setminus N(u)|\}$. The *distinctness* of a bipartite graph G, denoted by $DT(G)$, is defined as the minimum distinctness of any pair of vertices that belong to the same part of the bipartition. The following lemma relates the distinctness and the readability of graphs that are not matchings (for a matching, the readability is 1, provided that it has at least one edge, and 0 otherwise).

Lemma 6. *For every bipartite graph G that is not a matching, $r(G) \geq DT(G) + 1$.*

Proof. By Theorem 4, it suffices to show that $DT(G) \leq hub(G) - 1$. Let $h = hub(G)$, let $w : E(G) \to [h]$ be a minimum decomposition of G satisfying the HUB-rule, and consider the graphs $G_i = G_i^w$, for $i \in [h]$. We need to show that $DT(G) \leq h - 1$. Suppose first that each G_i is a matching. Then, since w is a decomposition of G, we have $\Delta(G) \leq h$. Moreover, since G is not a matching, it has a pair of distinct vertices, say u and v, with a common neighbor, which implies $DT(G) \leq DT(u, v) \leq \Delta(G) - 1 \leq h - 1$.

Suppose now that there exists an index $j \in [h]$ such that G_j is not a matching, and let j be the maximum such index. Then, there exist two distinct vertices in G, say u and v, that have a common neighbor in G_j, and therefore belong to the same biclique of G_j. It follows that u and v are non-isolated twins in G_j. Since w is satisfies the HUB-rule, this implies that u and v are twins in each G_i with $i \in [j - 1]$. Consequently, for each vertex x in G adjacent to u but not to v, the unique G_i with $(u, x) \in E(G_i)$ satisfies $i > j$. By the choice of j, each such G_i is a matching, and hence there can be at most $h - j$ such vertices x. Thus $|N(u) \setminus N(v)| \leq h - j$ and similarly $|N(v) \setminus N(u)| \leq h - j$, which implies the desired inequality $DT(G) \leq DT(u, v) \leq h - j \leq h - 1$. $\square$

While the distinctness is a much simpler graph parameter than the HUB number, simplicity comes with a price. Namely, the distinctness does not share the nice feature of the HUB number, that of being bounded on exactly the same sets of graphs as the readability. In Sect. 4.3, we show the existence of graphs (specifically, trees) of distinctness 1 and of arbitrary large readability.

We now introduce a family of graphs, inspired by the Hadamard error correcting code, and apply Lemma 6 to show that their readability is at least linear in the number of nodes. We define H_k as the bipartite graph with vertex sets $V_s = \{v_s \mid v \in \{0, 1\}^k \setminus \{0^k\}\}$ and $V_p = \{v_p \mid v \in \{0, 1\}^k \setminus \{0^k\}\}$ and edge set

$$E(H_k) = \left\{ (v_s, v_p) \in V_s \times V_p \mid \sum_{i=1}^{k} v_s[i]v_p[i] \equiv 1 \pmod{2} \right\}.$$

In other words, each vertex has a non-zero k-bit codeword vector associated with it and two vertices are adjacent if the inner product of their codewords is odd. Let $n = 2^k$. Graph H_k has $2(n - 1)$ vertices, all of degree $n/2$, and thus $(n - 1)n/2$ edges. Figure 1b illustrates H_3.

In the full version, we show that every pair of vertices in the same part of the bipartition of H_k has exactly $n/4$ common neighbors. This implies that the distinctness of H_k is $n/4$. Combining this with Lemma 6, we obtain the following theorem.

Theorem 6. $r(H_k) \geq n/4 + 1$.

This lower bound also translates to directed graphs: applying Theorem 1, there exists digraphs of readability $\Omega(n)$. A major open question is: Do there exist

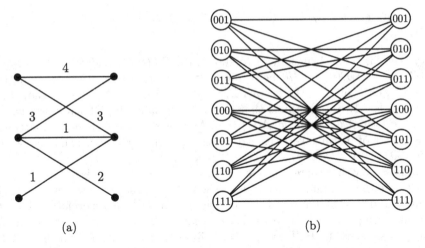

Fig. 1. (a) Illustration that Theorem 3 cannot be extended to graphs with a C_4. Example of a graph and decomposition that satisfies the strict P_4-rule, yet no overlap labeling ℓ exists that achieves it. (b) The graph H_3. The strings on the vertices correspond to the k-bit codeword vectors.

graphs that have exponential readability? We conjecture that they do, and that the graph family H_k has exponential readability. However, since distinctness is $O(n)$, we note that Lemma 6 is insufficient for proving stronger than $\Omega(n)$ lower bounds on the readability.

4.3   Trees

The purely graph theoretic characterization of readability given by Theorem 2 allows us to derive a sharp upper bound on the readability of trees. Recall that the *eccentricity* of a vertex u in a connected graph G is defined as $ecc_G(u) = \max_{v \in V(G)} dist_G(u, v)$, where $dist_G(u, v)$ is the number of edges in a shortest path from u to v. The *radius* of a graph G is defined as the minimum eccentricity of a vertex in G, that is $radius(G) = \min_{u \in V(G)} \max_{v \in V(G)} dist_G(u, v)$.

Theorem 7. *For every tree T, $r(T) \leq radius(T)$, and this bound is sharp. More precisely, for every $k \geq 0$ there exists a tree T such that $r(T) = radius(T) = k$.*

Proof. Let T be a tree. If $T = K_1$ (the one-vertex tree), then $radius(T) = r(T) = 0$ (note that assigning the empty string to the unique vertex of v results in an overlap labeling of T). Now, let T be of radius $r \geq 1$ and let $v \in V(T)$ be a vertex of T of minimum eccentricity (that is, $ecc_T(v) = r$). Consider the distance levels of T from v, that is, $V_i = \{w \in V(T) \mid dist_T(v, w) = i\}$ for $i \in \{0, 1, \ldots, r\}$. Also, for all $i \in [r]$, let E_i be the set of edges in T connecting a vertex in V_{i-1} with a vertex in V_i. Then $\{E_1, \ldots, E_r\}$ is a partition of $E(T)$ and the decomposition $w : E(T) \to [r]$ given by $w(e) = i$ if and only if $e \in E_i$ is well defined. We claim that w satisfies the P_4-rule. Let $P = (v_1, v_2, v_3, v_4)$ be an induced P_4 in T, and let $i = w(v_1, v_2)$, $j = w(v_2, v_3)$, $k = w(v_3, v_4)$. Suppose that $j = \max\{i, j, k\}$. We may assume without loss of generality that $v_2 \in V_{j-1}$ and $v_3 \in V_j$. Since T is a tree, v_2 is the only neighbor of v_3 in V_{j-1}, which implies that $v_4 \in V_{j+1}$ and consequently $k = j + 1$, contrary to the assumption $j = \max\{i, j, k\}$. Thus, the P_4-rule is trivially satisfied for w. By Theorem 2, we have $r(T) \leq \max_{e \in E(T)} w(e) = r = radius(T)$.

To show that for every $k \geq 0$ there exists a tree T with $r(T) = radius(T) = k$, we proceed by induction. We will construct a sequence $\{(T_i, v_i)\}_{i \geq 0}$ where T_i is a tree, v_i is a vertex in T_i with $ecc_{T_i}(v_i) \leq i$, the degree of v_i in T_i is i, and $r(T_i) = radius(T_i) = i$. For $i = 0$, take $(T_0, v_0) = (K_1, v_0)$ where v_0 is the unique vertex of K_1. This clearly has the desired properties. For $i \geq 1$, take i disjoint copies of (T_{i-1}, v_{i-1}), say (T_{i-1}^j, v_{i-1}^j) for $j \in [i]$, add a new vertex v_i, and join v_i by an edge to each v_{i-1}^j for $j \in [i]$. Let T_i be the so constructed tree. Clearly, the degree of v_i in T_i is i, and $ecc_{T_i}(v_i) \leq 1 + ecc_{T_i}(v_{i-1}) \leq 1 + (i - 1) = i$, which implies that $radius(T_i) \leq i$. On the other hand, we will show that $r(T_i) \geq i$, which together with inequality $r(T_i) \leq radius(T_i)$ will imply the desired conclusion $radius(T_i) = r(T_i) = i$. Suppose for a contradiction that $r(T_i) < i$. Then, by Lemma 1, there exists a decomposition w of T_i of size $i - 1$ satisfying the P_4-rule. In particular, this implies $i \geq 2$. Since the degree of v_i in T_i is i, there exist two edges incident with v_i, say (v_i, v_{i-1}^j) and (v_i, v_{i-1}^k) for some $j \neq k$ such that $w(v_i, v_{i-1}^j) = w(v_i, v_{i-1}^k)$. Let w_1 denote this common value. Let x be a neighbor of v_{i-1}^j in T_{i-1}^j. (Note that x exists since v_{i-1}^j is of degree $i - 1 \geq 1$ in T_{i-1}^j.) Then, $(x, v_{i-1}^j, v_i, v_{i-1}^k)$ is an induced P_4 in T_i. We claim that $w(x, v_{i-1}^j) > w_1$. Indeed, if $w(x, v_{i-1}^j) \leq w_1$ then we have $\max\{w(x, v_{i-1}^j), w(v_{i-1}^j, v_i), w(v_i, v_{i-1}^k)\} = \max\{w(x, v_{i-1}^j), w_1, w_1\} = w_1$, while $w_1 \not\geq w_1 + w(x, v_{i-1}^j)$, contrary to the P_4-rule. Since x was an arbitrary neighbor of v_{i-1}^j in T_{i-1}^j, we infer that every edge e in T_{i-1}^j incident with v_{i-1}^j satisfies $w(e) > w_1$. In particular, this leaves a set of at most $i - 2$ different values that can appear on these $i-1$ edges (the value w_1 is excluded), and hence again there must be two edges of the same weight, say w_2. Clearly, $w_2 > w_1$ and $i > 2$. Proceeding inductively, we construct a sequence of edges $e_1, e_2, \ldots, e_i$ forming a path in T_i from v_i to a leaf and satisfying $w_1 < w_2 < \ldots < w_i$, where $w_i = w(e_i)$. This implies that all the weights $w_1, \ldots, w_i$ are distinct, contrary to the fact that the range of w is contained in the set $[i - 1]$. This contradiction shows that $r(T_i) \geq i$ and completes the proof. $\square$

Note that for every $k \geq 2$, the tree T_k of radius k constructed in the proof of Theorem 2 has a pair of leaves in the same part of the bipartition and is therefore of distinctness 1. This shows that the readability of a graph cannot be upper-bounded by any function of its distinctness (cf. Lemma 6).

5   Conclusion

In this paper, we define a graph parameter called readability, and initiate a study of its asymptotic behavior. We give purely graph theoretic parameters (i.e., without reference to strings) that are exactly (respectively, asymptotically) equivalent to readability for trees (respectively, C_4-free graphs); however, for general graphs, the HUB number is equivalent to readability only in the sense that it is bounded on the same set of graphs. While an ℓ-decomposition always satisfies the HUB-rule, the converse is not true. For example, a decomposition of P_4 with weights $4, 5, 3$ satisfies the HUB-rule but cannot be achieved by an overlap labeling (by Lemma 1). For this reason, the upper bound given by Lemma 5 leaves a gap with the lower bound of Lemma 4. We are able to describe other properties that an ℓ-decomposition must satisfy (not included in the paper), however, we are not able to exploit them to close the gap. It is a very interesting direction to find other necessary rules that would lead to a graph theoretic parameter that would more tightly match readability on general graphs than the HUB number.

Consider $r(n) = \max\{r(D) \mid D$ is a digraph on n vertices$\}$. We have shown $r(n) = \Omega(n)$ and know from [BM02] that $r(n) = O(2^n)$. Can this gap be closed? Do there exist graphs with readability $\Theta(2^n)$ (as we conjecture), or, for example, is readability always bounded by a polynomial in n? Questions regarding complexity are also unexplored, e.g., given a digraph, is it NP-hard to compute its readability? For applications to bioinformatics, the length of reads can be said to be poly-logarithmic in the number of vertices. It would thus be interesting to further study the structure of graphs that have poly-logarithmic readability.

Acknowledgements. P.M. and M.M. would like to thank Marcin Kamiński for preliminary discussions. P.M. was supported in part by NSF awards DBI-1356529 and CAREER award IIS-1453527. M.M. was supported in part by the Slovenian Research Agency (I0-0035, research program P1-0285 and research projects N1-0032, J1-5433, J1-6720, and J1-6743). S.R. was supported in part by NSF CAREER award CCF-0845701, NSF award AF-1422975 and the Hariri Institute for Computing and Computational Science and Engineering at Boston University.

References

[BBT13] Bresler, G., Bresler, M., Tse, D.: Optimal assembly for high throughput shotgun sequencing. BMC Bioinform. **14**(Suppl 5), S18 (2013)

[BFK+02] Błażewicz, J., Formanowicz, P., Kasprzak, M., Schuurman, P., Woeginger, G.J.: DNA sequencing, eulerian graphs, and the exact perfect matching problem. In: Kučera, L. (ed.) WG 2002. LNCS, vol. 2573, pp. 13–24. Springer, Heidelberg (2002)

[BFKK02] Błażewicz, J., Formanowicz, P., Kasprzak, M., Kobler, D.: On the recognition of de Bruijn graphs and their induced subgraphs. Discrete Math. **245**(1), 81–92 (2002)

[BHKdW99] Blazewicz, J., Hertz, A., Kobler, D., de Werra, D.: On some properties of DNA graphs. Discrete Appl. Math. **98**(1), 1–19 (1999)

[BM02] Braga, M.D.V., Meidanis, J.: An algorithm that builds a set of strings given its overlap graph. In: Rajsbaum, S. (ed.) LATIN 2002. LNCS, vol. 2286, p. 52. Springer, Heidelberg (2002)

[BM08]   Bondy, J.A., Murty, U.S.R.: Graph Theory. Graduate Texts in Mathematics., vol. 244. Springer, New York (2008)

[GP14]   Gevezes, T.P., Pitsoulis, L.S.: Recognition of overlap graphs. J. Comb. Optim. **28**(1), 25–37 (2014)

[IW95]   Idury, R.M., Waterman, M.S.: A new algorithm for DNA sequence assembly. J. Comput. Biol. **2**(2), 291–306 (1995)

[LZ07]   Li, X., Zhang, H.: Characterizations for some types of DNA graphs. J. Math. Chem. **42**(1), 65–79 (2007)

[LZ10]   Li, X., Zhang, H.: Embedding on alphabet overlap digraphs. J. Math. Chem. **47**(1), 62–71 (2010)

[MGMB07]   Medvedev, P., Georgiou, K., Myers, G., Brudno, M.: Computability of models for sequence assembly. In: Giancarlo, R., Hannenhalli, S. (eds.) WABI 2007. LNCS (LNBI), vol. 4645, pp. 289–301. Springer, Heidelberg (2007)

[MKS10]   Miller, J.R., Koren, S., Sutton, G.: Assembly algorithms for next-generation sequencing data. Genomics **95**(6), 315–327 (2010)

[Mye05]   Myers, E.W.: The fragment assembly string graph. In: ECCB/JBI, pp. 85 (2005)

[NP09]   Nagarajan, N., Pop, M.: Parametric complexity of sequence assembly: theory and applications to next generation sequencing. J. Comput. Biol. **16**(7), 897–908 (2009)

[NP13]   Nagarajan, N., Pop, M.: Sequence assembly demystified. Nat. Rev. Genet. **14**(3), 157–167 (2013)

[PSW03]   Pendavingh, R., Schuurman, P., Woeginger, G.J.: Recognizing DNA graphs is difficult. Discrete Appl. Math. **127**(1), 85–94 (2003)

[Swe00]   Sweedyk, Z.: A $2\frac{1}{2}$-approximation algorithm for shortest superstring. SIAM J. Comput. **29**(3), 954–986 (2000)

[TU88]   Tarhio, J., Ukkonen, E.: A greedy approximation algorithm for constructing shortest common superstrings. Theor. Comput. Sci. **57**(1), 131–145 (1988)

Improved Algorithms for the Boxed-Mesh Permutation Pattern Matching Problem

Sukhyeun Cho[1], Joong Chae Na[2]([✉]), and Jeong Seop Sim[1]

[1] Department of Computer and Information Engineering,
Inha University, Incheon, Korea
csukhyeun@inha.edu, jssim@inha.ac.kr
[2] Department of Computer Science and Engineering,
Sejong University, Sejong, Korea
jcna@sejong.ac.kr

Abstract. Recently, various types of permutation patterns such as mesh patterns, boxed-mesh patterns, and consecutive patterns, have been studied where relative order between characters is considered rather than characters themselves. Among these, we focus on boxed-mesh patterns and study the problem of finding all boxed-subsequences of a text T of length n whose relative order between characters is the same as that of a pattern P of length m. Recently, it is known that this problem can be solved in $O(n^3)$ time. In this paper, we first propose an $O(n^2 m)$-time algorithm for the problem based on interesting properties of boxed subsequences. Then, we give a further improved algorithm which runs in $O(n^2 \log m)$ time using preprocessed information on P and order-statistics trees.

Keywords: Permutation pattern matching · Order-isomorphism · Boxed-mesh pattern

1 Introduction

Given a text T and a pattern P over a numeric alphabet Σ, the permutation pattern matching problem is to find every subsequence of T whose relative order between all characters (numbers) is the same as that of P [1,2]. For example, when $P = (5, 3, 4, 8, 9, 6, 7)$ and $T = (10, 6, 2, 7, 15, 16, 12, 19, 13, 11, 3)$ are given, P has the same relative order as those of two subsequences of T, i.e., $T' = (10, 6, 7, 15, 16, 12, 13)$ and $T'' = (10, 2, 7, 15, 16, 12, 13)$. The first character '10' in T' (resp. in T'') is the 3rd smallest character as '5' in P, the second character '6' in T' (resp. '2' in T'') is the smallest character as '3' in P, and so on. Due to their diverse applications in time series analysis, various types of permutation patterns have been studied such as mesh patterns [3], boxed-mesh patterns [4], and consecutive patterns [5–9].

Among the various types of permutation patterns, we focus on boxed-mesh patterns. In boxed-mesh permutation pattern matching (BPPM for short), the ith character c in a (numeric) string x can be represented as a point of coordinate

© Springer International Publishing Switzerland 2015
F. Cicalese et al. (Eds.): CPM 2015, LNCS 9133, pp. 138–148, 2015.
DOI: 10.1007/978-3-319-19929-0_12

(i, c) on a two-dimensional plane. Consider a rectangle R whose coordinates of four apexes are $(h_1, v_1), (h_1, v_2), (h_2, v_1)$ and (h_2, v_2). Assume R includes k number of points each of which represents a character of x. Then we can construct a subsequence x' of x ($|x'| = k$) by concatenating all the characters represented as points in R from left to right. In this case, we say R represents x' and x' is a *boxed subsequence* of x. Note that not every subsequence of x is a boxed subsequence. For the previous example, all characters of P and those of T can be represented in two-dimensional planes, respectively, as shown in Fig. 1. All the points of $T' = (10, 6, 7, 15, 16, 12, 13)$ are included in the shaded rectangle of the right plane and there are no other points in the shaded rectangle. Thus, T' is a boxed subsequence of T. But for another subsequence $T'' = (10, 2, 7, 15, 16, 12, 13)$, no rectangle can include all the points of T'' without including any other points of T. Thus, T'' is not a boxed subsequence of T.

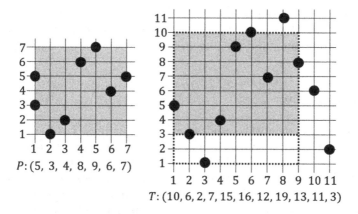

$P: (5, 3, 4, 8, 9, 6, 7)$

$T: (10, 6, 2, 7, 15, 16, 12, 19, 13, 11, 3)$

Fig. 1. Representations of $P = (5, 3, 4, 8, 9, 6, 7)$ and $T = (10, 6, 2, 7, 15, 16, 12, 19, 13, 11, 3)$ in two-dimensional planes, where the x-axis is labeled with indexes of characters and the y-axis is labeled with ranks of characters.

Given a text $T(|T| = n)$ and a pattern $P(|P| = m)$ over Σ, the boxed-mesh permutation pattern matching problem (BPPM problem for short) is to find every boxed subsequence of T whose relative order between characters is the same as that of P. In the previous example shown in Fig. 1, T' is an occurrence since T' is a boxed subsequence of T and its relative order between characters is the same as that of P. Though the relative order between characters of T'' is the same as that of P, T'' is not an occurrence since it is not a boxed subsequence of T. Avgustinovich et al. [4] firstly introduced the concept of boxed-mesh patterns and studied pattern avoidance, i.e., counting the number of permutations not containing a given pattern. Bruner et al. [2] introduced the BPPM problem and showed that it can be solved in $O(n^3)$ time as follows. First, they fix two characters c_1 and c_2 ($c_1 < c_2$) in T. Then, they check whether the boxed subsequence including c_1 and c_2 as the smallest character and the largest character, respectively, is an occurrence of P in T.

Table 1. μ_P and π for $P = (5, 3, 4, 8, 9, 6, 7)$.

i	1	2	3	4	5	6	7
$P[i]$	5	3	4	8	9	6	7
$\mu_P[i]$	1	1	2	4	5	4	5
$\pi[i]$	0	1	2	3	4	4	5

In this paper, we first propose an $O(n^2 m)$-time algorithm for the BPPM problem. Our algorithm fixes a position i and then finds all the occurrences of P in T whose first character is ith character of T based on interesting properties of boxed subsequences. Then using preprocessed information on P and order-statistics trees, we give a further improved algorithm which runs in $O(n^2 \log m)$ time.

This paper is organized as follows. In Sect. 2, we describe the previous works related to the BPPM problem. In Sect. 3, we present an $O(n^2 m)$-time algorithm for the BPPM problem, and then we present an $O(n^2 \log m)$-time algorithm in Sect. 4.

2 Preliminaries

We give some basic definitions and notations on strings. Let Σ denote the set of characters such that a comparison of two characters can be done in constant time. A string x over the alphabet Σ is a sequence of characters derived from the alphabet Σ. For simplicity, we assume that the characters of x are all distinct. We denote the length of x by $|x|$ and the ith character by $x[i]$ ($1 \leq i \leq |x|$). A sequence $(x[i_1], x[i_2], \ldots, x[i_k])$ of characters in x is called a *subsequence* of x when $1 \leq i_1 < i_2 < \cdots < i_k \leq |x|$. A *substring* of x denoted by $x[i..j]$ is a subsequence of consecutive characters $(x[i], x[i+1], \ldots, x[j])$ in x. A *prefix* of x is a substring of x starting at the first position. For a string x, we denote by $\min(x)$ (resp. $\max(x)$) the smallest (resp. largest) character in x. For a string x and a character c, we denote by $x \oplus c$ the concatenation of x and c. The *rank* of a character c for a string x is defined as $\mathrm{RANK}(x, c) = 1 + |\{i : x[i] < c, 1 \leq i \leq |x|\}|$.

We formally define the order-isomorphism [6]. Two strings x and y of the same length over Σ are called *order-isomorphic*, written $x \approx y$, if

$$x[i] \leq x[j] \Leftrightarrow y[i] \leq y[j] \text{ for all } 1 \leq i, j \leq |x|.$$

If two strings x and y are not order-isomorphic, we write $x \not\approx y$. For a string x, the *prefix representation* μ_x is defined as $\mu_x[i] = \mathrm{RANK}(x[1..i], x[i])$ ($1 \leq i \leq |x|$), which can be computed in $O(|x| \log |x|)$ time using order-statistics trees [10].

Lemma 1. *[7] For two strings x and y over Σ, $x \approx y$ if and only if $\mu_x = \mu_y$.*

In the lemma, it is assumed that there are no identical characters in each string [7] but this assumption can be avoided by extending a character $x[i]$ to the pair $(x[i], i)$ [11].

Given a text $T[1..n]$ and a pattern $P[1..m]$, we say that P occurs at (i, j) of T if there exists a boxed subsequence T' of T such that $T' = (T[i], \ldots, T[j])$ and $T' \approx P$. Also, we say T' is an occurrence of P at (i, j) in T. The following lemma shows the uniqueness of an occurrence at (i, j)

Lemma 2. *Given a pattern P and a text T, for fixed i and j $(1 \le i < j \le n)$, there exists at most one boxed subsequence T' of T such that $T' = (T[i], \ldots, T[j])$ and $T' \approx P$.*

Proof. It can be proven by contradiction. Let x and y be two distinct boxed subsequences whose first and last characters are $T[i]$ and $T[j]$, respectively. Assume both $x \approx P$ and $y \approx P$. Since x is a boxed subsequence whose first and last characters are $T[i]$ and $T[j]$, respectively, x includes all the characters c in $T[i..j]$ such that $\min(x) \le c \le \max(x)$. Similarly, y includes all the characters c in $T[i..j]$ such that $\min(y) \le c \le \max(y)$.

Now, we show that $\min(x) = \min(y)$ and $\max(x) = \max(y)$. Let l_x and l_y be the number of characters c in $T[i..j]$ such that $\min(x) \le c < T[i]$ and $\min(y) \le c < T[i]$, respectively. Then, $\text{RANK}(x, T[i]) = l_x + 1$ and $\text{RANK}(y, T[i]) = l_y + 1$. Since $x \approx P$ and $y \approx P$, $\text{RANK}(x, T[i]) = \text{RANK}(y, T[i])$ and thus $l_x = l_y$, which means $\min(x) = \min(y)$. Similarly, it can be shown $\max(x) = \max(y)$. Since $\min(x) = \min(y)$ and $\max(x) = \max(y)$, the boxed subsequences x and y consist of the same characters, which contradicts the assumption that x and y are distinct.   □

3 $O(n^2m)$-time Algorithm for the BPPM Problem

In this section, we present an $O(n^2m)$-time algorithm for the BPPM problem. Our algorithm consists of $n - m + 1$ phases. In each Phase i $(1 \le i \le n - m + 1)$, we find all the occurrences of P whose first character is $T[i]$. Since each phase is completely independent of the other phases, we devote our attention to a fixed phase (Phase i).

We first give some definitions and notations for description of Phase i. A subsequence x' of a string x is called a *full-width boxed* (for short, *f-boxed*) subsequence of x if x' includes only and all the characters c of x such that $\min(x') \le c \le \max(x')$. For example, a subsequence $T' = (10, 6, 7, 12)$ of T shown in Fig. 1 is an f-boxed subsequence of $T[1..9]$. However, T' is not an f-boxed subsequence of $T[1..10]$ since $\min(T') \le T[10] = 11 \le \max(T')$ but '11' is not a character in T'. For an f-boxed subsequence x' of a string x, we define the lower bound $lb(x, x')$ and the upper bound $ub(x, x')$ as follows:

- $lb(x, x')$ is the largest character in x smaller than $\min(x')$. If no such character exists, $lb(x, x')$ is $-\infty$.
- $ub(x, x')$ is the smallest character in x larger than $\max(x')$. If no such character exists, $ub(x, x')$ is ∞.

For the f-boxed subsequence $T' = (10, 6, 7, 12)$ of $T[1..9]$ in the previous example, $lb(T[1..9], T') = 2$ and $ub(T[1..9], T') = 13$. Notice that $lb(T[1..9], T') \neq 3$ since '3' $(= T[11])$ is not a character of $T[1..9]$.

Next, we define a set $\mathcal{Z}_j$ of f-boxed subsequences z of $T[i..j]$ $(1 \leq i \leq j \leq n)$ such that z includes the first character $T[i]$ and $z \approx P[1..|z|]$. Notice that the last character $T[j]$ may not be included in z. Moreover, $\mathcal{Z}_j$ always contains the subsequence $(T[i])$ of length one. The following lemma shows the strings in $\mathcal{Z}_j$ are of distinct lengths.

Lemma 3. *For a fixed k $(1 \leq k \leq m)$, there exists at most one string of length k in $\mathcal{Z}_j$.*

Proof. It can be proven by contradiction, similarly to the proof for Lemma 2. We omit the details. □

Lemma 4 shows the relation between $\mathcal{Z}_{j-1}$ and $\mathcal{Z}_j$.

Lemma 4. *For $z \in \mathcal{Z}_j$ $(i+1 \leq j \leq n)$, if z does not include $T[j]$, then $z \in \mathcal{Z}_{j-1}$; otherwise (if z includes $T[j]$), $z' \in \mathcal{Z}_{j-1}$ where z' is the string obtained by deleting $T[j]$ from z, i.e., $z = z' \oplus T[j]$.*

Proof. First, consider the case when z does not include $T[j]$. Since z is an f-boxed subsequence of $T[i..j]$ and z does not include $T[j]$, z is also an f-boxed subsequence of $T[i..j-1]$. Since $z \in \mathcal{Z}_j$, z includes $T[i]$ and $z \approx P[1..|z|]$. Hence, $z \in \mathcal{Z}_{j-1}$ by the definition of $\mathcal{Z}_{j-1}$.

Next, consider the case when z includes $T[j]$. Since $z = z' \oplus T[j]$ is an f-boxed subsequence of $T[i..j]$ and z' does not include $T[j]$, z' is an f-boxed subsequence of $T[i..j-1]$. Since $z' \oplus T[j] \in \mathcal{Z}_j$, z' includes $T[i]$ and $z' \approx P[1..|z'|]$. Hence, $z' \in \mathcal{Z}_{j-1}$ by the definition of $\mathcal{Z}_{j-1}$. □

By the definition of $\mathcal{Z}_j$, $\mathcal{Z}_j$ includes the occurrence of P (if exists) whose first and last characters are $T[i]$ and $T[j]$, respectively. Thus, we can get the following corollary from Lemma 4.

Corollary 1. *For an occurrence z of P at (i, j) in T, let z' be the string obtained by deleting $T[j]$ from z, i.e., $z = z' \oplus T[j]$. Then, $z' \in \mathcal{Z}_{j-1}$.*

Now we explain Phase i in details for finding all the occurrences of P whose first character is $T[i]$. Phase i consists of $n - i$ steps, from Step $i + 1$ to Step n. Let lz_j $(i \leq j \leq n)$ be the longest string in $\mathcal{Z}_j$ such that $|lz_j| < m$. (lz_i is simply $T[i]$ since $T[i]$ is the only string in $\mathcal{Z}_i$.) At each Step j $(i + 1 \leq j \leq n)$ of Phase i, we are given lz_{j-1}, and then we compute lz_j and decide whether P occurs at (i, j) or not. To decide whether P occurs at (i, j) or not, we check only whether $lz_{j-1} \oplus T[j] \approx P$ (Corollary 1). To compute lz_j, we make use of the strings in $\mathcal{Z}_{j-1}$, which can be computed using lz_{j-1}. (We will describe later how to compute $\mathcal{Z}_{j-1}$ using lz_{j-1}.) By Lemma 4, it is sufficient to consider $z \in \mathcal{Z}_{j-1}$ and $z \oplus T[j]$ to compute $\mathcal{Z}_j$. Furthermore, no two strings are of the same length in $\mathcal{Z}_{j-1}$ by Lemma 3. Therefore, to compute lz_j, for strings $z \in \mathcal{Z}_{j-1}$ in decreasing order of their lengths, we check if $z \oplus T[j] \in \mathcal{Z}_j$ and $z \in \mathcal{Z}_j$. The following two lemmas show conditions when z and $z \oplus T[j]$ will be elements in $\mathcal{Z}_j$.

Lemma 5. *For $z \in \mathcal{Z}_{j-1}$ $(i + 1 \leq j \leq n)$, $z \in \mathcal{Z}_j$ if and only if $T[j] < \min(z)$ or $T[j] > \max(z)$.*

Proof. If $T[j] < \min(z)$ or $T[j] > \max(z)$, z is an f-boxed subsequence in $T[i..j]$. Obviously, z includes $T[i]$ and $z \approx P[1..|z|]$. Thus, in this case, $z \in \mathcal{Z}_j$. If $\min(z) \leq T[j] \leq \max(z)$, z is not an f-boxed subsequence in $T[i..j]$ and thus, $z \notin \mathcal{Z}_j$. $\square$

Lemma 6. *For $z \in \mathcal{Z}_{j-1}$ $(i + 1 \leq j \leq n)$, let $B_{lb} = lb(T[i..j-1], z)$, $B_{ub} = ub(T[i..j-1], z)$, and $r = \mathrm{RANK}(z, T[j])$. Then, $z \oplus T[j] \in \mathcal{Z}_j$ if and only if $B_{lb} < T[j] < B_{ub}$ and $r = \mu_P[|z| + 1]$.*

Proof. We first prove $z \oplus T[j] \in \mathcal{Z}_j$ if $B_{lb} < T[j] < B_{ub}$ and $r = \mu_P[|z| + 1]$. Let $z' = z \oplus T[j]$, $l = |z'|$, and $P' = P[1..l]$. Obviously, z' includes $T[i]$. Moreover, z' is an f-boxed subsequence of $T[i..j]$ since z' includes only and all characters c of $T[i..j]$ such that $B_{lb} < c < B_{ub}$. We show that $\mu_{z'} = \mu_{P'}$ and thus $z' \approx P'$ (Lemma 1). By the definition of the prefix representation, $\mu_{z'}[1..l-1] = \mu_z[1..l-1]$ and $\mu_{P'}[1..l-1] = \mu_P[1..l-1]$. Since $z \in \mathcal{Z}_{j-1}$, $z \approx P[1..l-1]$, i.e., $\mu_z[1..l-1] = \mu_P[1..l-1]$ (by Lemma 1). Thus, $\mu_{z'}[1..l-1] = \mu_{P'}[1..l-1]$. Furthermore, $\mu_{z'}[l] = \mu_{P'}[l]$ by the condition $r = \mu_P[l]$, and thus $\mu_{z'} = \mu_{P'}$. Hence, $z \oplus T[j] \in \mathcal{Z}_j$.

Now, we prove $z \oplus T[j] \notin \mathcal{Z}_j$ if $T[j] \leq B_{lb}$ or $T[j] \geq B_{ub}$ or $r \neq \mu_P[|z| + 1]$. First, consider the case when $T[j] \leq B_{lb}$ or $T[j] \geq B_{ub}$. Without loss of generality, assume $T[j] \geq B_{ub}$. Let $T[k]$ be the character in $T[i..j-1]$ such that $T[k] = B_{ub}$. By definition of $B_{ub} = ub(T[i..j-1], z)$, $T[k]$ surely exists and $T[k] > \max(z)$. Since $\max(z) < T[k] \leq T[j]$ and $T[k]$ is not in $z \oplus T[j]$, $z \oplus T[j]$ is not an f-boxed subsequence of $T[i..j]$. Hence, $z \oplus T[j] \notin \mathcal{Z}_j$.

Next, consider the case when $r \neq \mu_P[|z| + 1]$. Let $z' = z \oplus T[j]$, $l = |z'|$, and $P' = P[1..l]$. We show $z' \not\approx P'$. Since the characters in T are all distinct, $\mathrm{RANK}(z', T[j]) = \mathrm{RANK}(z, T[j])$, i.e., $\mu_{z'}[l] = r$. By the definition of the prefix representation, $\mu_{P'}[l] = \mu_P[l]$. Thus, $\mu_{z'}[l] \neq \mu_{P'}[l]$ by the condition $r \neq \mu_P[l]$, which means $z' \not\approx P'$ by Lemma 1. Hence, $z \oplus T[j] \notin \mathcal{Z}_j$. $\square$

Algorithm 1 shows the pseudocode of our algorithm. The first for loop (line 2) represents the phases and the second for loop (line 4) represents the steps in each phase. While performing Phase i, we maintain a subsequence z and two variables B_{lb} and B_{ub} so that, at the beginning of Step j $(i + 1 \leq j \leq n)$, $z = lz_{j-1}$, $B_{lb} = lb(T[i..j-1], z)$ and $B_{ub} = ub(T[i..j-1], z)$. Initially, we set z to lz_i (i.e., $T[i]$) and B_{lb} and B_{ub} to $-\infty$ and ∞, respectively (line 3). At Step j of Phase i, we repeat the while loop of lines 5–14 until lz_j is computed as follows.

1. First, we check whether $z \oplus T[j] \in \mathcal{Z}_j$ (line 7). Let $r = \mathrm{RANK}(z, T[j])$. By Lemma 5, if $B_{lb} < T[j] < B_{ub}$ and $r = \mu_P[|z| + 1]$, then $z \oplus T[j] \in \mathcal{Z}_j$. If $|z| < m - 1$ (lines 8–9), then $lz_j = z \oplus T[j]$. Hence, we append $T[j]$ to z (the APPEND operation) and escape the while loop (break). In this case, B_{lb} and B_{ub} do not change. If $|z| = m - 1$ (lines 10–11), then $z \oplus T[j]$ is an occurrence at (i, j) but it is not lz_j due to the length restriction. Hence, we continue to compute lz_j.

Algorithm 1. Search for a boxed-mesh pattern

```
1: m ← |P|; n ← |T|;
2: for i ← 1 to n − m + 1 do
3:     z ← (T[i]); B_lb ← −∞; B_ub ← ∞;                          ▷ Initialization
4:     for j ← i + 1 to n do
5:         while true do
6:             r ← RANK(z, T[j]);
7:             if B_lb < T[j] < B_ub and r = μ_P[|z| + 1] then      ▷ if z ⊕ T[j] ∈ Z_j
8:                 if |z| < m − 1 then
9:                     APPEND(z, T[j]); break;                      ▷ z ← z ⊕ T[j]
10:                else
11:                    print "P occurs at (i, j)";
12:            if r = 1 then B_lb ← max(B_lb, T[j]); break;       ▷ if T[j] < min(z)
13:            else if r = |z| + 1 then B_ub ← min(B_ub, T[j]); break;  ▷ if T[j] > max(z)
14:            else NEXTCANDSEQ();                          ▷ if min(z) < T[j] < max(z)
```

```
1: procedure NEXTCANDSEQ
2:     do
3:         if P[|z|] > P[1] then B_ub ← EXTRACTMAX(z);     ▷ delete &return max(z)
4:         else B_lb ← EXTRACTMIN(z);                      ▷ delete &return min(z)
5:     while z ≉ P[1..|z|]
6: end procedure
```

2. Next, if lz_j is not computed yet, we check whether $z \in \mathcal{Z}_j$ (lines 12–14). If $r = 1$ or $r = |z| + 1$, i.e., $T[j] < \min(z)$ or $T[j] > \max(z)$ (lines 12–13), then $z \in \mathcal{Z}_j$ by Lemma 5. In this case, $lz_j = z$ since $|z| < m$. Due to $T[j]$, B_{lb} and B_{ub} may change. Hence, we set $B_{lb} = \max(B_{lb}, T[j])$ (if $r = 1$) and $B_{ub} = \min(B_{ub}, T[j])$ (if $r = |z| + 1$), and escape the while loop. If $1 < r \leq |z|$, i.e., $\min(z) < T[j] < \max(z)$ (line 14), then $z \notin \mathcal{Z}_j$ by Lemma 5. Since, for the current z, $z \notin \mathcal{Z}_j$ and $z \oplus T[j] \notin \mathcal{Z}_j$, we compute the longest string in $\mathcal{Z}_{j-1}$ whose length is less than $|z|$ by calling the NEXTCANDSEQ procedure and repeat the while loop for the new string in $\mathcal{Z}_{j-1}$.

The NEXTCANDSEQ procedure, from the current string z in $\mathcal{Z}_{j-1}$ at Step j, computes the longest string z' in $\mathcal{Z}_{j-1}$ such that $|z'| < |z|$. We compute z' by deleting characters one by one from z as follows. Let a (resp. b) be the number of characters in $P[1..|z|]$ larger (resp. smaller) than $P[1]$. Without loss of generality, assume that $P[|z|] > P[1]$. Consider an f-boxed subsequence x of $T[i..j-1]$ such that x includes the first character $T[i]$ and $a-1$ (resp. b) characters larger (resp. smaller) than $T[i]$. Note that x is the unique f-boxed subsequence of $T[i..j-1]$ such that $\text{RANK}(x, T[i]) = \text{RANK}(P[1..|z|-1], P[1])$ and thus x can be order-isomorphic to $P[1..|z| - 1]$. Thus, we repeat the following until z' is computed. We update B_{ub} to $\max(z)$ and delete it from z (EXTRACTMAX) if $P[|z|] > P[1]$ (line 3), or we update B_{lb} to $\min(z)$ and delete if from z (EXTRACTMIN) if $P[|z|] < P[1]$ (line 4). Then, we check whether $z' \approx P[1..|z'|]$ or not (line 5).

We analyze the time complexity of Algorithm 1. In the while loop in Algorithm 1, all the statements excepts lines 6 and 14 run at most once in each step. Moreover, the while loop is repeated only when the NEXTCANDSEQ procedure is called. Let us consider how many times EXTRACTMAX and EXTRACT-MIN in NEXTCANDSEQ are called in each phase. Whenever EXTRACTMAX or EXTRACTMIN is called, the length of z decreases by one. Moreover, the length of z increases at most by one in each step (by APPEND) and thus it increases at most by n in each phase. Therefore, all operations on z including RANK (line 6 of Algorithm 1) and order-isomorphism check (line 5 of NEXTCANDSEQ) runs $O(n)$ times in each phase. Operations RANK, APPEND, EXTRACTMAX, and EXTRACT-MIN can be performed in $O(m)$ time by maintaining z in an unsorted array. Moreover, we can check $z \approx P[1..|z|]$ (line 5 of NEXTCANDSEQ) in $O(m)$ time after preprocessing P in $O(m \log m)$ time [6–8]. The prefix representation for P can also be computed in $O(m \log m)$ time as mentioned in Sect. 2. Therefore, each phase takes $O(nm)$ time and thus Algorithm 1 takes in $O(n^2 m)$ time in total. Hence, we get the following theorem.

Theorem 1. *The BPPM problem can be solved in* $O(n^2 m)$ *time.*

4 $O(n^2 \log m)$-time Algorithm for the BPPM Problem

In this section, we improve Algorithm 1 to run in $O(n^2 \log m)$ time. In the time complexity of Algorithm 1, the factor m is due to the operations on string z, i.e., RANK, APPEND, EXTRACTMAX, EXTRACTMIN, and the order-isomorphism check with P (line 5 of NEXTCANDSEQ). First, we can easily reduce the time for all the operations except the order-isomorphism check to $O(\log m)$ by maintaining z in an order-statistics tree [10]. Second, we can avoid the order-isomorphism check by using preprocessed information on P.

We present in details how to avoid the order-isomorphism check. Let $\mathcal{Z}_q^P$ be the set for $P[1..q]$ defined equally to the set $\mathcal{Z}_j$ for $T[i..j]$. Precisely, $\mathcal{Z}_q^P$ $(1 \le q \le m)$ is the set of f-boxed subsequences x of $P[1..q]$ such that x includes the first character $P[1]$ and $x \approx P[1..|x|]$. Then, we define the function $\pi[q]$ $(1 \le q \le m)$ as the length of the longest string x in $\mathcal{Z}_q^P$ such that $|x| < q$. See Table 1 in Sect. 2 for an example. Since $P[1..q]$ is surely in $\mathcal{Z}_q^P$ and it is the longest string in $\mathcal{Z}_q^P$, $\pi[q]$ is the length of the second longest string in $\mathcal{Z}_q^P$.

1: **procedure** IMPROVEDNEXTCANDSEQ
2:     $l \leftarrow \pi[|z|]$;
3:     **do**
4:         **if** $P[|z|] > P[1]$ **then** $B_{ub} \leftarrow$ EXTRACTMAX(z);   ▷ delete &return max(z)
5:         **else** $B_{lb} \leftarrow$ EXTRACTMIN(z);   ▷ delete &return min(z)
6:     **while** $|z| > l$
7: **end procedure**

The IMPROVEDNEXTCANDSEQ procedure is an improved version of the NEXTCANDSEQ procedure. By using the π-function, we can skip checking the order-isomorphism between z and $P[1..|z|]$. The following lemma shows the correctness of IMPROVEDNEXTCANDSEQ.

Algorithm 2. Compute the π-function

```
1:  m ← |P|; π[1] ← 0;
2:  for q ← 2 to m do
3:      z ← P[1..q − 1]; B_lb ← −∞; B_ub ← ∞;              ▷ Initialization
4:      while true do
5:          r ← RANK(z, P[q]);
6:          if B_lb < P[q] < B_ub and r = μ_P[|z| + 1] then    ▷ if z ⊕ P[q] ∈ Z_q^P
7:              if |z| < q − 1 then
8:                  APPEND(z, P[q]); break;                     ▷ z ← z ⊕ P[q]
9:              if r = 1 or r = |z| + 1 then break;             ▷ if z ∈ Z_q^P
10:             else IMPROVEDNEXTCANDSEQ();                     ▷ if z ∉ Z_q^P
11:     π[q] = |z|;
```

Lemma 7. *For $z \in \mathcal{Z}_j$ ($i + 1 \le j \le n$), let z' be the longest string in $\mathcal{Z}_j$ (if exists) such that $|z'| < |z|$. Then, $|z'| = \pi[|z|]$.*

Proof. (Sketch) Let $q = |z|$ and $l = \pi[q]$. We first prove that there exists a string of length l in $\mathcal{Z}_j$. By the definition of $\pi[q]$, there exists an f-boxed subsequence P' of $P[1..q]$ of length l such that P' includes $P[1]$ and $P' \approx P[1..l]$. Assume $P' = (P[p_1], \ldots, P[p_l])$. Consider a subsequence $z' = (z[p_1], \ldots, z[p_l])$ of z of length l. We can show that $z'[1] = T[i]$, $z' \approx P[1..l]$, and z' is an f-boxed subsequence of $T[i..j]$ using the condition $z \in \mathcal{Z}_j$, i.e., $z[1] = T[i]$, $z \approx P[1..q]$, and z is an f-boxed subsequence of $T[i..j]$. Therefore, $z' \in \mathcal{Z}_j$.

Next, we prove by contradiction that there exists no string of length k ($l < k < q$) in $\mathcal{Z}_j$. Suppose that there exists a string x of length k in $\mathcal{Z}_j$. We can show that x consists of only the characters of z using conditions $x \in \mathcal{Z}_j$ and $z \in \mathcal{Z}_j$. Assume $x = (z[p_1], \ldots, z[p_k])$. Consider a subsequence $P' = (P[p_1], \ldots, P[p_k])$ of $P[1..q]$ of length k. Similarly to the above, we can show that $P' \in \mathcal{Z}_q^P$. It contradicts the definition of $\pi[q]$ since $|P'| > l$. Therefore, there is no such string x.   □

Lemma 8. *When the π-function is given, Algorithm 1 takes $O(n^2 \log m)$ time.*

Now we explain how to compute the π-function. Algorithm 2 shows the pseudocode of our algorithm, which is similar to Phase 1 of Algorithm 1. Algorithm 2 consists of $m - 1$ steps, from Step 2 to Step m. In Step q ($2 \le q \le m$), we compute $\pi[q]$ using the strings in $\mathcal{Z}_{q-1}^P$. (By definition, $\pi[1] = 0$.) That is, for strings $z \in \mathcal{Z}_{q-1}^P$ in decreasing order of their lengths, we check if $z \oplus P[q] \in \mathcal{Z}_q^P$ and $z \in \mathcal{Z}_q^P$. Let lz_q ($1 \le q \le m$) be the longest string in $\mathcal{Z}_q^P$ and lz_q' be the longest string in $\mathcal{Z}_q^P$ such that $|lz_q'| < q$. Differently from Algorithm 1, we do not compute lz_q for the next step since lz_q is simply $P[1..q]$. Thus, in Step q, we initially set $z = P[1..q - 1]$ (line 3) as lz_q and repeat the while loop (lines 4–10) until we compute lz_q'. Since the definition of $\mathcal{Z}_q^P$ is the same as that of $\mathcal{Z}_j$ in Sect. 3, we can check whether $z \oplus P[q]$ and z are in $\mathcal{Z}_q^P$ or not in the same way as in Algorithm 1. Notice that Algorithm 2 computes lz_q' while Algorithm 1 computes lz_q, which makes a difference in checking the length of z (line 8 of Algorithm 1 and line 7 of Algorithm 2). Moreover, we already know $\pi[k]$ $(1 \leq k \leq q-1)$ which is necessary to compute $\pi[q]$.

We analyze the time complexity of Algorithm 2. Similarly to the analysis of Algorithm 1, the running time of Algorithm 2 is bounded by the while loop in IMPROVEDNEXTCANDSEQ. In Step q, initially $|z| = q-1$ and it decreases whenever EXTRACTMAX or EXTRACTMIN is called. Thus, each step takes $O(m \log m)$ time when z is maintained in an order-statistics tree. Moreover, as mentioned in Sect. 2, the prefix representation for P can be computed in $O(m \log m)$ time. Since Algorithm 2 consists of $m-1$ steps, Algorithm 2 takes $O(m^2 \log m)$ time.

Lemma 9. *The π-function for P of length m can be computed in $O(m^2 \log m)$ time.*

From Lemmas 8 and 9 we can get the following theorem.

Theorem 2. *The BPPM problem can be solved in $O(n^2 \log m)$ time.*

Acknowledgements. Joong Chae Na was supported by Basic Science Research Program through the National Research Foundation of Korea (NRF) funded by the Ministry of Science, ICT &Future Planning (2014R1A1A1004901), and by the ICT R&D program of MSIP/IITP [10038768, The Development of Supercomputing System for the Genome Analysis]. Jeong Seop Sim was supported by the National Research Foundation of Korea (NRF) grant funded by the Korea government (MSIP) (No. 2012R1A2A2A01014892 & 2014R1A2A1A11050337), and by the ICT R&D program of MSIP/IITP [10041971, Development of Power Efficient High-Performance Multimedia Contents Service Technology using Context-Adapting Distributed Transcoding].

References

1. Bose, P., Buss, J.F., Lubiw, A.: Pattern matching for permutations. Inf. Process. Lett. **65**, 277–283 (1998)
2. Bruner, M.L., Lackner, M.: The computational landscape of permutation patterns. arXiv preprint arXiv:1301.0340 (2013)
3. Brändén, P., Claesson, A.: Mesh patterns and the expansion of permutation statistics as sums of permutation patterns. Electron. J. Combin **18**, P5 (2011)
4. Avgustinovich, S., Kitaev, S., Valyuzhenich, A.: Avoidance of boxed mesh patterns on permutations. Discrete Appl. Math. **161**, 43–51 (2013)
5. Elizalde, S., Noy, M.: Consecutive patterns in permutations. Adv. Appl. Math. **30**, 110–125 (2003)
6. Kubica, M., Kulczyński, T., Radoszewski, J., Rytter, W., Waleń, T.: A linear time algorithm for consecutive permutation pattern matching. Inf. Process. Lett. **113**, 430–433 (2013)
7. Kim, J., Eades, P., Fleischer, R., Hong, S.H., Iliopoulos, C.S., Park, K., Puglisi, S.J., Tokuyama, T.: Order-preserving matching. Theoret. Comput. Sci. **525**, 68–79 (2014)
8. Cho, S., Na, J.C., Park, K., Sim, J.S.: A fast algorithm for order-preserving pattern matching. Inf. Process. Lett. **115**, 397–402 (2015)
9. Crochemore, M., Iliopoulos, C.S., Kociumaka, T., Kubica, M., Langiu, A., Pissis, S.P., Radoszewski, J., Rytter, W., Waleń, T.: Order-preserving incomplete suffix trees and order-preserving indexes. In: Kurland, O., Lewenstein, M., Porat, E. (eds.) SPIRE 2013. LNCS, vol. 8214, pp. 84–95. Springer, Heidelberg (2013)

148   S. Cho et al.

10. Cormen, T.H., Leiserson, C.E., Rivest, R.L., Stein, C., et al.: Introduction to Algorithms. MIT press, Cambridge (2001)
11. Kim, J., Amir, A., Na, J.C., Park, K., Sim, J.S.: On representations of ternary order relations in numeric strings. In: Proceedings of the 2nd International Conference on Algorithms for Big Data, pp. 46–52, Palermo, 07–09 April (2014)

Range Minimum Query Indexes
in Higher Dimensions

Pooya Davoodi[1], John Iacono[1], Gad M. Landau[1,2], and Moshe Lewenstein[3(✉)]

[1] Polytechnic Institute of New York University, New York, USA
pooyadavoodi@gmail.com
[2] Haifa University, Haifa, Israel
landau@univ.haifa.ac.il
[3] Bar-Ilan University, Ramat Gan, Israel
moshe@cs.biu.ac.il

Abstract. Range minimum queries (RMQs) are essential in many algorithmic procedures. The problem is to prepare a data structure on an array to allow for fast subsequent queries that find the minimum within a range in the array. We study the problem of designing indexing RMQ data structures which only require sub-linear space and access to the input array while querying. The RMQ problem in one-dimensional arrays is well understood with known indexing data structures achieving optimal space and query time. The two-dimensional indexing RMQ data structures have received the attention of researchers recently. There are also several solutions for the RMQ problem in higher dimensions. Yuan and Atallah [SODA'10] designed a brilliant data structure of size $O(N)$ which supports RMQs in a multi-dimensional array of size N in constant time for a constant number of dimensions. In this paper we consider the problem of designing indexing data structures for RMQs in higher dimensions. We design a data structure of size $O(N)$ *bits* that supports RMQs in constant time for a constant number of dimensions. We also show how to obtain trade-offs between the space of indexing data structures and their query time.

1 Introduction

We consider the problem of preprocessing an array of elements into a data structure that supports range minimum queries (RMQs) asking for the position of the minimum element within a given range in the array. More formally, the d-dimensional range minimum query (d-RMQ) problem is:

J. Iacono—Research supported by NSF grant CCF-1018370 and BSF grant 2010437.
G.M. Landau—Research partially supported by the National Science Foundation Award 0904246, Israel Science Foundation grant 571/14, Grant No. 2008217 from the United States- Israel Binational Science Foundation (BSF) and DFG.
M. Lewenstein—Research supported by BSF grant 2010437, a Google Research Award and GIF grant 1147/2011.

F. Cicalese et al. (Eds.): CPM 2015, LNCS 9133, pp. 149–159, 2015.
DOI: 10.1007/978-3-319-19929-0_13

- **Input:** A d-dimensional array A over S with dimensions $(n_1 \times \cdots \times n_d)$ of size $N = n_1 n_2 \cdots n_d$, where S is a set of linearly ordered elements whose elements can be compared (for $\leq$) in constant time.
- **Output:** A data structure over A supporting the following queries.
- **Query:** Return the position of the minimum element in a range $q = [a_1..b_1] \times [a_2..b_2] \times \ldots \times [a_d..b_d]$ of A, that is,

$$d\text{-RMQ}(A, q) = \operatorname{argmin}_{[i_1, i_2, \ldots, i_d] \in q} A[i_1, i_2, \ldots, i_d]$$

This problem, in various dimensions, finds applications in databases, information retrieval, computational biology, etc. Also the well-known problem of range searching on points in higher dimensions can be reduced to range queries on arrays if the input point set is dense. In the realm of stringology its most well known use has been in the LCP (Longest Common Prefix) data structure, but many other uses have been made of this, e.g. [2].

Previous Work. The 1D-RMQ problem has been well studied. Initially, Gabow, Bentley and Tarjan [9] introduced the problem. They reduced the problem to the *Lowest Common Ancestor (LCA)* problem [11] on Cartesian Trees [13]. The *Cartesian Tree* is a binary tree defined on top of an array of n elements from a linear order. The root is the minimum element, say at location i of the array. The left subtree is recursively defined as the Cartesian tree of the sub-array of locations 1 to $i-1$ and the right subtree is defined likewise on the sub-array from $i+1$ to n. It is quite easy to see the connection between the RMQ problem and the Cartesian tree, which is what was utilized in [9]. The LCA problem was solved optimally in $\mathcal{O}(n)$ time and space allowing for $\mathcal{O}(1)$ time queries. This, in turn, yielded the result of $\mathcal{O}(n)$ preprocessing time and space for the 1D-RMQ problem with answers in $\mathcal{O}(1)$ time.

Fischer and Heun [8] showed that the space can be improved to $2n + o(n)$ *bits* and preprocessed in $\mathcal{O}(n)$ time for subsequent $\mathcal{O}(1)$ time queries. This space is close to the information-theoretic lower bound $2n - \Theta(\log n)$, which is derived by counting the number of Cartesian trees. Later Brodal et al. [5] showed how to reduce the size of the data structure to $\mathcal{O}(n/c)$ bits while supporting queries in optimal time $\mathcal{O}(c)$, for any parameter c that $1 \leq c \leq n$. This data structure is called an indexing data structure as it needs to read a number of elements from the input array of size $\mathcal{O}(n \log n)$ bits during querying. Such a model for designing data structures is called the *indexing model*, where we create an *index* which refers to A when answering queries, as opposed to the *encoding model*, where a data structure is able to answer queries *without* accessing A. Indexing data structures find applications in massive datasets where we would like to store the raw data in a slow memory and preprocess it into a small data structure stored in a fast memory.

For the 2D-RMQ problem the nice properties of 1D-RMQs do not seem to carry over. In Amir et al. [3] a solution for an $N = n_1 \times n_2$ sized array, and a parameter $k \geq 1$, was shown in $\mathcal{O}(Nk)$ word-space, $\mathcal{O}(N \log^{[k]} N)$ construction time with $\mathcal{O}(1)$ query time. Amir et al. [3] raised the question in what ways do the 2D RMQ problem and 1D RMQ problem differ. This turned out to have a complex answer. In the indexing model, Yuan and Atallah [14] showed an index of word-size $\mathcal{O}(N)$ with $\mathcal{O}(N)$ preprocessing time and $\mathcal{O}(1)$ query time. Brodal et al. [5] showed that an index with $\mathcal{O}(N)$ bits can also be constructed with the same preprocessing and query times. Furthermore, Brodal et al. [5] improved the space to a sub-linear number of bits, showing that for any index of size $\mathcal{O}(N/c)$ bits the query time will be $\Omega(c)$, for any parameter c that $1 \leq c \leq N$. They complemented this with an index of size $\mathcal{O}(N/c)$ bits that answers queries in time $\mathcal{O}(c \log^2 c)$. This query time was later improved to $\mathcal{O}(c \log c (\log \log c)^2)$ [4].

For the encoding model, Demaine et al. [7] showed that $o(N \log N)$ encodings are not possible in general for two dimensions. However, better solutions exist when $m << n$ [5] and when A is random [10].

For general dimension d, Chazelle and Rosenberg [6] proposed a solution which has a near constant query time (dependent on the inverse Ackerman function) and efficient space and time. Yuan and Atallah [14] gave the first constant query time RMQ algorithm for any constant sized dimension. Specifically, they show that there is an algorithm to preprocess a d-dimensional array A of size N using $\mathcal{O}((2.89)^d(d+1)!N)$ time and $\mathcal{O}(2^d d!N)$ space with query time $\mathcal{O}(3^d)$. Notice that the size of this data structure is $\mathcal{O}(2^d d!N \log N)$ in terms of bits. These results are in the indexing model as the input array A is needed by the data structures during querying.

Our Results. We consider higher dimensional arrays in the indexing model, as per the results of [6,14]. Our first result shows that for constant d we can create an index of size $\mathcal{O}(N)$ *bits* that supports queries in $\mathcal{O}(1)$ time. This improves the space usage of the data structure of [14] by a logarithmic factor. Our result works by a careful choice of a recursive algorithm utilizing a reduction technique that we define later. Our method makes a careful choice of use of certain calls to the algorithm of [14]. More specifically, our result constructs an index of size $\mathcal{O}((2.89)^d(d+1)!N)$ bits and answers queries in $\mathcal{O}(d!2^{\left(\binom{d+1}{2}\right)})$ time. For constant d, this translates to space $\mathcal{O}(N)$ bits and query time $\mathcal{O}(1)$.

Our second result allows for a parameter c that is $1 \leq c \leq N$. Here we construct an index with space $\mathcal{O}(d(2.89)^d(d+1)!N/c)$ bits which allows query time of $\mathcal{O}((d+1)!2^{\binom{d+1}{2}} + 2^d 2^{\binom{d+1}{2}} c(\log c)^{2(d-1)})$. This result does not achieve the state of the art 2D-RMQ indexes [4] which take space $\mathcal{O}(N/c)$ bits with query time $\mathcal{O}(c \log c (\log \log c)^2)$ since it is not clear how to extend the ideas of [4] which use properties of Fibonacci Lattices to higher dimensions. However it is the first d-RMQ index that gives a time-space trade-off for sub-linear space in bits.

The first result appears in Sect. 5 and is used as a sub-structure in the data structures of Sect. 6.

2   Preliminaries

The IndexEs that we present are recursive data structures that recurse on d, where the base of the recursion is when $d = 1$. It is known that RMQs on 1D arrays can be answered in constant time using a succinct representation of the Cartesian tree of the array (notice that an RMQ can be also answered naively in linear time with no preprocessing):

Lemma 1 *[8, 12]. For a 1D array of size N, there exists an Index of size $\mathcal{O}(N)$ bits which can answer 1D-RMQs on the array in $\mathcal{O}(1)$ time.*

This method is classical and there are several ways to solve this problem. The method most commonly known is a two-level partition with a tabulation scheme based upon encodings of Cartesian trees.

The Index that we present in Sect. 5 is based on a partitioning of the input array into blocks for two levels until the size of each block is small enough, and then we use a tabulation technique to create a lookup table storing the answer of queries within all possible blocks.

The tabulation technique that we use was introduced by Yuan and Atallah [14]. They designed an algorithm that preprocesses the input array of size N, using $\mathcal{O}(B(d)N)$ comparisons, into a data structure which then can be used to answer d-RMQs by comparing $\mathcal{O}(2^d)$ elements from the array, where $B(d) = O((2.89)^d \cdot (d+1)!)$ (the function is precisely described in [14] and is bounded using ordered Bell numbers). They used these comparisons to represent the array with a bit-sequence of size $\mathcal{O}(B(d)N)$. Applying this method to arrays of size G derives a lookup table that stores the answer of all G^2 queries within an array, for all possible arrays of size G which can be represented in $\mathcal{O}(B(d)G)$ bits:

Lemma 2 *[14]. For a d-dimensional array of size N which is partitioned into disjoint blocks of size G, there exists an index of size $O(2^{B(d)G}G^2 \log G + B(d)N)$ bits which can be used to answer RMQs within each block in time $O(2^d)$.*

2.1   Terminologies

- **Index:** We refer to an indexing data structure as simply index.
- **Rows and Directions:** For a dimension k, all elements in A that have the same coordinates in all dimensions except k form a row of A. More precisely, for fixed i_j for all $j \in \{1, \ldots, d\} - \{k\}$, all $A[i_1] \ldots [i_k] \ldots [i_d]$ form a row in A, for $i_k = 1 \cdots n_k$. We refer to this as a row that is in *direction* k. For example, in a 2D array which consists of rows and columns in dimensions 1 and 2 respectively, every row is in direction 2 and every column is in direction 1.
- **Block:** A block within a d-dimensional array $M[1..m_1, \ldots, 1..m_d]$ is defined by $M[i_1..j_1, \ldots, i_d..j_d]$ for any $i_k, j_k \in \{1, \ldots, m_k\}$, where $i_k \leq j_k$ and $k \in \{1, \ldots, d\}$. This block has dimensions (or size) $(j_1 - i_1 + 1 \times \cdots \times j_d - i_d + 1)$. The block is d'-dimensional if and only if the number of items $j_k - i_k + 1$ that are larger than 1 is d'.
- **Range Queries Spanning a Dimension:** We say a range (or a query) is row-spanning if and only if the range spans over one or several rows which means it *only* includes all the elements of the rows that are spanned by it. If the rows are in direction k, then the range is row-spanning in direction k. A block-spanning range is similarly a range that spans over a block.

  Observe that if a range spans a row that is in direction k, then all the other rows in direction k are either spanned by the range or do not have any element in the range. This is implied by the fact that ranges are axis-aligned. Such a range is also called a range that spans over dimension k.

3   Overview

We now give an overview of our approach. In Sect. 4 we describe a *reduction technique* which is a recursive partition of our d-dimensional input array. The reduction partitions q-dimensional arrays into a collection of $q - 1$-dimensional arrays for $q = d, d - 1, \ldots, 2$.

Upon receiving a query, if we are very lucky, the perimeter of the query may perfectly align with the partition of the d-dimensional array. In this case, we can answer the query easily using the reduction technique.

However, if the query does not align with the partition then we cannot use the result of the reduction technique directly and we need to handle our query more carefully. In this case we treat the internal aligned parts with the reduction technique. The leftover non-aligned parts of the query are now a collection of pieces within the "macro-blocks" of the array, which are of size ($\log n_1 \times \cdots \times \log n_d$). Each of these pieces is treated recursively for one more level down, which contains "micro-blocks" which are blocks of size ($\frac{\log \log n_1}{(2B(d))^{1/d}} \times \cdots \times \frac{\log \log n_d}{(2B(d))^{1/d}}$)). For leftover pieces within "micro-blocks" we use the tabulation technique of Lemma 2.

In Sect. 5.1 we explain this Index which we denote by $LBI(D)$. This Index is a recursive data structure on d, where the base of the recursion is $d = 1$, and we use the 1D-RMQ data structure of Lemma 1 at the base $LBI(1)$. In Sect. 5.2 we show how to make appropriate queries upon these data structures.

4   Reduction Technique

We present a method that, for a given input array M with dimensions ($m_1 \times \cdots \times m_d$) which can be partitioned into disjoint blocks with given dimensions $b = (b_1 \times \cdots \times b_d)$, generates an Index that supports d-RMQS spanning over the disjoint blocks. This Index is based on another RMQ Index, denoted by $\mathcal{I}$, which is given as part of the input. We denote by $R_{\mathcal{I}}(M, b)$ the RMQ Index generated by this method. In other words, the method constructs $R_{\mathcal{I}}(M, b)$ which is smaller than a given Index $\mathcal{I}$, but has less capability of supporting only block-spanning queries on M, while $\mathcal{I}$ supports all d-RMQS on M. We refer to this method as the *reduction technique*. This technique has been used in previous 2D-RMQ indexing data structures to reduce the space [4,5]. Here we use this technique for higher dimensions.

Now, we describe $R_{\mathcal{I}}(M, b)$. We partition M into disjoint blocks with dimensions b, and then we make an array D out of the minimum elements of the blocks, that is, $D[i_1, \ldots, i_d]$ contains the minimum element within the block $M[(i_1 - 1)b_1 + 1..i_1 b_1, \ldots, (i_d - 1)b_d + 1..i_d b_d]$. So D has dimensions $(m_1/b_1 \times \cdots \times m_d/b_d)$. Then we construct $\mathcal{I}$ for D, and we store it in $R_{\mathcal{I}}(M, b)$, and then we delete D.

A block-spanning d-RMQ q on M is reduced to an RMQ on D. To answer q, we first translate it to a query on D (by dividing the dimensions appropriately), and then use $\mathcal{I}$ to answer the query on D. However, since we do not have access

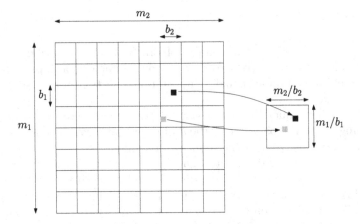

Fig. 1. Reduction technique for a 2D array: the two tiny squares denote the minimum elements within the corresponding blocks.

to D, whenever $\mathcal{I}$ wants to read an element in D, we will need to compute that element by reading the corresponding block from M and computing its minimum. Therefore, the query time of $R_{\mathcal{I}}(M,b)$ will be the query time of $\mathcal{I}$ multiplied by $b_1 b_2 \cdots b_d$ (the size of each block). The size of $R_{\mathcal{I}}(M,b)$ is obviously equal to the size of $\mathcal{I}$ (notice that we avoid storing D in $R_{\mathcal{I}}(M,b)$ to save space).

A block-spanning d-RMQ on M is reduced to a $(d-1)$-RMQ on D, if $b_k = m_k$ for some dimension k. In this case, we refer to the reduction technique as the *dimension reduction* technique. We note that the dimension reduction is a common technique in designing data structures for higher dimensions [1] (Fig. 1).

5   Space-Efficient Higher Dimensional RMQ

We present a recursive Index into a D-dimensional array A of size N with dimensions $(n_1 \times \cdots \times n_d)$ that supports d-RMQs on the array in time $\mathcal{O}(d! 2^{\binom{d+1}{2}})$ using $\mathcal{O}(B(d)N)$ bits. Specifically, this gives an Index of size $\mathcal{O}(N)$ bits with $\mathcal{O}(1)$ query time for $d = \mathcal{O}(1)$.

5.1   Data Structure

For each dimension $k \in \{1, \ldots, d\}$, we partition A into disjoint blocks with dimensions $b_k = (n_1 \times \cdots \times \log n_k \times \cdots \times n_d)$. We then make a full binary tree T_k on the blocks that we made for each dimension k, where each leaf corresponds to a block. We build a data structure for each of $T_1, \ldots, T_d$, that supports finding the lowest common ancestor (LCA) of two given query nodes in constant time using linear space on the size of the tree [11]. In the following, we explain the data structure that we store at T_k.

Let u be a left child at level h of T_k, where leaves are at level 0 (the number of blocks of dimensions b_k spanned by u is 2^h). We store the following data

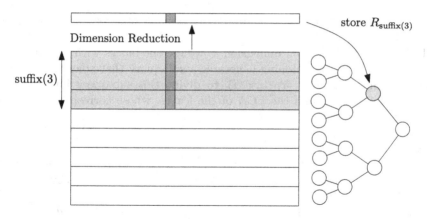

Fig. 2. Data structure for a 2D input array; The gray node corresponds to 4 blocks; The 3 gray blocks form suffix(3); suffix(3) consists of 3 blocks and is converted to a 1D array by the dimension reduction technique and $R_{\text{suffix}(3)}$ is constructed and stored at the gray node in the tree.

structure at u (we store no data structure at the root). Let suffix(i) denote i consecutive blocks of dimensions b_k in the sub-tree of u from right to left; that is, suffix(i) is a d-dimensional block in A with dimensions ($n_1 \times \cdots \times i \log n_k \times \cdots \times n_d$). We use the dimension reduction technique on suffix(i) with blocks of dimensions $b_{k,i} = (1 \times \cdots \times 1 \times i \log n_k \times 1 \times \cdots \times 1)$ that generates the Index $R_{\text{suffix}(i)}(b_{k,i}, LBI(d-1))$. This Index supports d-RMQS that span over blocks of dimensions $b_{k,i}$ in suffix(i). We make $R_{\text{suffix}(i)}(b_{k,i}, LBI(d-1))$ for all $i = 1, \ldots, 2^h$, and we store all of them at u. If u is a right child, we do the same thing symmetrically for consecutive blocks from left to right, and we make $R_{\text{prefix}(i)}(b_{k,i}, LBI(d-1))$ for all $i = 1, \ldots, 2^h$ similarly (Fig. 2).

The blocks with dimensions b_k used in partitioning A, altogether for all $k = 1, \ldots, d$, also partition A into disjoint d-dimensional *macro-blocks* with dimensions $b_{mac} = (\log n_1 \times \cdots \times \log n_d)$. Then we partition each macro-block into disjoint blocks with dimensions $b'_k = (\log n_1 \times \cdots \times (\frac{\log \log n_k}{(2B(d))^{1/d}}) \times \cdots \times \log n_d)$ for each dimension k, and we make $T_1 \ldots, T_d$ similarly for each macro-block.

Similarly the blocks with dimensions b'_k used in partitioning each macro-block, altogether for all $k = 1, \ldots, d$, also partition the macro-block into disjoint *micro-blocks* with dimensions $b_{mic} = (\frac{\log \log n_1}{(2B(d))^{1/d}} \times \cdots \times \frac{\log \log n_d}{(2B(d))^{1/d}})$. Now micro-blocks are small enough so that we can utilize the tabulation technique of Lemma 2 to support d-RMQs within micro-blocks ($G = b_{mic}$).

For each dimension $k = 1, \ldots, d$, we also store the 1D-RMQ Index of Lemma 1 built for each row in direction k of A. This data structure will be used to find the position of the minimum element within a range on a row in direction k.

5.2   Query Algorithm

Let $q = A[i_1..j_1 \times \cdots \times i_d..j_d]$ be a query range. If q is so small that fits within a micro-block, we use the tabulation technique of Lemma 2 to answer it. Otherwise, we partition q in two levels until sub-queries become small enough to fit within a micro-block.

In the first level for each dimension k, we use the blocks with dimensions b_k to partition q. This partitioning generates a sub-query that spans over several blocks of dimensions $b_{k,1}$. We use T_k to divide this sub-query into two sub-queries as follows: we find the LCA of the leaves, whose corresponding blocks contain i_k and j_k. We then use $R_{\text{suffix}(\ell_1)}(b_{k,\ell_1}, LBI(d-1))$ in the left child and $R_{\text{prefix}(\ell_2)}(b_{k,\ell_2}, LBI(d-1))$ in the right child of the LCA for the largest suffix(ℓ_1) and prefix(ℓ_2) that fit within q. These two IndexEs find the two rows in direction k which contain the minimum in suffix(ℓ_1) and prefix(ℓ_2) respectively. We then find the position of the minimum within each of these rows by answering appropriate 1D-RMQs on the rows.

The partitionings of the first level in all dimensions $k = 1, \ldots, d$, altogether generate at most 2^d sub-queries, where each sub-query is within a macro-block and contains a *corner* of q (a corner is a cell with maximum or minimum coordinates in all dimensions). We answer each of these sub-queries recursively using the data structures that we have stored for the corresponding macro-blocks. The partitionings of each macro-block similarly generate at most 2^d new sub-queries, where each new sub-query is within a micro-block and contains a corner of q. Now that these new sub-queries are within micro-blocks, we use the Index of Lemma 1 to answer them.

Theorem 1. *There exists an Index of size $\mathcal{O}(B(d)N)$ bits that supports d-RMQs on an array of size N in time $\mathcal{O}(d!2^{\binom{d+1}{2}})$, for $d = O(\log \log N / \log \log \log N)$.*

Proof. We prove the space and query time by induction on d. The base of the induction ($d = 1$) is clear due to Lemma 1. We first prove the space usage and then the query time. Let $(n_1 \times \cdots \times n_d)$ be the dimensions of the input array A.

The Index contains $T_1, \ldots, T_d$ in the first level of the partitioning. For each dimension k, the depth of T_k is $\mathcal{O}(\log(n_k / \log n_k)) = \mathcal{O}(\log n_k)$. The number of $R_{\text{suffix}(\cdot)}(b_{k,\cdot}, LBI(d-1))$ and $R_{\text{prefix}(\cdot)}(b_{k,\cdot}, LBI(d-1))$ that we store at each level is $n_k / \log n_k$. The size of each of these two IndexEs is $\mathcal{O}(B(d-1)N/n_k)$ bits by induction (due to the dimension reduction technique). Thus, $T_1, \ldots, T_d$ of the first level altogether take $\mathcal{O}(dB(d-1)N) = \mathcal{O}(B(d)N)$ bits, including the LCA data structure of each $T_1, \ldots, T_d$ [11]. There are $N/(\log n_1 \cdots \log n_d)$ macro-blocks. Each macro-block contains its own $T_1, \ldots, T_d$ which take $\mathcal{O}(B(d)(\log n_1 \cdots \log n_d))$ bits by similar argument. Thus, all the macro-blocks take $\mathcal{O}(B(d)N)$ bits. The size of the Index of Lemma 2 is $\mathcal{O}(B(d)N)$ bits for $G = b_{mic}$ if $d = O(\log \log N / \log \log \log N)$. The number of rows in direction k is n_k. Thus, the size of all the 1D-RMQ IndexEs that we make for each row is $\mathcal{O}(dN)$ bits for all dimensions. Therefore, the total space of $LBI(d)$ is $\mathcal{O}(B(d)N)$ bits.

In the first level, d sub-queries of q are generated that span over blocks of dimensions $b_{k,\ell}$ for each dimension $k = 1, \ldots, d$. To answer the sub-query in dimension k, an LCA query on T_k is performed in constant time [11]. Then the sub-query is answered using $R_{\text{suffix}(\cdot)}(b_{k,\cdot}, LBI(d-1))$ and

$R_{\text{prefix}(\cdot)}(b_{k,\cdot}, LBI(d-1))$, which take time $\mathcal{O}((d-1)!2^{\binom{d}{2}})$ (by induction) multiplied by the time to find the minimum in a block with dimensions $b_{k,i}$ (due to the dimension reduction technique), which can be done in constant time using the 1D-RMQ IndexEs stored for rows. Thus, each sub-query can be answered in time $\mathcal{O}((d-1)!2^{\binom{d}{2}})$, and all the sub-queries in time $\mathcal{O}(d(d-1)!2^{\binom{d}{2}}) = \mathcal{O}(d!2^{\binom{d}{2}})$. Then in the second level, q is reduced to at most 2^d queries within macro-blocks, which take time $\mathcal{O}(2^d d!2^{\binom{d}{2}}) = \mathcal{O}(d!2^{\binom{d+1}{2}})$ by similar argument. Finally, q is reduced to at most 2^d queries within micro-blocks, where each one can be answered in time $\mathcal{O}(2^d)$ due to Lemma 2, and thus time $\mathcal{O}(2^{2d})$ for all of them. Therefore, the overall query time is $\mathcal{O}(d!2^{\binom{d+1}{2}})$. □

6 Succinct High Dimensional RMQ

We present IndexEs that support d-RMQs on d-dimensional arrays, providing a trade-off between query time and the size of Index. The method that we use is to reduce the problem from d-RMQS on an array of size N to small d-RMQs on small arrays, and then create succinct IndexEs that support d-RMQs on small arrays.

6.1 Reduction to Small Arrays

We present an algorithm that transforms a given Index supporting d-RMQS on small arrays of size c^d with dimensions $(c \times \cdots \times c)$, to an Index that supports d-RMQS on an array of size N, where c is a parameter and $1 \leq c \leq N$.

The transformation algorithm is an extension of the algorithm of [4, Lemma 10] which was designed for two dimensions. Let A denote the input array of size N, $\mathcal{I}(d,c)$ denote the Index of size S bits for $(c \times \cdots \times c)$ arrays, and let $LBI(d)$ denote the Index of Theorem 1 for d dimensions.

Data Structure. For each dimension $k = 1, \ldots, d$, we use the reduction technique on A for blocks with dimensions $c_k = (1 \times \cdots \times 1 \times c \times 1 \times \cdots \times 1)$ (c is in dimension k), using the Index of Theorem 1. This provides an Index $R_{LBI(d-1)}(A, c_k)$ that supports d-RMQs that span over blocks with dimensions c_k on A. The blocks in all dimensions, altogether partition A into disjoint macro-blocks with dimensions $(c \times \cdots \times c)$. We construct $\mathcal{I}(d,c)$ for each macro-block.

Thus, the Index generated by the transformation algorithm contains $R_{LBI(d-1)}(A, c_k)$ for each $k = 1, \ldots, d$, plus $\mathcal{I}(d,c)$ for each macro-block. The size of each $R_{LBI(d-1)}(A, c_k)$ is $\mathcal{O}(B(d)N/c)$ bits due to Theorem 1 and the reduction technique. The number of macro-blocks is N/c^d. Therefore, the total space is $\mathcal{O}(dB(d)N/c + N/c^d S)$ bits.

Query Algorithm. If a given query is within a macro-block, it is answered using $\mathcal{I}(d,c)$ of the macro-block. Otherwise, the query is partitioned by the blocks with dimensions c_k in each dimension k. This generates a block-spanning sub-query in each dimension k, which can be answered using $R_{LBI(d-1)}(A, c_k)$. This can be done in time $\mathcal{O}(d!2^{\binom{d+1}{2}})$ multiplied by the time $\mathcal{O}(c)$ to find the minimum in a block (required by the reduction technique). Thus, all the sub-queries can be answered in time $\mathcal{O}(cdd!2^{\binom{d+1}{2}})$. These partitionings also generate at most 2^d sub-queries, where each sub-query is within a macro-block, and contains a corner of the query. Each of these sub-queries can be answered in time T using $\mathcal{I}(d, c)$ of the corresponding macro-block. Thus, all the sub-queries can be answered in time $\mathcal{O}(2^d T)$. Therefore, the total query time is $\mathcal{O}(c(d+1)!2^{\binom{d+1}{2}} + 2^d T)$.

Lemma 3. *If there exists an Index of size S bits with query time T that supports d-RMQS within an array of size c^d with dimensions $(c \times \cdots \times c)$, then for a d-dimensional array of size N, we can build an Index of size $\mathcal{O}(DB(d)N/c + N/c^d S)$ bits that supports d-RMQs in time $\mathcal{O}(c(d+1)!2^{\binom{d+1}{2}} + 2^d T)$, where c is a parameter and $1 \leq c \leq N$.*

6.2   Indexes on Small Arrays

We obtain an Index that supports d-RMQS on arrays with dimensions $(c \times \cdots \times c)$, which then immediately implies IndexEs on general arrays using Lemma 3. In the following, we let B denote an input array with dimensions $(c \times \cdots \times c)$. The Index is a recursive data structure that recurses on d, and we denote it by $\mathcal{I}(D, c)$.

   The general idea is to partition B into disjoint small blocks several times, each time with a different block size. These partitions divide a d-RMQ into block-spanning sub-queries which are then answered using the reduction technique. The full details of the data structure will appear in the journal version of this paper. The result that follows from the construction is:

Theorem 2. *There exists an Index of size $\mathcal{O}(DB(d)N/c)$ bits for a parameter c that $1 \leq c \leq N$, which supports d-RMQs on a d-dimensional array with N elements in time $\mathcal{O}((d+1)!2^{\binom{d+1}{2}}c + 2^d 2^{\binom{d+1}{2}}c(\log c)^{2(d-1)})$.*

References

1. Afshani, P., Arge, L., Larsen, K.D.: Orthogonal range reporting in three and higher dimensions. In: FOCS, pp. 149–158 (2009)
2. Amir, A., Apostolico, A., Landau, G.M., Levy, A., Lewenstein, M., Porat, E.: Range LCP. J. Comput. Syst. Sci. **80**(7), 1245–1253 (2014)
3. Amir, A., Fischer, J., Lewenstein, M.: Two-dimensional range minimum queries. In: Ma, B., Zhang, K. (eds.) CPM 2007. LNCS, vol. 4580, pp. 286–294. Springer, Heidelberg (2007)
4. Brodal, G.S., Davoodi, P., Lewenstein, M., Raman, R., Srinivasa Rao, S.: Two dimensional range minimum queries and fibonacci lattices. In: Epstein, L., Ferragina, P. (eds.) ESA 2012. LNCS, vol. 7501, pp. 217–228. Springer, Heidelberg (2012)

5. Brodal, G.S., Davoodi, P., Rao, S.S.: On space efficient two dimensional range minimum data structures. Algorithmica **63**(4), 815–830 (2012)
6. Chazelle, B., Rosenberg, B.: The complexity of computing partial sums off-line. Int. J. Comput. Geom. Appl. **1**(1), 33–45 (1991)
7. Demaine, E.D., Landau, G.M., Weimann, O.: On cartesian trees and range minimum queries. In: Albers, S., Marchetti-Spaccamela, A., Matias, Y., Nikoletseas, S., Thomas, W. (eds.) ICALP 2009, Part I. LNCS, vol. 5555, pp. 341–353. Springer, Heidelberg (2009)
8. Fischer, J., Heun, V.: Space-efficient preprocessing schemes for range minimum queries on static arrays. SIAM J. Comput. **40**(2), 465–492 (2011)
9. Gabow, H.N., Bentley, J.L., Tarjan, R.E.: Scaling and related techniques for geometry problems. In: Proceedings of the 16th Annual ACM Symposium on Theory of Computing, pp. 135–143. ACM Press (1984)
10. Golin, M., Iacono, J., Krizanc, D., Raman, R., Rao, S.S.: Encoding 2D range maximum queries. In: Asano, T., Nakano, S., Okamoto, Y., Watanabe, O. (eds.) ISAAC 2011. LNCS, vol. 7074, pp. 180–189. Springer, Heidelberg (2011)
11. Harel, D., Tarjan, R.E.: Fast algorithms for finding nearest common ancestors. SIAM J. Comput. **13**(2), 338–355 (1984)
12. Sadakane, K.: Succinct data structures for flexible text retrieval systems. J. Discrete Algorithms **5**(1), 12–22 (2007)
13. Vuillemin, J.: A unifying look at data structures. Commun. ACM **23**(4), 229–239 (1980)
14. Yuan, H., Atallah, M.J.: Data structures for range minimum queries in multidimensional arrays. In: Charikar, M. (ed.) SODA, pp. 150–160. SIAM (2010)

Alphabet-Dependent String Searching
with Wexponential Search Trees

Johannes Fischer[1][(✉)] and Paweł Gawrychowski[2]

[1] TU Dortmund, Dortmund, Germany
johannes.fischer@cs.tu-dortmund.de
[2] University of Warsaw, Warsaw, Poland
gawry@mimuw.edu.pl

Abstract. We consider finding a pattern of length m in a compacted (linear-size) trie storing strings over an alphabet of size σ. In static tries, we achieve $O(m + \lg \lg \sigma)$ deterministic time, whereas in dynamic tries we achieve $O(m + \frac{\lg^2 \lg \sigma}{\lg \lg \lg \sigma})$ deterministic time per query or update. One particular application of the above bounds (static and dynamic) are suffix trees, where we also show how to pre- or append letters in $O(\lg \lg n + \frac{\lg^2 \lg \sigma}{\lg \lg \lg \sigma})$ time. Our main technical contribution is a weighted variant of exponential search trees, which might be of independent interest.

1 Introduction

Text indexing is a fundamental problem in computer science. It requires storing a text of length n, composed of letters from an alphabet of size σ, such that subsequent pattern matching queries can be answered quickly. Typical such pattern matching queries are (1) *existential* queries (deciding whether or not the pattern occurs in the text), (2) *counting* queries (determining the number of occurrences), and (3) *enumeration* queries (listing all positions where the pattern occurs). Task (3) usually follows (2), and takes $O(occ)$ additional time for enumerating the occ occurrences. In our model, the text is either static, or can be modified by pre- or appending new letters [5].

Well-known static text indexes are *suffix trees* [12] and *suffix arrays* [11]. The former admit, in their plain form, $O(m \lg \sigma)$ existential and counting queries for a pattern of length m, while the latter achieve $O(m + \lg n)$ time. With perfect hashing [8], suffix trees can achieve $O(m)$ time, but then their $O(n)$ preprocessing time is only in *expectation*. Enjoying the best of both worlds, the *suffix tray* of Cole et al. [6] achieves $O(m + \lg \sigma)$ searching time, with a linear space data structure that can be constructed in $O(n)$ deterministic time for a static text. If superlinear construction time is allowed, then the proprocessing time rises to deterministic $O(n \lg^2 \lg n)$ time, yielding deterministic worst-case $O(m)$ search time [14]. Summing up, all current text indexes have either non-linear pattern

---

P. Gawrychowski—Currently holding a post-doc position at Warsaw Center of Mathematics and Computer Science.

F. Cicalese et al. (Eds.): CPM 2015, LNCS 9133, pp. 160–171, 2015.
DOI: 10.1007/978-3-319-19929-0_14

matching time, or super-linear size, or are randomized. In this paper, we focus on speeding up the pattern matching time for deterministic linear-space indexes.

In the dynamic setting, suffix trees can be updated in amortized expected constant time, where the amortization comes from the need to locate the node which should be updated, and expectation from the hashing used to store outgoing edges. If we insist on getting worst-case time bounds, a recent result of Kopelowitz [9] allows updates in $O(\lg \lg n + \lg \lg \sigma)$ worst-case (but still expected) time. The *suffix trists* of Cole et al. [6] achieve a deterministic bound of $O(m + \lg \sigma)$ for searching, with a linear space data structure that can be updated in $O(f(n, \sigma) + \lg \sigma)$ deterministic and worst-case time, where $f(n, \sigma)$ is the time required to locate the edge of the suffix tree which should be split, and the best bound on $f(n, \sigma)$ known so far for the general case is $O(\lg n)$ [1]. Also worth mentioning is the very recent result of Kucherov and Nekrich [10], who achieve deterministic worst-case constant update time and $O(m + occ)$ reporting time for constant size alphabets. It is not clear how to adapt their solution to larger alphabets, and they do not maintain the whole suffix tree, though.

In all of the tree based structures mentioned in the previous paragraph, the crucial point is how to implement the outgoing edges of the tree such that they can be searched efficiently for a given query character. This is the general setting of *trie search*, and in fact, we can and do formulate our results in terms of tries, and view suffix trees as one particular application. In this setting, it is worth mentioning that for static tries one can achieve $O(m + \lg n)$ query time after linear preprocessing by storing the edges outgoing from each node in a weighted binary search tree (see [13]; similar ideas are implicit in some other later papers). For dynamic tries, Andersson and Thorup [2] show how to update or search the trie in $O(m + \sqrt{\lg n / \lg \lg n})$ deterministic worst-case time, for n being the number of stored strings. In this article, however, we focus on alphabet-*dependent* running times, since the size of the alphabet is usually substantially smaller than the size of the whole input data. For instance, one often considers the case where the alphabet is of a polylogarithmic (in terms of, say, the input word) size.

1.1 Our Result and Outline

We formulate our results in the general setting of trie search (all our results are in the word RAM model). There it is usual to support stronger forms of existential queries, namely prefix queries (computing the longest prefix of the pattern which is a prefix of one of the stored strings), and predecessor queries (returning the largest string stored that is no larger than the query pattern):

Theorem 1. *A compacted trie storing n static strings over an integer alphabet of size σ can be stored in $O(n)$ space (in addition to the stored strings themselves) such that subsequent prefix- or predecessor queries can be answered in $O(m + \lg \lg \sigma)$ deterministic worst-case time, for patterns of length m. This data structure can be constructed in deterministic $O(n)$ time, for $\sigma = n^{O(1)}$.*

This improves the previously best deterministic solutions, either with $O(n)$ preprocessing time and $O(m + \lg \sigma)$ search time [6], or with $O(n \lg^2 \lg n)$ preprocessing time and $O(m)$ search time [14].

Theorem 2. *We can maintain a linear-size (in addition to the stored strings themselves) structure for a compacted trie storing strings over an integer alphabet of size σ under adding a new string in deterministic worst-case time $O(m + \frac{\lg^2 \lg \sigma}{\lg \lg \lg \sigma})$ so that subsequent on-line prefix- or predecessor queries can be answered in $O(m + \frac{\lg^2 \lg \sigma}{\lg \lg \lg \sigma})$ deterministic time, for strings of length m and $\sigma = n^{O(1)}$.*

This simultaneously improves the previously best deterministic worst-case $O(m + \sqrt{\lg n / \lg \lg n})$ bound by Andersson and Thorup [2], and the $O(m + \lg \sigma)$ bound by Cole et al. [6]. While these results are of mostly theoretical interest, they are not just improvements of the lower order terms, as the dominating part of $O(m + \lg \sigma)$ might actually be $\lg \sigma$. Given that "trie search" is a rather fundamental operation, we think that our improvements are substantial.

One particularly important application of Theorems 1 and 2 are suffix trees [15]. There, Theorem 2 allows us to perform updates (appending letters to the text) in $O(\frac{\lg^2 \lg \sigma}{\lg \lg \lg \sigma})$ time, assuming that we are given the edge of the suffix tree which should be split. It was already noted that the currently best bound [1] for a deterministic worst-case suffix tree oracle providing such information is $f(n, \sigma) = O(\lg n)$. Though not being the main goal of this article, in the full version [7] we show a faster suffix tree oracle with $f(n, \sigma) = O(\lg \lg n + \frac{\lg^2 \lg \sigma}{\lg \lg \lg \sigma})$, and thus we achieve truly superior times over [6].

1.2 Technical Contributions

Our main technical novelty is a weighted variant of exponential search trees [2], which we term *wexponential search trees*. The original exponential search tree achieves $O(\lg \lg n \cdot \frac{\lg \lg u}{\lg \lg \lg u})$ search and update times for n elements over a universe of size u. Our weighted variant generalizes this to $O(\lg \frac{\lg W}{\lg w} \cdot \frac{\lg \lg u}{\lg \lg \lg u})$, where w is the weight of the searched element, and W is the sum of all weights. The advantage of this is that in a sequence of t hierarchical accesses, where the old "w" is always the new "W", the sum telescopes to $O(\lg \lg n \cdot \frac{\lg \lg u}{\lg \lg \lg u})$ instead of $O(t \cdot \lg \lg n \cdot \frac{\lg \lg u}{\lg \lg \lg u})$. While this general idea is pervasive in data structures, we are not aware of any previous application in the doubly-logarithmic setting. We believe that this generalization might find other applications.

2 Preliminaries

We assume textbook knowledge of tries, suffix trees and suffix arrays.

Proposition 1 (Ružić, Theorem 3 [14]). *A static linear-size dictionary on a set of k keys can be deterministically constructed in time $O(k \lg^2 \lg k)$, so that lookups to the dictionary take worst-case time $O(1)$.*

Proposition 2 *A static linear-size data structure on a set of k sorted keys from a universe of size u can be deterministically constructed in time $O(k)$, so that subsequent* predecessor *queries can be answered in $O(\lg \lg u)$ worst-case time.*

Proposition 3 (Beame and Fich, Theorem 4.5 [3]). *A static data structure on a set S of k keys from a universe of size u can be deterministically constructed in $O(k^{1+\epsilon})$ time and space, so that subsequent* predecessor *queries can be answered in $O(\frac{\lg \lg u}{\lg \lg \lg u})$ time in the worst-case.*

Proposition 4 ([2]). *A dynamic linear-size data structure on a set S of k keys from a universe of size u can be maintained, so that subsequent* predecessor *queries can be answered in $O(\frac{\lg^2 \lg u}{\lg \lg \lg u})$ time, and new keys can be inserted in $O(\frac{\lg^2 \lg u}{\lg \lg \lg u})$ time, where both bounds are deterministic worst-case.*

3   New Static Data Structure

In this section we prove Theorem 1. We store *edges* of the trie as pairs of the form (v, a), where v is a pointer to the source of the edge, and $a \in \Sigma$ is the first character on the edge from v to its corresponding child w. As secondary information we also attach a pointer to w with the pair (v, a). This way, matching a pattern $P[1, m]$ reduces to repeatedly finding correct edges: assuming inductively that we have already matched $P[1, i]$ and are currently at node v, we check if the edge $(v, P[i + 1])$ exists, and move to that child if this is the case.

We now describe how the edges are stored. A naive storage with the data structures from Sect. 2 would result in superlinear construction time. Hence, we need to introduce several levels of indirection. Like in the suffix tray [6], we divide nodes into heavy and light, but this time with parameter $s := \Theta(\lg^2 \lg \sigma)$: a node with at least s leaves below it is called *heavy*, otherwise it is called *light*.

To continue, we classify the outgoing edges of heavy nodes into two different types: (heavy,heavy) and (heavy,light). Here, the first component of each tuple refers to the type of the parent node, and the second one to that of the child. For each *branching* heavy node (meaning it has more than 1 heavy child) we store the outgoing edges of type (heavy,heavy) in a dictionary using Proposition 1, with the key being the first character on the edge. Since there are at most $\frac{n}{s}$ heavy nodes with no heavy children, the total size of all dictionaries cannot exceed the sum of degrees in a tree on $2\frac{n}{s}$ nodes, which is $O(\frac{n}{s})$. Furthermore, the number of elements in each dictionary is at most σ, so constructing all those data structures takes $O(\frac{n}{s} \cdot \lg^2 \lg \sigma) = O(n)$ time. For nonbranching heavy nodes, we only have at most one edge of type (heavy,heavy), so each such node simply stores a special pointer to its only heavy child, which enables us to decide in $O(1)$ time if we need to continue matching there. Finally, for each (branching or not) heavy node we store all the outgoing edges of type (heavy,light) using the data structure from Proposition 2. Using the structure, we can locate the light child we should descend to in $O(\lg \lg \sigma)$ time. Then we are inside a small subtree of size at most s and binary search the remaining nodes in additional $O(m + \lg s)$ time.

The binary searching is very similar to the method used in pattern matching with suffix arrays. Each light node having a heavy parent stores a sorted list of all leaves in its subtree. We binary search over the list to locate the successor of the remaining part of the query among all strings corresponding to the leaves in $O(\lg s)$ time [11]. Summing up, the whole search takes $O(m + \lg\lg\sigma)$ time.

We described the structure to support prefix queries. For predecessor queries, we need to store additional predecessor data structures (again using Proposition 2) with *all* edges outgoing from a node, which is used only once if we terminate at a heavy node. Answering queries is straightforward.

4   New Dynamic Data Structure

In the full version [7] we prove that it is enough to maintain trees of size $O(\sigma)$ instead of $O(n)$. Here we first develop our *wexponential search trees*, which might be of independent interest. Then we show how to use it to build the new data structure, first in the amortized setting, and then making the bounds worst-case.

4.1   Weighted Exponential Trees

Theorem 3. *There is a linear-size data structure that allows us storing a collection of weighted sorted keys from an ordered universe of size u so that predecessor search takes $O(\lg\frac{\lg W}{\lg w}\frac{\lg\lg u}{\lg\lg\lg u})$ time, where W is the current total weight of all elements, and w is the weight of the predecessor found. Inserting a new element of weight 1 takes $O(\lg\lg W)$ time, and increasing by one the weight of an element of weight w (specified by a handle returned by the insertion procedure) takes $O(\lg\frac{\lg W}{\lg w})$ time. All bounds are deterministic worst-case.*

The proof of the theorem is based on the beautiful idea of Bender et al. [4], who have shown how to significantly simplify the deamortization presented by Andersson and Thorup [2]. We start with the amortized version of the theorem, and later show how to make it worst-case efficient.

Let $f(\ell) = \lfloor 2^{(\frac{3}{2})^\ell} \rfloor$, so $\ell = \Theta(\lg\lg f(\ell))$. We define a weighted multiway search tree with the degrees increasing doubly-exponentially along any leaf-to-root path, which we call a wexponential search tree of level ℓ, in the following recursive manner:

1. the (explicitly stored) current total weight W of all elements is less than $2f(\ell+1)$ ($2f(\ell+1) + \frac{1}{2}f(\ell)$ in the worst-case version), and if $W \geq 2f(\ell)$ the tree is *proper*,
2. we store a static predecessor structure (implemented using Proposition 3) containing a subset $S = \{e_1, \ldots, e_{|S|}\}$ of all elements (the "*splitters*")
3. the remaining elements are split into $X_0, \ldots, X_{|S|}$ such that e_i is between X_{i-1} and X_i,
4. the total weight of all elements in $\{e_i\} \cup X_i \cup \{e_{i+1}\}$ exceeds $f(\ell) - f(\ell-1)$ for all $i = 1, \ldots, |S| - 1$,

5. each X_i is stored in a wexponential search tree of level $\ell - 1$, which are called the *children*,
6. for each i the predecessor structure stores a bidirectional link to the child storing X_i, and additionally a link to the leftmost child storing X_0 is kept.

Observe that if a weight of an element is at least $2f(\ell)$, it must belong to S. Note also that the definition implies that some of the X_i's may be empty. Furthermore, we can bound the size of S as follows:

$$|S| \leq 2\frac{2f(\ell+1)}{f(\ell) - f(\ell-1)} = O(2^{(\frac{3}{2})^{\ell+1} - (\frac{3}{2})^\ell}) = O(2^{\frac{1}{2}(\frac{3}{2})^\ell}) = O(f^{\frac{1}{2}}(\ell)). \quad (1)$$

Updating the structure will be done in a bottom-up order. In other words, we will assume that the children are valid wexponential search trees of level at most $\ell-1$, and show how to ensure that their parent is a valid tree of level ℓ. Inserting a new element of weight one or increasing the weight of some element by one might cause the total weight to become $2f(\ell+1)$. As soon as we detect such a situation, we *split* the tree into two by choosing an element e_i from S. To choose this element we look at the sets of its predecessors $P_i = X_0 \cup \{e_1\} \cup \ldots \cup \{e_{i-1}\} \cup X_{i-1}$ and successors $S_i = X_i \cup \{e_{i+1}\} \cup \ldots \cup \{e_{|S|}\} \cup X_{|S|}$. As the total weight is $2f(\ell+1)$, and the weight of any X_i is less than $2f(\ell)$ (by conditions 1 and 5), we can always select an i so that the weight of both P_i and S_i is less than $f(\ell + 1) + f(\ell)$, but the weight of both $P_i \cup \{e_i\}$ and $\{e_i\} \cup S_i$ is at least $f(\ell + 1) - f(\ell)$; see Fig. 1. We construct two new wexponential search trees of level ℓ containing all elements in P_i and S_i, which is possible as their total weights are at most $f(\ell+1)+f(\ell) < 2f(\ell+1)$. This requires constructing static predecessor structures containing $e_1, e_2, \ldots, e_{i-1}$ and $e_{i+1}, e_{i+2}, \ldots, e_{|S|}$, respectively, and making each wexponential search tree of level $\ell - 1$ a child of the former or the latter new tree. Notice that we don't have to rebuild the smaller trees, as simply redirecting the pointers to already existing structures is enough. Then we look at the parent of the structure that we are splitting. If there is none, we simply create a new proper wexponential search tree of level $\ell+1$ with just one splitter e_i. Otherwise we add e_i to its set of splitters, and store pointers to the two newly created trees of level ℓ in its predecessor structure, which needs to be rebuilt, or *refreshed*. The whole splitting and refreshing process is a very local procedure, as it requires rebuilding the static predecessors structures only for the tree and its parent, while the descendants and further ancestors are kept intact.

Now we can describe how to query and update the wexponential search tree.

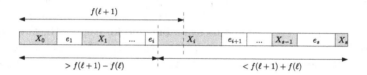

Fig. 1. Finding a suitable new splitter e_i. The smaller of the two parts has size at least $f(\ell + 1) - f(\ell)$, and hence the larger one at most $f(\ell + 1) + f(\ell)$.

Predecessor Search: First we use the static predecessor structure, which takes $O(\frac{\lg \lg u}{\lg \lg \lg u})$. If the query element belongs to S, we are done. Otherwise we recurse in the smaller structure.

Insert: Using the static predecessor structure we locate the smaller structure the new element belongs to, and insert it there recursively. Then we increase W by one and split the tree, if necessary. For each new element we allocate a record storing a link to the tree where we have used it as a splitter, and return a pointer to this record. Whenever some static predecessor structure is rebuilt, or an element is moved up to the parent, we update the record.

Increasing: We locate the tree where the element is a splitter using the record in constant time. Then we increase the total weight by one and split the tree, if necessary, and move to its parent.

Lemma 1. *Consider a proper wexponential search tree of total weight W and an element of weight w. The element is used as a splitter at depth $O(\lg \frac{\lg W}{\lg w})$.*

Proof. As the tree is proper, its level is $\Theta(\lg \lg W)$. On the other hand, if an element of weight w belongs to a subtree of level ℓ, but is not chosen as a splitter there, then $w < 2f(\ell)$, which is equivalent to $\ell = \Omega(\lg \lg w)$. Hence the maximum possible difference between the level of the whole tree and the level of the subtree where the element is used as a splitter is $O(\lg \frac{\lg W}{\lg w})$.   □

The above lemma shows that worst-case complexity of predecessor search in a wexponential search tree is $O(\lg \frac{\lg W}{\lg w} \frac{\lg \lg u}{\lg \lg \lg u})$, where W is the total current weight and w is the weight of the predecessor found. Indeed, consider this (unique) predecessor: it must be stored at depth $O(\lg \frac{\lg W}{\lg w})$, and traversing each level requires one query to the static predecessor structure. The complexity of both insert and increase is more tricky to estimate, as we might need to repeat the expensive splitting procedure a couple of times. Nevertheless, insert traverses all $O(\lg \lg W)$ levels, and increase traverses just $O(\lg \frac{\lg W}{\lg w})$ levels. At each of those levels we might need to split the tree, which takes a lot of time, but cannot happen very often. We start with an amortized bound.

For each wexponential tree we maintain an invariant that we have at least $\max(0, W - f(\ell+1))$ credits allocated there, where W is the current total weight and ℓ is the level. As long as we don't split, the invariant is easy to maintain: whenever we move from a tree to its parent during an insert or increase and add one to its total weight, we put an additional credit there. Now consider splitting a tree of total weight $W = 2f(\ell + 1)$. We need to rebuild the static predecessor structures at both the tree (or, more precisely, at the two new trees) and its parent, hence we need to apply Proposition 3 to a set of size which we bounded in (1) by $f^{\frac{1}{2}}(\ell+1)$, which takes $O(f(\ell+1))$ time. On the other hand, we have $W - f(\ell+1) = f(\ell+1)$ credits available, and for each of the two new trees we need to keep just $\max(0, f(\ell+1) + f(\ell) - f(\ell+1)) = f(\ell)$ of them (recall that the larger of the new trees has weight at most $f(\ell+1) + f(\ell)$). Hence we can use the remaining $f(\ell+1) - 2f(\ell) = \Theta(f(\ell+1))$ credits to pay for the reconstruction.

Deamortizing the running time requires more care. Fortunately, we can fairly closely follow the method of Bender et al. [4]. Instead of immediately splitting a tree as soon as its weight becomes $2f(\ell + 1)$, we perform it incrementally over $\frac{1}{2}f(\ell)$ updates concerning the tree or one of its descendants, starting when the weight becomes $2f(\ell + 1)$. Furthermore, instead of refreshing the parent as soon as we have two new trees, we use the bidirectional pointer to replace the link to the old tree kept there by a record containing the element e_i used for partitioning and the links to the new two trees. As long as the parent is not fully refreshed and the new trees are still linked from the same record, we call them *twins*. Similarly, refreshing is performed incrementally over $\frac{1}{2}f(\ell)$ updates concerning the tree or one of its descendants, starting whenever we notice that the weight is a multiple of $\frac{1}{2}f(\ell)$ not exceeding $2f(\ell + 1) - \frac{1}{2}f(\ell)$. This ensures that we never need to split a twin, as between splitting a child and splitting one of the two new children created as a result, the tree will be fully refreshed, hence we avoid a situation where we already keep a record containing two links, and now we would need to replace one of them by such a record again. As we never try to refresh and split a tree at the same time, there is no interference there.

Splitting requires first choosing a good splitter e_i, which can be done in a single left-to-right sweep through the contents of the static predecessor structure, and then building new static predecessor structures containing $\{e_1, e_2, \ldots, e_{i-1}\}$ and $\{e_{i+1}, \ldots, e_{|S|}\}$. As this is done in the background, it might happen that some updates concerning the already seen part occur. Hence we can only guarantee that the total weight of P_i and S_i is at most $f(\ell + 1) + \frac{3}{2}f(\ell)$, as there might be up to $\frac{1}{2}f(\ell)$ of such updates. An additional complication is that the weight of P_i or S_i might be so that we would expect the corresponding new tree to be undergoing refreshing at the moment, which is the reason we have chosen to refresh when the weight is a multiple of $\frac{1}{2}f(\ell)$ instead of simply taking a multiple of $f(\ell)$. We never skip two consecutive moments when we should start refreshing, as the next split starts not sooner than after $f(\ell + 1) - \frac{3}{2}f(\ell) \geq f(\ell)$ updates, hence it still never happens that we try to split a twin.

Rebuilding the static predecessor structures in the background requires storing two versions at the same time. More precisely, we have the old version, which we use for navigation and answering any query, and the new one that is being built. The corresponding elements in both versions are linked to each other, and any update is performed in both of them. When the new version is ready, we simply discard the old one in constant time. Discarding can be done by storing timestamps that can be used to determine which links are still valid, and which should be actually null at the moment. The definition of a wexponential tree must be relaxed so that the current total weight never exceeds $2f(\ell + 1) + \frac{1}{2}f(\ell)$, so the bounds on the complexity of search, and both insert and increase (excluding the cost of splitting) still hold, and the theorem follows.

4.2   Amortized Version of Theorem 2

For each node v we keep two separate structures. The first is a static dictionary implemented using Proposition 1 that stores edges leading to all children v_i of size "similar" to the size of v. The second is a dynamic predecessor data structure implemented using Theorem 3. To make the notion of "similar size" more precise, define the *weight* of a node to be the number of leaves in its subtree, and its *level* to be ℓ when the weight belongs to $[f(\ell), 2f(\ell+1))$ ($[f(\ell), 3f(\ell+1))$ in the worst-case version presented in the next section), where $f(\ell) = \left\lfloor 2^{(\frac{3}{2})^\ell} \right\rfloor$, again. Clearly, the weights and hence the levels along any root-to-leaf path are nonincreasing. We define the *fragment* of node v to be the maximal subtree containing v and consisting of nodes of the same level. The root r of a fragment containing v is its lowest ancestor of level ℓ, but with parent of level at least $\ell + 1$. For each such fragment we store the root r and a list of bidirectional links to all its nodes, and a counter with the number of leaves in the subtree rooted at r. For each node v we store all edges leading to children of v of the same level ℓ in a static dictionary. All edges leading to children of smaller levels are kept in a wexponential search tree. We would like the weights in this structure to be the same as the weights of the corresponding nodes, but for technical reasons we maintain a weaker condition, namely a node of weight w has weight from $[\sqrt{w}, w]$ in the wexponential search tree stored at its parent. First we show that this relaxation is enough to guarantee good bounds on the search time.

Lemma 2. *Traversing any path of length m takes $O(m + \frac{\lg^2 \lg \sigma}{\lg \lg \lg \sigma})$ time.*

Proof. At each node we first use the static dictionary to check if the next edge we would like to traverse is stored there. If so, we continue. Otherwise we query the wexponential search tree. In order to bound the complexity of the whole procedure, we only have to bound the total time of the latter steps, as the former sum to at most $O(m)$. Observe that whenever we query the wexponential search tree, we either terminate, or decrease the current level, which is already enough to get a bound of $O(\frac{\lg^3 \lg \sigma}{\lg \lg \lg \sigma})$. To get the claimed complexity, let $W_1, W_2, \ldots, W_k$ be the weights of nodes where we query a wexponential search tree. Similarly, let $w_1, w_2, \ldots, w_k$ be the corresponding weights of nodes that we find there (note that W_i is not necessarily stored in our implementation, but w_i certainly is). Let w_i' be the weight of the elements of the wexponential search tree corresponding to w_i, and W_i' the total weight of this structure. Then we have the following inequalities: $w_i' \in [\sqrt{w_i}, w_i]$, $W_i' \le W_i$, and $W_{i+1} \le w_i$. So the time of this part is order of $\frac{\lg \lg \sigma}{\lg \lg \lg \sigma} \sum_i \lg \frac{\lg W_i'}{\lg w_i'} \le \frac{\lg \lg \sigma}{\lg \lg \lg \sigma}(\lg \lg \sigma + \sum_i \lg \frac{\lg W_i}{\lg w_i}) \le \frac{\lg^2 \lg \sigma}{\lg \lg \lg \sigma}$. $\qquad\square$

Whenever we add a new leaf, the weights of all nodes on its path to the root increase by one. We iterate over all fragments above the new leaf and increase their counters. Iterating is done in $O(\lg \lg \sigma)$ time by starting at the leaf and repeatedly jumping to the root of the current fragment by following the bidirectional link stored at each node. To maintain the invariant that the weights on any path are nonincreasing, we actually first construct a list of all fragments, and then update their counters one-by-one in a top-down order. For each root that we consider we need to update its corresponding weight in the wexponential search tree at its parent, which takes at most $O(\lg \frac{\lg W}{\lg \sqrt{w}}) = O(1 + \lg \frac{\lg W}{\lg w})$ time, where w is the weight of this root, and W is at most the weight of its parent. Summing up over all roots, as in the proof of Lemma 2, we get a telescoping expression which is at most $O(\lg \lg \sigma)$.

During this procedure it might happen that we increase the weight of some root r to $2f(\ell+1)$, and hence need to increase its level. Maintaining the invariant in such situation is a very costly procedure, and we need to somehow amortize this cost. We start at $v := r$ and descend down to its (unique) child of weight exceeding $f(\ell + 1)$ as long as possible. Note that we don't actually store the weight of each node, but given the weight of v with exactly one child of level ℓ, we can compute the weight of this child by iterating through all other children, which are roots of their fragments, and hence have up-to-date counters available. Also, if there is more than one child of level ℓ, there cannot be any child of weight exceeding $f(\ell + 1)$. We call the traversed path the *tail*, and increase the level of all its nodes. Then maintaining the invariants requires four steps.

1. If the parent of r is of level $\ell + 1$, we must rebuild its static dictionary in order to include a new element there.
2. We move all nodes on the tail from the list of the current fragment to either the list of the fragment corresponding to the parent of r, if its level is $\ell + 1$, or a new fragment.
3. If the last node on the tail has more than one child of level ℓ, bumping its level to $\ell + 1$ splits the current fragment into more than one. We need to iterate through all nodes there, and partition them accordingly creating new fragments. Note that creating a new fragment requires computing the weights of their roots, which can be done by iterating through children of smaller levels for all nodes in the current fragment.
4. For each child of level ℓ of the last node on the tail we must add a new element to the wexponential search tree. Note that at this point all those nodes are roots of their fragments, hence have their weights computed.

Notice that all those steps are local in the sense that they modify only the nodes in the current fragment. To bound the time taken by the whole procedure, we allocate credits to fragments, making sure that a fragment of weight w and level ℓ has $\max(0, w - f(\ell+1))$ credits available. Whenever we split an edge and create a new leaf, we allocate one credit for each of the at most $O(\lg \lg \sigma)$ fragments above. Then when we are increasing the level of r, its weight is $2f(\ell + 1)$, so we have $f(\ell + 1)$ credits available, and because of the way we defined the tail, we can spend all of them, as all new fragments of level ℓ will be of weight at most $f(\ell + 1)$ after the update. We can bound the time required for maintaining the invariants, which we call *promoting* at r, as follows.

1. Rebuilding the static dictionary takes $O(\frac{f(\ell+2)}{f(\ell+1)} \log^2 \log \frac{f(\ell+2)}{f(\ell+1)}) = O(\frac{f^2(\ell+2)}{f^2(\ell+1)})$. We call this refreshing the parent of r.
2. There are at most $4f(\ell + 1) - 1$ nodes in the subtree of r, hence traversing the tail, including the time taken to compute the weight of all nodes there, takes $O(f(\ell + 1))$.
3. There are at most $2f(\ell + 1)$ nodes in the current fragment, hence the nodes in the current fragment can be partitioned into new fragments in $O(f(\ell+1))$ time. This includes the time to compute the weights of their roots, as in the worst-case we iterate through $4f(\ell + 1) - 1$ nodes in the subtree of r.
4. We must insert at most $\frac{2f(\ell+1)}{f(\ell)}$ elements into the wexponential tree stored at the last node of the tail, and inserting an element of weight w is done by first

adding a new element of weight one, and then increasing its weight repeatedly $\sqrt{w}$ times.

Thus, the total number of credits required by the promoting is order of $\frac{f^2(\ell+2)}{f^2(\ell+1)}+f(\ell+1)+\frac{f(\ell+1)}{f(\ell)}\sqrt{f(\ell)}\lg\lg f(\ell+1) = \frac{f^2(\ell+2)}{f^2(\ell+1)}+f(\ell+1)$. We therefore only need to ensure that $\frac{f^2(\ell+2)}{f^2(\ell+1)} \leq f(\ell+1)$, which is equivalent to $f^2(\ell+2) \leq f^3(\ell+1)$, and then $2(\frac{3}{2})^{\ell+2} \leq 3(\frac{3}{2})^{\ell+1}$, hence by the choice of f we always have enough credits to amortize the update.

4.3   Worst-Case Version of Theorem 2

The only non worst-case efficient part of the previous implementation is increasing the level of a root r. Instead of traversing the tail and updating all the structures as soon as its level reaches $2f(\ell+1)$, we will execute those operations incrementally over the next $f(\ell+1)$ insertions in the subtree rooted at r, and relax the condition on the weight of a node of level ℓ by saying that it shouldn't exceed $3f(\ell+1)$. As selecting the tail is done incrementally, we redefine it to be the maximal sequence of nodes of weight at least s at the moment we started the process, which requires storing at each edge a timestamp for its creation time.

First of all, we must make sure that there is at most one promoting process per fragment. Even though we have already chosen the tail, future insertions might increase the weight of some additional nodes to more than $f(\ell+1)$. Nevertheless, as we execute the procedure over just $f(\ell+1)$ insertions, no node in the current fragment (or in one of the new fragments) which doesn't belong to the tail can reach the weight of $2f(\ell+1)$ before we are done. Furthermore, there is some interaction between different fragments. While we are still promoting at r, new nodes might be added to the corresponding fragment. Also, different children of a node might need to be refreshing it at overlapping periods of time.

We choose the speed of the simulation so that there is enough time to process a fragment consisting of $3f(\ell+1)$ nodes over $f(\ell+1)$ insertions. When splitting the current fragment into more than one, we run a depth-first search to determine the new fragments. As soon as we reach a node, we set the link to its (new) fragment. Then when a new node appears, its parent is either already processed and hence has the correct link set, or will be seen later, and we will notice the new child then. The same reasoning works for inserting the new elements into the wexponential tree stored at the last node of the tail. Refreshing a node is a quite different issue, though, as we must somehow deal with the problem that many children might need to refresh the same parent. Each node of level ℓ stores a list of all its children of weight at least $f(\ell)$, where we simply append a child as soon as its weight becomes large enough. As a part of our simulation we run the refreshing process. In other words, each child of a node can be potentially running a deferred process refreshing its parent. The first step of such process is making a read-only copy of the current list. As this list only grows, instead of making a physical copy we can simply store the current last element, which can be done in constant time. Then we build a new static dictionary containing all elements on this read-only copy over the insertions in the subtree rooted at the child. As soon as the construction finishes, we substitute the old dictionary in constant time by replacing one pointer. There is just additional detail: we should first check if we are really replacing an older version, i.e., we can simply look at the number of elements stored there. This is because it might have happened that there was a refreshing process which started later, yet finished earlier (as there were more insertions to the corresponding subtree). This ensures that when the weight of r reaches $3f(\ell+1)$, the static dictionary stored at its parent surely contains its edge (and, potentially, also some more recent edges).

References

1. Amir, A., Franceschini, G., Grossi, R., Kopelowitz, T., Lewenstein, M., Lewenstein, N.: Managing unbounded-length keys in comparison-driven data structures with applications to online indexing. SIAM J. Comput. **43**(4), 1396–1416 (2014)
2. Andersson, A., Thorup, M.: Dynamic ordered sets with exponential search trees. J. ACM 53(3) (2007). Article No. 13
3. Beame, P., Fich, F.E.: Optimal bounds for the predecessor problem and related problems. J. Comput. Syst. Sci. **65**(1), 38–72 (2002)
4. Bender, M.A., Cole, R., Raman, R.: Exponential structures for efficient cache-oblivious algorithms. In: Widmayer, P., Triguero, F., Morales, R., Hennessy, M., Eidenbenz, S., Conejo, R. (eds.) ICALP 2002. LNCS, vol. 2380, pp. 195–207. Springer, Heidelberg (2002)
5. Breslauer, D., Italiano, G.F.: Near real-time suffix tree construction via the fringe marked ancestor problem. J. Discrete Algorithms **18**, 32–48 (2013)
6. Cole, R., Kopelowitz, T., Lewenstein, M.: Suffix trays and suffix trists: structures for faster text indexing. In: Bugliesi, M., Preneel, B., Sassone, V., Wegener, I. (eds.) ICALP 2006. LNCS, vol. 4051, pp. 358–369. Springer, Heidelberg (2006)
7. Fischer, J., Gawrychowski, P.: Alphabet-dependent string searching with wexponential search trees. CoRR abs/1302.3347 (2013). http://arxiv.org/abs/1302.3347
8. Fredman, M.L., Komlós, J., Szemerédi, E.: Storing a sparse table with $O(1)$ worst case access time. J. ACM **31**(3), 538–544 (1984)
9. Kopelowitz, T.: On-line indexing for general alphabets via predecessor queries on subsets of an ordered list. In: Proceedings of the FOCS, pp. 283–292. IEEE Computer Society (2012)
10. Kucherov, G., Nekrich, Y.: Full-fledged real-time indexing for constant size alphabets. In: Fomin, F.V., Freivalds, R., Kwiatkowska, M., Peleg, D. (eds.) ICALP 2013, Part I. LNCS, vol. 7965, pp. 650–660. Springer, Heidelberg (2013)
11. Manber, U., Myers, E.W.: Suffix arrays: a new method for on-line string searches. SIAM J. Comput. **22**(5), 935–948 (1993)
12. McCreight, E.M.: A space-economical suffix tree construction algorithm. J. ACM **23**(2), 262–272 (1976)
13. Mehlhorn, K.: Dynamic binary search. SIAM J. Comput. **8**(2), 175–198 (1979)
14. Ružić, M.: Constructing efficient dictionaries in close to sorting time. In: Aceto, L., Damgård, I., Goldberg, L.A., Halldórsson, M.M., Ingólfsdóttir, A., Walukiewicz, I. (eds.) ICALP 2008, Part I. LNCS, vol. 5125, pp. 84–95. Springer, Heidelberg (2008)
15. Weiner, P.: Linear pattern matching algorithms. In: Proceedings of the Annual Symposium on Switching and Automata Theory, pp. 1–11. IEEE Computer Society (1973)

Lempel Ziv Computation in Small Space (LZ-CISS)

Johannes Fischer, Tomohiro I, and Dominik Köppl[(✉)]

Department of Computer Science, TU Dortmund, Dortmund, Germany
{johannes.fischer,tomohiro.i}@cs.tu-dortmund.de,
dominik.koeppl@tu-dortmund.de

Abstract. For both the Lempel Ziv 77- and 78-factorization we propose factorization algorithms using $(1 + \epsilon)n \lg n + \mathcal{O}(n)$ bits (for any positive constant $\epsilon \leq 1$) working space (including the space for the output) for any text of size n over an integer alphabet in $\mathcal{O}(n/\epsilon^2)$ time.

1 Introduction

Two of the most important algorithms for text compression are the methods by Lempel and Ziv, LZ77 [19] and LZ78 [20]. While there are naive algorithms that take $\mathcal{O}(1)$ working space with quadratic running time (for both LZ77 and LZ78), linear time algorithms with very restricted space emerged only in recent years.

For LZ77, the bound of $3n \lg n$ bits set by [7] was very soon lowered to $2n \lg n$ by [10]. For small alphabet size σ, the upper bound of $n \lg n + \mathcal{O}(\sigma \lg n)$ bits by [8] is also very compelling. In [12], a practical variant having the worst case performance guarantees of $(1+\epsilon)n \lg n + n + \mathcal{O}(\sigma \lg n)$ bits of working space and $\mathcal{O}(n \lg \sigma/\epsilon^2)$ time was proposed.

Wrt. LZ78, by using a naive trie implementation, the factorization is computable with $\mathcal{O}(z \lg z)$ bits of space and $\mathcal{O}(n \lg \sigma)$ overall running time, where z is the size of LZ78 factorization. More sophisticated trie implementations [6] improve this to $\mathcal{O}(n + z \lg^2 \lg \sigma/\lg \lg \lg \sigma)$ time using the same space. Jansson et al. [9] proposed a compressed dynamic trie based on word packing, and showed an application to LZ78 trie construction that runs in $\mathcal{O}(n(\lg \sigma + \lg \lg_\sigma n)/\lg_\sigma n)$ bits of working space and $\mathcal{O}(n \lg^2 \lg n/ (\lg_\sigma n \lg \lg \lg \lg n))$ time, which is superlinear in the worst case. For an integer alphabet, Nakashima et. al [15] recently presented a linar time algorithm using $\mathcal{O}(n \lg n)$ bits of space, but the use of the (complicated) dynamic marked ancestor queries [1] seems to prevent them from achieving a small constant factor.

2 Preliminaries

Let Σ denote an integer alphabet of size $\sigma = |\Sigma| = n^{\mathcal{O}(1)}$. An element w in Σ^* is called a **string**, and $|w|$ denotes its length. The empty string of length 0 is called ε. For any $1 \leq i \leq |w|$, $w[i]$ denotes the i-th character of w. When w is

F. Cicalese et al. (Eds.): CPM 2015, LNCS 9133, pp. 172–184, 2015.
DOI: 10.1007/978-3-319-19929-0_15

represented by the concatenation of $x, y, z \in \Sigma^*$, i.e., $w = xyz$, then x, y and z are called a **prefix**, **substring** and **suffix** of w, respectively. In particular, a suffix starting at position i of w is called the **i-th suffix** of w. For any $1 \leq j \leq |w|$, let $S_j(w)$ denote the set of substrings of w that start strictly before j.

In the rest of this paper, we take a string T of length $n > 0$, which is subject to LZ77 or LZ78 factorization. For convenience, let $T[n]$ be a special character that appears nowhere else in T, so that no suffix of T is a prefix of another suffix of T. Our computational model is the word RAM model with word size $\Omega(\lg n)$. Further, we assume that T is read-only; accessing a word costs $\mathcal{O}(1)$ time (e.g., T is stored in RAM using $n \lg \sigma$ bits).

The **suffix trie** of T is the trie of all suffixes of T. The **suffix tree** of T, denoted by ST, is the tree obtained by compacting the suffix trie of T. ST has n leaves and at most n internal nodes. We denote by V the nodes and by E the edges of ST. For any edge $e \in E$, the string stored in e is denoted by $c(e)$ and called the **label** of e. Further, the **string depth** of a node $v \in V$ is defined as the length of the concatenation of all edge labels on the path from the root to v. The leaf corresponding to the i-th suffix is labeled with i. SA and ISA denote the suffix array and the inverse suffix array of T, respectively [13]. For any $1 \leq i \leq n$, SA[i] is identical to the label of the *lexicographically i-th* leaf in ST. *LCP* and *RMQ* are abbreviations for *longest common prefix* and *range minimum query*, respectively. LCP is a DS (data structure) on SA such that LCP[i] is the LCP of the *lexicographically i-th* smallest suffix with its lexicographic predecessor for $i = 2, \ldots, n$.

For any bit vector B with length $|B|$, $B.\text{rank}_1(i)$ counts the number of '1'-bits in $B[1..i]$, and $B.\text{select}_1(i)$ gives the position of the i-th '1' in B. Given B, a DS that uses additional $o(|B|)$ bits of space and supports any rank/select query on B in constant time can be built in $\mathcal{O}(|B|)$ time [3].

2.1   Lempel Ziv Factorization

A **factorization** partitions T into z substrings $T = f_1 \cdots f_z$. These substrings are called **factors**. In particular, we have:

Definition 1. A factorization $f_1 \cdots f_z = T$ is called the **LZ77 factorization** of T iff $f_x = \text{argmax}_{S \in S_j(T) \cup \Sigma} |S|$ for all $1 \leq x \leq z$ with $j = |f_1 \cdots f_{x-1}| + 1$.

The classic LZ77 factorization involving the innovative character at the end is treated similarly, and is part of the full version of this paper [4].

Definition 2. A factorization $f_1 \cdots f_z = T$ is called the **LZ78 factorization** of T iff $f_x = f'_x \cdot c$ with $f'_x = \text{argmax}_{S \in \{f_y : y < x\} \cup \{\varepsilon\}} |S|$ and $c \in \Sigma$ for all $1 \leq x \leq z$.

We identify factors by text positions, i.e., we call a text position j the **factor position** of f_x ($1 \leq x \leq z$) iff factor f_x starts at position j. A factor f_x may refer to either (LZ77) a previous text position j (called f_x's **referred position**), or (LZ78) to a previous factor f_y (called f_x's **referred factor**—in this case y is also called the **referred index** of f_x). If there is no suitable reference found for a given factor f_x with factor position j, then f_x consists of just the single letter $T[j]$. We call such a factor a **free letter**. The other factors are called **referencing factors**. Our final data structures allow us to access arbitrary factors (factor position and referred position (LZ77)/referred index (LZ78)) in constant time.

2.2   Data Structures

Common to both our algorithms is the construction of a succinct ST representation. It consists of SA with $n \lg n$ bits, LCP with $2n + o(n)$ bits, and a $2|V| + o(|V|)$-bit representation of the topology of ST, for which we choose the DFUDS [3]. The latter is denoted by SucST. We make use of several construction algorithms from the literature:

- SA can be constructed in $\mathcal{O}(n/\epsilon^2)$ time and $(1+\epsilon)n \lg n$ bits of space, including the space for SA itself [11].
- Given SA, LCP can be computed in $\mathcal{O}(n)$ time with no extra space [18]. Note that LCP can only answer LCP$[i]$ in constant time if SA$[i]$ is also available. This is an important remark, because we will discard SA at several occasions in order to free space, and this discarding causes additional difficulties.
- Given both SA and LCP, a space economical construction of SucST was discussed in [17, Alg. 1]. The authors showed that the DFUDS of ST can be built in $\mathcal{O}(n)$ time with $n + o(n)$ bits of working space.

We identify a node $v \in V$ with its pre-order number, which is also the order in which the opening parentheses occur in the DFUDS. So we implicitly identify every node $v \in V$ with its pre-order number (enumerated by $1, \ldots, |V|$).

Since our ST is static, we can perform various operations on the tree topology in constant time (see, e.g., [16,17]). Among them, we especially use the following operations (for any $v \in V$ and $i \in \mathbb{N}$): parent(v) returns the parent of v; and level_anc(v, i) returns the i-th ancestor of v. By building the *min-max tree* [16] on the DFUDS of ST in $\mathcal{O}(n)$ time (using $\mathcal{O}(n)$ bits of space), we can get SucST supporting these operations in constant time. Additionally, we are interested in answering str_depth(v) on ST; str_depth(v) returns the string depth of $v \in V$. As noted in [17], an RMQ data structure on LCP can be built in $\mathcal{O}(n)$ time and $n + o(n)$ bits of working space to support str_depth in constant time. Note that the operation str_depth becomes unavailable when SA is discarded.

Our algorithms in Sects. 3 and 4 make use of two arrays: A_1 of size $n \lg n$ bits, and a small helper array A_2 of size $\epsilon n \lg n$ bits. (We chose such generic names since the contents of these arrays will change several times during the LZ-computation.)

Node-Marking Vectors. In our algorithms, we sometimes deal with subsets V' of V. Pre-order numbers enumerating only the nodes in V' can naturally be used to map nodes in V' to the range $[1..|V'|]$. For this purpose, we use a **node-marking vector** $M_{V'}$, which is a bit vector of length $|V|$, such that $M_{V'}[v] = 1$

iff $v \in V'$ for any $1 \leq v \leq |V|$. We write $\rho_{V'}(v) := M_{V'}.\mathrm{rank}_1(v)$ for any node $v \in V'$.

3 LZ77

The main idea is to perform leaf-to-top traversals accompanied by the marking of visited nodes. The marked nodes are indicated by a '1' in a bit vector of size $|V|$. Starting from the situation where only the root is marked, in the j-th leaf-to-top traversal for any $1 \leq j \leq n$, we traverse ST from the leaf labeled with j towards the root, while marking visited nodes until we encounter an already marked node. Observe that right before the j-th leaf-to-top traversal, each string of $S_j(T)$ can be obtained by following the path from the root to some marked node. Hence, the LZ77 factorization can be determined during these leaf-to-top traversals: If j is a factor position of a factor f, the last accessed node v during the j-th leaf-to-top traversal reveals f's referred position. More precisely, v is either the root, or a node that was already marked in a former traversal. If v is the root, f is a free letter. Otherwise, we call v the **referred node** of f. Then, the factor length is str_depth(v), and the referred position is the minimum leaf label in the subtree rooted at v (retrieved, e.g., by an RMQ on SA). Since every visited node will be marked, and a marked node will never be unmarked, the total number of parent($\cdot$)-operations is upper bounded by the number of nodes in ST, i.e., $\mathcal{O}(n)$.

We start with SA stored in $A_1[1..n]$, and some $\mathcal{O}(n)$-bit DS to provide SucST, RMQs on SA, and RMQs on LCP. Note that the LZ77 computation via leaf-to-top traversals, as explained above, accesses ISA n times to fetch suffix leaves that are starting nodes of the traversals, and accesses SA $\mathcal{O}(z)$ times to compute the factor lengths and the referred positions. Then, if we have both SA and ISA, the LZ77 factorization can be easily done in $\mathcal{O}(n)$ time by the leaf-to-top traversals. However, allowing only $(1 + \epsilon)n \lg n + \mathcal{O}(n)$ bits for the entire working space, it is no longer possible to store both SA and ISA completely at the same time.

With Extra Output Space. Let us first consider the easier case where the result of the factorization can be output *outside* the working space. We can then use the array+inverse DS of Munro et al. [14, Sect. 3.1], which allows us to access inverse array's values in $\mathcal{O}(1/\epsilon)$ time by spending additional $\epsilon n \lg n$ bits (on top of the array's size). Since ISA is accessed more often than SA, we first convert SA on A_1 into ISA and then create its array+inverse DS so that accessing ISA and SA can be done in $\mathcal{O}(1)$ and $\mathcal{O}(1/\epsilon)$ time, respectively. Although it is not explicitly mentioned in [14], the DS can be constructed in $\mathcal{O}(n)$ time. Then, the leaf-to-top traversals can be smoothly conducted, leading to $\mathcal{O}(z/\epsilon + n) = \mathcal{O}(n)$ running time.

Although this is already an improvement over the currently best linear-time algorithm using $2n \lg n$ bits [10], using Munro et al.'s DS as a black box would prevent us from also storing the *output* of the LZ77 factorization in the working space. Solving this is exactly what is explained in the remainder of this section.

Without Extra Space – Outline. It is difficult to find space for writing the referred positions; the former algorithm already uses $(1+\epsilon)n \lg n$ bits of working space for the array+inverse DS. Overwriting it would corrupt the DS and cause a problem when accessing SA or ISA. We evade this problem by performing several rounds of leaf-to-top traversals during which we build an array that registers every visit of a referred node. (A minor remark is that this approach does not even need RMQs on SA.) Our algorithm is divided into three rounds of leaf-to-top traversals and a final matching phase, all of which will be discussed in detail in the following:

First Round: Construct a bit vector $B_f[1..n]$ marking all factor positions in T, and a bit vector $B_r[1..z]$ marking the referencing factors. Determine the set of referred nodes $V_r \subset V$, and mark them with a node-marking vector M_{V_r}.
Second Round: Construct a bit vector B_D counting (in unary) the number of *referred* nodes from V_r visited during each traversal.
Third Round: Construct an array D storing the pre-order numbers of all referred nodes visited during each traversal (as counted in the second round).
Matching: Convert the pre-order numbers in D to referred positions.

Details. In the **first round**, we compute the factor lengths as before by leaf-to-top traversals, which are used to construct B_f. Since the set of referred nodes can be identified during the leaf-to-top traversals, M_{V_r} can be easily constructed. We also compute B_r by setting $B_r[x] \leftarrow 1$ for every referencing factor f_x with $1 \leq x \leq z$. For the rest of the algorithm, the information of SA is not needed any longer.

We now aim at generating the array D storing a sequence of pre-order numbers of referred nodes, which will finally enable us to determine the referred positions of each referencing factor. D is formally defined as a sequence obtained by outputting the pre-orders of referred nodes whenever they are marked or referred to during the leaf-to-top traversals. Hence, each referred node appears in D for the first time when it is marked, and after that it occurs whenever it is the last accessed node of the j-th traversal, where $1 \leq j \leq n$ coincides with a factor position. To see how D will be useful for obtaining the referred positions, consider a node $v \in V$ that was marked during the k-th traversal. If we stumble upon v during the j-th traversal (for any factor position $j > k$) we know that k is the referred position for the factor with factor position j (because v had not been marked before the k-th traversal).

Alas, just D alone does not tell us *which* referred nodes are found during *which* traversal. We want to partition D by the n text positions, s.t. we know the traversal numbers which the referred nodes belong to. This is done by a bit vector B_D that stores a '1' for each text position j, and intersperses these '1's with '0's counting the number of referred nodes written to D during the j-th traversal. The size of the j-th partition $(1 \leq j \leq n)$ is determined by the number of referred nodes accessed during the j-th traversal. Hence the number of '0's between the $(j-1)$-th and j-th '1' represents the number of entries in

D for the j-th suffix. Formally, B_D is a bit vector such that $D[j_b..j_e]$ represents the sequence of referred nodes that are written to D during the j-th leaf-to-top traversal, where, for any $1 \leq j \leq n$, $j_b := B_D.\mathrm{rank}_0(B_D.\mathrm{select}_1(j-1))+1$ and $j_e := B_D.\mathrm{rank}_0(B_D.\mathrm{select}_1(j))$. Note that for each factor position j of a referencing factor f we encountered its referred node during the j-th traversal; this node is the last accessed node during that traversal, and was stored in $D[j_e]$, which we call the **referred entry** of f. Note that we do not create a rank_0 nor a select_1 DS on B_D because we will get by with sequential scans over B_D and D.

Finally, we show the actual computation of B_D and D. Unfortunately, the computation of D cannot be done in a single round of leaf-to-top traversals; overwriting A_1 naively with D would result in the loss of necessary information to access the suffix tree's leaves. This is solved by performing *two* more rounds of leaf-to-top traversals, as already outlined above: In the **second round**, with the aid of M_{V_r}, B_D is generated by counting the number of referred nodes that are accessed during each leaf-to-top traversal. Next, according to B_D, we sparsify ISA by discarding values related to suffixes that will not contribute to the construction of D (i.e., those values i for which there is no '0' between the $(i-1)$-th and the i-th '1' in B_D). We align the resulting sparse ISA to the right of A_1. Afterwards, we overwrite A_1 with D from left to right in a **third round** using the sparse ISA. The fact that this is possible is proved by the following

Lemma 1. $|D| \leq n$.

Proof. First note that the size of D is $|V_r| + z_r$, where z_r is the number of referencing factors (number of '1's in B_r). Hence, we need to prove that $|V_r| + z_r \leq n$. Let z_r^1 (resp. $z_r^{>1}$) denote the number of referencing factors of length 1 (resp. longer than 1), and let V_r^1 (resp. $V_r^{>1}$) denote the referred nodes whose string depth is 1 (resp. longer than 1). Also, z_f denotes the number of free letters. Clearly, $|V_r| = |V_r^1| + |V_r^{>1}|$, $z_r = z_r^1 + z_r^{>1}$, $|V_r^1| \leq z_f$, and $|V_r^{>1}| \leq z_r^{>1}$. Hence $|V_r| + z_r = |V_r^1| + |V_r^{>1}| + z_r^1 + z_r^{>1} \leq z_f + z_r^1 + 2z_r^{>1} \leq n$. The last inequality follows from the fact that the factors are counted disjointly by z_f, z_r^1 and $z_r^{>1}$, and the sum over the lengths of all factors is bounded by n, and every factor counted by $z_r^{>1}$ has length at least 2. $\square$

By Lemma 1, D fits in A_1. Since each suffix having an entry in the sparse ISA has at least one entry in D, overwriting the remaining ISA values before using them will never happen. Once we have D on A_1, we start **matching** referencing factors with their referred positions. Recall that each referencing factor has one referred entry, and its referred position is obtained by matching the leftmost occurrence of its referred node in D.

Let us first consider the easy case with $|V_r| \leq \lfloor n\epsilon \rfloor$ such that all referred positions fit into A_2 (the helper array of size $\epsilon n \lg n$ bits). By B_D we know the leaf-to-top traversal number (i.e., the leaf's label) during which we wrote $D[i]$ (for any $1 \leq i \leq |D|$). For $1 \leq m \leq |V_r|$, the zero-initialized $A_2[m]$ will be used to store the smallest suffix number at which we found the m-th referred node (i.e., the m-th node of V_r identified by pre-order).

Let us consider that we have set $A_2[m] = k$, i.e., the m-th referred node was discovered for the first time by the traversal of the suffix leaf labeled with k.

Whenever we read the referred entry $D[i]$ of a factor f with factor position larger than k and $\rho_{V_r}(D[i]) = m$, we know by $A_2[m] = k$ that the referred position of f is k. Both the filling of A_2 and the matching are done in one single, sequential scan over D (stored in A_1) from left to right: While tracking the suffix leaf's label with a counter $1 \le k \le n$, we look at $t := \rho_{V_r}(D[i])$ and $A_2[t]$ for each array position $1 \le i \le |D|$: if $A_2[t] = 0$, we set $A_2[t] \leftarrow k$. Otherwise, $D[i]$ is a referred entry of the factor f with factor position k, for which $A_2[t]$ stores its referred position. We set $A_1[i] \leftarrow A_2[t]$. By doing this, we overwrite the referred entry of every referencing factor f in D with the referred position of f.

If $|V_r| > \lfloor n\epsilon \rfloor$, we run the same scan multiple times, i.e., we partition $\{1, \ldots, |V_r|\}$ into $\lceil |V_r|/(n\epsilon) \rceil$ equi-distant intervals (pad the size of the last one) of size $\lfloor n\epsilon \rfloor$, and perform $\lceil |V_r|/(n\epsilon) \rceil$ scans. Since each scan takes $\mathcal{O}(n)$ time, the whole computation takes $\mathcal{O}(|V_r|/\epsilon) = \mathcal{O}(z/\epsilon)$ time.

Now we have the complete information of the factorization: The length of the factors can be obtained by a select-query on B_f, and A_1 contains the referred positions of all referencing factors. By a left shift we can restructure A_1 such that $A_1[x]$ tells us the referred position (if it exists, according to $B_r[x]$) for each factor $1 \le x \le z$. Hence, looking up a factor can be done in $\mathcal{O}(1)$ time.

4   LZ78

Common implementations use a trie for storing the factors, which we call the **LZ78 trie**. Recall that all trie implementations have a (log-)logarithmic dependence on σ for top-down-traversals (see the Introduction); one of our tricks is using level_anc queries starting from the leaves in order to get rid of this dependence. For this task we need ISA to fetch the correct suffix leaf; hence, we first overwrite SA by its inverse.

4.1   Algorithm

The LZ78 trie structure can be represented by ST, with an additional DS storing the number of LZ78 trie nodes that lie on each edge of ST. Each trie node v is called **explicit** iff it is not discarded during the compactification of the suffix trie towards ST; the other trie nodes are called **implicit**.

For every edge e of ST we use a counting variable $0 \le n_e \le |c(e)|$ that keeps track of how far e is explored. If $n_e = 0$, then the factorization has not (yet) explored this edge, whereas $n_e = |c(e)|$ tells us that we have already reached the ending node $v \in V$ of $e =: (u, v)$. We defer the question how the n_e- and $|c(e)|$-values are stored in $\epsilon n \lg n$ bits to Sect. 4.2, as those technicalities might not be of interest to the general audience.

Because we want to have a representative node in ST for *every* LZ78-factor, we introduce the concept of witnesses: For any $1 \le x \le z$, the **witness** of f_x is the ST node that is either the explicit representation of f_x, or, if such an explicit representation does not exist, the ending node in ST of the edge on which f_x lies.

Our next task is therefore the creation of an array $W[1..z]$ s.t. $W[x]$ stores the pre-order number of f_x's witness. With W it will be easy to find the referred index y of any referencing factor f_x. That is because f_y will either share the witness with f_x, or $W[y]$ is the parent node of $W[x]$. Storing W will be done by overwriting the first z positions of the array A_1.

We start by computing $W[x]$ for all $1 \leq x \leq z$ in increasing order. Suppose that we have already processed $x - 1$ factors, and now want to determine the witness of f_x with factor position j. $\mathsf{ISA}[j]$ tells us where to find the ST leaf labeled with j. Next, we traverse ST from the root towards this leaf (navigated by level_anc queries in deterministic constant time per edge) until we find the first edge e with $n_e < |c(e)|$, namely, e is the edge on which we would insert a new LZ78 trie leaf. It is obvious that the ending node of e is f_x's witness, which we store in $W[x]$. We let the LZ78 trie grow by incrementing n_e. The length of f_x is easily computed by summing up the $|c(\cdot)|$-values along the traversed path, plus n_e's value. Having processed f_x with factor position $j \in [x..n]$, ISA's values in $A_1[1..j]$ are not needed anymore. Thus, it is eligible to overwrite $A_1[x]$ by $W[x]$ for $1 \leq x \leq z$ while computing f_x. Finally, $A_1[1..z]$ stores W. Meanwhile, we have marked the factor positions in a bit vector $B_f[1..n]$. Matching the factors with their references can now be done in a top-down-manner by using W. Let us consider a referencing factor f_x with referred factor f_y. We have two cases: Whenever f_y is explicitly represented by a node v (i.e., by f_y's witness), v is the parent of f_x's witness. Otherwise, f_y has an implicit representation and hence has the same witness as f_x. Hence, if W stores at position x the *first* occurrence of $W[x]$ in W, f_y is determined by the largest position $y < x$ for which $W[y] = \mathsf{parent}(v)$; otherwise ($W[x]$ is *not* the first occurrence of $W[x]$ in W), then the referred factor of f_x is determined by the largest $y < x$ with $W[x] = W[y]$.

Now we hold W in $A_1[1..z]$, leaving us $A_1[z+1..n]$ as free working space that will be used to store a new array R, storing for each witness w the index of the most recently processed factor whose witness is w. However, reserving space in R for *every* witness would be too much (there are potentially z many of them); we will therefore have to restrict ourselves to a carefully chosen subset of witnesses. This is explained next. First, let us consider a witness w that is witnessed by a single factor f_x whose LZ78 trie node is a leaf. Because no other factor will refer to f_x, we do not have to involve w in the matching. Therefore, we can neglect all such witnesses during the matching. The other witnesses (i.e., those being witnessed by at least one factor that is not an LZ78 trie leaf) are collected in a set V_Ξ and marked by a bit vector M_{V_Ξ}. $|V_\Xi|$ is at most the number z_i of internal nodes of the LZ78 trie, which is bounded by $n - z$, due to the following

Lemma 2. $z + z_i \leq n$.

Proof. Let α (resp. β) be the number of free letters that are internal LZ78 trie nodes (resp. LZ78 trie leaves). Also, let γ (resp. δ) be the number of referencing factors that are internal LZ78 trie nodes (resp. LZ78 trie leaves). Obviously, $\alpha + \beta + \gamma + \delta = z$. Wrt. the factor length, each referencing factor has length of at least 2, while each free letter is exactly one character long. Hence $2(\gamma + \delta) + \alpha + \beta = z + \gamma + \delta \leq n$. Since each LZ78 leaf that is counted by δ has an LZ78 internal node of depth one as ancestor (counted by α), $\alpha \leq \delta$ holds. Hence, $z + z_i \leq z + \alpha + \gamma \leq z + \gamma + \delta \leq n$.   □

By Lemma 2, if we let R store only the indices of factors whose witnesses are in V_Ξ, it fits into $A_1[z + 1..n]$, and we can use M_{V_Ξ} to address R.

We now describe how to convert W (stored in $A_1[1..z]$) into the referred indices, such that in the end $A_1[x]$ contains the referred index of f_x for $1 \leq x \leq z$. We scan $W = A_1[1..z]$ from left to right while keeping track of the index of the most recently visited factor that witnesses v, for each witness $v \in V_\Xi$ at $R[\rho_{V_\Xi}(v)]$. Suppose that we are now processing f_x with witness $v = W[x]$.

- If $v \notin V_\Xi$ or $R[\rho_{V_\Xi}(v)]$ is empty, we are currently processing the first factor that witnesses v. Further, if f_x is not a free letter, its referred factor is explicitly represented by the parent of v. We can find its referred index at position $\rho_{V_\Xi}(\mathsf{parent}(v))$ in R.
- Otherwise, $v \in V_\Xi$, and $R[\rho_{V_\Xi}(v)]$ has already stored a factor index. Then $R[\rho_{V_\Xi}(v)]$ is the referred index of f_x.

In either case, if $v \in V_\Xi$, we update R by writing the current factor index x to $R[\rho_{V_\Xi}(v)]$. Note that after processing f_x, the value $A_1[x]$ is not used anymore. Hence we can write the referred index of f_x to $A_1[x]$ (if it is a referring factor) or set $A_1[x] \leftarrow 0$ (if it is a free letter). In the end, $A_1[1..z]$ stores the referred indices of every referring factor.

Now we have the complete information about the LZ78 factorization: For any $1 \leq x \leq z$, f_x is formed by $f_y c$, where $y = A_1[x]$ is the referred index and $c = T[B_f.\mathsf{select}_1(x + 1) - 1]$ the additional letter (free letters will refer to $f_0 := \varepsilon$). Hence, looking up a factor can be done in $\mathcal{O}(1)$ time.

4.2   Bookkeeping the LZ78 Trie Representation

Basically, we store both n_e and $|c(e)|$ for each edge e so as to represent the LZ78 trie construction in each step. A naive approach would spend $2\lg(\max_{e \in E} |c(e)|)$ bits for every edge, i.e., $4n \lg n$ bits in the worst case. In order to reduce the space consumption to $\epsilon n \lg n + o(n)$ bits, we will exploit two facts: (1) the superimposition of the LZ78 trie on ST takes place only in the *upper* part of ST, and (2) most of the needed $|c(e)|$- and n_e-values are actually small.

More precisely, we will introduce an upper bound for the n_e values, which shows that the necessary memory usage for managing the n_e and $|c(e)|$ values is, without a priori knowledge of the LZ78 trie's shape, actually very low.

Note that although we do not know the LZ78 trie's shape, we will reason about those nodes that might be created by the factorization. For a node $v \in V$, let $\mathsf{height}(v)$ denote the height of v in the LZ78 trie if v is the explicit representation of an LZ78 trie node; otherwise we set $\mathsf{height}(v) = 0$.

For any node $v \in V$, let $l(v)$ denote the number of descendant leaves of v. The following lemma gives us a clue on how to find an appropriate upper bound:

Lemma 3. *Let $u, v \in V$ with $e := (u, v) \in E$. Further assume that u is the explicit representation of an LZ78 trie node. Then $\mathsf{height}(v)$ is upper bounded by $l(v) - |c(e)|$.*

Proof. Let π be a longest path from u to some descendant leaf of v, and $d := \mathsf{height}(v) + |c(e)|$ (i.e., the number of LZ78 trie edges along π). By construction of the LZ78 trie, the ST node v must have at least d leaves, for otherwise the (explicit or implicit) LZ78 trie nodes on π will never get explored by the factorization. So $d \leq l(v)$, and the statement holds. $\qquad \square$

Further, let root denote the root node of the suffix *trie*. In particular, root is an explicit LZ78 trie node. Consider two arbitrary nodes $u, v \in V$ with $e := (u, v) \in E$. Obviously, the suffix *trie* node of v is deeper than the suffix *trie* node of u by $|c(e)|$. Putting this observation together with Lemma 3, we define $h : V \to \mathbb{N}_0$, which upper bounds $\mathsf{height}(\cdot)$:

$$h(v) = \begin{cases} n & \text{if } v = \mathsf{root}, \\ \max\left(0, \min\left(h(u), l(v)\right) - |c(e)|\right) & \text{if there is an } e := (u, v) \in E. \end{cases}$$

Since the number of LZ78 trie nodes on an edge below any $v \in V$ is a lower bound for $\mathsf{height}(v)$, we conclude with the following lemma:

Lemma 4. *For any edge $e = (v, w) \in E$, $n_e \leq \min\left(|c(e)|, h(v)\right)$.*

Let us remark that Lemma 4 does not yield a tight bound. For example, the height of the LZ78 trie is indeed bounded by $\sqrt{2n}$ (see, e.g., [2, Lemma 1]). But we do not use this property to keep the analysis simple.

Instead, we classify the edges $e \in E$ into two sets, depending on whether $n_e \leq \Delta := \lfloor n^{\epsilon/4} \rfloor$ holds for sure or not. By Lemma 4, this classification separates E into $E_{\leq \Delta} := \{(u, v) \in E : \min\left(|c((u, v))|, h(u)\right) \leq \Delta\}$ and $E_{>\Delta} := E \setminus E_{\leq \Delta}$. Since $2 \lg \Delta$ bits are enough for bookkeeping any edge $e \in E_{\leq \Delta}$, the space needed for these edges fits in $2 |E_{\leq \Delta}| \lg \Delta \leq n\epsilon \lg n$ bits. Thus, our focus lies now on the edges in $E_{>\Delta}$; each of them costs us $2 \lg n$ bits. Fortunately, we will show that $|E_{>\Delta}|$ is so small that the space of $2 |E_{>\Delta}| \lg n$ bits needed by these edges is in fact $o(n)$ bits.

We call any $e \in E_{>\Delta}$ a Δ-*edge* and its ending node a Δ-*node*. The set of all Δ-nodes is denoted by V_Δ. As a first task, let us estimate the number of Δ-edges on a path from a node $v \in V_\Delta$ to any of its descendant leaves; because v is a Δ-node with $\mathsf{height}(v) \leq h(v)$, this number is upper bounded by $\left\lfloor \frac{h(v)}{\Delta} \right\rfloor \leq \left\lfloor \frac{l(v) - \Delta}{\Delta} \right\rfloor = \left\lfloor \frac{l(v)}{\Delta} \right\rfloor - 1$. For the purpose of analysis, we introduce $\hat{h} : (V_\Delta \cup \{\mathsf{root}\}) \to \mathbb{N}_0$, which upper bounds the number of Δ-edges that occur on a path from a node to any of its descendant leaves:

$$\hat{h}(v) = \begin{cases} \left\lfloor \frac{n}{\Delta} \right\rfloor & \text{if } v = \mathsf{root}, \\ \min\left(\hat{h}(\hat{p}(v)) - 1, \left\lfloor \frac{l(v)}{\Delta} \right\rfloor - 1\right) & \text{otherwise,} \end{cases}$$

where $\hat{p} : V_\Delta \rightarrow (V_\Delta \cup \{root\})$ returns for a node v either its deepest ancestor that is a Δ-node, or the root if such an ancestor does not exist. Note that $\hat{h}$ is non-negative by the definition of V_Δ.

For the actual analysis, $\alpha(v)$ shall count the number of Δ-edges in the subtree rooted at $v \in V_\Delta \cup \{root\}$.

Lemma 5. *For any node* $v \in (V_\Delta \cup \{root\})$, $\alpha(v) \leq \frac{l(v)}{\Delta} \sum_{i=1}^{\hat{h}(v)} \frac{1}{i}$.

Proof. We proceed by induction over the values of $\hat{h}(v)$ for every $v \in V_\Delta$. For $\hat{h}(v) = 0$ the subtree rooted at v has no Δ-edges; hence $\alpha(v) = 0$. If $\hat{h}(v) = 1$, any Δ-node w of the subtree rooted at v holds the property $\hat{h}(w) = 0$. Hence, none of those Δ-nodes are in ancestor-descendant relationship to each other. By the definition of Δ-nodes, for any Δ-node u, we have $0 \leq \left\lfloor \frac{l(u)}{\Delta} \right\rfloor - 1$, and hence, $\Delta \leq l(u)$. By $\Delta \alpha(v) \leq \sum_{u \in V_\Delta, \hat{p}(u)=v} l(u) \leq l(v)$ we get $\alpha(v) \leq \frac{l(v)}{\Delta}$.

For the induction step, let us assume that the induction hypothesis holds for every $u \in V_\Delta$ with $\hat{h}(u) < k$. Let us take a $v \in V_\Delta$ with $\hat{h}(v) = k$. Further, let $V_{k'} := \left\{ u \in V_\Delta : \hat{p}(u) = v \text{ and } \hat{h}(u) = k' \right\}$ for $0 \leq k' \leq k - 1$ denote the set of Δ-nodes that have the same $\hat{h}$ value and are descendants of v, without having a Δ-node as ancestor that is a descendant of v. These constraints ensure that there does not exist any $u \in \bigcup_{0 \leq k' \leq k-1} V_{k'} =: \mathcal{V}$ that is ancestor or descendant of some node of $\mathcal{V}$. Thus the sets of descendant leaves of the nodes of $\mathcal{V}$ are disjoint. So it is eligible to denote by $L_{k'} := \sum_{u \in V_{k'}} l(u)$ the number of descendant leaves of all nodes of $V_{k'}$. It is easy to see that $\sum_{k'=0}^{k-1} L_{k'} \leq l(v)$. Now, by the hypothesis, and the fact that each $u \in \mathcal{V}$ is the highest Δ-node on every path from v to any leaf below u, we get $\alpha(v) \leq |V_0| + \sum_{k'=1}^{k-1} \left(\sum_{u \in V_{k'}} \frac{l(u)}{\Delta} \sum_{i=1}^{k'} \frac{1}{i} + |V_{k'}| \right) = |V_0| + \sum_{k'=1}^{k-1} \left(\frac{L_{k'}}{\Delta} \sum_{i=1}^{k'} \frac{1}{i} + |V_{k'}| \right)$. By definition of $V_{k'}$ and $\hat{h}$, we have $\hat{h}(u) = k' \leq \left\lfloor \frac{l(u)}{\Delta} \right\rfloor - 1$ and hence $(k'+1)\Delta \leq l(u)$ for any $u \in V_{k'}$. This gives us $\frac{L_{k'}}{(k'+1)\Delta} = \sum_{u \in V_{k'}} \frac{l(u)}{(k'+1)\Delta} \geq |V_{k'}|$. In sum, we get $\alpha(v) \leq \frac{L_0}{\Delta} + \sum_{k'=1}^{k-1} \frac{L_{k'}}{\Delta} \sum_{i=1}^{k'+1} \frac{1}{i} = \sum_{k'=0}^{k-1} \frac{L_{k'}}{\Delta} \sum_{i=1}^{k'+1} \frac{1}{i} \leq \frac{l(v)}{\Delta} \sum_{i=1}^{k} \frac{1}{i}$. $\qquad \square$

By Lemma 5, $|E_{>\Delta}| = \alpha(root) \leq \frac{n}{\Delta} \sum_{i=1}^{\frac{n}{\Delta}} \frac{1}{i}$. Since $\sum_{i=1}^{\frac{n}{\Delta}} \frac{1}{i} \leq 1 + \ln \frac{n}{\Delta}$, we have $\alpha(root) \leq \frac{n}{\Delta} + \frac{n}{\Delta} \ln \frac{n}{\Delta} = \mathcal{O}\left(\frac{n}{\Delta} \lg \frac{n}{\Delta}\right) = \mathcal{O}\left(n \lg n / \left(n^{\epsilon/4}\right)\right)$. We conclude that the space needed for $E_{>\Delta}$ is $2 |E_{>\Delta}| \lg n = \mathcal{O}\left(\frac{n \lg^2 n}{n^{\epsilon/4}}\right) = o(n)$ bits.

Finally, we explain how to implement the data structures for bookkeeping the LZ78 trie representation. By an additional node-marking vector M_{V_Δ} that marks the V_Δ-nodes, we divide the edges into $E_{\leq \Delta}$ and $E_{>\Delta}$. rank / select on M_{V_Δ} allows us to easily store, access and increment the n_e values for all edges in constant time. M_{V_Δ} can be computed in $\mathcal{O}(n)$ time when we have SA on A_1: since str_depth allows us to compute every $|c(e)|$ value in constant time, we can traverse ST in a DFS manner while computing $h(v)$ for each node v, and hence, it is easy to judge whether the current edge belongs to $E_{>\Delta}$. In order to store the h values for all ancestors of the current node we use a stack. Observe that the h values on the stack are monotonically increasing; hence we can implement it using a DS with $\mathcal{O}(n)$ bits [5, Sect. 4.2].

References

1. Amir, A., Farach, M., Idury, R.M., Poutré, J.A.L., Schäffer, A.A.: Improved dynamic dictionary matching. Inf. Comput. **119**(2), 258–282 (1995)
2. Bannai, H., Inenaga, S., Takeda, M.: Efficient LZ78 factorization of grammar compressed text. In: Calderón-Benavides, L., González-Caro, C., Chávez, E., Ziviani, N. (eds.) SPIRE 2012. LNCS, vol. 7608, pp. 86–98. Springer, Heidelberg (2012)
3. Benoit, D., Demaine, E.D., Munro, J.I., Raman, R., Raman, V., Rao, S.S.: Representing trees of higher degree. Algorithmica **43**(4), 275–292 (2005)
4. Fischer, J, I, T., Köppl, D.: Lempel Ziv computation in small space (LZ-CISS) (2015). arXiv:1504.02605
5. Fischer, J.: Optimal succinctness for range minimum queries. In: López-Ortiz, A. (ed.) LATIN 2010. LNCS, vol. 6034, pp. 158–169. Springer, Heidelberg (2010)
6. Fischer, J., Gawrychowski, P.: Alphabet-dependent string searching with wexponential search trees. In: Cicalese, F., Porat, E., Vaccaro, U. (eds.) CPM 2015. LNCS, vol. 9133, pp. 160–171. Springer, Heidelberg (2015)
7. Goto, K., Bannai, H.: Simpler and faster Lempel Ziv factorization. In: Proceedings of the 2013 Data Compression Conference, DCC 2013, pp. 133–142. IEEE Computer Society, Washington (2013)
8. Goto, K., Bannai, H.: Space efficient linear time Lempel-Ziv factorization for small alphabets. In: Data Compression Conference, DCC 2014, Snowbird, UT, USA, 26–28 March 2014, pp. 163–172 (2014)
9. Jansson, J., Sadakane, K., Sung, W.: Linked dynamic tries with applications to LZ-compression in sublinear time and space. Algorithmica **71**(4), 969–988 (2015)
10. Kärkkäinen, J., Kempa, D., Puglisi, S.J.: Linear time Lempel-Ziv factorization: simple, fast, small. In: Fischer, J., Sanders, P. (eds.) CPM 2013. LNCS, vol. 7922, pp. 189–200. Springer, Heidelberg (2013)
11. Kärkkäinen, J., Sanders, P., Burkhardt, S.: Linear work suffix array construction. J. ACM **53**(6), 918–936 (2006)
12. Kempa, D., Puglisi, S.J.: Lempel-Ziv factorization: simple, fast, practical. In: ALENEX, pp. 103–112 (2013)
13. Manber, U., Myers, E.W.: Suffix arrays: a new method for on-line string searches. SIAM J. Comput. **22**(5), 935–948 (1993)
14. Munro, J.I., Raman, R., Raman, V., Rao, S.S.: Succinct representations of permutations and functions. Theor. Comput. Sci. **438**, 74–88 (2012)
15. Nakashima, Y., I, T., Inenaga, S., Bannai, H., Takeda, M.: Constructing LZ78 tries and position heaps in linear time for large alphabets. Inform. Process. Lett. **115**(9), 655–659 (2015)
16. Navarro, G., Sadakane, K.: Fully functional static and dynamic succinct trees. ACM Trans. Algorithms **10**(3), 16:1–16:39 (2014)
17. Ohlebusch, E., Fischer, J., Gog, S.: CST++. In: Chavez, E., Lonardi, S. (eds.) SPIRE 2010. LNCS, vol. 6393, pp. 322–333. Springer, Heidelberg (2010)
18. Välimäki, N., Mäkinen, V., Gerlach, W., Dixit, K.: Engineering a compressed suffix tree implementation. ACM J. Exp. Algorithmics **14** (2009)

19. Ziv, J., Lempel, A.: A universal algorithm for sequential data compression. IEEE Trans. Inf. Theory **23**(3), 337–343 (1977)
20. Ziv, J., Lempel, A.: Compression of individual sequences via variable-rate coding. IEEE Trans. Inf. Theory **24**(5), 530–536 (1978)

Succinct Non-overlapping Indexing

Arnab Ganguly[1]([✉]), Rahul Shah[1], and Sharma V. Thankachan[2]

[1] School of Electrical Engineering and Computer Science,
Louisiana State University, Baton Rouge, USA
`agangu4@lsu.edu, rahul@csc.lsu.edu`
[2] School of Computational Science and Engineering,
Georgia Institute of Technology, Atlanta, USA
`sharma.thankachan@gatech.edu`

Abstract. Given a text T having n characters, we consider the *non-overlapping indexing* problem defined as follows: pre-process T into a data-structure, such that whenever a pattern P comes as input, we can report a maximal set of non-overlapping occurrences of P in T. The best known solution for this problem takes linear space, in which a suffix tree of T is augmented with $O(n)$-word data structures. A query P can be answered in optimal $O(|P| + \mathsf{nocc})$ time, where nocc is the output size [Cohen and Porat, ISAAC 2009]. We present the following new result: let CSA (not necessarily a compressed suffix array) be an index of T that can compute (i) the suffix range of P in $\mathsf{search}(P)$ time, and (ii) a suffix array or an inverse suffix array value in t_{SA} time; then by using CSA alone, we can answer a query P in $O(\mathsf{search}(P) + \mathsf{nocc} \cdot \mathsf{t}_{\mathsf{SA}})$ time. Additionally, we present an improved result for a generalized version of this problem called *range non-overlapping indexing*.

1 Introduction and Related Work

Indexing a text, so as to facilitate efficient pattern matching queries in the future, is a fundamental problem in the domain of information retrieval. The objective is to pre-process a text T, such that whenever a pattern P comes as query, all start positions (or simply, occurrences) of P in T can be reported efficiently. In most of the earlier works [5,12,14], both text and pattern were provided at query time. However, in most cases the text is static, and patterns come in as online query. This motivated the development of full text indexes such as (compressed) suffix arrays/trees.

Suffix tree [19,20] (resp. *suffix array* [16]) are the most well known *full-text* indexes supporting $O(|P| + occ)$ (resp. $O(|P| + \log n + occ)$) query time; here, $|P|$ is the length of P, and occ is the output size (i.e., the number of occurrences of P in T). The query time for suffix arrays can be reduced to $O(|P| + occ)$ by using a modified form of it, called *enhanced suffix arrays* [1]. Both suffix trees and suffix arrays require $\Theta(n \log n)$ bits of space, which is too large for most practical

This research is funded in part by National Science Foundation (NSF) Grant CCF 1218904.

© Springer International Publishing Switzerland 2015
F. Cicalese et al. (Eds.): CPM 2015, LNCS 9133, pp. 185–195, 2015.
DOI: 10.1007/978-3-319-19929-0_16

purposes. Grossi and Vitter [9], and Ferragina and Manzini [8] addressed this by presenting space efficient indexes named compressed suffix arrays and FM-Index respectively. Subsequently, an exciting field of compressed text indexing was established; see [17] for an excellent survey.

In some applications [6] (such as data-compression, speech recognition, linguistics), we are interested in only those occurrences of the pattern P that do not overlap with each other. This problem is a variation of the pattern matching problem, and can be formally stated as follows.

Problem 1 (Non-overlapping Indexing). *Given a text* T *of n characters, pre-process* T *into a data-structure, such that for any input pattern P, we can report a set of maximal starting positions of P, such that any two distinct starting positions are at least $|P|$ characters apart.*

Cohen and Porat [6] presented the first optimal time solution to this problem. Their data structure consists of a suffix tree of T and additional $O(n)$-word data structures. We show that Problem 1 can be solved efficiently using any index of T alone. The result is summarized in the following theorem.

Theorem 1. *Let* CSA *(not necessarily a compressed suffix array) be a full-text-index of* T*, that can compute (i) the suffix range of a pattern P in* search(P) *time and (ii) suffix array or an inverse suffix value in* t_{SA} *time. Then, without using any additional data structures, we can report a maximal set of* nocc *non-overlapping occurrences of P in* T *in* $O(\text{search}(P) + \text{nocc} \cdot t_{SA})$ *time.*

By avoiding use of any additional data-structures, various space time trade-off can be easily obtained. For example, if we use a suffix tree of T, optimal $O(|P| + \text{nocc})$ query time can be obtained (same as the result by Cohen and Porat [6]). On the other hand, an $n \log \sigma + o(n \log \sigma)$ bits of space and $O(|P| + \text{nocc} \log^{1+\epsilon} n)$ query time result can be obtained by using a recent version of compressed suffix array by Belazzougui and Navarro [4]. Here σ is the size of the alphabet from which the characters in T are drawn from and $\epsilon > 0$ is an arbitrary small constant. We remark that our solution is conceptually much simpler than the previous solution.

The second problem addressed in this paper can be seen as a variation of the well known *position restricted substring searching* problem of Makinen and Navarro [15]. The problem can be stated as follows.

Problem 2 (Range Non-overlapping Indexing). *Given a text* T *of n characters, pre-process* T *into a data-structure, such that whenever a pattern P, and a range $[a, b]$, $1 \leq a \leq b \leq n$, are provided as input, a maximal set of starting positions of P in the range $[a, b]$ are reported such that any two distinct positions are at least $|P|$ characters apart.*

For Problem 2, Keller et al. [13] presented an $O(n \log n)$ space and $O(|P| + \text{nocc}_{a,b} \log \log n)$ time data-structure, where $\text{nocc}_{a,b}$ is the number of non-overlapping occurrences of P in $[a, b]$. Iliopoulos et al. [7] presented an $O(n^{1+\epsilon})$ space and $O(|P| + \text{nocc}_{a,b})$ time solution. Nekrich and Navarro [18] presented a linear space and $O(|P| + \mathsf{nocc}_{a,b} \log^\epsilon n)$ time solution. Cohen and Porat [6] presented an $O(n \log^\epsilon n)$ space and $O(|P| + \log \log n + \mathsf{nocc}_{a,b})$ time solution. The following theorem summarizes our solution to Problem 2.

Theorem 2. *Given a text* T *of* n *characters, we can pre-process* T *to create data-structures of total size* $O(n \log^\epsilon n)$ *words, such that for any pattern* P, *and range* $[a, b]$, $1 \leq a \leq b \leq n$, *we can find a maximal set of non-overlapping occurrences of* P *in* $[a, b]$ *in optimal* $O(|P| + \mathsf{nocc}_{a,b})$ *time, where* $\mathsf{nocc}_{a,b}$ *is the output size.*

Organization of the Paper. The rest of the paper is dedicated for proving Theorem 1 and Theorem 2. In Sect. 2, we introduce notations and terminologies. We prove Theorem 1 in Sect. 3 and Theorem 2 in Sect. 4. Finally, we conclude the paper in Sect. 5.

2 Preliminaries and Notations

We refer the reader to [10] for standard definitions and terminologies. Throughout this paper, T is a text having n characters, and P is a pattern having $|P|$ characters. We assume that T terminates in a special character that does not appear at any other position in the document. We denote by $\mathsf{T}[t, t']$, the substring of T from t to t' (both inclusive). Further, ϵ is any arbitrarily small positive constant. A pattern P is said to occur at a position t in T if P starts at t. We denote by nocc, the number of non-overlapping occurrences of P in T. Likewise, $\mathsf{nocc}_{a,b}$ is the number of non-overlapping occurrences in the range $[a, b]$ i.e., in the substring $\mathsf{T}[a, b]$.

Suffix Tree. A suffix tree, denoted by ST, is a compact trie that stores all the (non-empty) suffixes of T. Leaves in the suffix tree are numbered in the lexicographic order of the suffix they represent. The locus of a pattern P is the highest node u such that P is a prefix of the string formed by the concatenation of the edge labels from the root to u. The suffix range of P is denoted by $[\mathsf{sp}, \mathsf{ep}]$, where sp (resp. ep) is the leftmost (resp. rightmost) leaf in the sub-tree of ST rooted at the locus of P. Using suffix trees, the suffix range of any pattern P can be determined in $O(|P|)$ time.

Suffix Array. A suffix array, denoted by SA, is an array of size n that maintains the lexicographic arrangement of all the suffixes of the text. More specifically, if the ith smallest suffix of T starts at j, we let $\mathsf{SA}[i] = j$ and $\mathsf{SA}^{-1}[j] = i$. Using suffix arrays, the locus node of any pattern P, or equivalently, the suffix range of P, can be found in $O(|P| + \log n)$ time. Using enhanced suffix arrays [1], this can be done in $O(|P|)$ time. Moving forward, we use the term suffix array to denote enhanced suffix arrays. The suffix value $\mathsf{SA}[\cdot]$ and the inverse suffix value $\mathsf{SA}^{-1}[\cdot]$ can be found in constant time.

In general, suffix trees (arrays) require $O(n)$ words for storage. *Compressed Suffix Arrays* reduce this space to $O(n \log \sigma)$ bits (or close to the size of the text) with a slowdown in query time. In what follows, we use CSA to denote a full-text-index of T (not necessarily a compressed index) that can compute the suffix range of P in search(P) time, and can compute a suffix array or inverse suffix array value in t_{SA} time. In most cases, search(P) is proportional to the length of the pattern P.

Lemma 1. *Given the suffix range* [sp, ep] *of pattern P, using* CSA, *we can verify in time* $O(t_{SA})$ *whether P appears at a text-position t, or not.*

Proof. The lexicographic position ℓ of the suffix $T[t, n]$ (i.e., $SA^{-1}[t]$) can be found in $O(t_{SA})$ time. By checking if ℓ lies in the range [sp, ep], we can determine whether P appears at t, or not.   ∎

3   Non-overlapping Indexing

In this section, we present our solution to Problem 1. We start with a few definitions.

Definition 1 (Period of a Pattern). *The period of a pattern P is its shortest prefix Q, such that P can be written as the concatenation of several (say $\alpha \geq 0$) number of copies of Q and a prefix Q' of Q. Specifically, $P = Q^\alpha Q'$.*
For e.g., if $P = abcabcab$, then $Q = abc$, $Q' = ab$ and $\alpha = 2$.

Definition 2 (Critical Occurrence of a Pattern). *A position t_c in the text* T *is called a critical occurrence of a pattern $P = Q^\alpha Q'$ if and only if t_c is an occurrence of P but the position $t_c + |Q|$ is not. In continuation with our example above, let the text* T *be $xyzabcabcabcabxyz$. Then $t_c = 7$ is a critical occurrence of P, but $t_c = 4$ is not.*

Observation 1. *Every critical occurrence of P in* T *corresponds to at least one non-overlapping occurrence of P in* T.

Definition 3 (Range of a Critical Occurrence). *Let t_c be a critical occurrence of the pattern P in the text* T. *Let $t' \leq t_c$ be the maximal position such that $t', t' + |Q|, t' + 2|Q|, \cdots, t_c$ are occurrences of P but the position $t' - |Q|$ is not. The range of the critical occurrence t_c is* range$(t_c) = [t', t_c + |P| - 1]$. *Continuing with our example above,* range$(7) = [4, 14]$.

Observation 2. *Let t_c be any critical occurrence of P in* T. *Then t_c is the rightmost occurrence of P in* range(t_c). *Furthermore, the ranges of any two distinct critical occurrences are disjoint.*

It follows from Observations 1 and 2 above that in order to find all non-overlapping positions of P in the text T, it suffices to find the non-overlapping occurrences of P in the range of every critical occurrence of P. We show how to achieve this in the following lemma.

t			t_c				t'		t		t_c						t'		t		t_c		
x	z	b	a	b	a	y	z	a	b	a	b	a	b	a	x	y	z	z	b	a	b	a	b	a
1	2	3	4	5	6	7	8	9	10	11	12	13	14	15	16	17	18	19	20	21	22	23	24	25

Fig. 1. Illustration of Lemma 2. Bottom row shows the text-position of every character, and the shaded text-positions mark the critical occurrences of the pattern $P = aba$ for which $Q = ab$ and $Q' = a$. Top row shows the text, and the shaded region shows the range of the critical occurrences t_c; t and t' have the same meaning as in Lemma 2.

Lemma 2. *Given a pattern $P = Q^\alpha Q'$, the suffix range of P, and a critical occurrence t_c of P in T, we can find all maximal non-overlapping occurrences of P in $\mathsf{range}(t_c)$ in time $O(nocc' \cdot t_{\mathsf{SA}})$, where $nocc'$ is the output size.*

Proof. First, we report t_c as a non-overlapping occurrence of P. Now, we consider the position $t = t_c - \alpha|Q|$. If $t \leq 0$ or if P does not appear at t (which can be verified in $O(t_{\mathsf{SA}})$ time using Lemma 1), then terminate. Otherwise, t belongs to $\mathsf{range}(t_c)$, and let $t' = t - |Q|$. If $t' \leq 0$ or if P does not appear at t', then terminate. Otherwise t' belongs to $\mathsf{range}(t_c)$, and is another non-overlapping occurrence of P. See Fig. 1 for an illustration. We repeat the process by letting $t_c = t'$. Clearly, the reported t''s are the maximal non-overlapping occurrences of P in $\mathsf{range}(t_c)$, and reporting them takes $O(nocc' \cdot t_{\mathsf{SA}})$ time. ∎

From Lemma 2, we conclude that given the suffix range of P (which can be found in $\mathsf{search}(P)$ time), and every critical occurrence of P in T, we can find all maximal non-overlapping occurrences of P in time $\mathsf{search}(P) + O(nocc \cdot t_{\mathsf{SA}})$, thereby proving Theorem 1. Therefore, our task is to find all critical occurrences of P in T. The following lemma shows how to achieve this.

Lemma 3. *Given a pattern $P = Q^\alpha Q'$, we can find all critical occurrences of P in T in time bounded by $search(P) + O(nocc \cdot t_{\mathsf{SA}})$.*

Proof. The proof of the lemma relies on the following observation.

Observation 3. *A critical occurrence of a pattern P is same as the text-position of a leaf which belongs to the suffix range of P, but not of QP.*

First observe that since Q' is a prefix of Q, the suffix range of QP is contained within that of P. Now, assume the contrary to Observation 3 above. Then, there is a critical occurrence, say t_c in text-order, of P in T, such that $\mathsf{SA}^{-1}[t_c]$ lies in the suffix range of $QP = Q^{\alpha+1}Q'$. Clearly, there is an occurrence of P at the position $t = t_c + |Q|$ which presents a contradiction. Therefore, our objective translates to locating the suffix range of P, say $[\mathsf{sp}, \mathsf{ep}]$, and of QP, say $[\mathsf{sp}', \mathsf{ep}']$; this can be done in time $\mathsf{search}(QP)$ which can be bounded by $\mathsf{search}(P)$. (We assume that $\mathsf{search}(P)$ is proportional to $|P|$.) See Fig. 2 for an illustration.

Note that for each leaf ℓ in the suffix ranges $[\mathsf{sp}, \mathsf{sp}' - 1]$ and $[\mathsf{ep}' + 1, \mathsf{ep}]$ (taken together), the text position $\mathsf{SA}[\ell]$ is a (distinct) critical occurrence of P. Thus, the number of leaves in the suffix range of P but not QP is same as the number of critical occurrences of P in T, and by Observation 1, the number of

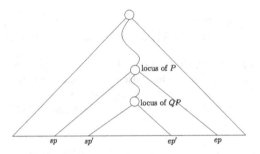

Fig. 2. Illustration of Lemma 3. Since, Q' is a prefix of Q, the locus of $P = Q^\alpha Q'$ lies on the path from root to the locus of QP. For each leaf ℓ in $[\mathsf{sp}, \mathsf{sp}' - 1] \cup [\mathsf{ep}' + 1, \mathsf{ep}]$, the text position $\mathsf{SA}[\ell]$ is a critical occurrence of P in the text T.

critical occurrences is at most the output size nocc. For every leaf, we can find the corresponding critical occurrence (i.e., its text-position) in time $O(\mathsf{t_{SA}})$ using $\mathsf{SA}[\cdot]$. Therefore, once the suffix ranges of P and QP are located, all the critical occurrences are found in time bounded by $O(\mathsf{nocc} \cdot \mathsf{t_{SA}})$.   ∎

4   Range Non-overlapping Indexing

In the range non-overlapping indexing problem, a range $[a, b]$ is provided as input in addition to the pattern P, and we are required to report all non-overlapping occurrences of P that start within the range $[a, b]$. For this problem, we use suffix trees or (enhanced) suffix arrays as the full-text indexes for T. Therefore, $\mathsf{search}(P) = O(|P|)$ and $\mathsf{t_{SA}} = O(1)$.

If $|P| \geq b - a$, then by using Lemma 1, we first find all occurrences of P (in sorted-order) in $[a, b]$ in $O(|P|)$ time. Then, we consider the first occurrence, skip over all occurrences that occur within the next P characters, and then report the next one, and so forth. Clearly, total time can be bounded by $O(|P|)$. Moving forward, we assume $|P| < b - a$. For our purposes, we will need a slightly modified form of Lemma 2 presented below.

Lemma 4. *Given a pattern $P = Q^\alpha Q'$, the suffix range of P, an occurrence t of P in T, and a range $[t', t]$, let t_c be the critical occurrence such that t belongs to $\mathsf{range}(t_c)$. Then, we can find all maximal non-overlapping occurrences of P in $\mathsf{range}(t_c)$ within $[t', t]$ in time $O(\mathsf{nocc'})$, where $\mathsf{nocc'}$ is the output size.*

Proof. First, we report t as a non-overlapping occurrence of P. Now, we consider the position $t'' = t - \alpha|Q|$. If $t'' < t'$ or if P does not appear at t'', then terminate. Otherwise, t'' belongs to $\mathsf{range}(t_c)$, and let $t''' = t'' - |Q|$. If $t''' < t'$ or if P does not appear at t''', then terminate. Otherwise, t''' belongs to $\mathsf{range}(t_c)$, and is another non-overlapping occurrence of P. We repeat the process by letting $t = t'''$. Clearly, the reported t''''s are the desired maximal non-overlapping occurrences of P, and reporting them takes $O(\mathsf{nocc'})$ time.   ∎

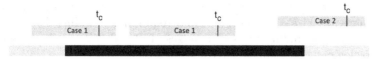

Fig. 3. Illustration of different cases for long patterns. Bottom row shows the text T in light shade, and the substring T[a, b] in dark shade. Case 1 shows the range of critical occurrences t_c, where $t_c \in [a, b]$. Case 2 shows range(t_c) for a critical occurrence $t_c > b$, where range(t_c) begins in [a, b].

Based on the length of the pattern P, we divide the proof of Theorem 2 into two cases. We say that P is a *long pattern* if $|P| > \log \log n$, and *short pattern*, otherwise. The following couple of lemmas summarize our result for the range non-overlapping indexing problem. We arrive at Theorem 2 by combining them.

Lemma 5. *There exists a data-structure that takes $O(n \log^\epsilon n)$ space and finds all non-overlapping occurrences of pattern P in the range $[a, b]$ in time $O(|P| + \mathsf{nocc}_{a,b})$, where $\mathsf{nocc}_{a,b}$ is the output size and $|P| > \log \log n$.*

Lemma 6. *There exists a data-structure that takes $O(n(\log \log n)^2)$ space and finds all non-overlapping occurrences of pattern P in the range $[a, b]$ in time $O(|P| + \mathsf{nocc}_{a,b})$, where $\mathsf{nocc}_{a,b}$ is the output size and $|P| \leq \log \log n$.*

4.1 Proof of Lemma 5

For answering queries for a long pattern $P = Q^\alpha Q'$, and a range $[a, b]$, we start by first finding the suffix range $[\mathsf{sp}, \mathsf{ep}]$ of P, and the suffix range $[\mathsf{sp}', \mathsf{ep}']$ of QP. Since, Q' is a prefix of Q, this can be achieved in $O(|P|)$ time. We show how to obtain all non-overlapping occurrences of P in $[a, b]$ by considering the following two cases. See Fig. 3 for an illustration.

Case 1. Consider those critical occurrences of P which lie in the range $[a, b]$. We find them by using the data-structure of the following lemma.

Lemma 7. [2]. *Given a set $\mathcal{P}$ of n two-dimensional points, we can pre-process them to create a data-structure which takes $O(n \log^\epsilon n)$ space, such that when a rectangular box $B = [x_1, x_2] \times [y_1, y_2]$ comes as query, we can report all k points of $\mathcal{P}$ that lie in B in $O(\log \log n + k)$ time.*

In addition to the full-text index over T, we maintain the above 2-dimensional range reporting data-structure of Alstrup et al. [2] over the suffix array SA of the text T. Specifically, the points for this data-structure are $(i, \mathsf{SA}[i])$, $1 \leq i \leq n$. For the range $[a, b]$, by using this data-structure, we first find all the points which lie within the bounding box $[\mathsf{sp}, \mathsf{sp}' - 1] \times [a, b]$, say $\mathcal{P}_1$, and those within the bounding box $[\mathsf{ep}' + 1, \mathsf{ep}] \times [a, b]$, say $\mathcal{P}_2$. Since, ranges of critical occurrences do not overlap, any point $(i, \mathsf{SA}[i]) \in \mathcal{P}_1 \cup \mathcal{P}_2$ corresponds to a unique critical occurrence that lies in the range $[a, b]$, which in turn corresponds to at least one non-overlapping occurrence of P in $[a, b]$ (refer to Observations 1, 2 and 3); thus, we have $|\mathcal{P}_1 \cup \mathcal{P}_2| \leq nocc_{a,b}$. Time required for finding $\mathcal{P}_1$ and $\mathcal{P}_2$, in addition to that for finding the suffix ranges of P and QP, can be bounded by $O(|P| + \log\log n + |\mathcal{P}_1 \cup \mathcal{P}_2|) = O(|P| + \mathsf{nocc}_{a,b})$.

For finding the non-overlapping occurrences of P in the range of every critical occurrence t_c reported above, we follow the procedure described in Lemma 4 for the range $[a, t_c]$. Since, each non-overlapping occurrence of P in the range of a critical occurrence can be found in constant time, total time required for finding all non-overlapping occurrences of P can be bounded by $O(|P| + \mathsf{nocc}_{a,b})$. Clearly, the reported occurrences are those of P that start at or after a, and ends before or at b, as desired.

Case 2. Apart from the non-overlapping occurrences reported in Case 1 above, the other non-overlapping occurrences that we are interested in are those which lie in $[a, b]$ and also in the range of a critical occurrence t_c, where $t_c > b$. Note that there is at most one such critical occurrence, and with abuse of notation, we use t_c to denote it. Consider the rightmost occurrence t_r of P in range(t_c) such that $t_r \leq b$. Note that $t_r \in [b - |P|, b]$. We can therefore verify the existence of such a t_c and t_r pair, and also find t_r (if any) in $O(|P|)$ time (refer to Lemma 1). (For our purposes, it is not necessary to locate the exact location t_c but only to verifying its existence.) Assume that there exists such a pair t_r and t_c. By our assumption $|P| < b - a$; therefore, $t_r \in [a, b]$. Now, for the range $[a, t_r]$, and occurrence t_r, we report all the desired non-overlapping occurrences of P as described in Lemma 4. Total time can again be bounded by $O(|P| + \mathsf{nocc}_{a,b})$.

4.2   Proof of Lemma 6

It is to be observed that the procedure described in the previous section will report non-overlapping occurrences in $[a, b]$ even for short patterns i.e., when $|P| \leq \log\log n$. In this case, however, the query complexity will incur an extra $\log\log n$ factor (because $|P|$ does not necessarily cascade $\log\log n$). This $\log\log n$ factor is incurred due to the use of the 2-dimensional range reporting data-structure of Alstrup et al. [2] for finding all critical occurrences in the range $[a, b]$ (refer to Lemma 7) The motivation, therefore, is to find the critical occurrences, once the suffix range of P is found, in time bounded by $O(\mathsf{nocc}_{a,b})$. In light of this, we present the following result of Alstrup et al. [3].

Lemma 8 [3]. *Given a set $\mathcal{P}$ of n one-dimensional points that are drawn from a set $\{0, 1, 2, \cdots, 2^w - 1\}$, where $w \geq \log n$ is the word size, there exists a data-structure of $O(n)$ words, such that when a range $R = [x_1, x_2]$ comes as query, we can report all k points of $\mathcal{P}$ that lie in R in $O(k)$ time.*

Observation 4. *Denote by* leaf(u), *the set of leaves in the sub-tree of* ST *rooted at u, and by $\mathcal{U}_k$, the set of all nodes at depth k. Since, the number of leaves in* ST *is n, it follows that $\sum_{u \in \mathcal{U}_k} |\mathsf{leaf}(u)| \leq n$.*

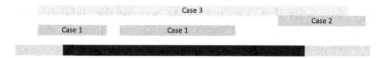

Fig. 4. Illustration of different cases for short patterns. Bottom row shows the text T in light shade, and the substring $T[a, b]$ in dark shade. In Case 1, $\mathsf{range}(t_c)$ for any critical occurrence t_c terminates in $[a, b]$. In Case 2, $\mathsf{range}(t_c)$ for the unique critical occurrence t_c (if any) starts in $[a, b]$ and terminates after b. Case 3 can occur only if none of Case 1 and Case 2 occurs. For the relevant Case 3, the entire range of a critical occurrence overlaps $T[a, b]$.

Building the Index. Now we show how to create a data-structure that is capable of answering queries for short patterns P in time $O(|P| + \mathsf{nocc}_{a,b})$. Let d be a parameter to be defined later. Consider a string P_d of length d formed by concatenating edge-labels on any path from *root*. For every prefix P' of P_d, we first find all the critical occurrences $\mathsf{C}_{P'}$ of P' in T. We maintain two data-structures, as described in Lemma 8, one each for the start and the end of the range of every critical occurrence in $\mathsf{C}_{P'}$. Based on Observations 2 and 4, the total space required for storing this structure for every prefix of every d length string P_d can be bounded by $O(nd)$ words. We do this for every $d \in [1, \lceil \log \log n \rceil]$. Total space can be bounded by $O(n(\log \log n)^2)$ words.

Querying the Index. For a short pattern P, we first retrieve the data-structures maintaining the start and end of the range of every critical occurrence in C_P. This can be done in $O(|P|)$ time. Depending on the critical occurrences, and the range $[a, b]$, we have the following three cases. See Fig. 4 for an illustration.

Case 1. Retrieve all critical occurrences t_c in C_P such that the end of $\mathsf{range}(t_c)$ lies in the interval $[a, b]$. By Lemma 8, this can be done in time $O(nocc')$, where $nocc'$ is the number of such critical occurrences. By definition, if the range of any critical occurrence t_c is $[t^-, t^+]$, then $t_c = t^+ - |P| + 1$. Given the range of a critical occurrence t_c, we first compute t_c using this formula. Consider any critical occurrence t_c. If $t_c \geq a$, we report all non-overlapping occurrences of P in the range $[a, t_c]$ for the occurrence t_c as described in Lemma 4. If $t_c < a$, we ignore it. Note that there is at most one critical occurrence such that its range starts before a and ends in $[a, b]$. Every other critical occurrence will result in at least one non-overlapping occurrence of P. Therefore, $nocc' \leq \mathsf{nocc}_{a,b} + 1$, and total time for finding all occurrences can be bounded by $O(\mathsf{nocc}_{a,b})$.

Case 2. Among the critical occurrences reported in Case 1 above, let t_{max} be the critical occurrence that appears last in the text and let t_{max}^+ be end of $\mathsf{range}(t_{max})$. If no critical occurrence was reported, then let $t_{max}^+ = a-1$. Observe that there is at most one critical occurrence, say t_c, such that $t_{max}^+ < t^- \leq b$, where $[t^-, t^+] = \mathsf{range}(t_c)$. Also, if t_c exists, then $t^+ > b$.

To find t_c (or verify if it does not exist), we use the data-structure which contains the start of range(t_c') for every t_c' in C_P and query it with the range $[t_{max}^+ + 1, b]$. Assume that t_c exists. We find the last occurrence $t_r \in [t_{max}^+ + 1, b]$ of P in range(t_c). This can be done in $O(|P|)$ time (refer to Case 2 in Sect. 4.1). We retrieve all the non-overlapping occurrences in $[t_{max}^+ + 1, t_r]$ for the occurrence t_r as described in Lemma 4. Total time can be bounded by $O(\mathsf{nocc}_{a,b})$.

Case 3. Suppose there is no critical occurrence retrieved in the above two cases. This may happen in the following scenarios.

- Range of critical occurrences in C_P either terminate before a, or start after b, or both. Clearly, these are of no interest, and can be safely ignored.
- There exists a single critical occurrence t_c whose range starts before a and terminates after b. By our assumption $|P| < b - a$, and if t_c exists, there must be an occurrence of P in $[a, b]$. We locate the last occurrence t_r of P in the range $[a, b]$. Note that t_r must lie within $|P|$ characters of b, and hence can be located in $O(|P|)$ time (refer to Lemma 1). If t_r does not exist, then neither does t_c. Otherwise, as described in Lemma 4, we locate all other non-overlapping occurrences of P in the range $[a, t_r]$ for the occurrence t_r. Total time is again bounded by $O(\mathsf{nocc}_{a,b})$.

5   Conclusion

In this paper, we revisit the problem of reporting all maximal non-overlapping occurrences of a pattern P in a text T. We show that by maintaining only a full-text index on T, we can find all nocc occurrences in optimal time $O(|P| + \mathsf{nocc})$. Further, the space can be made succinct by maintaining a self index of T, and the occurrences can be found in $\mathsf{search}(P) + O(\mathsf{nocc} \cdot \mathsf{t}_{\mathsf{SA}})$, where $\mathsf{search}(P)$ is the time required to find the suffix range of T, and t_{SA} is the time required to find suffix value or inverse suffix value. For the range-reporting version of the problem, where a range $[a, b]$ is provided as input in addition to the pattern P, we present an $O(n \log^\epsilon n)$ space index which can report all $\mathsf{nocc}_{a,b}$ occurrences in this range in optimal time $O(|P| + \mathsf{nocc}_{a,b})$. We remark that it is highly unlikely to have an efficient succinct data-structure for this problem, based on the hardness result of the position restricted substring searching problem [11].

References

1. Abouelhoda, M.I., Kurtz, S., Ohlebusch, E.: Replacing suffix trees with enhanced suffix arrays. J. Discret. Algorithms **2**(1), 53–86 (2004)
2. Alstrup, S., Brodal, G.S., Rauhe, T.: New data structures for orthogonal range searching. In: 41st Annual Symposium on Foundations of Computer Science, FOCS 2000, 12–14 November 2000, Redondo Beach, California, USA, pp. 198–207 (2000)
3. Alstrup, S., Brodal, G.S., Rauhe, T.: Optimal static range reporting in one dimension. In: Proceedings on 33rd Annual ACM Symposium on Theory of Computing, 6–8 July 2001, Heraklion, Crete, Greece, pp. 476–482 (2001)

4. Belazzougui, D., Navarro, G.: Alphabet-independent compressed text indexing. ACM Trans. Algorithms **10**(4), 23 (2014)
5. Boyer, R.S., Moore, J.S.: A fast string searching algorithm. Commun. ACM **20**(10), 762–772 (1977)
6. Cohen, H., Porat, E.: Range non-overlapping indexing. In: Dong, Y., Du, D.-Z., Ibarra, O. (eds.) ISAAC 2009. LNCS, vol. 5878, pp. 1044–1053. Springer, Heidelberg (2009)
7. Crochemore, M., Iliopoulos, C.S., Kubica, M., Rahman, M.S., Walen, T.: Improved algorithms for the range next value problem and applications. In: STACS 2008, Proceeding of the 25th Annual Symposium on Theoretical Aspects of Computer Science, Bordeaux, France, 21–23 February 2008, pp. 205–216 (2008)
8. Ferragina, P., Manzini, G.: Indexing compressed text. J. ACM **52**(4), 552–581 (2005)
9. Grossi, R., Vitter, J.S.: Compressed suffix arrays and suffix trees with applications to text indexing and string matching (extended abstract). In: Proceedings of the Thirty-Second Annual ACM Symposium on Theory of Computing, 21–23 May 2000, Portland, OR, USA, pp. 397–40 (2000)
10. Gusfield, D.: Algorithms on Strings, Trees, and Sequences : Computer Science and Computational Biology. Cambridge University Press, New York (1997)
11. Hon, W., Shah, R., Thankachan, S.V., Vitter, J.S.: On position restricted substring searching in succinct space. J. Discret. Algorithms **17**, 109–114 (2012)
12. Karp, R.M., Rabin, M.O.: Efficient randomized pattern-matching algorithms. IBM J. Res. Dev. **31**(2), 249–260 (1987)
13. Keller, O., Kopelowitz, T., Lewenstein, M.: Range non-overlapping indexing and successive list indexing. In: Dehne, F., Sack, J.-R., Zeh, N. (eds.) WADS 2007. LNCS, vol. 4619, pp. 625–636. Springer, Heidelberg (2007)
14. Knuth, D.E., Jr, J.H.M., Pratt, V.R.: Fast pattern matching in strings. SIAM J. Comput. **6**(2), 323–350 (1977)
15. Mäkinen, V., Navarro, G.: Position-restricted substring searching. In: Correa, J.R., Hevia, A., Kiwi, M. (eds.) LATIN 2006. LNCS, vol. 3887, pp. 703–714. Springer, Heidelberg (2006)
16. Manber, U., Myers, E.W.: Suffix arrays: a new method for on-line string searches. SIAM J. Comput. **22**, 935–948 (1993)
17. Navarro, G., Mäkinen, V.: Compressed full-text indexes. ACM Comput. Surv., vol. 39(1) (2007)
18. Nekrich, Y., Navarro, G.: Sorted range reporting. In: Fomin, F.V., Kaski, P. (eds.) SWAT 2012. LNCS, vol. 7357, pp. 271–282. Springer, Heidelberg (2012)
19. Ukkonen, E.: On-line construction of suffix trees. Algorithmica **14**(3), 249–260 (1995)
20. Weiner, P.: Linear pattern matching algorithms. In: 14th Annual Symposium on Switching and Automata Theory, Iowa City, Iowa, USA, 15–17 October 1973, pp. 1–11 (1973)

Encodings of Range Maximum-Sum Segment Queries and Applications

Paweł Gawrychowski[1] and Patrick K. Nicholson[2]([⊠])

[1] Institute of Informatics, University of Warsaw, Warsaw, Poland
[2] Max-Planck-Institut für Informatik, Saarbrücken, Germany
pnichols@mpi-inf.mpg.de

Abstract. Given an array A containing arbitrary (positive and negative) numbers, we consider the problem of supporting *range maximum-sum segment queries* on A: i.e., given an arbitrary range $[i,j]$, return the subrange $[i',j'] \subseteq [i,j]$ such that the sum $\sum_{k=i'}^{j'} A[k]$ is maximized. (We use the terms segment and subrange interchangeably, but only use segment when referring to the name of the problem, for consistency with prior work.) Chen and Chao [Disc. App. Math. 2007] presented a data structure for this problem that occupies $\Theta(n)$ words, can be constructed in $\Theta(n)$ time, and supports queries in $\Theta(1)$ time. Our first result is that if only the indices $[i',j']$ are desired (rather than the maximum sum achieved in that subrange), then it is possible to reduce the space to $\Theta(n)$ bits, regardless the numbers stored in A, while retaining the same construction and query time. Our second result is to improve the trivial space lower bound for any encoding data structure that supports range maximum-sum segment queries from n bits to $1.89113n - \Theta(\lg n)$, for sufficiently large values of n. Finally, we also provide a new application of this data structure which simplifies a previously known linear time algorithm for finding k-covers: given an array A of n numbers and a number k, find k disjoint subranges $[i_1, j_1], ..., [i_k, j_k]$, such that the total sum of all the numbers in the subranges is maximized. As observed by Csürös [IEEE/ACM TCBB 2004], k-covers can be used to identify regions in genomes.

1 Introduction

Many core data structure problems involve supporting *range queries* on arrays of numbers: see the surveys of Navarro [14] and Skala [18] for numerous examples. Likely the most heavily studied range query problem of this kind is that of supporting *range maximum queries* (resp. *range minimum queries*): given an array A of n numbers, preprocess the array such that, for any range $[i,j] \subseteq [1,n]$ we can return the index $k \in [i,j]$ such that $A[k]$ is maximum (resp. minimum). These kinds of queries have a large number of applications in the area of text indexing [8, Sect. 3.3]. Solutions have been proposed to this problem that achieve $\Theta(n)$

P. Gawrychowski—Currently holding a post-doc position at Warsaw Center of Mathematics and Computer Science.

© Springer International Publishing Switzerland 2015
F. Cicalese et al. (Eds.): CPM 2015, LNCS 9133, pp. 196–206, 2015.
DOI: 10.1007/978-3-319-19929-0_17

space (in terms of number of machine words[1]), and constant query time [1,7]. At first glance, one may think this to be optimal, since the array A itself requires n words to be stored. However, if we only desire the index of the maximum element, rather than the value of the element itself, it turns out that it is possible to reduce the space [9].

By a counting argument, it is possible to show that $2n - o(n)$ bits are necessary to answer range maximum queries on an array of n numbers [9, Sect. 1.1.2]. On the other hand, rather surprisingly, it is possible to achieve this space bound, to within lower order terms, while still retaining constant query time [9]. That is, regardless of the number of bits required to represent the individual numbers in A, we can encode a data structure in such a way as to support range maximum queries on A using $2n + o(n)$ bits. The key point is that we need not access A during any part of the query algorithm. In a more broad sense, results of this type are part of the area of succinct data structures [11], in which the aim is to represent a data structure using space matching the information theoretic lower bound, to within lower order terms.

In this paper, we consider *range maximum-sum segment queries* [5], where, given a range $[i,j]$, the goal is to return a subrange $[i',j'] \subseteq [i,j]$ such that $\sum_{k=i'}^{j'} A[k]$ is maximized. Note that this problem only becomes non-trivial if the array A contains negative numbers. With a bit of thought it is not difficult to see that the well-studied problem of supporting range maximum queries in an array A can be reduced to supporting range maximum-sum segment queries on a modified version of A that we get by padding each element of A with a sufficiently large negative number (see [5] for the details of the reduction). However, Chen and Chao [5] showed that a reduction holds in the other direction as well: range maximum-sum segment queries can be answered using a combination of range minimum and maximum queries on several different arrays, easily constructible from A. Specifically, they show that these queries can be answered in constant time with a data structure occupying $\Theta(n)$ words, that can be constructed in linear time.

A natural question one might ask is whether it is possible to improve the space of their solution to $\Theta(n)$ bits rather than $\Theta(n)$ words, while still retaining the constant query time. On one hand, we were aware of no information theoretic lower bound that ruled out the possibility of achieving $\Theta(n)$ bits. On the other hand, though Chen and Chao reduce the problem to several range maximum and range minimum queries, they still require comparisons to be made between various word-sized elements in arrays of size $\Theta(n)$ words in order to make a final determination of the answer to the query; we review the details of their solution in Sect. 4.1.

Our first result is summarized in the following theorem:

Theorem 1. *There is a data structure that occupies $12n + o(n)$ bits for supporting range maximum-sum segment queries. That is, the data structure returns the indices $[i',j']$ of the maximum-sum segment $\mathrm{RMAxSSQ}(A,i,j)$ for any $1 \le i \le j \le n$ in constant time, but not the sum of the numbers in the subrange $[i',j']$. Furthermore, the data structure is constructible in linear time.*

---

[1] In this paper we assume the word-RAM model with word size $\Theta(\log n)$ bits.

We present a high level overview of the steps of the proof of Theorem 1 in Sect. 4.2. Proofs of technical lemmas are omitted due to lack of space, but can be found in the in the full version of this paper [15]. The main idea is to sidestep the need for explicitly storing the numeric arrays required by Chen and Chao by storing two separate graphs that are judiciously defined so as to be embeddable in one page. By a well known theorem of Jacobson [11], combined with later improvements [10,13], it is known that one-page graphs—also known as outerplanar graphs—can be stored in a number of bits that is linear in the total number of vertices and edges, while still supporting constant time navigation operations. Navigating these graphs allows us to implicitly simulate comparisons between certain numeric array elements, thus avoiding the need to store the arrays themselves.

In terms of lower bounds, in the full version of the paper [15] we have also proved the following:

Theorem 2. *For an array A of length n, any data structure that encodes the solution to range maximum-sum segment queries must occupy at least $1.89113n - \Theta(\lg n)$ bits, if n is sufficiently large.*

We note that it is quite trivial to show that one requires n bits to support range maximum-sum segment queries by observing that if an array contains only numbers from the set $\{1, -1\}$ we can recover them via $\Theta(n)$ queries. In contrast, the main idea of the proof of Theorem 2 is to enumerate a combinatorial object which we refer to as maximum-sum segment trees, and then bound the number of trees of this type using generating functions.

Our final result, presented in Sect. 5, is a new application for maximum-sum segment data structures. Given an array and a number k, we want to find a k-cover: i.e., k disjoint subranges with the largest total sum. This problem was first studied by Csűrös [6], who was motivated by an application in bioinformatics, and constructed an $\mathcal{O}(n \log n)$ time algorithm. Later, an optimal $\mathcal{O}(n)$ time solution was found by Bengtsson and Chen [3]. We provide an alternative $\mathcal{O}(n)$ time solution, which is an almost immediate consequence of any constant time maximum-sum segment data structure that can be constructed in linear time. An advantage of our algorithm is that it can be also used to preprocess the array just once, and then answer the question for any $k \in [1, n]$ in $\mathcal{O}(k)$ time. We remark that this is related, but not equivalent, to finding k non-overlapping maximum-sum segments, and finding k maximum-sum segments. In the latter, one considers all $\binom{n}{2} + n$ subranges ordered non-increasingly according to their corresponding sums, and wants to select the k-th largest [12]. In the former, one repeats the following operation k times: find a maximum-sum segment disjoint from all that were previously chosen, and add it to the current set [17].

2   Notation and Definitions

We follow the notation of Chen and Chao [5] with a few minor changes. Let A be an array of n numbers. Let $S(i, j)$ denote the sum of the values in the range

$[i, j]$: i.e., $S(i, j) = \sum_{k=i}^{j} A[k]$. Let C be an array of length n such that $C[i]$ stores the cumulative sum $S(1, i)$. Note that $S(i, j) = C[j] - C[i-1]$ if $i > 1$.

Given an arbitrary array B, a range maximum query $\text{RMaxQ}(B, i, j)$ returns the index of the rightmost maximum value in the subarray $B[i, j]$; the query $\text{RMinQ}(B, i, j)$ is defined analogously. A range maximum-sum segment query $\text{RMaxSSQ}(A, i, j)$ returns a subrange $A[i', j']$ such that $i \leq i' \leq j' \leq j$, and $S(i', j')$ is maximum. If there is a tie, then our data structure will return the shortest range with the largest value of j'; i.e., the rightmost one.[2] Note that the answer is a range, specified by its endpoints, rather than the sum of the values in the range. Also note that if the range $A[i, j]$ contains only non-positive numbers, we return an empty range as the solution (for a discussion of alternatives we refer the reader to the full version of this paper [15]).

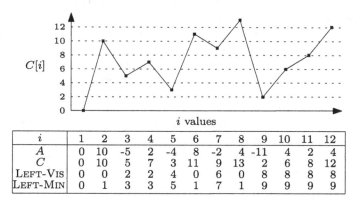

i	1	2	3	4	5	6	7	8	9	10	11	12
A	0	10	-5	2	-4	8	-2	4	-11	4	2	4
C	0	10	5	7	3	11	9	13	2	6	8	12
Left-Vis	0	0	2	2	4	0	6	0	8	8	8	8
Left-Min	0	1	3	3	5	1	7	1	9	9	9	9

Fig. 1. Example array A and the values of the various definitions presented in this section that are induced by A. The list of candidates for this array are: $(1, 2)$, $(3, 4)$, $(1, 6)$, $(1, 8)$, $(9, 10)$, $(9, 11)$, $(9, 12)$.

The *left visible region* $\text{Left-Vis}(i)$ of array C at index i is defined to be the maximum index $1 \leq j < i$ such that $C[j] \geq C[i]$, or 0 if no such index exists. The *left minimum* $\text{Left-Min}(i)$ of array C is defined to be $\text{RMinQ}(C, \text{Left-Vis}(i) + 1, i)$ for $1 < i \leq n$. See Fig. 1 for an illustration of these definitions. The pairs $(\text{Left-Min}(i), i)$ where $\text{Left-Min}(i) < i$ are referred to as *candidates*. Thus, candidates are such that the sum in $A[\text{Left-Min}(i) + 1..i]$ is positive. One issue is that the pair $(\text{Left-Min}(1), 1)$ might have a positive sum, but not be a candidate. Without loss of generality we can ignore this case by assuming $A[1] = 0$, as in Fig. 1. Define the *candidate score array* D such that $D[i] = S(\text{Left-Min}(i) + 1, i)$ if $(\text{Left-Min}(i), i)$ is a candidate, and $D[i] = 0$ otherwise, for all $i \in [1, n]$. Thus, for *non-candidates*, the candidate score is 0. Let $x' = \text{RMaxQ}(D, 1, n)$ and $t' = \text{Left-Min}(x')$. From the definitions it is not too difficult to see that $\text{RMaxSSQ}(A, 1, n)$ is $[t' + 1, x']$ if $t' \neq x'$, and the empty range otherwise.

---

[2] Alternatively, we can return the leftmost such range by symmetry.

3   Preliminary Data Structures

We make heavy use of the following result:

Lemma 1 [9]. *Given an array B of n numbers, we can store a data structure of size $2n+o(n)$ bits such that $\text{RMAXQ}(B, i, j)$ can be returned for any $1 \le i \le j \le n$ in $\mathcal{O}(1)$ time. Similarly, we can also answer $\text{RMINQ}(B, i, j)$ for $1 \le i \le j \le n$ using the same space bound. These data structures can be constructed in linear time so as to return the rightmost maximum (resp. minimum) in the case of a tie.*

We also require the following succinct data structure result for representing one-page graphs. A one-page (or outerplanar) graph G has the property that it can be represented by a sequence of balanced parentheses [11]. Equivalently, there exists a labelling $1, ..., n$ of the vertices in G in which there is no pair of edges (u_1, u_2) and (u_3, u_4) in G such that $1 \le u_1 < u_3 < u_2 < u_4 \le n$. That is, if we refer to vertices by their labels, then we have that the set of ranges represented by edges are either nested or disjoint: we refer to this as the *nesting property*. Note that our definitions of the navigation operations differ (slightly) from the original source so as to be convenient for our application.

Lemma 2 (Based on [10,13]). *Let G be a one-page multigraph with no self-loops: i.e., G has vertex labels $1, ..., n$, and m edges with the nesting property. There is a data structure that can represent G using $2(n + m) + o(n + m)$ bits, and be constructed in $\Theta(n + m)$ time from the adjacency list representation of G, such that the following operations can be performed in constant time:*

1. $\text{DEGREE}(G, u)$ *returns the degree of vertex u.*
2. $\text{NEIGHBOUR}(G, u, i)$ *returns the index of the vertex which is the endpoint of the i-th edge incident to u, for $1 \le i \le \text{DEGREE}(u)$. The edges are sorted in non-decreasing order of the indices of the endpoints: $\text{NEIGHBOUR}(G, u, 1) \le \text{NEIGHBOUR}(G, u, 2) \le ... \le \text{NEIGHBOUR}(G, u, \text{DEGREE}(G, u))$.*
3. $\text{ORDER}(G, u, v)$ *returns the order of the edge (u, v) among those incident to u: i.e., return an i such that $\text{NEIGHBOUR}(G, u, i) = v$.*[3]

In all of the operations above, a vertex is referred to by its label, which is an integer in the range $[1, n]$.

4   Supporting Range Maximum-Sum Segment Queries

In this section we present our solution to the range maximum-sum segment query problem which occupies linear space in bits. First we begin by summarizing the solution of Chen and Chao [5]. Then, in Sect. 4.2 we describe an alternative data structure that occupies $\Theta(n)$ words of space. Finally, we reduce the space of our alternative data structure to linear in bits.

---

[3] Since G may be a multigraph the value of i may be arbitrary among all possible values that satisfy the equation. This is more general than we require, as we will not execute this type of query on a multigraph, so the answer will always be unique.

4.1 Answering Queries using $\Theta(n)$ Words

In the solution of Chen and Chao the following data structures are stored: the array C; a data structure to support $\text{RMINQ}(C, i, j)$ queries; a data structure for supporting $\text{RMAXQ}(D, i, j)$ queries; and, finally, an array P of length n where $P[i] = \text{LEFT-MIN}(i)$. Thus, the overall space is linear in words.

The main idea is to examine the candidate $(P[x], x)$ whose right endpoint achieves the maximum sum in the range $[i, j]$. If $P[x] + 1 \in [i, j]$ then Chen and Chao proved that $[P[x] + 1, x]$ is the correct answer. However, if $P[x] + 1 \notin [i, j]$ then they proved that there are two possible ranges which need to be examined to determine the answer. In this case we check the sum for both ranges and return the range with the larger sum. The pseudocode for their solution to answering the query $\text{RMAXSSQ}(A, i, j)$ is presented in Algorithm 1:

Algorithm 1. Computing $\text{RMAXSSQ}(A, i, j)$.

1: $x \leftarrow \text{RMAXQ}(D, i, j)$
2: **if** $P[x] = x$ **then** $\triangleright$ In this case x is a non-candidate, so $D[x] = 0$
3:      **return** the empty range
4: **else if** $P[x] + 1 \geq i$ **then** $\triangleright$ In this case $[P[x] + 1, x] \subseteq [i, j]$
5:      **return** $[P[x] + 1, x]$
6: **else** $\triangleright$ In this case $[P[x] + 1, x] \not\subseteq [i, j]$
7:      $y \leftarrow \text{RMAXQ}(D, x + 1, j)$
8:      $t \leftarrow \text{RMINQ}(C, i - 1, x - 1)$
9:      **if** $S(t + 1, x) > S(P[y] + 1, y)$ **then**
10:          **return** $[t + 1, x]$
11:      **else**
12:          **return** $[P[y] + 1, y]$
13:      **end if**
14: **end if**

Items (1), (2) and (3) of the following collection of lemmas by Chen and Chao imply that the query algorithm is correct. We use item (4) later.

Lemma 3 [5]. *The following properties hold (using the notation from Algorithm 1):*

1. *If $[P[x] + 1, x] \subseteq [i, j]$ then $\text{RMAXSSQ}(A, i, j)$ is $[P[x] + 1, x]$.*
2. *The following inequalities hold: $x < P[y] \leq y$.*
3. *If $[P[x] + 1, x] \not\subseteq [i, j]$ then $\text{RMAXSSQ}(A, i, j)$ is $[P[y] + 1, y]$ or $[t + 1, x]$.*
4. *If $1 \leq i < j \leq n$ then it cannot be the case that $P[i] < P[j] \leq i$. That is, the ranges $[P[i], i]$ and $[P[j], j]$ have the nesting property for all $1 \leq i < j \leq n$.*

On line 9 of Algorithm 1 the sums can be computed in constant time using the array C. All other steps either defer to the range maximum or minimum structures, or a constant number of array accesses. Thus, the query algorithm takes constant time to execute.

4.2   Reducing the Space to $\Theta(n)$ Bits

Observe that the data structure for answering RMAXQ (resp. RMINQ) queries on D (resp. C) only requires $2n+o(n)$ bits by Lemma 1; $4n+o(n)$ bits in total for both structures. Thus, if we can reduce the cost of the remaining data structures to $\Theta(n)$ bits, while retaining the correctness of the query algorithm, then we are done. There are two issues that must be overcome in order to reduce the overall space to linear in bits:

1. The array P occupies n words, so we cannot store it explicitly.
2. In the case where $[P[x] + 1, x]$ is not contained in $[i, j]$, we must compare $S(t + 1, x)$ and $S(P[y] + 1, y)$ without explicitly storing the array C.

The first issue turns out to be easy to deal with: we instead encode the graph $G = ([n], \{(P[x], x) | 1 \leq x \leq n, P[x] < x\})$, which we call the *candidate graph* using Lemma 2 combined with item 4 of Lemma 3, which implies the following:

Lemma 4. *The candidate graph G can be represented using $4n + o(n)$ bits of space, such that given any $x \in [1, n]$ we can return* LEFT-MIN(x) *in $\mathcal{O}(1)$ time.*

From here onward, we can assume that we have access to the array P, which we simulate using Lemma 4. Unfortunately, the second issue turns out to be far more problematic. We overcome this problem via a two step approach. In the first step, we define another array Q which we will use to avoid directly comparing the sums $S(t + 1, x)$ and $S(P[y] + 1, y)$. This eliminates the need to store the array C. We then show how to encode the array Q using $\Theta(n)$ bits.

Left Siblings and the Q Array: Given candidate $(P[x], x)$, we define the *left sibling* LEFT-SIB$((P[x], x))$ to be the largest index $\ell \in [1, P[x] - 1]$, such that there exists an $\ell' \in [\ell + 1, P[x]]$ with $S(\ell + 1, \ell') > S(P[x] + 1, x)$, if such an index exists. Moreover, when discussing ℓ' we assume that ℓ' is the smallest such index. If no such index ℓ exists, or if $(P[x], x)$ is a non-candidate, we say LEFT-SIB$((P[x], x))$ is undefined. We define the array Q such that $Q[x] =$ LEFT-SIB$((P[x], x))$ for all $x \in [1, n]$; if $Q[x]$ is undefined, then we store the value 0 to denote this. By case analysis, we have proved the following lemma which shows that we can compare $S(t + 1, x)$ to $S(P[y] + 1, y)$ using the Q array:

Lemma 5. *If $P[y] = y$ or $Q[y] \geq t$ then* RMAXSSQ$(A, i, j) = [t + 1, x]$. *Otherwise,* RMAXSSQ$(A, i, j) = [P[y] + 1, y]$

Note that in the previous lemma, we need not know the value of ℓ' in order to make the comparison: only the value $Q[y]$ is required.

Encoding the Q Array: Unfortunately, the graph defined by the set of edges $(Q[x], x)$ does not have the nesting property. Instead, we construct an n-vertex graph H using the pairs $(Q[x], P[x])$ as edges, for each $x \in [1, n]$ where LEFT-SIB(x) is defined (i.e., $Q[x] \neq 0$). We call H the *left sibling graph*. We give an example illustrating both the graphs G and H in Fig. 2. Note in the figure that each edge in G has a corresponding edge in H unless its left sibling is undefined. We have proved that we can navigate G and H to find the left sibling of a candidate, as summarized in the following lemma:

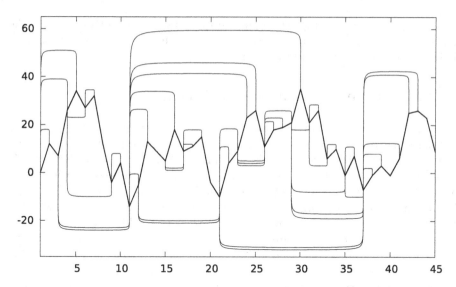

Fig. 2. A graph of the cumulative sums C (thick middle line) for a randomly generated instance with $n = 45$ and floating point numbers drawn uniformly from the range $[-20, 20]$. The x-axis represents the number i, and the y-axis represents $C[i]$. The edges drawn above the line for C represent the candidate graph G and the edges below represent the left sibling graph H: note that H is a multigraph.

Lemma 6. *Let $(P[x], x)$ be a candidate, and suppose $i = \text{DEGREE}(H, P[x]) - \text{ORDER}(G, P[x], x) + 1$. If $i > 0$ then it is the case that $\text{LEFT-SIB}((P[x], x)) = \text{NEIGHBOUR}(H, P[x], i)$. Otherwise, $\text{LEFT-SIB}((P[x], x))$ is undefined.*

Finally, we have proved the property that can be observed in Fig. 2: namely, that we can apply Lemma 2 to H, since the edges are nested.

Lemma 7. *The left-sibling graph H can be represented using no more than $4n + o(n)$ bits of space, such that given any $x \in [1, n]$ we can return $Q[x]$ in constant time, assuming access to the data structure of Lemma 4.*

Thus, to simulate the query algorithm of Chen and Chao we need: the range maximum structure for the array D (Lemma 1); the range minimum structure for the array C (Lemma 1); the representation of the graph G (Lemma 2); the representation of the graph H (Lemma 2). Since the sum of the sizes of these data structures is $12n + o(n)$ bits, we have completed the proof of Theorem 1, except for the claim about the linear construction time. The details of the construction can be found in the full version [15].

Remark 1. We note that the constant factor of 12 in Theorem 1 is suboptimal. As Rajeev Raman [16] has pointed out, the space for Lemma 4 can be reduced to $2n + o(n)$ bits. Furthermore, we have also noted that an additional n bits can be saved by combining the range maximum and minimum encodings for D and C.

However, both of these improvements are quite technical and we suspect the optimal constant factor is much lower than 9. As such, we leave determination of this optimal constant as future work.

5   Application to Computing k-Covers

Given an array A of n numbers and a number k, we want to find k disjoint subranges $[i_1, j_1], \ldots, [i_k, j_k]$, called a k-cover, such that the total sum of all numbers inside, called the score, is maximized. For $k = 1$ (the RMAxSSQ problem on the entire array) this is a classic exercise, often used to introduce dynamic programming. For larger values of k, it is easy to design an $\mathcal{O}(nk)$ time dynamic programming algorithm, but an interesting question is whether we can do better. As shown by Csurös [6], one can achieve $\mathcal{O}(n \log n)$ time complexity. This was later improved to $\mathcal{O}(n\alpha(n,n))$ [2] (where $\alpha(n,n)$ is the inverse Ackermann function), and finally to optimal $\mathcal{O}(n)$ time [3]. In this section we show that, assuming a constant time range maximum-sum segment structure, which can be constructed in linear time, we can preprocess the array in time $\mathcal{O}(n)$, so that given any k, we can compute a maximum k-cover in $\mathcal{O}(k)$ time. This improves the previous linear time algorithm, which needs $\mathcal{O}(n)$ time to compute a maximum k-cover regardless of how small k is, so our algorithm is more useful when there are multiple different values of k for which we want to compute a maximum k-cover.

We iteratively construct a maximum score k-cover for $k = 0, 1, 2, \ldots, n$. This is possible due to the following property already observed by Csurös.

Lemma 8 [6]. *A maximum score $(k + 1)$-cover can be constructed from any maximum score k-cover consisting of intervals $[i_1, j_1], \ldots, [i_k, j_k]$ in one of the two ways:*

1. *adding a new interval $[i_{k+1}, j_{k+1}]$ disjoint with all $[i_1, j_1], \ldots, [i_k, j_k]$,*
2. *replacing some $[i_\ell, j_\ell]$ with two intervals $[i_\ell, j'], [i', j_\ell]$.*

As any such transformation results in a valid $(k + 1)$-cover, we can construct a maximum score $(k + 1)$-cover by simply choosing the one increasing the score the most. In other words, we can iteratively select the best transformation. Now the question is how to do so efficiently.

We will first show that the best transformation of each type can be found in $\mathcal{O}(1 + k)$ time using the range maximum-sum queries. Assume that we have both a range maximum-sum and a range minimum-sum query structure available. Recall that out of all possible transformations of every type, we want the find the one increasing the score the most.

1. To add a new interval $[i_{k+1}, j_{k+1}]$ disjoint with all $[i_1, j_1], \ldots, [i_k, j_k]$ increasing the score the most, we guess an index ℓ such that the new interval is between $[i_\ell, j_\ell]$ and $[i_{\ell+1}, j_{\ell+1}]$ (if $\ell = 0$ we ignore the former and if $\ell = k$ the latter condition). Then $[i_{k+1}, j_{k+1}]$ can be found with RMAxSSQ$(A, j_\ell + 1, i_{\ell+1} - 1)$.

2. To replace some $[i_\ell, j_\ell]$ with two intervals $[i_\ell, j'], [i', j_\ell]$ increasing the score the most, we observe that the score increases by $-S(j' + 1, i' - 1)$, hence we can guess ℓ and then find $(j' + 1, i' - 1)$ with $\mathrm{RMINSSQ}(A, i_\ell, j_\ell)$.

For every type, we need $1 + k$ calls to one of the structures. If each call takes constant time, the claimed $\mathcal{O}(1 + k)$ complexity follows.

We will now show that, because we repeatedly apply the best transformation, the situation is more structured and the best transformation of each type can be found faster. To this end we define a *transformation tree* as follows. Its root corresponds to the maximum-sum segment $[i, j]$ of the whole A, meaning that its weight is $S(i, j)$, and has up to three children. If A is empty or consists of only negative numbers, the transformation tree is empty.

1. The left child is the transformation tree recursively defined for $A[1..i - 1]$.
2. The middle child is the transformation tree recursively defined for $-A[i..j]$, i.e., for a copy of $A[i..j]$ with all the numbers multiplied by -1.
3. The right child is the transformation tree recursively defined for $A[j + 1..n]$.

If any of these ranges is empty, we don't create the corresponding child. Now the transformation tree is closely related to the maximum score k-covers.

Lemma 9. *For any $k \geq 1$, a k-cover constructed by the iterative method corresponds to a subtree of the transformation tree containing the root.*

This suggests that a maximum k-cover can be found by computing a maximum weight subtree of the transformation tree containing the root and consisting of k nodes. Indeed, any such subtree corresponds to a k-cover, and by Lemma 9 a maximum k-cover corresponds to some subtree. To find a maximum weight subtree efficiently, we observe the following property of the transformation tree.

Lemma 10. *The transformation tree has the max-heap property, meaning that the weight of every node is at least as large as the weight of its parent.*

Therefore, to find a maximum weight subtree consisting of k nodes, we can simply choose the k nodes with the largest weight in the whole tree (we assume that the weights are pairwise distinct, and if not we break the ties by considering the nodes closer to the root first). This can be done by first explicitly constructing the transformation tree, which takes $\mathcal{O}(n)$ time assuming a constant time maximum and minimum range-sum segment structures. Then we can use the linear time selection algorithm [4] to find its k nodes with the largest weight. This is enough to solve the problem for a single value of k in $\mathcal{O}(n)$ time.

If we are given multiple values of k, we can process each of them in $\mathcal{O}(k)$ time assuming the following linear time and space preprocessing. For every $i = 0, 1, 2, \ldots, \log n$ we select and store the 2^i nodes of the transformation tree with the largest weight. This takes $\mathcal{O}(n + n/2 + n/4 + \ldots) = \mathcal{O}(n)$ total time and space. Then, given k, we find i such that $2^i \leq k < 2^{i+1}$ and again use the linear time selection algorithm to choose the k nodes with the largest weight out of the stored 2^{i+1} nodes.

References

1. Bender, M.A., Farach-Colton, M.: The LCA problem revisited. In: Gonnet, G.H., Viola, A. (eds.) LATIN 2000. LNCS, vol. 1776, pp. 88–94. Springer, Heidelberg (2000)
2. Bengtsson, F., Chen, J.: Computing maximum-scoring segments in almost linear time. In: Chen, D.Z., Lee, D.T. (eds.) COCOON 2006. LNCS, vol. 4112, pp. 255–264. Springer, Heidelberg (2006)
3. Bengtsson, F., Chen, J.: Computing maximum-scoring segments optimally. Technical report, Research Report, Luleå University of Technology (2007)
4. Blum, M., Floyd, R., Pratt, V., Rivest, R., Tarjan, R.: Time bounds for selection. J. Comput. Syst. Sci. **7**, 448–461 (1972)
5. Chen, K.Y., Chao, K.M.: On the range maximum-sum segment query problem. Discrete Appl. Math. **155**(16), 2043–2052 (2007)
6. Csűrös, M.: Maximum-scoring segment sets. IEEE/ACM Trans. Comput. Biol. Bioinform. **1**(4), 139–150 (2004)
7. Durocher, S.: A simple linear-space data structure for constant-time range minimum query. In: Brodnik, A., López-Ortiz, A., Raman, V., Viola, A. (eds.) Ianfest-66. LNCS, vol. 8066, pp. 48–60. Springer, Heidelberg (2013)
8. Fischer, J.: Data structures for efficient string algorithms. Ph.D. thesis, Ludwig-Maximilians-Universität München, October 2007
9. Fischer, J., Heun, V.: Space-efficient preprocessing schemes for range minimum queries on static arrays. SIAM J. Comput. **40**(2), 465–492 (2011)
10. Geary, R.F., Rahman, N., Raman, R., Raman, V.: A simple optimal representation for balanced parentheses. Theor. Comput. Sci. **368**(3), 231–246 (2006)
11. Jacobson, G.: Space-efficient static trees and graphs. In: Proceedings of the 30th Annual Symposium on Foundations of Computer Science, FOCS 1989, pp. 549–554. IEEE Computer Society, Washington, DC (1989)
12. Liu, H.F., Chao, K.M.: Algorithms for finding the weight-constrained k longest paths in a tree and the length-constrained k maximum-sum segments of a sequence. Theor. Comput. Sci. **407**(1–3), 349–358 (2008)
13. Munro, J.I., Raman, V.: Succinct representation of balanced parentheses and static trees. SIAM J. Comput. **31**(3), 762–776 (2001)
14. Navarro, G.: Spaces, trees, and colors: the algorithmic landscape of document retrieval on sequences. ACM Comput. Surv. **46**(4), 52 (2013)
15. Nicholson, P.K., Gawrychowski, P.: Encodings of range maximum-sum segment queries and applications. CoRR abs/1410.2847 (2014). http://arxiv.org/abs/1410.2847
16. Raman, R.: Personal communication
17. Ruzzo, W.L., Tompa, M.: A linear time algorithm for finding all maximal scoring subsequences. In: Proceedings of the Seventh International Conference on Intelligent Systems for Molecular Biology, pp. 234–241. AAAI Press (1999)
18. Skala, M.: Array range queries. In: Brodnik, A., López-Ortiz, A., Raman, V., Viola, A. (eds.) Ianfest-66. LNCS, vol. 8066, pp. 333–350. Springer, Heidelberg (2013)

Compact Indexes for Flexible Top-k Retrieval

Simon Gog[1]([⊠]) and Matthias Petri[2]

[1] Karlsruhe Institute of Technology, 76131 Karlsruhe, Germany
gog@kit.edu
[2] The University of Melbourne, Parkville, VIC 3010, Australia
matthias.petri@unimelb.edu.au

Abstract. We design and engineer a self-index based retrieval system capable of rank-safe evaluation of top-k queries. The framework generalizes the GREEDY approach of Culpepper et al. (ESA 2010) to handle multi-term queries, including over phrases. We propose two techniques which significantly reduce the ranking time for a wide range of popular Information Retrieval (IR) relevance measures, such as TF × IDF and BM25. First, we reorder elements in the document array according to document weight. Second, we introduce the repetition array, which generalizes Sadakane's (JDA 2007) document frequency structure to document subsets. Combining document and repetition array, we achieve attractive functionality-space trade-offs. We provide an extensive evaluation of our system on terabyte-sized IR collections.

1 Introduction

Calculating the k most relevant documents for a multi-term query Q against a set of documents $\mathcal{D}$ is a fundamental problem – the top-k document retrieval problem – in Information Retrieval (IR). The relevance of a document d to Q is determined by evaluating a similarity function $\mathcal{S}$ such as BM25. Exhaustive evaluation of $\mathcal{S}$ generates scores for all d in $\mathcal{D}$. The top-k scored documents in the list are then reported. Algorithms which guarantee production of the same top-k results list as the exhaustive process are called *rank-safe*.

The *inverted index* is a highly-engineered data structure designed to solve this problem. The index stores, for each unique term in $\mathcal{D}$, the list of documents containing that term. Queries are answered by processing the lists of all the query terms. Advanced query processing schemes [2] process lists only partially while remaining rank-safe. However, additional work during construction time is required to avoid scoring non-relevant documents at query time. Techniques used to speed up query processing include sorting lists in decreasing score order, or pre-storing score upper bounds for sets of documents which can then safely be skipped during query processing. These pre-processing steps introduce a dependency between $\mathcal{S}$ and the stored index. Changing the $\mathcal{S}$ requires at least partial reconstruction of the index, which in turn reduces the flexibility of the retrieval system.

Another family of retrieval systems is based on self-indexes [16]. These systems support functionality not easily provided by inverted indexes, such as

© Springer International Publishing Switzerland 2015
F. Cicalese et al. (Eds.): CPM 2015, LNCS 9133, pp. 207–218, 2015.
DOI: 10.1007/978-3-319-19929-0_18

efficient phrase search, and text extraction. Systems capable of single-term top-k queries have been proposed [17,20] and work well in practice [8,13].

Hon et al. [12] investigate top-k indexes which support different scoring schemes such as term frequency or static document scores. They also extended their framework to multi-term queries and term proximity. While the space of the multi-term versions is still linear in the collection size the query time gets dependent on the root of the collection size. At CPM 2014, Larson et al. [14] showed that it is not expected to improve their result significantly by reducing the boolean matrix multiplication problem to a relaxed version of Hon et al.'s problem, i.e. just answering the question whether there is a document which contains both of the query times. While the single-term version of Hon et al.'s framework was implemented and studied by several authors [5,19] there is not yet an implementation of a rank-safe multi-term version.

Our Contributions. We propose, to the best of our knowledge, the first flexible self-index based retrieval framework capable of rank-safe evaluation of multi-term top-k queries for complex IR relevance measures. It is based on a generalization of GREEDY [4]. We suggest two techniques to decrease the number of evaluated nodes in the GREEDY approach. The first is reordering of documents according to their length (or other suitable weight), the second is a new structure called the *repetition array*, R. The latter is derived from Sadakane's [25] document frequency structure, and is used to calculate the document frequency for subsets of documents. We further show that it is sufficient to use only R and a subset of the document array if query terms, which can also be phrases, are length-restricted. Finally, we evaluate our proposal on two terabyte-scale IR collections. This is, to our knowledge, three orders of magnitudes larger than previous self-index based studies. Our source code and experimental setup is publicly available.

2   Notation and Problem Definition

Let $\mathcal{D}' = \{d_1, \ldots, d_{N-1}\}$ be a collection of $N-1$ documents. Each d_i is a string over an alphabet (words or characters) $\Sigma' = [2, \sigma]$ and is terminated by the sentinel symbol '1', also represent as '#'. Adding the one-symbol document $d_0 = 0$ results in a collection $\mathcal{D}$ of N documents. The concatenation $\mathcal{C} = d_{\pi(0)}d_{\pi(1)} \ldots d_{\pi(N-1)}$ is a string over $\Sigma = [0, \sigma]$, where π is a permutation of $[0, N-1]$ with $\pi(N-1) = 0$. We denote the length of a document d_i with $|d_i| = n_{d_i}$, and $|\mathcal{C}| = n$. See Fig. 1 for a running example. In the "bag of words" model a query $Q = \{q_0, q_1, \ldots, q_{m-1}\}$ is an unordered set of length m. Each element q_i is either a *term* (chosen from Σ') or a *phrase* (chosen from Σ'^p for $p > 1$).

Top-k Document Retrieval Problem. Given a collection $\mathcal{D}$, a query Q of length m, and a similarity measure $\mathcal{S}: \mathcal{D} \times \mathcal{P}_{=m}(\Sigma') \to \mathbb{R}$. Calculate the top-$k$ documents of $\mathcal{D}$ with regard to Q and $\mathcal{S}$, i.e. a sorted list of document identifiers $\mathrm{T} = \{\tau_0, \ldots, \tau_{k-1}\}$, with $\mathcal{S}(d_{\tau_i}, Q) \geq \mathcal{S}(d_{\tau_{i+1}}, Q)$ for $i < k$ and $\mathcal{S}(d_{\tau_{k-1}}, Q) \geq \mathcal{S}(d_j, Q)$ for $j \notin \mathrm{T}$.

$i =$	0	1	2	3	4	5	6	7	8	9	10	11	12	13
$\mathcal{C}^{word} =$	LA	O	LA	#	O	LA	LA	LA	#	O	O	LA	#	$
$\mathcal{C} =$	2	3	2	1	3	2	2	2	1	3	3	2	1	0

$$\underbrace{}_{d_1} \quad \underbrace{}_{d_3} \quad \underbrace{}_{d_2} \quad \underbrace{}_{d_0}$$

Fig. 1. $\mathcal{C}$ is the concatenation of a document collection $\mathcal{D}$ for $\pi = [1, 3, 2, 0]$. We use both words (as in $\mathcal{C}^{word}$) or integer identifiers (as in $\mathcal{C}$) to refer to document tokens.

A basic similarity measure used in many self-index based document retrieval systems (see [16]), is the *frequency measure* $\mathcal{S}^{freq}$. It scores d by accumulating the *term frequency* of each term. Term frequency $f_{d,q}$ is defined as the number of occurrences of term q in d; e.g. $f_{d_1,\text{LA}} = 2$ in Fig. 1. In IR, more complex TF × IDF measures also include two additional factors. The first is the inverse of the *document frequency* (DF), which is the number of documents in $\mathcal{D}$ that contain q, defined $F_{\mathcal{D},q}$; e.g. $F_{\mathcal{D},\text{LA}} = 3$. The second is the length of the document n_d. Due to space limitations, we only present the popular Okapi BM25 function:

$$S_{Q,d}^{\text{BM25}} = \sum_{q \in Q} \underbrace{\frac{(k_1 + 1) f_{d,q}}{k_1 \left(1 - b + b \frac{n_d}{n_{\text{avg}}}\right) + f_{d,q}}}_{= w_{d,q}} \cdot \underbrace{f_{Q,q} \cdot \ln \left(\frac{N - F_{\mathcal{D},q} + 0.5}{F_{\mathcal{D},q} + 0.5}\right)}_{= w_{Q,q}} \tag{1}$$

where n_{avg} is the mean document length, and $w_{d,q}$ and $w_{Q,q}$ refer to components that we address shortly. Parameters k_1 and b are commonly set to 1.2 and 0.75 respectively. Note that the $w_{Q,q}$ part is negative for $F_{\mathcal{D},q} > \frac{N}{2}$. To avoid negative scores, real-world systems, such as Vigna's MG4J [1] search engine, set $w_{Q,q}$ to a small positive value (10^{-6}), in this case. We refer to Zobel and Moffat [29] for a survey on IR similarity measures including TF × IDF, BM25, and LMDS.

3   Data Structure Toolbox

We briefly describe the two most important building blocks of our systems, and refer the reader to Navarro's survey [16] for detailed information. A *wavelet tree* (WT) [10] of a sequence $X[0, n-1]$ over alphabet $\Sigma[0, \sigma-1]$ is a perfectly balanced binary tree of height $h = \lceil \log \sigma \rceil$, referred to as WT-X. The i-th node of level $\ell \in [0, h-1]$ is associated with symbols c such that $\lceil c/2^{h-1-\ell} \rceil = i$. Node v, corresponding to symbols $\Sigma_v = [c_b, c_e] \subseteq [0, \sigma - 1]$, represent the subsequence X_v of X filtered by symbols in Σ_v. Only the bitvector which indicates if an element will move to the left or right subtree is stored at each node; that is, WT-X is stored in $n \lceil \log \sigma \rceil$ bits. Using only sub-linear extra space it is possible to efficiently navigate the tree. Let v be the i-th node on level $\ell < h-1$, then method *expand(v)* returns in constant time a node pair $\langle u, w \rangle$, where u is the $(2 \cdot i)$-th and w the $(2 \cdot i + 1)$-th node on level $\ell + 1$. A range $[l, r] \subseteq [0, n-1]$ in X can be mapped to range $[l, r]_v$ in node v such that the sequence $X_v[l, r]_v$ represents $X[l, r]$ filtered by Σ_v. Operation *expand($v, [l, r]_v$)* then returns in constant time

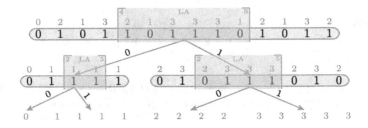

Fig. 2. Wavelet tree over document array D. Method $expand(v_{root}, [4, 9])$ maps range $[4, 9]$ (locus of LA) to range $[2, 3]$ in the left and range $[2, 5]$ in the right child.

a pair of ranges $\langle [l, r]_u, [l, r]_w \rangle$ such that the sequence $X_u[l, r]_u$ (resp. $X_w[l, r]_w$) represents $X[l, r]$ filtered by Σ_u (resp. Σ_w). Figure 2 provides an example.

The *binary suffix tree* (BST) of string $X[0, n-1]$ is the compact binary trie of all suffixes of X. For each path p from the root to a leaf, the concatenation of the edge labels of p, corresponds to a suffix. The children of a node are ordered lexicographically by their edge labels. Each leaf is labeled with the starting position of its suffix in X. Read from left to right, the leaves form the *suffix array* (SA), which is a permutation of $[0, n-1]$ such that $X[SA[i], n-1] <_{lex} X[SA[i+1], n-1]$ for all $0 \le i < n-1$. We refer to Fig. 3 for an example. Compressed versions of SA and ST – the compressed SA (CSA) and compressed ST (CST) – use space essentially equivalent to that of the compressed input, while efficiently supporting the same operations [23, 24]. For example, given a pattern P of length m, the range $[l, r]$ in SA containing all suffixes start with P or the corresponding node, that is the *locus* of P, in the BST is calculated in $\mathcal{O}(m \log \sigma)$.

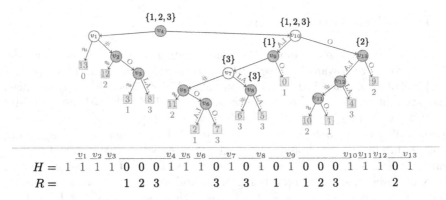

Fig. 3. Top: BST of the example in Fig. 1. The leaves form SA, the gray numbers below form D. Bottom: Bitvector $H[0, 2n - N - 1]$ and repetition array R.

4   Revisiting and Generalizing the GREEDY Framework

The GREEDY framework [4] consists of two parts: a CSA built over $\mathcal{C}$, and a WT over the *document array* $D[0..N-1]$; with each $D[i]$ specifying the document in which suffix $SA[i]$ starts. A top-k query using $\mathcal{S}^{freq}$ with $m=1$ is answered as follows. For term $q = q_0$ the CSA returns a range $[l, r]$, such that all suffixes in $SA[l, r]$ are prefixed by q. The size of the range corresponds to $f_{\mathcal{D},q}$, the number of occurrences of q in $\mathcal{D}$. In WT-D the alphabet Σ_v of each node represents a subset $\mathcal{D}_v \subseteq \mathcal{D}$ of documents of $\mathcal{D}$; and the size of the mapped interval $[l, r]_v$ equals $f_{\mathcal{D}_v,q}$, the number of occurrences of q in the subset $\mathcal{D}_v$. Each leaf v in WT-D corresponds to a $d \in \mathcal{D}$, such that the size of $[l, r]_v$ equals term frequency $f_{d,q}$.

To calculate the documents with maximal $f_{d,q}$, i.e. maximizing $\mathcal{S}^{freq}_{q,d}$, a max priority queue stores $\langle v, [l, r]_v \rangle$-tuples sorted by interval size. Initially, WT-D's root node and $[l, r]$ is enqueued. The following process is repeated until k documents are reported or the queue is empty: dequeue the top element $\langle v, [l, r]_v \rangle$. If v is a leaf, the corresponding document is reported. Otherwise the interval is expanded and the two tuples $\langle u, [l, r]_u \rangle$ and $\langle w, [l, r]_w \rangle$ containing the expanded ranges are enqueued.

This process returns the correct result if $f_{\mathcal{D},q}$ at a parent is never smaller than that of a child ($f_{\mathcal{D}_u,q}$ or $f_{\mathcal{D}_w,q}$). The interval size $f_{\mathcal{D},q}$ is never smaller than the maximum $f_{d,q}$ value in the subtree. Thus, in general the algorithm is correct, if (1) the score estimate s_v at any node v is larger than or equal to the maximum document score in v's subtree and (2) the score estimates s_u and s_w of the children of v are not larger than s_v. For many similarity measures (e.g. TF × IDF, BM25, and LMDS) theses conditions hold true if s_v is computed as follows: first, all document-independent parts, such as query weight $w_{Q,q}$ are determined. Then n_d is estimated with the smallest document length n_{min} in $\mathcal{D}$ if v is an inner node. Last, the maximal term frequency $f_{d,q}$ of each term q_i is set to $f_{\mathcal{D},q_i}$, the size of interval $[l_i, r_i]_v$. Since each interval size is non-increasing when traversing down WT-D the algorithm is correct, but not necessarily efficient. Instead of processing only one range, wavelet tree based algorithms can be process multiple ranges simultaneously [6]. In this case, the queue stores states $\langle s_v, v, \{[l_0, r_0]_v, \ldots, [l_{m-1}, r_{m-1}]_v\} \rangle$ sorted by s_v. Processing a state takes $\mathcal{O}(m)$ time as m intervals are expanded.

5   Improving Score Estimation

The runtime of GREEDY is dependent on the process time of a state and the number of states evaluated, which is determined by the quality of the score estimations.

Length Estimation by Document Relabelling. We improve document length estimation in $\mathcal{D}_v$ by replacing the collection-wide value n_{min} by the smallest document length $n_{\tilde{d}}$ in the sub-collection $\mathcal{D}_v$. The computation of $n_{\tilde{d}}$ can be performed in constant time if the document identifiers are assigned in documents length order. Thus, the smallest document corresponds to smallest symbol in $\mathcal{D}_v$ which is $\Sigma_v[0]$ which can be computed in constant time. Let v be the i-th node of level ℓ in WT-D^n then $\Sigma_v[0] = i \cdot 2^{\lceil \log N \rceil - \ell - 1}$. The document lengths are stored in an array $L[0, N-1]$. In Figs. 1 and 2 the documents are reordered using the permutation $\pi = [1, 3, 2, 0]$. The additional space of $N \log N + N \log n_{max}$ bits is negligible compared to the size of the CSA and D.

Improved Term Frequency Estimation. Until now we use the range size $f_{\mathcal{D}_v, q}$ of term q in v to estimate an upper bound for the maximal term frequency in a document $d \in \mathcal{D}_v$. Knowing the number of distinct documents in $\mathcal{D}_v$, called $F_{\mathcal{D}_v, q}$, helps to improve the upper bound to the number of repetitions plus one: $\delta_{\mathcal{D}_v, q} = f_{\mathcal{D}_v, q} - F_{\mathcal{D}_v, q} + 1$. In this section, we present a method that computes $\delta_{\mathcal{D}_v, q}$ in constant time during WT-D traversal. The solution is built on top of Sadakane's [25] document frequency structure (DF), which solves the problem solely for $\mathcal{C}_v = \mathcal{C}$. We briefly revisit the structure: first, a BST is built over $\mathcal{C}$, see Fig. 3. The leaves are labeled with the corresponding documents, i.e. from left to right D is formed. The inner nodes are numbered from 1 to $n-1$ in-order. Each node w_i holds a list $\mathcal{R}_i$, containing all documents which occur in both subtrees of w_i. We refer to elements in $\mathcal{R}_i$ as *repetitions*. Let w_i be the locus of a term q in the BST and let $[l, r]$ be w_i's interval. Then the total number of repetitions in $D[l, r]$ can be calculated by accumulating the length of all repetition lists in w_i's subtree. To achieve this, Sadakane generated a bitvector H that concatenates the unary coding of the lengths of all $\mathcal{R}_i$: $H = 10^{|\mathcal{R}_0|}10^{|\mathcal{R}_1|}1 \ldots 0^{|\mathcal{R}_{n-1}|}1$. The subtree interval $[l, r]$ can be mapped into H via select operations: $[l', r'] = [select_1(l, H), select_1(r, H)]$, since the accumulation of the list lengths equals the number of zeros in $[l', r']$. The following example illustrates the process: interval $[4, 9]$ corresponds to term $q = \text{LA}$ and is mapped to $[l', r'] = [select_1(4, H), select_1(9, H)] = [7, 15]$ in H. It follows that there are $z_l = l' - l = 3$ zeros in $H[0, l']$ and $z_r = r' - r = 6$ in $H[0, r']$; thus there are $6 - 3 = 3$ repetitions in $D[4, 9]$. We can overestimate the maximal term frequency by assuming that all repetitions belong to the same document d_x and add one for d_x itself. So $\delta_{\mathcal{D}_v, q} = 4$ in this case. This overestimates the maximal term frequency, which is $f_{d_3, q} = 3$, by one. The interval size estimate would have been 6.

We now extend Sadakane's solution to work on all subsets $\mathcal{D}_v$. First, we concatenate all $\mathcal{R}_i$ and form the *repetition array* $R[0, n - N - 1]$ (again, see Fig. 3), containing the actual repetition value for each zero in H. As above, using H and $select_1$, we can map $[l, r]$ to the corresponding range $[l'', r''] = [z_l, z_r - 1]$ in R. To calculate $\delta_{\mathcal{D}_v, q}$ for $\mathcal{D}_v$ we represent R as a WT. Now, we can traverse WT-D and WT-R simultaneously, mapping $[l'', r'']$ to $[l'', r'']_v$ in WT-R. The size of $[l'', r'']_v + 1$ equals $\delta_{\mathcal{D}_v, q}$ since node v contains only repetitions of $\mathcal{D}_v$.

6   Space Reduction

The space of R can be reduced to array $\hat{R}$ by omitting all elements belonging to the root v_{ST} of the non-binary ST since we will never query the empty string. In Fig. 3 all nodes with empty path labels correspond to v_{ST}, i.e. $v_1, v_4,$ and v_{10}.

Hence $\hat{R} = \{3, 3, 1, 2\}$ and we use a bitvector to map from the index domain of R into $\hat{R}$.

Second, we note that the space of WT-D and WT-$\hat{R}$ can be reduced if the length of query phrases is restricted to length ℓ. In this case, we can sort ranges in $\hat{R}$ which belong to nodes v_i, where v_i are the loci of patterns of length ℓ. Since all query ranges are aligned at borders of sorted ranges, the interval sizes during processing will not be affected. In our example, if $\ell = 1$, we sort the elements of v_9's subtree, resulting in $\hat{R}^1 = \{1, 3, 3, 2\}$. The sorting will result in $H_0(T)$-compression of WT-$\hat{R}^\ell$ for $\ell = 1$.

Third, we observe that when using WT-$\hat{R}^\ell$ *only a part of* WT-D is necessary to calculate $\delta_{\mathcal{D}_v, q}$. If q occurs more than once in $\mathcal{D}_v$, WT-$\hat{R}^\ell$ can be used to get $\delta_{\mathcal{D}_v, q}$. Hence, WT-D is only used to determine the existence of q in $\mathcal{D}_v$, and we only need to store the unique values inside ranges corresponding to loci of ℓ-length patterns. In addition, values in these ranges can be sorted, since this does not change the result of the existence queries. In our example we get $D^1 = \{3, 0, 1, 2, 0, 1, 2, 0, 1, 2\}$; one increasing sequence per symbol. A bitvector is again used to map into D^ℓ.

7 Experimental Study

Indexes and Implementations. To evaluate our proposals we created the SUccinct Retrieval Framework (SURF) which implements document retrieval specific components, like Sadakane's DF structure. These components can be parametrized by structures provided by the SDSL library [7]. We assembled three self-index based systems, corresponding to different functionality-space trade-offs. All systems use the same CSA and DF structure. The CSA is a FM-index using a WT. The WT as well as DF use RRR bitvectors [18, 22] to minimize space.

Our first index (I-D^n) adds WT-D^n, which uses plain bitvectors to allow fast WT traversal. Our second structure ($\text{I-D}^n\hat{\text{R}}^n$) adds WT-$\hat{R}^n$. A RRR bitvectors compresses the increasing sequences in $\hat{R}^n$. A variant of the latter index is $\text{I-D}^1\hat{\text{R}}^1$, which restricts the phrase length to one, and will show a functionality-space trade-off. In this version WT-D^1 is also compressed by using RRR vectors.

As a reference point we also implemented a competitive inverted index (INVIDX) which stores block-based postings lists compressed using OPTPFD [15, 28]. For each block, a representative is stored to allow efficient skipping. The top-k documents are calculated using two processing schemes. The first scheme – INVIDX-W – uses the efficient WAND list processing algorithm [2]. However, WAND requires pre-computation specific to $\mathcal{S}$ at construction time. A more flexible, but less efficient algorithm – INVIDX-E – exhaustively evaluates all postings in document-at-a-time order without either the burden or benefit of score pre-computation.

Data Sets and Environment. We use two standard IR test collections: (1) the GOV2 test collection of the TREC 2004 Terabyte Track and (2) the CLUEWEB09 collection consists of "Category B" subset of the ClueWeb09 dataset. To ensure reproducibility we extract the integer token sequence $\mathcal{C}$ from Indri [26] using

Collection	n	N	n_{avg}	σ	$|\mathcal{C}^{raw}|$	$|\mathcal{C}^{word}|$
Gov2	23,468,782,575	25,205,179	931	39,177,922	$\approx$ 426 GiB	72 GiB
ClueWeb09	40,579,891,952	50,220,423	808	90,411,635	$\approx$ 1.2 TiB	128 GiB

Fig. 4. Collection statistics for Gov2 and ClueWeb09.

default parameters. We selected 1000 randomly sampled queries from both the TREC 2005 and 2006 Terabyte tracks efficiency queries, ensuring all query terms are present in both collections. Statistics of our datasets (see Fig. 4) are in line with other studies [27]. We support ranked disjunctive (Ranked-OR, at least one term must occur) and ranked conjunctive (Ranked-AND, all terms must occur) retrieval. All indexes were loaded into RAM prior to query processing. Our machine was equipped with 256 GiB RAM and one Intel E5-2680 CPU.

Space Usage. The space usage of our indexes is summarized in Fig. 5 (right). All indexes are much larger than our inverted index, which uses 7.3 GiB for Gov2 and 22.8 GiB for ClueWeb09. The space of the compressed docid and frequency representations is 5.1 GiB and 17.6 GiB respectively which is comparable to other recent studies [21]. However, an inverted index supporting phrase queries would require additional positional information, which would significantly increase its size. The size of our integer parsing of size $n\lceil \log \sigma \rceil$ is shown as a horizontal line. The CSA for both collections compresses to roughly 30 % of the size of the integer parsing. The space for DF is negligible. The WT-D^n has the size of

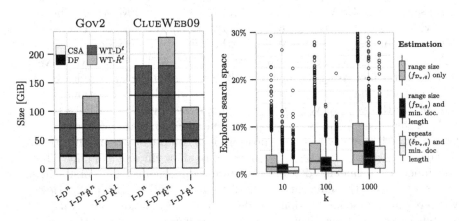

Fig. 5. Left: Memory breakdown of our indexes. $|\mathcal{C}^{raw}|$ denotes the original size of the collection, while $|\mathcal{C}^{word}|$ denotes the size after parsing it. A more detailed space breakdown of the indexes is available at http://go.unimelb.edu.au/6a4n. Right: Percentage of states evaluated for $k = 10, 100$, and 1000 during Ranked-OR retrieval using BM25 for queries on Gov2.

the integer parsing plus 5% overhead for a rank structure. The size reduction from R to $\hat{R}$ is substantial. For example, storing R for CLUEWEB09 requires 123 GiB, whereas $\hat{R}$ requires only 74 GiB. Restricting the phrase length to one ($\text{I-D}^1\hat{\text{R}}^1$), which makes it equivalent to a non-positional inverted index, shrinks the structure below the original input size.

Processed States. First, we measure the quantitative effects of our improved score estimation during GREEDY processing. We compare the range size ($f_{\mathcal{D}_v,q}$)-only estimation to (a) range size estimation including document length estimation and (b) repeats estimation ($\delta_{\mathcal{D}_v,q}$) including document length estimation. Figure 5 shows the percentage of processed states for all methods and $k = \{10, 100, 1000\}$ for both query sets on GOV2 using BM25 Ranked-OR processing. The percentage is calculated as the fraction of states processed compared to the exhaustive process-ing of each query ($k = N$). For all k, range size only estimation evaluates the most states on average. For $k = 10$, the median percentage of evaluated states for range size only estimation is 1.6%. Adding document length estimation reduces the number of evaluated states to 0.8%. Using $\delta_{\mathcal{D}_v,q}$ instead of $f_{\mathcal{D}_v,q}$ to estimate the frequency further improved the percentage of evaluated states to 0.06%. Similar effects can be observed for $k = 100$ and $k = 1000$. For $k = 1000$, docu-ment length estimation reduces the percentage from 5.1% to 3.2%. Frequency estimation using $\delta_{\mathcal{D}_v,q}$ again marginally improves the number of evaluated nodes to 2.8%. Overall, document length estimation has a larger impact on GREEDY than better frequency estimation via $\delta_{\mathcal{D}_v,q}$.

Disjunctive Ranked Retrieval. Next we evaluate the performance of I-D^n, $\text{I-D}^n\hat{\text{R}}^n$, $\text{I-D}^1\hat{\text{R}}^1$ for BM25 Ranked-OR query processing. Figure 6 (left) shows run-time on GOV2 and both query sets for $k = \{10, 100, 1000\}$. We additionally included INVIDX-W as a reference point for an efficient inverted index. The latter uses additional similarity measure dependent information and clearly outper-forms all self-index based indexes. For $k = 10$, it achieves a median runtime performance of less than 20 ms, and performs well for other test cases. How-ever, if an additional $k + 1$-th item is to be retrieved with the inverted index, the computation has to be restarted, whereas returning additional results using GREEDY is efficient. Our fastest index, I-D^n, is roughly 15 times slower, achiev-ing a median runtime of 300 ms for $k = 10$. The indexes $\text{I-D}^n\hat{\text{R}}^n$ and $\text{I-D}^1\hat{\text{R}}^1$ are approximately two times slower than I-D^n. This can be explained by the fact that I-D^n uses an uncompressed WT, whereas the other indexes use compressed WTs to save space. Also note that $\text{I-D}^1\hat{\text{R}}^1$ is faster than $\text{I-D}^n\hat{\text{R}}^n$ as ranges in $\hat{R}^1$ can be sorted, which creates runs in the WT which in turn allows faster state processing. The mean time per processed state – depicted in Fig. 6 (right) – highlights this observation. For I-D^n, the time linearly increases from 2 to 5 microseconds. While there is a correlation to the number of query terms, rank operations occur in close proximity – cache friendly – within WT-D^n, which increases performance. For the other indexes, we simultaneously access two WTs to evaluate a single state. This doubles the processing time per state.

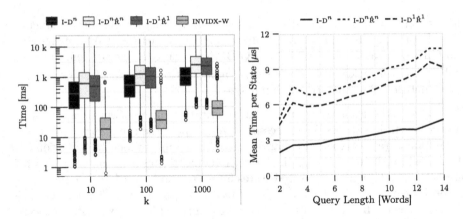

Fig. 6. Runtime (left) and WT state process time for $k = 100$ (right) for BM25 Ranked-OR.

Efficient Retrieval Using Multi-Word Expressions. Multi-word queries often contain mutli word expressions (MWE), i.e. sequences of words which describe one concept; e.g. the terms "saudi" and "arabia" are *strongly associated* [3] in our collections and would be recognized as one concept "saudi arabia". Using our index we can efficiently parse a query into MWE [9] (the problem of parsing MWE is also know as the *query segmentation* problem in IR; see e.g. [11]). Figure 7 (left) explores the runtime of MWE queries generated from TREC2006 for GOV2 using I-D^n. The runtime is reduced by an order of magnitude. This experiment shows how our system would support retrieval tasks where the vocabulary does not consist of words but a large number of entities. Supporting MWE does not increase the size of our index, but vastly increases the size of an inverted index.

Flexible Document Retrieval. Our indexes efficiently support a wide range of similarity measures, which can be changed and tuned after the index is built, while optimized inverted indexes require pre-computation depending on S at construction time [2]. If ranking functions are only chosen at query time, inverted indexes require exhaustive list processing. Figure 7 (right) shows the benefits of scoring flexibility. We compare our index structures to INVIDX-E using three ranking formulas: TF × IDF, BM25, and LMDS on GOV2. Our index structures significantly outperform the exhaustive inverted index for TF × IDF. This can be attributed to the influence of the document length n_d on $S^{TF \times IDF}$. Unlike BM25 or LMDS, the final document score is linearly proportional to the actual size of the document, thus document length estimation significantly reduces the number of evaluated states. For BM25, the document length contribution to the final document score is normalized by the average document length, and thus has a smaller effect on the overall score of each document.

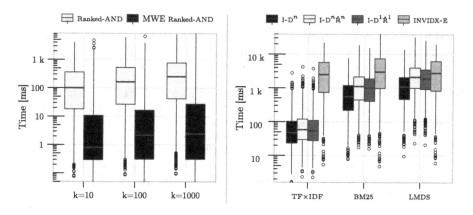

Fig. 7. BM25 runtime for MWE queries (left) Ranked-OR runtime for different $\mathcal{S}$ (right).

8   Conclusions

We presented a self-index based retrieval framework which allows rank-safe top-k retrieval on multi-term queries using complex scoring functions. The proposed estimation methods have improved the query speed compared to frequency-only score estimation. We found that top-k document retrieval is still solved more efficiently by inverted indexes, if augmented by similarity measure-dependent pre-computations. However, self-index based systems provide can be used in scenarios where the inverted index is not applicable or slower such as phrase retrieval or query segmentation.

Acknowledgments. We are grateful to Paul Cook, who pointed us to [3], and Alistair Moffat and Andrew Turpin for fixing our grammar. This research was supported by a Victorian Life Sciences Computation Initiative (VLSCI) grant number VR0052 on its Peak Computing Facility at the University of Melbourne, an initiative of the Victorian Government. Both authors were funded by ARC DP grant DP110101743.

References

1. Boldi, P., Vigna, S.: MG4J at TREC 2005. In: Proceedings of the TREC (2005)
2. Broder, A.Z., Carmel, D., Herscovici, H., Soffer, A., Zien, J.: Efficient query evaluation using a two-level retrieval process. In: Proceedings of the CIKM, pp. 426–434 (2003)
3. Church, K.W., Hanks, P.: Word association norms, mutual information, and lexicography. Comput. Linguist. **16**(1), 22–29 (1990)
4. Culpepper, J.S., Navarro, G., Puglisi, S.J., Turpin, A.: Top-k ranked document search in general text databases. In: de Berg, M., Meyer, U. (eds.) ESA 2010, Part II. LNCS, vol. 6347, pp. 194–205. Springer, Heidelberg (2010)
5. Culpepper, J.S., Petri, M., Scholer, F.: Efficient in-memory top-k document retrieval. In: Proceedings of the SIGIR, pp. 225–234 (2012)
6. Gagie, T., Navarro, G., Puglisi, S.J.: New algorithms on wavelet trees and applications to information retrieval. Theoret. Comput. Sci. **426**, 25–41 (2012)

7. Gog, S., Beller, T., Moffat, A., Petri, M.: From theory to practice: plug and play with succinct data structures. In: Gudmundsson, J., Katajainen, J. (eds.) SEA 2014. LNCS, vol. 8504, pp. 326–337. Springer, Heidelberg (2014)
8. Gog, S., Navarro, G.: Improved single-term top-k document retrieval. In: Proceedings of the ALENEX, pp. 24–32 (2015)
9. Gog, S., Moffat, A., Petri, M.: On identifying phrases using collection statistics. In: Hanbury, A., Kazai, G., Rauber, A., Fuhr, N. (eds.) ECIR 2015. LNCS, vol. 9022, pp. 278–283. Springer, Heidelberg (2015)
10. Grossi, R., Gupta, A., Vitter, J.S.: High-order entropy-compressed text indexes. In: Proceedings of the SODA, pp. 841–850 (2003)
11. Hagen, M., Potthast, M., Beyer, A., Stein, B.: Towards optimum query segmentation: in doubt without. In: Proceedings of the DIR, pp. 28–29 (2013). http://ceur-ws.org/Vol-986/paper_8.pdf
12. Hon, W.K., Shah, R., Thankachan, S.V., Vitter, J.S.: Space-efficient frameworks for top-k string retrieval. J. ACM 61(2), 1–36 (2014)
13. Konow, R., Navarro, G.: Faster compact top-k document retrieval. In: Proceedings of the DCC, pp. 351–360 (2013)
14. Larsen, K.G., Munro, J.I., Nielsen, J.S., Thankachan, S.V.: On hardness of several string indexing problems. In: Kulikov, A.S., Kuznetsov, S.O., Pevzner, P. (eds.) CPM 2014. LNCS, vol. 8486, pp. 242–251. Springer, Heidelberg (2014)
15. Lemire, D., Boytsov, L.: Decoding billions of integers per second through vectorization. Soft. Prac. Exp. 45, 1–29 (2013)
16. Navarro, G.: Spaces, trees and colors: the algorithmic landscape of document retrieval on sequences. ACM Comp. Surv. 46(4), 1–47 (2014)
17. Navarro, G., Nekrich, Y.: Top- k document retrieval in optimal time and linear space. In: Proceedings of the SODA, pp. 1066–1077 (2012)
18. Navarro, G., Providel, E.: Fast, small, simple rank/select on bitmaps. In: Klasing, R. (ed.) SEA 2012. LNCS, vol. 7276, pp. 295–306. Springer, Heidelberg (2012)
19. Navarro, G., Puglisi, S.J., Valenzuela, D.: General document retrieval in compact space. J. Experimental Alg. 19(2), 1–46 (2014)
20. Navarro, G., Thankachan, S.V.: New space/time tradeoffs for top-k document retrieval on sequences. Theor. Comput. Sci. 542, 83–97 (2014)
21. Ottaviano, G., Venturini, R.: Partitioned Elias-Fano indexes. In: Proceedings of the SIGIR, pp. 273–282 (2014)
22. Raman, R., Raman, V., Rao, S.S.: Succinct indexable dictionaries with applications to encoding k-ary trees and multisets. In: Proceedings of the SODA, pp. 233–242 (2002)
23. Sadakane, K.: New text indexing functionalities of the compressed suffix arrays. J. Alg. 48(2), 294–313 (2003)
24. Sadakane, K.: Compressed suffix trees with full functionality. Theory Comput. Syst. 41(4), 589–607 (2007)
25. Sadakane, K.: Succinct data structures for flexible text retrieval systems. J. Discrete Alg. 5(1), 12–22 (2007)
26. Strohman, T., Metzler, D., Turtle, H., Croft, W.B.: Indri: a language model-based search engine for complex queries. In: Proceedings of the International Conference on Intelligent Analysis (2005)
27. Vigna, S.: Quasi-succinct indices. In: Proceedings of the WSDM, pp. 83–92 (2013)
28. Yan, H., Ding, S., Suel, T.: Inverted index compression and query processing with optimized document ordering. In: Proceedings of the WWW, pp. 401–410 (2009)
29. Zobel, J., Moffat, A.: Inverted files for text search engines. ACM Comp. Surv. 38(2), 1–56 (2006)

LZD Factorization: Simple and Practical Online Grammar Compression with Variable-to-Fixed Encoding

Keisuke Goto$^{(\boxtimes)}$, Hideo Bannai, Shunsuke Inenaga, and Masayuki Takeda

Department of Informatics, Kyushu University, Fukuoka, Japan
keisuke.gotou@gmail.com, {bannai,inenaga,takeda}@inf.kyushu-u.ac.jp

Abstract. We propose a new variant of the LZ78 factorization which we call the LZ Double-factor factorization (LZD factorization). Each factor of the LZD factorization of a string is the concatenation of the two longest previous factors, while each factor of the LZ78 factorization is that of the longest previous factor and the following character. Interestingly, this simple modification drastically improves the compression ratio in practice. We propose two online algorithms to compute the LZD factorization in $O(m(M + \min(m, M) \log \sigma))$ time and $O(m)$ space, or in $O(N \log \sigma)$ time and $O(N)$ space, where m is the number of factors to output, M is the length of the longest factor(s), N is the length of the input string, and σ is the alphabet size. We also show two versions of our LZD factorization with variable-to-fixed encoding, and present online algorithms to compute these versions in $O(N + \min(m, 2^L)(M + \min(m, M, 2^L) \log \sigma))$ time and $O(\min(2^L, m))$ space, where L is the bit-length of each fixed-length code word. The LZD factorization and its versions with variable-to-fixed encoding are actually grammar-based compression, and our experiments show that our algorithms outperform the state-of-the-art online grammar-based compression algorithms on several data sets.

1 Introduction

Large-scale, highly repetitive texts such as collections of genomes of the same or similar species or the edit history of version controlled documents, have been increasing. Grammar compression algorithms, which are compression algorithms that output a compressed representation of the input text in the form of a context free grammar (CFG), have recently been gaining renewed interest since they are effective for such text collections [3], and also since CFGs are a convenient compressed representation that allows for various efficient processing on the strings without explicit decompression, e.g. pattern matching [13], q-gram frequencies [4], and edit-distance [5] computation.

While many previous grammar compression algorithms such as RE-PAIR [6] or SEQUITUR [10] give good compression ratios and run in linear time and working space, smaller working space is essential in order to compress large-scale data that does not fit in main memory. Maruyama et al. [7] proposed a fast and space efficient algorithm OLCA, which uses a simple strategy to

© Springer International Publishing Switzerland 2015
F. Cicalese et al. (Eds.): CPM 2015, LNCS 9133, pp. 219–230, 2015.
DOI: 10.1007/978-3-319-19929-0_19

determine the priority in selecting pairs of consecutive characters to form a production rule. Their algorithm runs online, and the working space depends only on the output, i.e., the compressed size of the input string. The working space was further reduced to the information theoretic lower bound of the output size in [9]. Maruyama and Tabei [8] proposed a variant that uses only constant working space, at the cost of some degradation in the compression ratio. Sekine et al. [12] proposed a modified version of RE-PAIR, called ADS, that splits the input string into blocks and compresses each block. In order to maintain a good compression ratio, they devised a technique to reuse non-terminal variables that are created and used frequently in each block, to the next block. Each non-terminal variable is encoded as a fixed-length code word, and since the length of the decompressed string that a code represents may vary, it is a variable-to-fixed code. The algorithm runs in $O(N)$ time and $O(B + 2^L)$ working space, where N is the length of the input string, B is the block size, and L is the bit-length of each fixed-length code word.

In this paper, we propose a new grammar-based compression based on the LZ78 factorization, which we call the *LZ Double-factor factorization* (*LZD factorization*). While each factor of the LZ78 factorization of a string is the longest previous factor and the following character, each factor of the LZD factorization is the concatenation of the two longest previous factors. We propose two online algorithms to compute the LZD factorization in $O(m(M + \min(m, M) \log \sigma))$ time and $O(m)$ space, or in $O(N \log \sigma)$ time and $O(N)$ space, where m is the number of factors to output, M is the length of the longest factor(s), N is the length of the input string, and σ is the alphabet size. We also show two versions of our LZD factorization with variable-to-fixed encoding, and present online algorithms to compute these versions in $O(N + \min(m, 2^L)(M + \min(m, M, 2^L) \log \sigma))$ time and $O(\min(2^L, m))$ space, where L is the bit-length of each fixed-length code word. When L can bee seen as a constant, these algorithms run in $O(N)$ time and $O(1)$ space. Computational experiments show that, in practice, our algorithms run fast and compress well while requiring small working space, outperforming the state-of-the-art online grammar-based compression algorithms on several data sets.

2   Preliminaries

Let Σ be a finite *alphabet*, and let $\sigma = |\Sigma|$. An element of Σ^* is called a *string*. The length of a string T is denoted by $|T|$. The empty string ε is the string of length 0, namely, $|\varepsilon| = 0$. For a string $T = XYZ$, X, Y and Z are called a *prefix, substring,* and *suffix* of T, respectively. If a prefix X (resp. substring Y, suffix Z) is of a string T is shorter than T, then it is called a *proper prefix* (resp. *proper substring, proper suffix*) of T. The set of suffixes of T is denoted by **Suffix**(T).

The i-th character of a string T is denoted by $T[i]$ for $1 \leq i \leq |T|$, and the substring of T that begins at position i and ends at position j is denoted by $T[i..j]$ for $1 \leq i \leq j \leq |T|$. For convenience, let $T[i..j] = \varepsilon$ if $j < i$. For convenience, we assume that $T[|T|] = \$$, where $\$$ is a special delimiter character that does not occur elsewhere in the string.

The Patricia tree of a set S of k strings, denoted $\mathbf{PT}_S$, is a rooted tree satisfying the following: (1) each edge is labeled with a non-empty substring of a string in S, (2) the labels of any two distinct out-going edges of the same node must begin with distinct characters; (3) for each string $s \in S$ there exists a node v such that $\mathbf{str}(v) = s$, where $\mathbf{str}(v)$ represents the concatenation of the edge labels from the root to v; (4) a string p is a non-empty prefix of a string $s \in S$ iff there are nodes u, v such that u is the parent of v, $\mathbf{str}(u)$ is a proper prefix of p, and p is a prefix of $\mathbf{str}(v)$. Because of conditions (2)-(4), there are at most k non-branching nodes (including leaves) and at most $k - 1$ branching nodes in $\mathbf{PT}_S$. Also, if we represent each edge label ℓ by a pair of the beginning and ending positions of an occurrence of ℓ in one of the strings in S, then $\mathbf{PT}_S$ can be stored in $O(k)$ space (excluding the string S). For a node u of $\mathbf{PT}_S$, let $\mathbf{depth}(u) = |\mathbf{str}(u)|$. If the string p of Condition (4) is $\mathbf{str}(v)$ itself, then we say that p is represented by an *explicit* node of $\mathbf{PT}_S$. Otherwise (if p is a proper prefix of $\mathbf{str}(v)$), then we say that it is represented by an *implicit* node of $\mathbf{PT}_S$.

The suffix tree of a string T, denoted $\mathbf{ST}_T$, is the Patricia tree of $\mathbf{Suffix}(T)$, namely $\mathbf{ST}_T = \mathbf{PT}_{\mathbf{Suffix}(T)}$. Since we have assumed T terminates with a special character \$, there is a one-to-one correspondence between the suffixes of T and the leaves of $\mathbf{ST}_T$. $\mathbf{ST}_T$ has at most $2N - 1$ nodes, and can be stored in $O(N)$ space, where $N = |T|$. For a string T of length N over an alphabet of size σ, $\mathbf{ST}_T$ can be constructed in $O(N \log \sigma)$ time and $O(N)$ space in an online manner [14].

3 LZD Factorization

We propose a new greedy factorization of a string inspired by the LZ78 factorization [16], which is able to achieve better compression ratios. We simply change the definition of a factor f_i, from the pair of the longest previously occurring factor and the immediately following character, to the pair of the longest previously occurring factor f_{j_1} and the longest previously occurring factor f_{j_2} which also appears at position $|f_1 \cdots f_{i-1}| + |f_{j_1}| + 1$. We call this new factorization the *LZ Double-factor factorization (LZD)*, and its formal definition is the following:

Definition 1 (LZD Factorization). The LZD factorization of a string T of length N, denoted $\mathbf{LZD}_T$, is the factorization $f_1, \ldots, f_m$ of T such that $f_0 = \varepsilon$, and for $1 \leq i \leq m$, $f_i = f_{i_1} f_{i_2}$ where f_{i_1} is the longest prefix of $T[k..N]$ with $f_{i_1} \in \{f_j \mid 1 \leq j < i\} \cup \Sigma$, f_{i_2} is the longest prefix of $T[k + |f_{i_1}|..N]$ with $f_{i_2} \in \{f_j \mid 0 \leq j < i\} \cup \Sigma$, and $k = |f_1 \cdots f_{i-1}| + 1$.

Note that for any $1 \leq i < m$ the length of f_i is at least 2, while f_m can be of length 1. This happens only when $|f_1 \cdots f_{m-1}| = N - 1$.

$\mathbf{LZD}_T = f_1, \ldots, f_m$ can be represented by a sequence of m integer pairs, where each pair (i_1, i_2) represents the ith factor $f_i = f_{i_1} f_{i_2}$. For example, the LZD factorization of string abaaababababaabbbbbabab\$ is $f_1 = $ ab, $f_2 = $ aa, $f_3 = $ abab, $f_4 = $ abaa, $f_5 = $ bb, $f_6 = $ bbabab, $f_7 = $ \$, and can be represented by (a, b), (a, a), $(1, 1)$, $(1, 2)$, (b, b), $(5, 3)$, and $(\$, 0)$.

One can regard $\mathbf{LZD}_T$ as a context-free grammar which only generates string T, with $m + 1$ production rules $S \to f_1 \cdots f_m$, $f_i \to f_{i_1} f_{i_2}$ for $1 \leq i \leq m$, where the set of rules $f_i \to f_{i_1} f_{i_2}$ $(1 \leq i \leq m)$ is called the dictionary.

Lemma 1. *For any string T, all factors of $\mathbf{LZD}_T$ are different.*

Proof. Let $\mathbf{LZD}_T = f_1, \ldots, f_m$. Since $f_m[|f_m|] = \$$, f_m is different from any other factors. Assume on the contrary that $f_h = f_i$ for some $1 \leq h < i < m$. Since both f_{i_1} and f_{i_2} are of length at least 1, $|f_{i_1}| < |f_i|$. However, we have assumed $f_h = f_i$, and this contradicts that f_{i_1} is the longest prefix of $T[|f_1 \cdots f_{i-1}| + 1..N]$ which belongs to $\{f_j \mid 1 \leq j < i\} \cup \Sigma$. Hence each factor f_i is distinct. □

Using the idea of [16] and Lemma 1, we get the following lemma:

Lemma 2. *For any string T of length N, the number of factors in $\mathbf{LZD}_T$ is $O(N/\log_\sigma N)$.*

Let $F = \{f_0, \ldots, f_m\}$ be the set of factors of $\mathbf{LZD}_T$. In a similar way to the case of LZ78 factorization, computing $\mathbf{LZD}_T$ reduces to computing $\mathbf{PT}_F$, the Patricia tree of F. We call $\mathbf{PT}_F$ the *LZD* tree of T. Figure 1 illustrates the LZD tree of the example string abaaabababaabbbbbabab$.

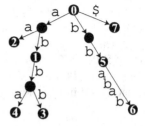

Fig. 1. The LZD tree for string abaaabababaabbbbbabab$. Each node numbered i represents the ith factor f_i of the LZD factorization of the string

In what follows, we will propose two algorithms to compute $\mathbf{LZD}_T$ for a given string T of length N in an online manner. The first one is space-efficient, namely, its extra space usage is linear in the number of factors in $\mathbf{LZD}_T$. The second one is fast, namely, it runs in $O(N \log \sigma)$ time.

3.1   Space-Efficient Online Algorithm for LZD Factorization

We present an online algorithm to compute $\mathbf{LZD}_T$ for a string T of length N in $O(m(M + \min\{M, m\} \log \sigma))$ time using $O(m)$ working space, where m is the number of factors in $\mathbf{LZD}_T$ and M is the length of the longest factor in $\mathbf{LZD}_T$.

The LZD tree of a given string T can be computed incrementally, in quite a similar way to the LZ78 trie [16], as follows: We first construct a tree only with the root. To compute a factor $f_i = f_{i_1} f_{i_2}$ starting at a position $k = |f_1 \cdots f_{i-1}| + 1$, we assume that the LZD tree contains nodes which represent all previous factors f_1 to f_{i-1}, and these nodes are marked. We also assume that the LZD tree contains nodes which represent all characters occurring in $T[1..|f_1 \cdots f_{i-1}|]$, and these nodes are marked. Let $T[k..q]$ be the longest prefix of $T[k..N]$ that is represented by the LZD tree, where $k \leq q \leq N$. This string $T[k..q]$ can be computed by traversing the tree from the root. If k is the first occurrence of character $c = T[k]$ (namely $q = 0$), then we create a new child of the root representing c, and mark this node. The first element f_{i_1} is c in this case. Otherwise, since there are at most $\min(m-1, M-1)$ branching nodes in any path of the LZD tree, and since $\mathbf{depth}(v) \le M$ for any leaf v, the number of character comparisons to compute $T[k..q]$ is $O(M + \min(m, M) \log \sigma)$. Then, the lowest marked node in the path which spells out $T[k..q]$ is exactly the first element f_{i_1} of f_i. The second element f_{i_2} of f_i can be computed analogously, traversing the LZD tree with $T[k + |f_{i_1}|..N]$ in $O(M + \min(m, M) \log \sigma)$ time. After computing f_i, we update the LZD tree so that f_i is represented by an explicit marked node in the tree. Recall that in the LZD tree there always exists a path spelling out f_{i_1} from the root. We traverse f_{i_2} from the end of this path, to compute the longest prefix y of f_i that is represented by the current LZD tree. There are four cases to consider:

1. If $y = f_i$ and f_i is represented by an explicit node u (i.e., $\mathbf{str}(u) = f_i$), then we simply mark u. Note that, by Lemma 1, u was always unmarked before computing f_i.
2. If $y = f_i$ and f_i is represented by an implicit node, then we create a new internal non-branching node v such that $\mathbf{str}(v) = f_i$ by splitting the edge on which the path spelling out f_i ends. We then mark v.
3. If $|y| < |f_i|$ and f_i is represented by an explicit node u, then we create a new leaf node v such that $\mathbf{str}(v) = f_i$, with a new edge from u to v. We then mark v.
4. If $|y| < |f_i|$ and f_i is represented by an implicit node, then we first create a new internal node u such that $\mathbf{str}(u) = y$, by splitting the edge on which the path spelling out y ends. Next, we create a new leaf node v such that $\mathbf{str}(v) = f_i$, with a new edge from u to v. We finally mark v.

Since we repeat the above procedure m times, it takes a total of $O(m(M + \min(m, M) \log \sigma))$ time to compute the LZD tree for all the factors. Notice that $N \le mM$, and hence N is hidden in the above time complexity. Since the LZD tree is the Patricia tree for the set of m factors of $\mathbf{LZD}_T$, the size of the tree (and hence the extra space requirement of this algorithm) is $O(m)$.

Since $\mathbf{LZD}_T$ is a kind of context-free grammar which only generates string T, we can obtain the original string T in $O(N)$ time from $\mathbf{LZD}_T$.

The following theorem summarizes this subsection.

Theorem 1 (Space-Efficient Online LZD Factorization). *Given a string T of length N, we can compute $\mathbf{LZD}_T = f_1 \cdots f_m$ in $O(m(M + \min(m, M) \log \sigma))$ time and $O(m)$ space in an online manner, where M is the length of the longest factor in $\mathbf{LZD}_T$.*

Since $m = O(N/\log_\sigma N)$ and $M = O(N)$[1], the space-efficient algorithm takes $O(N^2/\log_\sigma N)$ time. However, we have not found an instance which gives the above bound. As we will see in Sect. 5, this algorithm runs fast in practice.

---

[1] The bound $M = O(N)$ can be achieved with string $a^{N-1}\$$ with $N - 1 = 2^k$ for some k. Observe that $f_1 = aa$, $f_2 = f_1 f_1 = aaaa$, $\ldots$, $f_{m-1} = a^{\frac{N-1}{2}}$, and $f_m = \$$.

3.2   Fast Online Algorithm for LZD Factorization

Here, we present a fast online algorithm to compute $\mathbf{LZD}_T$ for a given string T of length N. Our algorithm uses the suffix tree $\mathbf{ST}_T$ of a given string T. Since every factor f_i of $\mathbf{LZD}_T = f_1, \ldots, f_m$ is a substring of T, it is also represented by either an implicit or explicit node of $\mathbf{ST}_T$. Hence we have the following observation: For any string T, the LZD tree for $\mathbf{LZD}_T$ can be superimposed on $\mathbf{ST}_T$, by possibly introducing some non-branching internal nodes. Due to this observation, we can compute $\mathbf{LZD}_T$ in $O(N)$ time and space in an *offline* manner for integer alphabets, using the offline algorithm of [2] which computes the LZ78 factorization of T from the suffix tree of T. In what follows, we show how to compute $\mathbf{LZD}_T$ in $O(N \log \sigma)$ time using $O(N)$ space in an *online* manner.

The basic strategy of our online algorithm is as follows. We first build the suffix tree of T incrementally, using Ukkonen's online suffix tree construction algorithm [14]. Then, for each $1 \le i \le m$, we find f_{i_1} and f_{i_2} on the suffix tree, and then mark the node which represents f_i (if there is no such node in the tree, then we create a new node and mark it).

We modify Ukkonen's algorithm as follows. As soon as we find the first occurrence of each character c at some position r in the string, we create a marked non-branching node v representing c, i.e., $\mathbf{str}(v) = c$. A new leaf for the suffix starting at position r is then created as a child of v. This permits us to superimpose the children of the root of the LZD tree onto the suffix tree.

We construct the suffix tree of $T[1..j]$ online, for increasing $j = 1, \ldots, N$. For each position $1 \le j \le N$, Ukkonen's algorithm maintains the following invariant: the longest suffix $T[s_j..j]$ of $T[1..j]$ that has an occurrence in $T[1..j-1]$. For convenience, when the longest suffix is the empty string ε, then let $s_j = j + 1$. Also, let $s_0 = 0$. We will use this suffix (and its location in the suffix tree) to determine the first and second elements of each LZD factor.

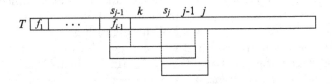

Fig. 2. $T[s_{j-1}..j-1]$ (resp. $T[s_j..j]$) is the longest suffix of $T[1..j-1]$ (resp. $T[1..j]$) that has an occurrence in $T[1..j-2]$ (resp. $T[1..j-1]$). We have computed $f_1, \ldots, f_{i-1}$ for the minimum i satisfying $s_{j-1} \le |f_1 \cdots f_{i-1}| + 1 < s_j$.

Assume that we have constructed the suffix tree for $T[1..j]$ for some $1 \le j < N$ such that $s_{j-1} < s_j$. Also, assume that we have computed $f_1, \ldots, f_{i-1}$ for the minimum integer i satisfying $s_{j-1} \le |f_1 \cdots f_{i-1}| + 1 < s_j$ (see also Fig. 2). For any $s_{j-1} \le k < s_j$, let P_k be the path spelling out $T[k..j-1]$ from the root. While we update the suffix tree of $T[1..j-1]$ to that of $T[1..j]$ by Ukkonen's algorithm, the ending position of path P_k in the tree can be found in amortized constant time for each k, in increasing order. Let $f_i, \ldots, f_{i'}$ be the consecutive

LZD factors such that i' is the minimum integer with $|f_1 \cdots f_{i'}| + 1 \geq s_j$. Since a node of the suffix tree is marked iff it represents one of the previous LZD factors or a single character, for any k ($s_{j-1} \leq k < s_j$) the lowest marked node v_k in the path P_k represents the longest prefix $T[k..k + \mathbf{depth}(v_k) - 1]$ of $T[k..N]$ which is also a previous LZD factor or a single character. This allows us to efficiently compute f_ℓ for each $\ell = i, \ldots, i' - 1$ in increasing order. As soon as we finish computing each f_ℓ, we maintain the suffix tree so that it contains a marked node which represents f_ℓ. Since we already know the location of the node which represents f_{ℓ_1}, we can find the ending position of the path spelling out $f_\ell = f_{\ell_1} f_{\ell_2}$ simply by traversing f_{ℓ_2} from the node representing f_{ℓ_1}. If f_ℓ is represented by an explicit node in the current tree, we mark the node. Otherwise, we insert a new marked node representing f_ℓ into the tree. Since $\sum_{\ell=i}^{i'-1} |f_{\ell_2}| < |f_i \cdots f_{i'-1}|$, this takes a total of $O(|f_i \cdots f_{i'-1}| \log \sigma)$ time for all $i \leq \ell < i'$.

In the sequel, we show how to compute the first element $f_{i'_1}$ of $f_{i'}$. If $s_j = j+1$ (i.e., j is the first occurrence of character $T[j]$ in $T[1..j]$), then $f_{i'_1} = T[j]$. After computing this, we mark the node representing $T[j]$. Otherwise, let z be the lowest marked node in the path from the root which spells out $T[|f_1 \cdots f_{i'-1}| + 1..j]$. By definition, it holds that $|f_1 \cdots f_{i'-1}| + \mathbf{depth}(z) \leq j$. If $|f_1 \cdots f_{i'-1}| + \mathbf{depth}(z) < j$, then $f_{i'_1}$ is computed in the same way as above, namely $f_{i'_1} = \mathbf{str}(z)$. If $|f_1 \cdots f_{i'-1}| + \mathbf{depth}(z) = j$, then we update the suffix tree of $T[1..j]$ to that of $T[1..j']$, where $j' > j$ is the minimum integer such that $s_j = s_{j'-1} \leq |f_1 \cdots f_{i'-1}| + 1 < s_{j'}$. Then, we can compute $f_{i'_1}$ in the same way as above, on the suffix tree for $T[1..j']$. The second element $f_{i'_2}$ can be computed analogously, and the node representing $f_{i'}$ can be found and marked in $O(|f_{i'}| \log \sigma)$ time. We repeat this procedure till we obtain all LZD factors for T (Fig. 3).

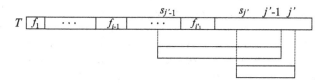

Fig. 3. When computing $f_{i'_1}$, if $|f_1 \ldots f_{i'-1}| + \mathbf{depth}(z) > j$, then we update the suffix tree of $T[1..j]$ to that of $T[1..j']$ with the minimum $j' > j$ such that $s_j = s_{j'-1} \leq |f_1 \ldots f_{i'-1}| + 1 < s_{j'}$. Then, $f_{i'_1}$ is represented by the lowest marked node in the path $P_{|f_1 \ldots f_{i'-1}|+1}$.

What remains is how to efficiently compute the lowest marked node in each path P_k. We use the following result:

Lemma 3 ([1,15]). *A semi-dynamic rooted tree can be maintained in linear space in its size so that the following operations are supported in amortized $O(1)$ time: (1) find the nearest marked ancestor of any node; (2) insert an unmarked node; (3) mark an unmarked node.*

By semi-dynamic we mean that insertions of new nodes to the tree are allowed, while deletions of existing nodes from the tree are not allowed. Since Ukkonen's algorithm does not delete any existing nodes, we can use the above lemma in our algorithm. If path P_k ends on an edge (i.e., if $T[k..j-1]$ is represented by an implicit node), then we can use the lowest explicit node in the path P_k to find the desired nearest marked ancestor.

After computing all LZD factors, we can discard the suffix tree. Ukkonen's algorithm constructs the suffix tree $\mathbf{ST}_T$ of string T in $O(N \log \sigma)$ time and $O(N)$ space. Since we can find all LZD factors in $O(\sum_{i=1}^{m} |f_i| \log \sigma) = O(N \log \sigma)$ time and $O(N)$ space, we obtain the following theorem:

Theorem 2 (Fast Online LZD Factorization). *Given a string T of length N, we can compute $\mathbf{LZD}_T = f_1, \ldots, f_m$ in $O(N \log \sigma)$ time and $O(N)$ space in an online manner, where σ is the alphabet size.*

4   LZD Factorization with Variable-to-Fixed Encoding

This section proposes an extension of LZD factorization of Sect. 3 to a variable-to-fixed encoding that runs in $O(N + \min(m, 2^L)(M + \min(m, M, 2^L) \log \sigma))$ time and $O(\min(2^L, m))$ space, where L is the fixed bit-length of code words representing factors, m is the number of factors, and M is the length of the longest factor. We call this variant the *LZDVF* factorization.

Since we are allowed to use only 2^L codes to represent the factors, we can store at most 2^L previous factors to compute new factors. A naïve solution would be to compute and store the first 2^L factors for the prefix $T[1..|f_1 \ldots f_{2^L}|]$, and then factorize the remaining suffix $T[|f_1 \ldots f_{2^L}| + 1..N]$ using the existing dictionary, without introducing new factors to it. We store these factors in a Patricia tree, and hence this algorithm uses $O(\min(2^L, m))$ space. Since there are at most $\min(m, M, 2^L) - 1$ branching nodes in the trie, this algorithm runs in $O(N + \min(m, 2^L)(M + \min(m, M, 2^L) \log \sigma))$ time. However, when the content of the remainder $T[|f_1 \ldots f_{2^L}| + 1..N]$ is significantly different from that of the prefix $T[1..|f_1 \ldots f_{2^L}|]$, then the naïve algorithm would decompose the remainder into many short factors, resulting in a poor compression ratio.

To overcome the above difficulties, our algorithms reuse limited encoding space by deleting some factors, and store new factors there. We propose two kinds of replacement strategies which we call LZDVF Count and LZDVF Pre respectively. The first one counts the number of factors appearing in the derivation trees of the factors that are currently stored in the dictionary, and deletes factors with low frequencies. This method is similar to the ones used in [8,12]. The second one deletes the least recently used factor in the dictionary in a similar way to [11] which uses an LRU strategy for LZ78 factorization.

In both strategies, there are at most 2^L entries in the dictionary and thus each factor is encoded by an L-bit integer. Since code words are reused as new factors are inserted and old factors are deleted from the dictionary, one may think that this introduces difficulties in decompression. However, since the procedure is deterministic, the change in assignment can be recreated during decompression, and thus will not cause problems.

4.1   Counter-Based Strategy

We define the derivation tree of each factor $f_i = f_{i_1} f_{i_2}$ recursively, as follows. The root of the tree is labeled with f_i, with two children such that the subtree rooted at the left child is the derivation tree of f_{i_1}, and the subtree rooted at the right child is the derivation tree of f_{i_2}. If f_{i_1} is a single character a, then its derivation tree consists only of the root labeled with a. The same applies to f_{i_2}. Let $\mathbf{vOcc}_i(f_j)$ denote the number of nodes in the derivation tree of f_i which are labeled with f_j. For all factors f_j that appear at least once in the derivation tree of f_i, we can compute $\mathbf{vOcc}_i(f_j)$ in a total of $O(|f_i|)$ time by simply traversing the derivation tree. Let $\mathbf{count}(f_j)$ be the sum of $\mathbf{vOcc}_q(f_j)$ for all factors f_q that are currently stored in the dictionary.

Assume that we have just computed a new factor $f_i = f_{i_1} f_{i_2}$. For each factor f_j with $\mathbf{vOcc}_i(f_j) > 0$, we first add $\mathbf{vOcc}_i(f_j)$ to $\mathbf{count}(f_j)$. If 2^L factors are already stored, then we do the following to delete factors from the dictionary. Depending on whether f_{i_1} and f_{i_2} are single characters or not, at least one (just f_i), and at most 3 (f_i and both f_{i_1}, f_{i_2}) new factors are introduced. For all factors f_h that are currently stored in the dictionary, we decrease $\mathbf{count}(f_h)$ one by one, until for some factor f_k, $\mathbf{count}(f_k) = 0$. We delete all such factors and repeat the procedure until enough factors have been deleted.

As the number of nodes in the derivation tree of each factor f_j is $O(|f_j|)$, the sum of counter values for all factors is $O(N)$. Hence, the total time required to increase and decrease the counter values is $O(N)$. Thus, the counter-based algorithm takes $O(N + \min(m, 2^L)(M + \min(m, M, 2^L) \log \sigma))$ time and $O(\min(2^L, m))$ space. When L can be seen as a constant, the algorithm runs in $O(N + M + \log \sigma) = O(N)$ time and uses $O(1)$ space.

4.2   Prefix-Based Strategy

Assume that we have computed the first i factors $f_1, \ldots, f_i$. In the prefix-based strategy, we consider a factor to be *used* at step i if it is a prefix of f_i. If $f_{h_1} (= f_i), f_{h_2}, \ldots, f_{h_k}$ are the sequence of all k factors in the dictionary which are prefixes of f_i in decreasing order of their lengths, then we consider that these factors are used in this chronological order. Hence, f_{h_k} will be the most recently used factor for step i. We use a doubly-linked list to maintain the factors, with the most recently used factor at the front and the least recently used factor at the back of the list. At each step i, we update the information for the factors $f_{h_1}, \ldots, f_{h_k}$. For any $1 \leq j \leq k$, if f_{h_j} is already in the list, we simply move it to the front of the list. Since the list is doubly linked, this can be done in $O(1)$ time. Otherwise, we simply insert a new element for f_{h_j} to the front of the list, and delete the LRU factor at the back of the list if the size of the list exceeded 2^L. This can also be done in $O(1)$ time.

The factors $f_{h_1}, \ldots, f_{h_k}$ can easily be found by maintaining the existing factors in a trie. Note that in each step of the algorithm, the LRU factor to be deleted is always a leaf of the trie since we have inserted the most recently used factors in decreasing order of their lengths. Hence, it takes $O(1)$ time to remove the LRU factor from the trie. Overall, the prefix-based algorithm also takes $O(N + \min(m, 2^L)(M + \min(m, M, 2^L) \log \sigma))$ time and $O(\min(2^L, m))$ space, which are respectively $O(N)$ and $O(1)$ when L is a constant.

5   Computational Experiments

All computations were conducted on a Mac Xserve (Early 2009) (Mac OS X 10.6.8) with 2 x 2.93 GHz Quad Core Xeon processors and 24 GB Memory, but only running a single process/thread at once. Each core has L2 cache of 256 KB and L3 cache of 8 MB. The programs were compiled using LLVM C++ compiler (`clang++`) 3.4.2 with the `-Ofast` option for optimization.

We implemented the space efficient on-line LZD algorithm described in Sect. 3.1, and the algorithms LZDVF Count and Pre with variable-to-fixed encoding described in Sect. 4[2], and compared them with the state-of-the art of grammar compression algorithms OLCA [7] and FOLCA [9]. For LZD, the resulting grammar is first transformed to a Straight line program (SLP) by transforming the first rule $S \rightarrow f_1 \cdots f_m$; replacing consecutive factors with non-terminal variables iteratively until the number of non-terminal variables equals to 1, and then the SLP is encoded in the same way as [7]. The output of LZDVF is a sequence of pairs of fixed-length code words that describes each LZD factor.

We evaluated the compression ratio, compression and decompression speed[3] of each algorithm for data (non highly-repetitive[4] and highly-repetitive[5]) taken from the Pizza & Chili Corpus. The running times are measured in seconds, and includes the time reading from and writing to the disk. The disk and memory caches are purged before every run using the `purge` command. The average of three runs is reported. The results are shown in Fig. 4 (a)-(d). We can see that compared to LZ78, LZD improves the compression ratio for all cases, as well as compression/decompression times in almost all most cases. The compression ratio of LZD is roughly comparable to OLCA, but the compression time of LZD slightly outperforms that of OLCA for highly repetitive texts, though not for the non-highly repetitive texts.

We also evaluated the performance of our algorithms for large-scale highly repetitive data, using 10 GB of English Wikipedia edit history data[6] (See Fig. 4 (e) and (f)). In this experiment, we modified LZDVF Pre and Count so that they do not read the whole input text into memory, and to explicitly store the edge labels of the Patricia tree that represents the factors. This modification increases the required working space from $O(\min(2^L, m))$ to $O(\min(2^L, m)M)$, but allows us to process large-scale data which does not fit in main memory. We compared the modified version of LZDVF Pre and Count with Freq and Lossy

---

[2] Source codes are available at https://github.com/kg86/lzd.

[3] The number of characters the algorithm can process a second.

[4] http://pizzachili.dcc.uchile.cl/texts.html.

[5] http://pizzachili.dcc.uchile.cl/repcorpus.html.

[6] The first 10 GB of enwiki-20150112-pages-meta-history1.xml-p000000010p00000 2983.7z, downloaded from http://dumps.wikimedia.org/backup-index.html.

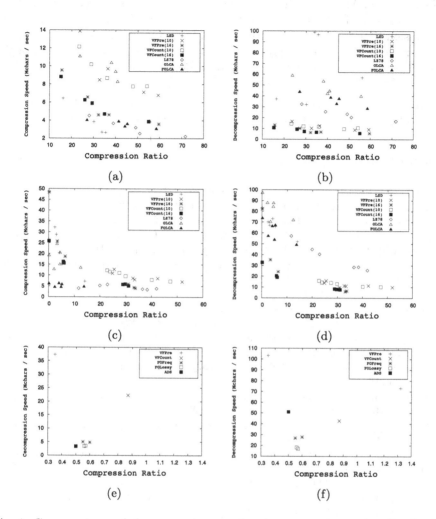

Fig. 4. Compression and decompression speed wrt. compression ratios. Results for LZD, LZDVF Pre and Count (VFPre and VFCount), OLCA [7] and FOLCA [9] on: (a), (b) non highly repetitive texts (DNA, English, Proteins, Sources, XML) of size 200 MB and (c), (d) highly repetitive texts (einstein.en, Escherichia_Coli, influenza, kernel, para, world_leaders). (e), (f): Results for LZDVF Pre and Count (VFPre and VFCount), Freq and Lossy FOLCA [8] (FOFreq and FOLossy), and ADS [12], which are grammar compression algorithms that do not store the whole input text in RAM, on 10 GB of English Wikipedia edit history. The parameters that determine the maximum number of non-terminal variables that VFPre, VFCount, FOFreq, ADS can store are varied between 2^{12}, 2^{14}, 2^{16} respectively. The block size parameter is varied 100 MB and 500 MB for ADS, and 100 MB, 500 MB, 1000 MB for FOLossy. Note that the points out of the frame are not plotted for visibility.

230 K. Goto et al.

FOLCA [8], and ADS [12] which use constant space. In this experiment, LZDVF Pre with bit-size of 16 shows the best performance. Surprisingly, it reduces the compression time to about a seventh of that of FOLCA Freq, which is the fastest of the previous grammar compression algorithms applicable to such large-scale data, while achieving a better compression ratio.

Acknowledgements. We would like to thank Shirou Maruyama and Takuya Kida for providing source codes of their compression programs FOLCA and ADS.

References

1. Amir, A., Farach, M., Idury, R.M., Poutré, J.A.L., Schäffer, A.A.: Improved dynamic dictionary matching. Inf. Comput. **119**(2), 258–282 (1995)
2. Bannai, H., Inenaga, S., Takeda, M.: Efficient LZ78 factorization of grammar compressed text. In: Calderón-Benavides, L., González-Caro, C., Chávez, E., Ziviani, N. (eds.) SPIRE 2012. LNCS, vol. 7608, pp. 86–98. Springer, Heidelberg (2012)
3. Claude, F., Navarro, G.: Self-indexed grammar-based compression. Fundamenta Informaticae **111**(3), 313–337 (2011)
4. Goto, K., Bannai, H., Inenaga, S., Takeda, M.: Speeding up q-gram mining on grammar-based compressed texts. In: Kärkkäinen, J., Stoye, J. (eds.) CPM 2012. LNCS, vol. 7354, pp. 220–231. Springer, Heidelberg (2012)
5. Hermelin, D., Landau, G.M., Landau, S., Weimann, O.: Unified compression-based acceleration of edit-distance computation. Algorithmica **65**(2), 339–353 (2013)
6. Larsson, N.J., Moffat, A.: Offline dictionary-based compression. In: DCC 1999, 296–305 (1999)
7. Maruyama, S., Sakamoto, H., Takeda, M.: An online algorithm for lightweight grammar-based compression. Algorithms **5**(2), 214–235 (2012)
8. Maruyama, S., Tabei, Y.: Fully online grammar compression in constant space. In: DCC 2014, pp. 173–182 (2014)
9. Maruyama, S., Tabei, Y., Sakamoto, H., Sadakane, K.: Fully-online grammar compression. In: Kurland, O., Lewenstein, M., Porat, E. (eds.) SPIRE 2013. LNCS, vol. 8214, pp. 218–229. Springer, Heidelberg (2013)
10. Nevill-Manning, C.G., Witten, I.H., Maulsby, D.L.: Compression by induction of hierarchical grammars. In: DCC 1994. pp. 244–253 (1994)
11. Peter, T.: A modified LZW data compression scheme. In: Australian Computer Science Communications, pp. 262–272 (1987)
12. Sekine, K., Sasakawa, H., Yoshida, S., Kida, T.: Adaptive dictionary sharing method for re-pair algorithm. In: DCC 2014, p. 425 (2014)
13. Shibata, Y., Kida, T., Fukamachi, S., Takeda, M., Shinohara, A., Shinohara, T., Arikawa, S.: Speeding up pattern matching by text compression. In: Bongiovanni, G., Petreschi, R., Gambosi, G. (eds.) CIAC 2000. LNCS, vol. 1767, pp. 306–315. Springer, Heidelberg (2000)
14. Ukkonen, E.: On-line construction of suffix trees. Algorithmica **14**(3), 249–260 (1995)
15. Westbrook, J.: Fast incremental planarity testing. In: Kuich, W. (ed.) ICALP 1992. LNCS, vol. 623, pp. 342–353. Springer, Heidelberg (1992)
16. Ziv, J., Lempel, A.: Compression of individual sequences via variable-length coding. IEEE Trans. Inf. Theory **24**(5), 530–536 (1978)

Combinatorial RNA Design: Designability and Structure-Approximating Algorithm

Jozef Haleš[1], Ján Maňuch[1,3], Yann Ponty[1,2($\boxtimes$)], and Ladislav Stacho[1]

[1] Department of Mathematics, Simon Fraser University, Burnaby, Canada
[2] Pacific Institute for Mathematical Sciences, CNRS UMI3069, Vancouver, Canada
yann.ponty@lix.polytechnique.fr
[3] Department of Computer Science,
University of British Columbia, Vancouver, Canada

Abstract. In this work, we consider the *Combinatorial RNA Design problem*, a *minimal* instance of the RNA design problem in which one must return an RNA sequence that admits a given secondary structure as its unique base pair maximizing structure.

First, we fully characterize designable structures using restricted alphabets. Then, under a classic four-letter alphabet, we provide a complete characterization for designable structures without unpaired bases. When unpaired bases are allowed, we characterize extensive classes of (non-)designable structures, and prove the closure of the set of designable structures under the stutter operation. Membership of a given structure to any of the classes can be tested in $\Theta(n)$ time, including the generation of a solution sequence for positive instances. Finally, we consider a structure-approximating version of the problem that allows to extend bands (stems). We provide a $\Theta(n)$ algorithm which, given a structure S avoiding two trivially non-designable motifs, transforms S into a designable structure by adding at most one base-pair to each of its stems, and returns a solution sequence.

1 Introduction

RiboNucleic Acids (RNAs) are biomolecules which act in almost every aspect of cellular life, and can be abstracted as a sequence of nucleotides, i.e., a string over the alphabet $\{A, U, C, G\}$. Due to their versatility, and the specificity of their interactions, they are increasingly being used as therapeutic agents [21], and as building blocks for the emerging field of synthetic biology [16,18]. A substantial proportion of the functional roles played by RNA rely on interactions with other molecules to activate/repress dynamical properties of some biological system, and ultimately require the adoption of a specific conformation. Accordingly, RNA bioinformatics has dedicated much effort to developing energy models [13,20] and algorithms [14,24] to predict the secondary structure of RNA, a combinatorial description of the conformation adopted by an RNA which only retains interacting positions, or base pairs. Historically, structure prediction has been addressed as an optimization problem, whose expected output is

© Springer International Publishing Switzerland 2015
F. Cicalese et al. (Eds.): CPM 2015, LNCS 9133, pp. 231–246, 2015.
DOI: 10.1007/978-3-319-19929-0_20

a secondary structure which minimizes some notion of free-energy [14,24]. The performances of the RNA folding prediction problem have now reached a point where *in silico* predictions have become generally reliable [13], allowing for large scale studies and fueling the discovery of an increasing number of functional families [8].

Due to the existence of expressive, yet tractable, energy models, coupled with promising applications in multiple fields (pharmaceutical research, natural computing, biochemistry...), a wide array of computational methods [1–5,7,9–12,15,19,22,23] have been proposed to tackle the natural inverse version of the structure prediction, the RNA design problem. In this problem, one attempts to perform the *in silico* synthesis of artificial RNA sequences, performing a predefined biological function *in vitro* or *in vivo*. Given the prevalence of structure in the function of an RNA, one of the foremost goal of RNA design (sometimes named inverse folding in the literature) is that the designed sequence should fold into a predefined secondary structure. In other words, it should not be challenged by alternative stable structures having similar or lower free-energy (Fig. 2).

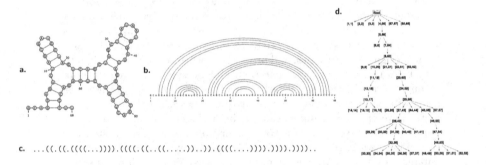

Fig. 1. Four equivalent representations for an RNA secondary structure of length 68, consisting of 20 base pairs forming 7 bands: outer-planar graph (a.), arc-annotated representation (b.), parenthesized expression (c.), and tree representation (d.)

Despite a rich, fast-growing, body of literature dedicated to the problem, there is currently no exact polynomial-time algorithm for the problem. Moreover, the complexity of the problem remains open (see Sect. 5 for details). It can be argued that this situation, quite exceptional in the field of computational biology, partly stems from the intricacies of the Turner free-energy model [20] which associates experimentally-determined energy contributions to $\sim 2.4 \times 10^4$ structure/sequence motifs. This motivates a reductionist approach, where one studies an idealized version of the RNA design problem, lending itself to algorithmic intuitions, while hopefully retaining the presumed difficulty of the original problem.

In this work, we introduce the *Combinatorial RNA Design problem*, a *minimal* instance of the RNA design problem which aims at finding a sequence that admits the target structure as its unique base pair maximizing structure. After this short introduction, Sect. 2 states definitions and problems. In Sect. 3, we summarize our results, some of which are proven in Sect. 4. Finally, we conclude in Sect. 5 with some remarks, open problems and future extensions of this work.

2   Definitions and Notations

RNA secondary structure. An RNA can be encoded as a sequence of nucleotides, i.e., a string $w = w_1 \cdots w_{|w|} \in \{A, U, C, G\}^\star$. The prefix of w of length i is denoted as $w_{[1,i]}$ and $|w|_b$ denotes the number of occurrences of b in w. A (pseudoknot-free) secondary structure S on an RNA of length n is a pair (n, P), where P is a set of base pairs $\{(l_i, r_i)\}_{i=1}^p \subset [1, n]^2$ such that:

- $\forall i \in [1, p]$, $l_i < r_i$;
- $\forall i, j \in [1, p]$, $l_i \neq l_j$, $l_i \neq r_j$, $r_i \neq r_j$ (each position is involved in at most one base pair);
- $\nexists i, j \in [1, p]$, $l_i < l_j < r_i < r_j$ (base pairs (l_i, r_i) and (l_j, r_j) do not cross).

The set of all secondary structures is denoted by $\mathcal{S}$, and its restriction to structures of length n by $\mathcal{S}_n$. The unpaired positions U_S in a secondary structure $S = (n, P)$ is the set of indices $k \in [1, n]$ that are not involved in a base pair. A structure S is saturated if $U_S = \emptyset$. Given a sequence w and a structure $S = (|w|, P)$, let $u_i = \varepsilon$ if $i \in U_S$ and $u_i = w_i$, otherwise, where ε is the empty sequence. Define the S-paired restriction of w (paired restriction of S), denoted as $\mathsf{Paired}(w, S)$ ($\mathsf{Paired}(S)$), as $u = u_1 \cdots u_{|w|}$ (respectively, $(|u|, \{(|u_1 \cdots u_i|, |u_1 \cdots u_j|) \mid (i, j) \in P\})$). A maximal subset $B = \{(i, j), (i+1, j-1), \ldots, (i+\ell, j-\ell)\}$ of P for some integer i, j, ℓ is called a band (sometimes referred to as helix or stem) of size $\ell = |B|$, of $S = (n, P)$. Note that every base pair belongs to exactly one band.

Dot-parentheses notation. A well-parenthesized sequence $s \in \{(,), .\}^*$ can be used to represent a secondary structure. There is one-to-one correspondence between secondary structures and such well-parenthesized sequences: any base pair $(l, r) \in P$ becomes a pair of corresponding opening and closing parentheses in s at position l and r respectively ($s_l = ($ and $s_r =)$), and any unpaired position i corresponds to a dot ($s_i = .$). A concatenation of two structures S and S', denoted by SS', is the structure corresponding to the well-parenthesized sequence obtained by concatenating the well-parenthesized sequences of S and S'.

k-stutter. The k-stutter of a sequence s, denoted by $s^{[k]}$ is the result of an independent copy k-times of each of the characters in s. This operation can be applied to both RNA sequences and structures in the dot-parentheses notation.

Tree representation. Alternatively, the tree representation, denoted by T_S, for $S = (n, P)$ is a rooted ordered tree whose vertex set V_S consists of intervals $[l, r]$ for any base pair $(l, r) \in P$, and $[k, k]$ for every $k \in U_S$. A virtual root $[0, n+1]$ is added for convenience. Each $[k, k]$ node is called unpaired node, all other nodes (including the root) are called **paired** nodes. The **children** of an interval $I \in V_S$ are the maximal proper subintervals $I' \in V_S$ of I ordered by the left points of the intervals. The **degree** of a vertex $I \in V_S$ is the total number of its paired neighbors, including its parent (if any). We denote by $D(S)$ the maximal degree of nodes in T_S.

Proper, greedy and separated coloring of the tree representation. Consider the tree representation T_S of structure S. Color every paired node of T_S different from the root by black, white or grey color. This coloring is called **proper** if:

1. every node has at most one black, at most one white and at most two grey children;
2. a node with color c has at most one child with color c;
3. a black node does not have a white child and a white node does not have a black child.

A **greedy** coloring of T_S is the coloring obtained by recursive application of the following rule starting from the root and continuing towards leaves: if the node is black, color the first paired child black and the remaining paired children grey, if the node is white, color the first paired child white and the remaining paired children grey, otherwise (the grey node or the root), color the first paired child black, second white and the remaining paired children grey. It is easy to check that if the degree of each node is at most four then the greedy coloring is a proper coloring.

Given a proper coloring of T_S, let the **level** of each node be the number of black nodes minus the number of white nodes on the path from this node to the root. A proper coloring is called **separated** if the two sets of levels, associated with grey and unpaired nodes respectively, do not overlap.

2.1   Statement of the Generic RNA Design Problem

Consider an energy model $\mathcal{M}$, which associates a free-energy $E_{\mathcal{M}}(w, S) \in \mathbb{R}^- \cup \{+\infty\}$ to each secondary structure $S \in \mathcal{S}_{|w|}$ for a given RNA sequence w. The minimum free-energy (MFE) structure prediction problem is typically defined as follows:

> **RNA-FOLD$_{\mathcal{M}}$ problem**
> **Input:** RNA sequence w
> **Output:** $S^\star_{\mathcal{M}}(w) := \operatorname{argmin}_{S' \in \mathcal{S}_{|w|}} E_{\mathcal{M}}(w, S')$.

The existence of competing structures, having comparable or lower free-energy for a given RNA, impacts the well-definedness of the folding process. The detection of such situations is therefore of interest, and can be rephrased as follows:

> **UNIQUE-FOLD$_{\mathcal{M}}$ problem**
> **Input:** Sequence w + Energy distance $\Delta > 0$
> **Output:** True if, for every $S' \in \mathcal{S}_{|w|} \setminus \{S^\star_{\mathcal{M}}(w)\}$, $E_{\mathcal{M}}(w, S') \geq E_{\mathcal{M}}(w, S^\star_{\mathcal{M}}(w)) + \Delta$.
> False otherwise.

a. Target sec. str. S **b.** Invalid sequence for S **c.** Design for S

Fig. 2. The combinatorial RNA design problem: Starting from a secondary structure S (a.), our goal is to design an RNA sequence which uniquely folds, with maximum number of base pairs, into S. The sequence proposed in b. is invalid due to the existence of an alternative structure (lower half-plane, red) having the same number of base pairs as S. The right-most sequence (c.) is a design for S.

We can now define the combinatorial RNA Design problem as:

RNA-DESIGN$_{\mathcal{M},\Sigma}$ problem
Input: Secondary structure S + Energy distance $\Delta > 0$
Output: RNA sequence $w \in \Sigma^*$ — called an $(\mathcal{M}, \Sigma, \Delta)$-design for S — such that:

RNA-FOLD$_{\mathcal{M}}(w) = S$ and UNIQUE-FOLD$_{\mathcal{M}}(w, \Delta)$,

or $\varnothing$ if no such sequence exists.

Structures for which there exists an $(\mathcal{M}, \Sigma, \Delta)$-design are called $(\mathcal{M}, \Sigma, \Delta)$-designable. Let Designable$(\mathcal{M}, \Sigma, \Delta)$ be the set of all such structures. If it is clear from the context, we will usually drop $\mathcal{M}$, Σ and/or Δ (Fig. 2).

2.2 Combinatorial Design in a Simple Base Pair Energy Model

In this work, we adopt a Watson-Crick energy model $\mathcal{W}$, which only allows pairs involving complementary letters, i.e., in $\{C, G\}$ and $\{A, U\}$.

Definition 1 (Watson-Crick energy model $\mathcal{W}$).

$$E_{\mathcal{W}}(w, S) = \begin{cases} -|S| & \text{if } \forall (l, r) \in S, w_l \text{ is complementary with } w_r, \\ +\infty & \text{otherwise.} \end{cases}$$

We say that the structure is compatible with a sequence w, if $E_{\mathcal{W}}(w, S) < +\infty$.

Minimizing $E_{\mathcal{W}}(w, S)$ is equivalent to maximizing $|S|$, thus RNA $-$ FOLD$_{\mathcal{W}}$ is a classic base pair maximization problem. It can be solved by dynamic programming, historically in $\mathcal{O}(n^3)$ complexity [14], or in $\mathcal{O}(n^3/\log(n))$ current best time complexity [6]. A backtracking procedure reconstructs the MFE structure, and can be easily adapted to assess the uniqueness of the MFE structure.

3 Statement of the Results

We consider the design problem in a base pairing energy model $\mathcal{W}$ restricted to Watson-Crick base pairs $\{C, G\}$ and $\{A, U\}$. We set $\Delta = 1$, which forbids designed

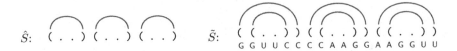

Fig. 3. An example of undesignable (left) and designable structure (right).

sequence to adopt alternative structures having greater or equal number of base pairs than the target structure. Let us first characterize the sets $\mathsf{Designable}(\Sigma)$ of designable structures over partial alphabets Σ. Let $\Sigma_{c,u}$ be an alphabet with c pairs of complementary bases and u bases without a complementary base.

Designability over restricted alphabets.

R1: For every $u \in \mathbb{N}^+$, $\mathsf{Designable}(\Sigma_{0,u}) = \{(n, \emptyset) \mid \forall n \in \mathbb{N}\}$;

R2: $\mathsf{Designable}(\Sigma_{1,0}) = \{S \in \mathcal{S} \mid S \text{ is saturated and } D(S) \leq 2\} \cup \{(n, \emptyset) \mid \forall n \in \mathbb{N}\}$;

R3: $\mathsf{Designable}(\Sigma_{1,1}) = \{S \in \mathcal{S} \mid D(S) \leq 2\}$.

Designability over the complete alphabet $\Sigma_{2,0} = \{\mathsf{A}, \mathsf{U}, \mathsf{C}, \mathsf{G}\}$.

R4: $\mathsf{Designable}(\Sigma_{2,0}) \cap \{S \in \mathcal{S} \mid S \text{ is saturated}\} = \{S \in \mathcal{S} \mid D(S) \leq 4\} \cap \{S \in \mathcal{S} \mid S \text{ is saturated}\}$.

When unpaired positions are allowed in the target structure, our characterization is only partial:

R5: Let m_5 represent "a node having degree more than four", and $m_{3\circ}$ be "a node having one or more unpaired children, and degree greater than two", then
$$\mathsf{Designable}(\Sigma_{2,0}) \cap \{S \in \mathcal{S} \mid S \text{ contains } m_5 \text{ or } m_{3\circ}\} = \emptyset\,;$$

R6: Let Sep be the set of structures for which there exists a separated (proper) coloring of the tree representation, then $\mathsf{Sep} \subset \mathsf{Designable}(\Sigma_{2,0})$;

R7: The set of $\Sigma_{2,0}$-designable structures is closed under the k-stutter operations:
$$\forall S \in \mathcal{S}, \forall k \in \mathbb{N}^+ : \quad S \in \mathsf{Designable}(\Sigma_{2,0}) \implies S^{[k]} \in \mathsf{Designable}(\Sigma_{2,0})\,.$$

We note that $S^{[k]} \in \mathsf{Designable}(\Sigma_{2,0})$ does not imply that $S \in \mathsf{Designable}()$ $\Sigma_{2,0}$. For instance in Fig. 3, it can be verified that $\hat{S}^{[2]}$ is $\Sigma_{2,0}$-designable, while $\hat{S}$ is not. Membership to the classes described in **R1-R5** can be tested by trivial linear-time algorithms, which can also be adapted into linear-time algorithms for the $\mathsf{RNA-DESIGN}_{\mathcal{M},\Sigma}$ problem.

Structure-approximating algorithm. Unfortunately, the absence of m_5 or $m_{3\circ}$, while necessary, is generally not sufficient to ensure designability. For instance, $\hat{S}$ in Fig. 3 clearly does not contain m_5 or $m_{3\circ}$, yet cannot be designed. In such cases, the unwanted interactions can be penalized by the duplication of some base pairs. For instance, duplicating the base pairs in the above example yields $\Sigma_{2,0}$-designable structure $\tilde{S}$.

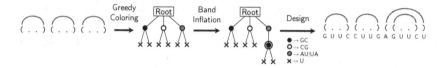

Fig. 4. Application of the structure-approximating algorithm to the non-designable structure $\hat{S}$ in Fig. 3: A base pair (circled black node) is inserted in the greedily colored tree, offsetting the levels of white and unpaired nodes (crosses) to even and odd levels respectively, so that the resulting tree is proper/separated, representing a designable structure.

R8: Any structure S without m_5 and $m_{3\circ}$ can be transformed in $\Theta(n)$ time into a $\Sigma_{2,0}$-designable structure S', by inflating a subset of its base pairs (at most one per band) so that the greedy coloring of the resulting structure is proper and separated, as illustrated by Fig. 4.

4 Proofs

R1 is trivial since, in the absence of complementary letters, the structures without base pairs are the only structures whose energy is not infinite.

Theorem 1 (Result R4). *A saturated sec. str. S is $\Sigma_{c,0}$-designable if and only if $D(S) \leq 2c$.*

Proof. First, we will show that the degree condition is necessary. Assume to the contrary that $D(S) > 2c$ and S has a design w. Let $[a, b]$ be a vertex with degree $d \geq 2c + 1$ in T_S. Let $\{[l_i, r_i]\}_{i=1}^d$ be the (paired) children of $[a, b]$ and the node $[a, b]$ if $[a, b]$ is not the root. Let $L_i = l_i$ and $R_i = r_i$ if $[l_i, r_i]$ is a child of $[a, b]$, and $L_i = r_i$ and $R_i = l_i$ if it is $[a, b]$. Then among bases $w_{L_1}, \ldots, w_{L_d}$ must be a pair of repeated letters. Let $w_{L_i} = w_{L_j}$ be such a pair with $L_i < L_j$. It is easy to check that $S \setminus \{(l_i, r_i), (l_j, r_j)\} \cup \{(L_i, R_j), (R_i, L_j)\}$ is a structure compatible with w with the same number of base pairs as S, a contradiction with the assumption that w is a design for S.

To show that the degree condition is also sufficient, we need further definitions and claims. First, we say that a sequence $w \in \Sigma^*$ is **saturable** if there is a saturated structure compatible with w. Note that the concatenation of two saturable sequences is also saturable. Then the following claim characterizes the cases when a saturable sequence can be split into saturable sequences.

Claim 1.1. *Let $w = uv$ be a saturable sequence of length k. If u is saturable, then so is v.*

Proof. Consider a saturated structure S compatible with sequence w and saturated structure S_u compatible with u. We will construct a saturated structure S_v compatible with v.

Consider a graph G with vertex set $\{1, \ldots, k\}$ and edge set defined by pairs in $S \cup S_u$. Obviously, this graph is a collection of alternating paths (alternating between pairs from S and from S_u, starting and ending with positions in v) and alternating cyclic paths, and it has a planar embedding such that all vertices lie on a line in their order: pairs in S are drawn as non-crossing arcs above the line and pairs in S_u as non-crossing arcs below the line. Note that every position in v is an end-point of exactly one path in the collection.

Define set of base pairs S_v by pairing the end-points of the paths in G, cf. Fig. 5. We will show that S_v is a structure. Consider a graph G' constructed by adding pairs in S_v to G. This graph is a collection of cyclic paths. Consider an embedding of G' into plane that extends the planar embedding of G by adding arcs corresponding to the pairs in S_v below the line containing all the vertices. If two base pairs $b, b' \in S_v$ cross then the cyclic path containing b and the cyclic path containing b' intersect in exactly one point. By Jordan's curve theorem, this is a contradiction. It follows that S_v is a saturated structure, and hence v is also saturable. $\qquad \square$

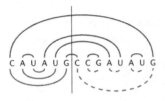

Fig. 5. Construction the saturated structure compatible with the suffix v. The vertical line splits the sequence into a prefix u and a suffix v. Blue and red arcs depict saturated structures compatible with w and u respectively. Dashed red arcs represent the induced saturated structure compatible with v.

We define w to be an atomic saturable sequence if no proper prefix of w is saturable. Clearly, every saturated structure compatible with an atomic saturable sequence w contains the base pair $(1, |w|)$. On the other hand, by Claim 1.1, if every saturated structure compatible with w contains the pair $(1, |w|)$, then w is an atomic saturable sequence. A design w that is also an atomic saturable sequence will be called an atomic saturable design. A concatenation of two or more atomic saturable designs is obviously not an atomic saturable sequence and it is not necessarily a design. However, we have the following claims.

Claim 1.2. *The concatenation of t atomic saturable designs $w^1, \ldots, w^t$ for structures $S^1, \ldots, S^t$, such that $w_1^i \neq w_1^j, \forall 1 \leq i < j \leq t$, is a design for the concatenated (saturated) structure $S = S^1 \cdots S^t$.*

Proof. Assume that $W := w^1 \cdots w^t$ is not a design, then there exist a saturated structure $S' \neq S$ for W. We show that positing such an alternative structure leads to a contradiction. Recall that each S^i is saturated and contains a pair

$(1, |w^i|)$. If S' pairs the first and last letters in each w^i, $i \in [1, t]$, then $S' = S$ since each w^i is a design, a contradiction. Let w_i be the leftmost sequence such that w_1^i is not paired with $w_{|w^i|}^i$ in S'. Since S' must be also saturated, w_1^i must be paired. Let w_k^j, $j \geq i$, be the partner of w_1^i in S', and let $u := w^i \cdots w^{j-1} w_{[1,k]}^j$. If $k = |w^j|$, then $j > i$ and, by complementarity, $w_1^i = w_1^j$ which contradicts the preconditions. Hence, we can assume that $k < |w^j|$. Since u and each of the $w^i, \ldots, w^{j-1}$ are saturable, by iterated application of Claim 1.1, we conclude that $v = w_{[1,k]}^j$ is saturable as well. This contradicts the precondition that w^j is an atomic saturable design, since v is a proper prefix of w^j. We conclude that no alternative saturated folding exists for W, i.e., W is a design for S.   □

Claim 1.3. *Consider t atomic saturable designs $w^1 = w_1^1 \cdots w_{|w^1|}^1, \ldots, w^t = w_1^t \cdots w_{|w^t|}^t$ and a pair a, b of complementary letters such that $w_1^i \neq b$ for every $1 \leq i \leq t$ and $w_1^i \neq w_1^j$ for every $1 \leq i < j \leq t$. Then $W = aw^1 \cdots w^t b$ is an atomic saturable design.*

Proof. We will first show that W is an atomic saturable sequence. Assume to the contrary that there is a proper prefix of W that is saturable. Consider the shortest such prefix $aw^1 \cdots w^i w_{[1,j]}^{i+1}$. Obviously, a has to be paired with w_j^{i+1}, otherwise we can find a shorter saturable prefix. This implies that $b = w_j^{i+1}$ and that $w^1 \cdots w^i w_{[1,j-1]}^{i+1}$ is saturable as well. By repeated application of Claim 1.1, we have that $w_{[1,j-1]}^{i+1}$ is saturable. Since it is a prefix of atomic saturable sequence w^{i+1}, it must be the empty sequence, i.e., $j = 1$. Therefore, $b = w_1^{i+1}$, a contradiction with the assumptions of the claim. Thus, W is an atomic saturable sequence.

Now we will show that W is a design. Consider any MFE (saturated) structure S for W. Since W is atomic saturable, a is paired with b in S. By Claim 1.2, $w^1 \cdots w^t$ is a design. It follows that W is a design as well.   □

To prove the sufficiency of the degree condition, consider the following algorithm, which takes as input a saturated structure S with $D(S) \leq 2c$, and returns a design w for S:

– Let $\{[l_i, r_i]\}_{i=1}^d$ be the children of the root. Assign to each w_{l_i}, w_{r_i} complementary bases such that $\forall 1 \leq i < j \leq d : w_{l_i} \neq w_{l_j}$.
– While there exists an unprocessed internal node $[a, b]$ whose parent has been processed (if there is no such node, stop and return w). Let $\{[l_i, r_i]\}_{i=1}^d$ be the children of $[a, b]$. Assign to each w_{l_i}, w_{r_i} complementary bases such that $\forall 1 \leq i \leq d : w_{l_i} \neq w_a$ and $\forall 1 \leq i < j \leq d : w_{l_i} \neq w_{l_j}$.

Note that since the alphabet contains c pairs of complementary bases, the assignment at each step of the algorithm is possible. We will show that the returned sequence w is a design for S. We will show by tree induction on the size subtrees that $w_i \cdots w_j$ is an atomic saturable design for every internal node $[i, j]$. It is easy to check that this is satisfied at the leaves. Consider an internal node u.

By the induction hypothesis, sequences for each child subtree of u are atomic saturable designs. Furthermore, by the choice of bases at children nodes of u, all assumptions of Claim 1.3 are satisfied, hence, the sequence for node u is also an atomic saturable design. The claim holds. Finally, we can apply Claim 1.2 at the root, which yields that w is a design.   □

Corollary 2 (Result R2). *A structure S is $\Sigma_{1,0}$-designable if and only if it does not contain any base pairs, or it is saturated and $D(S) \leq 2$.*

Proof. If S contains a base pair and an unpaired position, then it can be easily checked that S is not $\Sigma_{1,0}$-designable. Hence, any $\Sigma_{1,0}$-designable structure is either empty, and trivially designable using a single letter, or saturated. In the latter case, by Theorem 1, we know that designable structures are exactly those that are saturated, and such that $D(S) \leq 2$. The claim follows.   □

Corollary 3 (Result R3). *A structure S is $\Sigma_{1,1}$-designable if and only if $D(S) \leq 2$.*

Proof. First, suppose S is $\Sigma_{1,1}$-designable and let w be a design for S. Then $\mathsf{Paired}(w, S)$ is a design for $\mathsf{Paired}(S)$. Since the paired restriction $\mathsf{Paired}(S)$ is saturated, it is over alphabet $\Sigma_{1,0} \subset \Sigma_{1,1}$, and by Theorem 1, $D(\mathsf{Paired}(S)) \leq 2$. Hence, $D(S) = D(\mathsf{Paired}(S)) \leq 2$.

Conversely, suppose that $D(S) \leq 2$. Construct a design for S as follows. Since $\mathsf{Paired}(S)$ is saturated, by Theorem 1, there is a design $\bar{w}$ for $\mathsf{Paired}(S)$ over $\Sigma_{1,0} \subset \Sigma_{1,1}$. Construct w from $\bar{w}$ by inserting the base without a complementary base at every unpaired position of S. Let S' be an MFE structure for w. Obviously, all unpaired positions in S are also unpaired in S'. We must have $\mathsf{Paired}(S') = \mathsf{Paired}(S)$, otherwise we have an alternative structure for $\bar{w}$, a contradiction. Hence, $S' = S$, i.e., w is a design for S.   □

Result **R4** follows readily from Theorem 1 by taking $c = 2$.

Lemma 4 (Result R5). *Any structure that contains m_5 or m_{30} is not $\Sigma_{2,0}$-designable.*

Proof. Assume that S is $\Sigma_{2,0}$-designable and let w be a design for S. Then $\mathsf{Paired}(w, S)$ is a design for $\mathsf{Paired}(S)$. Since $\mathsf{Paired}(S)$ is saturated, by Theorem 1, $D(S) = D(\mathsf{Paired}(S)) \leq 4$, hence, S cannot contain motif m_5. Now, assume to the contrary that S contain motif m_{30} appearing at node $[a, b]$ of T_S. Let $\{[l_i, r_i]\}_{i=1}^3$ be some paired children of $[a, b]$ and the node $[a, b]$ if $[a, b]$ is not the root, and $[u, u]$ an unpaired child of $[a, b]$. Let $L_i = l_i$ and $R_i = r_i$ if $[l_i, r_i]$ is a child of $[a, b]$, and $L_i = r_i$ and $R_i = l_i$ if it is $[a, b]$. If among bases $w_{L_1}, \ldots, w_{L_3}$ there is a pair of repeated letters, then we can construct an alternative MFE structure for w (see the first paragraph in the proof of Theorem 1). Assume that these three bases are different. Then for some $i = 1, 2, 3$, w_u equals either w_{l_i} or w_{r_i}, say it equals w_{l_i}. Then $S \setminus \{(l_i, r_i)\} \cup \{(u, r_i)\}$ is an MFE structure for S, a contradiction with the assumption that w is a design for S.   □

Theorem 5 (Result R6). *If the tree representation of a structure S admits a separated coloring then S is $\Sigma_{2,0}$-designable.*

Proof. Given a sequence w, we define the level $L(i)$ of position i as $L(i) = |w_{[1,i]}|_G - |w_{[1,i]}|_C$.

Claim 5.1. *Consider any structure compatible with sequence w that contains some A − U base pair between positions at different levels, then some G or C is left unpaired.*

Proof. Consider that the A − U base pair occurs at position (a, b), and note that the bases of the substring $w_{[a+1,b-1]}$ can only base pair among themselves without introducing crossings. We will show that G's and C's are not balanced in this substring. Since $w_b \in \{A, U\}$, $L(b) = L(b - 1)$. Hence, by the definition of L, we have that

$$|w_{[a+1,b-1]}|_G - |w_{[a+1,b-1]}|_C = L(b - 1) - L(a) = L(b) - L(a) \neq 0 .$$

Therefore, at least one G or C in the substring remains unpaired in this structure.

□

Consider a separated coloring of the tree representation of S. We will use this coloring to construct a design w for S, by specifying a nucleotide at each position of w. First, for each unpaired position i, set $w_i = U$. Second, apply a modified version of the algorithm described in Theorem 1 to set the bases of paired positions in which black nodes are assigned to base pair G − C, white nodes to C − G and grey nodes to A − U or U − A. The algorithm ignores unpaired nodes in the tree representation of S. Since the coloring is proper such assignment is always possible at every step of the algorithm. We claim that for any node $[i, j]$ (paired or unpaired), the level of position i is the same as the level of the node $[i, j]$. To verify this, observe that the substring of w corresponding to any subtree has the same number of G's and C's. Hence, for any node $[i, j]$, the level of position i depends only on nodes on the path from this node to the root. It is easy to check that the level of i is equal to the level the node. Note that if $[i, j]$ is a grey node then the level of position j is the same as the level of i, i.e., the same as the level of $[i, j]$.

We will show that the constructed w is a design for S. Since all C's and A's of w are paired in S, S is an MFE structure for w. We need to show that it is the only MFE structure for w. Consider an MFE structure S' for w different from S. Since w has the same number of G's and C's, S' must pair all G's, C's and A's of w. We will show that all unpaired positions in S are also unpaired in S'. Assume to the contrary that position i is unpaired in S, but it is paired to j in S'. We must have $w_i = U$ and $w_j = A$. Since the coloring is separated, the unpaired node $[i, i]$ has a different level than the grey node containing j, and hence, the level of i is different from the level of j. It follows by Claim 5.1 that some G or C is unpaired in S', a contradiction. Consider paired restrictions of S, S' and w. Both Paired(S) and Paired(S') are saturated and compatible with

Paired(w, S) and they are different since S and S' are different and agree on the unpaired positions. Furthermore, Paired(w, S) can be produced by the algorithm described in Theorem 1 for the input structure Paired(S), and hence, by Theorem 1, Paired(w, S) is a design for Paired(S), which contradicts the existence of Paired(S'). Hence, w is a design for S.   □

Theorem 6 (Result R7). *If w is a design for a structure S, then for any integer $k \geq 1$, $w^{[k]}$ is a design for $S^{[k]}$. In particular, if a structure S is $\Sigma_{2,0}$-designable, then so is $S^{[k]}$.*

Proof. Consider a designable structure S and let $w = w_1 \cdots w_n$ be a design for S. We will show that $w^{[k]}$ is a design for $S^{[k]}$. Let the i-th k positions in S be called the *region i*. Note that the positions in region i of $S^{[k]}$ correspond to the i-th position in S.

First, we will show that $S^{[k]}$ is an MFE structure for $w^{[k]}$. Consider an MFE structure S' of $w^{[k]}$. Define an *interaction graph* of S', denoted by $I(S') = (V_{I(S')}, E_{I(S')})$, as follows: the vertex set $V_{I(S')}$ is the set of positions in w, i.e., $\{1, \ldots, n\}$, and there is an edge between i and j in $I(S')$ if there exists a pair between a position in region i and a position in region j in S'. Note that $I(S')$ is a bipartite graph: indeed, vertices of any cycle in $I(S')$ are positions in w that alternate between A and U, or between C and G. Also note that $I(S')$ is an outer-planar graph: base pairs are pairwise non-crossing and can therefore be drawn without crossings on the upper half-plane, leaving the lower half-plane on the outer face. Assign each edge e in $E_{I(S')}$ a weight $c(e)$ equal to the number of pairs between regions i and j in S'. Note that the sum of all weights in $I(S')$, denoted as $\|E_{I(S')}\|$, equals $|S'|$. We have the following claim.

Claim 6.1. *If M is a maximum matching in $I(S')$ then $|S'| \leq k|M|$. Moreover, if $|S'| = k|M|$ then every minimum vertex cover of $I(S')$ covers every edge exactly once.*

Proof. Note that for any vertex i in $V_{I(S')}$, the sum of the weights of edges incident with i is at most k. Consider a smallest vertex cover C of $I(S')$, and take the sum of these inequalities over all vertices i in the cover C:

$$\sum_{i \in C} \sum_{e \text{ incident with } i} c(e) \leq k|C|. \tag{1}$$

Since C is a vertex cover, the weight of every edge in $E_{I(S')}$ appears at least once on the left side of (1), hence $|S'| = \|E_{I(S')}\| \leq k|C|$. By König's Theorem, the maximum matching M in $I(S')$ has the same number of edges as C, i.e., $|S'| \leq k|M|$. The equality implies that the weight of every edge in $E_{I(S')}$ appears exactly once on the left side of (1), i.e., that vertex cover C covers every edge exactly once.   □

Given a matching M in $I(S')$, we can construct a structure S_M for w with $|M|$ pairs as follows: for every edge $\{i, j\}$ in M, add pair (i, j). This is a valid (pseudoknot-free) structure, since M is a subgraph of outer-planar graph $I(S')$. It follows that $|M| \leq |S|$. If M is a maximum matching on $I(S')$, we have by Claim 6.1 that $|S'| \leq k|M| \leq k|S| = |S^{[k]}|$ i.e., $S^{[k]}$ is an MFE structure for $w^{[k]}$. It also follows that $|S'| = k|M|$ and that $|M| = |S|$. Since S is a unique structure for w and $|S_M| = |M| = |S|$, we have that $S_M = S$, i.e., there is only one maximum matching in $I(S')$. We need the following claim to show that all connected components in $I(S')$ have at most 2 vertices.

Claim 6.2. *Let G be a connected bipartite graph on at least three vertices with unique maximum matching M. Then there exists a minimum vertex cover of G that covers some edge twice.*

Proof. First, we will show that every vertex in G is incident to an edge in matching M. Assume the contrary and consider all vertices in G which are incident to only non-matching edges. If two of these vertices are incident then the matching is not maximal. Otherwise, let u be such a vertex and v its neighbor. Vertex v must be incident to a matching edge. We can construct a new matching by removing this edge and adding edge uv, which contradicts the assumption that M is a unique maximal matching.

Take a maximal path P alternating between matching and non-matching edges in G. Let u be an endpoint of P and e the edge on P incident to u. If e is a non-matching edge then u must be incident to a matching edge, say f. By maximality of P, the other endpoint v of f must be on P. Since every internal vertex of P is incident to a matching on P, v must be the other endpoint of P and the edge incident to v on P must be a non-matching edge. Hence, we have an alternating cycle $P + f$ which contradicts the uniqueness of the maximal matching. Thus, P starts and ends with matching edges. Next, we show that u is a pendant vertex (has degree one). Assume to the contrary u is incident to another (non-matching) edge $f = uv$. By maximality of P, v is on P, which yields a cycle. If this cycle is even, we have an alternating cycle, which contradicts the uniqueness of the matching, and if it is odd, we have a contradiction with the fact that G is bipartite. Hence, both endpoints of P are pendant.

Consider a minimum vertex cover C of G. By well-known König's theorem, every minimum vertex cover in a bipartite graph uses exactly one endpoint of every edge of a maximum matching and no other vertices. Since the endpoints of P are pendant, and G is connected and has ≥ 3 vertices, P must have at least three edges. Since endpoints of P are pendant and incident to matching edges, we can assume that C does not contain endpoints of P, i.e., contains the second and last by one vertex of P. It is easy to see that at least one non-matching edge is covered twice by C. □

Consider a connected component K of $I(S')$. Since $I(S')$ has a unique maximum matching, so does K. If K has more than two vertices, it contains a minimum vertex cover of K that covers some edge twice. It follows that there is a minimum vertex cover of $I(S')$ that covers some edge twice. Hence, by Claim 6.1, $|S'| \leq k|M|$, a contradiction. It follows that every connected component of $I(S')$ has at most two vertices, hence, either S' is not MFE or $S' = S^{[k]}$. □

Theorem 7 (Result R8). *Each structure S without m_5 and m_{30} can be transformed into a $\Sigma_{2,0}$-designable structure S' by inflating a subset of its base pairs (at most one per band). Furthermore, this transformation can be done in $\Theta(n)$ time.*

Proof. We start with the greedy coloring of T_S. Since S does not contain m_5 and m_{30}, it is a proper coloring and there is no node having both a grey child and an unpaired child. We will insert base pairs within S so that the grey nodes and any unpaired node end up at levels of different parities. If the root has a grey child, assign even parity to the grey nodes, otherwise (if the root has an unpaired child, or no grey and no unpaired children), assign even parity to the unpaired nodes.

Now we proceed from the children of the root towards leaves adjusting parity level for grey and unpaired nodes to keep one type even and the other one odd. We repeatedly apply the following simple operation on T_S: If the node N does not match its intended parity level. Denote N_P the parent of N (N_P is not the root as all children of the root already have the correct parity level) and N_{PP} the parent of N_P. Insert a new paired node N_N between N_{PP} and N_P, assign it with the color of N_P, and apply the greedy algorithm on N_N. Observe that N_P always takes either black or white color changing the parity level of all its descendants (including N). Note that the children of N_P may get recolored, we can even get one more grey child but after this operation the parity levels of all children of N are correct and we do not change parity levels outside the subtree rooted at N. After fixing all nodes, we get a separated proper coloring (which is actually the greedy coloring) of $T_{S'}$. Hence, by Theorem 5, S' is designable. Figure 4 illustrates this process.   □

5   Conclusion, Discussion and Perspectives

In this work, we introduced the *Combinatorial RNA Design problem*, a *minimal* instance of the RNA design problem which aims at finding a sequence that admits the target structure as its unique base pair maximizing structure. First, we provided complete characterizations for the structures that can be designed using restricted alphabets. Then we considered the RNA design under a four-letter alphabet, and provide a complete characterization of designable saturated structures, i.e., free of unpaired positions. Turning to those target structures that contain unpaired positions, we provided partial characterizations for classes of designable/undesignable structures, and showed that the set of designable structures is closed under the stutter operation. Finally, we introduced structure-approximating version of the problem and, assuming that the input structure avoids two motifs, provided a structure approximating algorithm of ratio 2 for general structures.

An important question that is left open by this work is the computational complexity of the RNA design problem. Schnall-Levin *et al.* [17] established the NP-hardness of a more general problem, called the inverse Viterbi algorithm, which takes as input a stochastic grammar (representing the energy model) and a targeted parse tree (representing the structure), and outputs a sequence (design) whose most probable parsing should match the target. However this result does not settle the complexity of the RNA design, essentially because the proposed reduction relies critically on an encoding of 3-SAT instances within the input grammar. While the hypothetical *perfect* grammar/energy model for RNA folding probably differs from the currently accepted Turner model, it should ultimately reflect the laws of physics and should certainly not depend on the instance. As the reduction [17] requires a different grammar (i.e., energy model) for each instance, it does not seem easily adaptable into a proof that holds for a fixed energy model. Consequently, despite two decades of work on the subject, the computational tractability of RNA design is still open, either in its general instance and in our combinatorial version.

Besides complexity issues, natural extensions of this work may include the consideration of more general base pairing models, more realistic energy models (ideally, the Turner energy model [20]), or the design under other objectives, such as the Boltzmann probability [22]. However, even the simplest of modifications, allowing $G - U$ base pairs, would invalidate parity properties that are critical to the proofs of some of our results and algorithms. More precise bounds for the ratio of the structure-approximating could be established. Finally, better algorithms could be designed for the problem, attempting to minimize the number of modifications so that a given structure becomes designable (or, more modestly, belongs to an identified class of designable structures).

References

1. Aguirre-Hernández, R., Hoos, H.H., Condon, A.: Computational RNA secondary structure design: empirical complexity and improved methods. BMC Bioinform. **8**, 34 (2007)
2. Avihoo, A., Churkin, A., Barash, D.: RNAexinv: an extended inverse RNA folding from shape and physical attributes to sequences. BMC Bioinform. **12**(1), 319 (2011)
3. Busch, A., Backofen, R.: INFO-RNA–a fast approach to inverse RNA folding. Bioinformatics **22**(15), 1823–1831 (2006)
4. Dai, D.C., Tsang, H.H., Wiese, K.C.: RNADesign: local search for RNA secondary structure design. In: IEEE Symposium on Computational Intelligence in Bioinformatics and Computational Biology (CIBCB) (2009)
5. Esmaili-Taheri, A., Ganjtabesh, M., Mohammad-Noori, M.: Evolutionary solution for the RNA design problem. Bioinformatics **30**(9), 1250–1258 (2014)
6. Frid, Y., Gusfield, D.: A simple, practical and complete $o(n^3/\log n)$-time algorithm for RNA folding using the Four-Russians speedup. Algorithms Mol. Biol. **5**, 13 (2010)
7. Garcia-Martin, J.A., Clote, P., Dotu, I.: RNAiFOLD: a constraint programming algorithm for RNA inverse folding and molecular design. J. Bioinform. Comput. Biol. **11**(2), 1350001 (2013)
8. Griffiths-Jones, S., Bateman, A., Marshall, M., Khanna, A., Eddy, S.R.: RFAM: an RNA family database. Nucleic Acids Res. **31**(1), 439–441 (2003)

9. Höner Zu Siederdissen, C., Hammer, S., Abfalter, I., Hofacker, I.L., Flamm, C., Stadler, P.F.: Computational design of RNAs with complex energy landscapes. Biopolymers **99**(12), 1124–1136 (2013)

10. Hofacker, I.L., Fontana, W., Stadler, P., Bonhoeffer, L., Tacker, M., Schuster, P.: Fast folding and comparison of RNA secondary structures. Monatshefte für Chemie/Chem. Monthly **125**(2), 167–188 (1994)

11. Levin, A., Lis, M., Ponty, Y., O'Donnell, C.W., Devadas, S., Berger, B., Waldispühl, J.: A global sampling approach to designing and reengineering RNA secondary structures. Nucleic Acids Res. **40**(20), 10041–10052 (2012)

12. Lyngsø, R.B., Anderson, J.W., Sizikova, E., Badugu, A., Hyland, T., Hein, J.: FRNAkenstein: multiple target inverse RNA folding. BMC Bioinform. **13**, 260 (2012)

13. Mathews, D.H., Sabina, J., Zuker, M., Turner, D.H.: Expanded sequence dependence of thermodynamic parameters improves prediction of RNA secondary structure. J. Mol. Biol. **288**(5), 911–940 (1999)

14. Nussinov, R., Jacobson, A.: Fast algorithm for predicting the secondary structure of single-stranded RNA. Proc. Natl. Acad. Sci. USA **77**, 6903–6913 (1980)

15. Reinharz, V., Ponty, Y., Waldispühl, J.: A weighted sampling algorithm for the design of RNA sequences with targeted secondary structure and nucleotide distribution. Bioinformatics **29**(13), i308–i315 (2013)

16. Rodrigo, G., Landrain, T.E., Majer, E., Daròs, J.-A., Jaramillo, A.: Full design automation of multi-state RNA devices to program gene expression using energy-based optimization. PLoS Comput. Biol. **9**(8), e1003172 (2013)

17. Schnall-Levin, M., Chindelevitch, L., Berger, B.: Inverting the Viterbi algorithm: an abstract framework for structure design. In: Machine Learning, Proceedings of the Twenty-Fifth International Conference (ICML 2008), Helsinki, Finland, June 5–9, 2008, pp. 904–911 (2008)

18. Takahashi, M.K., Lucks, J.B.: A modular strategy for engineering orthogonal chimeric RNA transcription regulators. Nucleic Acids Res. **41**(15), 7577–7588 (2013)

19. Taneda, A.: MODENA: a multi-objective RNA inverse folding. Adv. Appl. Bioinform. Chem. **4**, 1–12 (2011)

20. Turner, D.H., Mathews, D.H.: NNDB: the nearest neighbor parameter database for predicting stability of nucleic acid secondary structure. Nucleic Acids Res. **38**, D280–D282 (2010). (Database issue)

21. Wu, S.Y., Lopez-Berestein, G., Calin, G.A., Sood, A.K.: RNAi therapies: drugging the undruggable. Sci. Transl. Med. **6**(240), 240ps7 (2014)

22. Zadeh, J.N., Wolfe, B.R., Pierce, N.A.: Nucleic acid sequence design via efficient ensemble defect optimization. J. Comput. Chem. **32**(3), 439–452 (2011)

23. Zhou, Y., Ponty, Y., Vialette, S., Waldispuhl, J., Zhang, Y., Denise, A.: Flexible RNA design under structure and sequence constraints using formal languages. In: Proceedings of the International Conference on Bioinformatics, Computational Biology and Biomedical Informatics, BCB 2013, pp. 229–238. ACM (2013)

24. Zuker, M., Stiegler, P.: Optimal computer folding of large RNA sequences using thermodynamics and auxiliary information. Nucleic Acids Res. **9**, 133–148 (1981)

Dictionary Matching with Uneven Gaps

Wing-Kai Hon[1], Tak-Wah Lam[2], Rahul Shah[3]($\boxtimes$), Sharma V. Thankachan[4],
Hing-Fung Ting[2], and Yilin Yang[1]

[1] Department of Computer Science, National Tsing Hua University, Hsinchu, Taiwan
{wkhon,yilinyang}@cs.nthu.edu.tw

[2] Department of Computer Science, University of Hong Kong, Hong Kong, China
{twlam,hfting}@cs.hku.hk

[3] Division of Computer Science, Louisiana State University, Baton Rouge, USA
rahul@csc.lsu.edu

[4] School of Computational Science and Engineering,
Georgia Institute of Technology, Atlanta, USA
sharma.thankachan@gatech.edu

Abstract. A gap-pattern is a sequence of sub-patterns separated by bounded sequences of don't care characters (called gaps). A one-gap-pattern is a pattern of the form $P[\alpha, \beta]Q$, where P and Q are strings drawn from alphabet Σ and $[\alpha, \beta]$ are lower and upper bounds on the gap size g. The gap size g is the number of don't care characters between P and Q. The dictionary matching problem with one-gap is to index a collection of one-gap-patterns, so as to identify all sub-strings of a query text T that match with any one-gap-pattern in the collection. Let $\mathcal{D}$ be such a collection of d patterns, where $\mathcal{D} = \{P_i[\alpha_i, \beta_i]Q_i \mid 1 \le i \le d\}$. Let $n = \sum_{i=1}^{d} |P_i| + |Q_i|$. Let γ and λ be two parameters defined on $\mathcal{D}$ as follows: $\gamma = |\{j \mid j \in [\alpha_i, \beta_i], 1 \le i \le d\}|$ and $\lambda = |\{\alpha_i, \beta_i \mid 1 \le i \le d\}|$. Specifically γ is the total number gap lengths possible over all patterns in $\mathcal{D}$ and λ is the number of distinct gap boundaries across all the patterns. We present a linear space solution (i.e., $O(n)$ words) for answering a dictionary matching query on $\mathcal{D}$ in time $O(|T|\gamma \log \lambda \log d + occ)$, where occ is the output size. The query time can be improved to $O(|T|\gamma + occ)$ using $O(n + d^{1+\epsilon})$ space, where $\epsilon > 0$ is an arbitrarily small constant. Additionally, we show a compact/succinct space index offering a space-time trade-off. In the special case where parameters α_i and β_i's for all the patterns are same, our results improve upon the work by Amir et al. [CPM, 2014]. We also explore several related cases where gaps can occur at arbitrary locations and where gap can be induced in the text rather than pattern.

Keywords: Dictionary matching · Point enclosure queries

This research is funded in part by US National Science Foundation (NSF) Grant CCF–1218904 and Taiwan MOST Grant 102-2221-E-007-068. Part of this work was done during Y. Yang's visit at the University of Hong Kong. This paper is a merger of two independent similar works.

© Springer International Publishing Switzerland 2015
F. Cicalese et al. (Eds.): CPM 2015, LNCS 9133, pp. 247–260, 2015.
DOI: 10.1007/978-3-319-19929-0_21

1   Introduction

Pattern Matching is a fundamental research field in Computer Science. In pattern matching, we are given a text T and a pattern P both drawn from same alphabet set Σ and the task is to find all occurrences of the pattern P in the text T. In earlier times, the focus was to develop algorithms to achieve this goal efficiently [8,22,24]. In many applications, the text would be known in advance and queries in the form of pattern would arrive in online manner. This motivated the development of data structures like suffix trees, suffix arrays and compressed indexes [13,15,26,30]. There are also applications (like virus scanning, packet routing etc.) where the patterns are known in advanced and are to be indexed and then text T comes as an online stream. This variant is called *dictionary matching* and solutions for this in many cases are based on suffix trees.

In many useful applications the pattern match may not be exact but we are allowed few character mismatches. In some formulations, text and/or patterns could have wild-cards or don't care characters. We consider the variant where the patterns have don't care characters which are clustered in one location as one-gap. Moreover, we allow this gap length to be variable but within some bounds. Thus, our pattern would look like $P\phi^g Q$ where P and Q are strings and between are g don't care symbols ϕ. The gap length g has to be at least α characters long and at most β characters long. We denote such one-gap-patterns by $P[\alpha, \beta]Q$. Thus, pattern $abc[2,4]cba$ will match with $abcdedcba$ (with gap of size 3) but not with $abcdcba$ or $abcdefedcba$ (because gap of 1 or 5 is not allowed).

More formally, let $\mathcal{D} = \{P_i[\alpha_i, \beta_i]Q_i \mid 1 \leq i \leq d\}$ be a collection of d one-gap-patterns over an alphabet set Σ of size σ. Let the total number of characters excluding any don't care character be $n = \sum_{i=1}^{d} |P_i| + |Q_i|$. We associate two additional parameters with $\mathcal{D}$, namely γ and λ, where γ is the size of the set $cover(\mathcal{D}) = \{j \mid j \in [\alpha_i, \beta_i], 1 \leq i \leq d\}$ and $\lambda = |\{\alpha_i, \beta_i \mid 1 \leq i \leq d\}|$. Specifically $cover(\mathcal{D})$ is the set of all possible gap lengths over all patterns in $\mathcal{D}$ and λ is number of distinct gap boundaries across all the patterns.

Problem 1. *Our task is to index $\mathcal{D}$, such that whenever a text T comes as a query, we can report all sub-strings of T that match with any pattern in $\mathcal{D}$.*

As noted by Amir et al. [6], many problems in computational biology can be modeled as above [14,16,27,28].

1.1   Previous Work

For the dictionary matching problem with exact pattern matching, [2] presented a classic Aho-Corasick (AC) automata which requires linear space $O(n)$, where n is the total size of all patterns in number of characters. The AC automata supports matching in the query text T in optimal $O(|T| + occ)$ time, occ being the number of matches produces as the output. Amir et al., [3,4] used a suffix tree based approach to solve the dynamic version of this problem. The state transitions in AC automata can be viewed as following suffix links in a generalized suffix tree of all patterns. Recently, there has been interest in making the index space succinct or compact [7,12,18–21].

Approximate dictionary matching has also had a long line of research. For 1-error matching, Amir et al. [5] presented a solution comprising of forward and reverse suffix trees. This solution was later made succinct by using sparsification (suffix sampling) techniques [17]. In this approach, the pattern is matched in three parts. The prefix of pattern is matched in the reverse suffix tree, the middle character (wild-card or gap or mismatching character) and the suffix of the pattern which is matched in the forward suffix tree. Based on these matches, a geometric range searching query is issued to find all matches. A framework for approximate dictionary matching with k-errors (or wild cards) was provided by Cole et al. [11]. For further reading, we refer to the recent survey by Lewenstein [25].

For the particular case of dictionary matching with one-gap, Amir et al. [6] recently studied the case where all the patterns have same bounds on the gap length allowed. Thus, in their case for all i, we have $\alpha_i = \alpha$ and $\beta_i = \beta$. Our model is generalization of theirs. Moreover, in their particular case, our results directly improve theirs in space and time complexities.

We list the comparison next.

1.2   Our Results and Comparison

Our main results are summarized in the following theorems.

Theorem 1. *The collection $\mathcal{D}$ can be indexed in $O(n)$ space and can answer any dictionary matching query T in $O(|T|\gamma \log \lambda \log d + occ)$ time, where occ is the output size.*

Theorem 2. *A dictionary matching query T on $\mathcal{D}$ can be answered in time $O(|T|\gamma + occ)$ using an index of space $O(n + d^{1+\epsilon})$, where $\epsilon > 0$ is an arbitrarily small constant.*

Amir et al.'s work [6] focused on a special case where $\alpha_i = \alpha, \beta_i = \beta, \forall i$ (i.e., $\gamma = \beta - \alpha + 1$ and $\lambda = 2$). Two solutions offered by them require either (i) $O(n + d\log^\epsilon d)$ space and $O(|T|(\beta - \alpha + 1)\log^2 n \log \log d + occ)$ time or (ii) $O(n + d^2)$ space and $O(|T|(\beta - \alpha + 1) + occ)$ time. The first solution is based on the ideas from the classical solutions for dictionary matching with one error, where as the second result uses look up tables. In contrast, our results are (i) $O(n)$ space and $O(|T|(\beta - \alpha + 1)\log d + occ)$ time or (ii) $O(n + d^{1+\epsilon})$ space and $O(|T|(\beta - \alpha + 1) + occ)$ time, both using a one-shot framework. Thus, on first count we get substantial improvement in times and some improvement in space and on the second count we get some improvement in space. Clearly, our results not only improve, but also generalize these result. Moreover, we obtain the following succinct result.

Theorem 3. *There exists a $n \log \sigma + O(d \log n) + o(n \log \sigma)$ bits structure to solve dictionary matching query T over the collection $\mathcal{D}$, in time $O(|T|\gamma \log \lambda \log^{2+\epsilon} n + occ)$.*

Apart from this main problem involving one-gap-patterns, we consider several generalizations and variants. In particular, we consider the case where the gap of an acceptable length can occur at more than one locations in the pattern. We show that our framework can answer such queries easily. We also consider an orthogonal variant called *dictionary matching with one missing substring* problem. In this variant, a gap of acceptable length can be created in the text, for matching puposes. More, precisely we denote missing substring pattern as $P_i[\beta_i]Q_i$, the pattern can be interpreted as P_i followed by a suffix of Q_i with anywhere between 0 to β_i characters missing from Q_i. We give a $O(n)$ space index, taking $O(|T|\log n + occ)$ query time. In this case also, we further show how to generalize it to the case of missing string occuring at arbitrary location in the pattern.

2   Notation and Preliminaries

For a string X, we use $|X|$ to denote its length, $X[i], 1 \le i \le |X|$, to denote its ith character, $X[i \ldots j]$ to denote its sub-string starting from position i and ending at position j. For simplicity, we may use $X[i \ldots]$ to denote $X[i \ldots |X|]$ and $X[\ldots i]$ to denote $X[1 \ldots i]$. The reverse of X is denote by $\overleftarrow{X}$. Specifically, the ith character of $\overleftarrow{X}$ is the $(|X| - i + 1)$th character of X. Throughout this paper, ϵ denotes an arbitrary small positive constant. For two strings (which may consist of a single character) X and Y, XY represents the concatenation. Always interpret $\log x$ as 1 for $x < 2$.

2.1   Suffix Tree and Loci

Let $\mathcal{S} = \{S_1, S_2, \ldots, S_d\}$ be a collection of d strings and of total n characters. The (generalized) suffix tree of $\mathcal{S}$ is a compact trie storing all the suffixes of each S_i in $\mathcal{S}$ [30]. For our purpose, we slightly change[1] the definition of suffix tree: first create a new set $\mathcal{S}' = \{S_1\$, S_1\#, S_2\$, S_2\#, \ldots, S_d\$, S_d\#\}$ by duplicating each $S_i \in \mathcal{S}$ into $S_i\$$ and $S_i\#$, where $\$$ and $\#$ are two special characters that does not appear in Σ. We now create a compact trie of all suffixes (except $\$$ and $\#$) of all strings in $\mathcal{S}'$ and call it the suffix tree $\mathcal{T}$ of $\mathcal{S}$. Following are the characteristics of $\mathcal{T}$.

1. $\mathcal{T}$ is a tree of exactly $2n$ leaves. There are no internal nodes with degree 1.
2. Edges are labeled with strings and for any node u, $\mathsf{path}(u)$ is the string obtained by the concatenation of edge labels on the path from root u.
3. Corresponding to each $S_i\$$ (resp., $S_i\#$), there exists a unique leaf node in $\mathcal{T}$ with its path equals $S_i\$$ (resp., $S_i\#$).

---

[1] This is just for the ease ensuring some properties.

4. For each i, there exists a unique node u in $\mathcal{T}$, where $\mathsf{path}(u) = S_i$. Specifically, u is the lowest common ancestor (lca) of the leaves corresponding to $S_i\$$ and $S_i\#$. We mark node u with id i.
5. For any node u there always exists a node v such that $\mathsf{path}(v)$ is the string obtained by deleting the first character from the string $\mathsf{path}(u)$. Node v is called the suffix link of u, denoted $\mathsf{slink}(u)$.
6. Each node in $\mathcal{T}$ maintains its pre-order id, its suffix link, and a pointer to its nearest marked ancestor.
7. The locus of a string T in $\mathcal{T}$, denoted by $\mathsf{locus}(T)$, is a node u, such that u is the lowest (farthest from root) where $\mathsf{path}(u)$ is a prefix of T.

We now define a problem, which can be efficiently solved.

Problem 2 (Prefix Matching). *Let S be collection of d strings of total length n, where the characters are drawn from an alphabet set $[\sigma]$. Then, pre-process S into a suffix tree data structure $\mathcal{T}$, such that we can answer the following: given a text T, take all the suffixes $T[i\ldots]$ of this text and report all $\mathsf{locus}(T[i\ldots])$ for all possible starting locations i of the suffixes.*

The following result is by Amir et al. [4].

Lemma 1 ([4]). *There exists an $O(n)$ space and $O(|T|)$ query time data structure for Problem 2.*

This is obtained by following process, first we match T as much as we can and find the locus then we follow suffix link and match further until we can (thus finding locus for $T[2\ldots]$) and so on. For more precise and detailed description, see [4,21].

In the usual dictionary matching problem, once the locus node v is obtained, every marked node on the path from v to root forms an output. This is done by following marked ancestors pointers starting from the locus node v. This is repeated for all the loci.

3   Forward and Reverse Suffix Trees

Based on the ideas used for 1-error dictionary matching [5] we show the framework of using forward and reverse suffix trees for our problem. We define two string collections $\overleftarrow{S}$ and $\overrightarrow{S}$ w.r.t. $\mathcal{D}$ as follows.

$$\overleftarrow{S} = \{\overleftarrow{P_i} \mid P_i[\alpha_i, \beta_i]Q_i \in \mathcal{D}, 1 \leq i \leq d\}$$

$$\overrightarrow{S} = \{Q_i \mid P_i[\alpha_i, \beta_i]Q_i \in \mathcal{D}, 1 \leq i \leq d\}$$

For the string collection $\overleftarrow{S}$, construct a suffix tree $\overleftarrow{\mathcal{T}}$ as described in Sect. 2.1. Similarly, for $\overrightarrow{S}$, construct the corresponding trie $\overrightarrow{\mathcal{T}}$. For the string $\overleftarrow{T[\ldots i]}$ (reverse of the prefix of T ending at location i), let u_i be the node corresponding to its locus in $\overleftarrow{\mathcal{T}}$. Similarly, let v_i be the node corresponding to the locus of

$T[i \ldots]$ in $\overrightarrow{T}$. Both $\overleftarrow{T}$ and $\overrightarrow{T}$ can be maintained in $O(n)$ space and whenever T comes as a query, we can compute u_i's and v_i's for all values of i in $O(|T|)$ time (refer to Lemma 1). Following is a crucial observation.

Observation 1. *Let $j \in [1, |T|]$ and g be two integers. Then, a pattern $P_i[\alpha_i, \beta_i]Q_i$ match with a sub-string $T[j - |P_i| + 1, j + g + |Q_i|]$ if and only if*

1. *$\overleftarrow{P_i}$ is a prefix of $\mathsf{path}(u_j)$ in $\overleftarrow{T}$.*
2. *Q_i is a prefix of $\mathsf{path}(v_{j+g+1})$ in $\overrightarrow{T}$.*
3. *$\alpha_i \leq g \leq \beta_i$.*

Therefore, dictionary matching problem where $\alpha_i = \alpha, \beta_i = \beta$, for all i can be reduced to the following problem.

Problem 3. *For all $j \in [1, |T|]$ and $g \in [\alpha, \beta]$, find all tuples (j, g, i), where $\overleftarrow{P_i}$ is a prefix of $\mathsf{path}(u_j)$ in $\overleftarrow{T}$ and Q_i is a prefix of $\mathsf{path}(v_{j+g+1})$ in $\overrightarrow{T}$. A tuple (j, g, i) can be interpreted as a match of $P_i[\alpha_i, \beta_i]Q_i$ with $T[j - |P_i| + 1, j + g + |Q_i|]$.*

Given j and g, the main task here is to find all i's which go with them. For finding such i's, we can first find loci u_j and v_{j+g+1}. Then make the list of marked nodes above u_j in $\overleftarrow{T}$ and the list of marked nodes above v_{j+g+1} in $\overrightarrow{T}$. Now intersect these to list to find the common one-gap-patterns. However, here lies the catch. These lists individually might be much bigger than the size of their intersection. Thus, we are trying out many more things than what the final output contains. Consequently, we cannot bound our query time in terms of the size of the output.

To overcome this issue, Amir et al. [6] took the approach of heavy path decomposition [29]. Both the forward and reverse trees are decomposed into heavy paths. Now, any locus to root path (from which the marked nodes need to be considered) will overlap with at most $\log n$ heavy paths. Between every pair of heavy path – one from forward and one from reverse suffix tree – they maintain a 2D range reporting data structure if their intersection is non-zero. In this structure, pre-order ids are assigned in a such a manner that all the ids on any heavy path form contiguous numbers. So, 2D range reporting data structure will have points corresponding to intersection of both heavy paths (one in forward and one in reverse). The coordinates of this point are its pre-order id in forward tree and pre-order id in reverse tree. Now, when we query based on u_j and v_{j+g+1} we consider $\log n$ heavy paths in forward and $\log n$ in reverse, and based their cross product, we look at $\log^2 n$ 2D range reporting data structures. In each structure, we issue an orthogonal range reporting query which takes $O(\log \log d + output)$. (The structure giving such query times take super linear $O(d \log^\epsilon d)$ space [9].) Based on this Amir et al. [6] got the $O(|T|(\beta - \alpha + 1) \log^2 n \log \log d + occ)$ time bound.

4   Our Framework

Unlike the heavy path decomposition along with 2D orthogonal range searching approach of Amir et al., we take an orthogonal approach to find the intersection of marked ancestors. We formulate this marked node intersection problem as 3D **rectangular stabbing** problem instead. In our framework, the bound of gaps are not the same for all patterns. Then by combining the definition of $cover(\mathcal{D})$ with Observation 1, the dictionary matching problem can be reduced to the following.

Problem 4. For all $j \in [1, |T|]$ and $g \in cover(\mathcal{D})$, find all tuples (j, g, i), where

1. $\overleftarrow{P_i}$ is a prefix of $\mathsf{path}(u_j)$ in $\overleftarrow{T}$.
2. Q_i is a prefix of $\mathsf{path}(v_{j+g+1})$ in $\overrightarrow{T}$.
3. $\alpha_i \leq g \leq \beta_i$.

As described before, we interpret tuple (j, g, i) as a match of $P_i[\alpha_i, \beta_i]Q_i$ with $T[j - |P_i| + 1, j + g + |Q_i|]$. We now present the way we handle a sub-query in Problem 4. i.e., for a specific j and g. We map each pattern $P_i[\alpha_i, \beta_i]Q_i$ to a rectangular region R_i in 3D as follows: $R_i = [x_i', x_i''] \times [y_i', y_i''] \times [\alpha_i, \beta_i]$, where

- x_i' is the (pre-order rank of the) locus of $\overleftarrow{P_i}$ in $\overleftarrow{T}$
- x_i'' is the rightmost leaf node in the sub-tree of x_i'
- y_i' is the (pre-order rank of the) locus of Q_i in $\overrightarrow{T}$
- y_i'' is the rightmost leaf node in the sub-tree of y_i'

Clearly the three conditions in Problem 4 are satisfied iff the rectangle R_i is stabbed by the point (u_j, v_{j+g+1}, g) as illustrated in Fig. 1. Therefore by maintaining structures for answering stabbing queries over the rectangles $R_1, R_2, \ldots, R_d$, the dictionary matching problem can be reduced to $O(|T|\gamma)$ number of 3D rectangle stabbing queries. Following lemma summarizes one of the best known space-time trade-offs for rectangular stabbing problem.

Lemma 2 [1,10]. A set $\mathcal{R}^k$ of d k-dimensional rectangles ($k \geq 2$ is a constant) can be pre-processed into an $O(d \log^{k-2} d)$ space data structure and can answer rectangle stabbing queries in time $O(\log^{k-1} d + output)$.

Therefore by using the result in Lemma 2 as a black box (with $k = 3$), we obtain an $O(n + d \log d)$ space and $O(|T|\gamma \log^2 d + occ)$ query time data structure for our problem. The set of rectangles $R_1, R_2, \ldots, R_d$ in our case have bounded number (specifically λ) of distinct corner points in the third dimension. Therefore by using the result in Lemma 4, we obtain the following improved result.

Lemma 3. The collection $\mathcal{D}$ of d one-gap patterns can be indexed in $O(n + d \log \lambda)$ space and can answer any dictionary matching query T in $O(|T|\gamma \log \lambda \log d + occ)$ time, where occ is the output size.

Lemma 4. Let $\mathcal{R}^3$ be a set of d rectangles in three-dimension with the number of distinct corner points in the third dimension is at most λ, we can pre-process $\mathcal{R}^3$ into an $O(d \log \lambda)$ space data structure and can answer stabbing queries in $O(\log d \log \lambda + output)$ time.

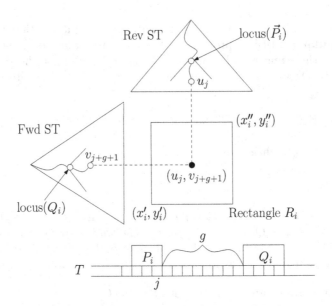

Fig. 1. Rectangle Stabbing

Proof. (sketch) The result follows from the proof of Lemma 2 in [1] for $k \geq 3$. Essentially, they showed that any $k \geq 3$ dimensional rectangle stabbing query can be decomposed into $O(\log \lambda)$ $(k-1)$-dimensional rectangle stabbing queries, where λ is the number of distinct corner points in k-th dimension. The space of the new structure will be $O(\log \lambda)$ times the space of the structure for $(k-1)$th dimensional stabbing queries. By using the structure by Chazelle [10] as the base case (i.e., Lemma 2 with $d = 2$), we obtain the result. ∎

4.1   Achieving Linear Space

In Lemma 3, we achieve the desired query time as in Theorem 1. However the space is not linear when d is large. In order to achieve linear space, we categorize the patterns into *long* and *short* based on a threshold $\tau = \log \lambda$. Specifically, a pattern $P_i[\alpha_i, \beta_i]Q_i$ is long if $|P_i| + |Q_i| \geq \tau$, and is short otherwise. We now treat the collection of long patterns ($\mathcal{S}_l$) and the collection of short patterns ($\mathcal{S}_s$) independently. Notice that the number of long patterns $|\mathcal{S}_l| \leq n/\log \lambda$. Therefore using Lemma 3, we can maintain $\mathcal{S}_l$ in $O(n + |\mathcal{S}_l| \log \lambda) = O(n)$ space and a dictionary matching query on $\mathcal{S}_l$ can be answered in $O(|T|\gamma \log \lambda \log d + occ_l)$ time, where occ_l is the number of long patterns that appear as a sub-string of T.

We now show how to handle dictionary matching queries on $\mathcal{S}_s$. As described before, we create forward suffix tree $\overrightarrow{\mathcal{T}}$ and reverse suffix tree $\overleftarrow{\mathcal{T}}$ of $\mathcal{S}_s$. Space can be bounded by $O(n)$ words. Now we associate a set $\mathcal{R}(w)$ of two-dimensional rectangles with each node w in $\overleftarrow{\mathcal{T}}$ as follows:

$$\mathcal{R}(w) = \{[y_i', y_i''] \times [\alpha_i, \beta_i] \mid P_i[\alpha_i, \beta_i]Q_i \in \mathcal{D} \text{ and } \mathsf{path}(w) = P_i\}$$

Recall that y_i' is the (pre-order rank of the) locus of Q_i in $\overrightarrow{\mathcal{T}}$ and y_i'' is the rightmost leaf node in the sub-tree of y_i'. Each $\mathcal{R}(w)$ is maintained as a data structure for answering two-dimensional stabbing queries (using Lemma 2 with $d = 2$). As each rectangle belongs to a unique set $\mathcal{R}(\cdot)$, the total space can be bounded by $O(n)$. Now a specific instance (j, g, i) of Problem 4 can be answered by issuing a two-dimensional stabbing query (v_{j+g+1}, g) on $\mathcal{R}(w)$ over all ancestors w of u_j. Notice that the number of ancestor of any node in $\overleftarrow{\mathcal{T}}$ is at most τ (which is $\log \lambda$), and each two-dimensional range stabbing query require time $O(\log d)$ plus the number of outputs. By combining all, the time for dictionary matching query on $\mathcal{S}_s$ can also be bounded by $O(|T|\gamma \log \lambda \log d + occ_s)$, where occ_s is the number of short patterns that appear as a sub-string of T.

Finally by combining both the short pattern and the long pattern cases, we obtain Theorem 1.

5   More Space-Time Trade-Offs

We present two additional space-time trade-offs in this section.

5.1   Achieving Faster Query Time

In this section we present the details of Theorem 2. Similar to the solution in Theorem 1, we have two phases (i) prefix matching phase and (ii) stabbing query phase. For Phase (i), we use the result in Lemma 1, where as for Phase (ii), we use the following structure for handling stabbing queries.

Lemma 5. *A given set $\mathcal{R}^3$ of d rectangles in an $[n]^3$ grid can be pre-processed into an $O(n + d^{1+\epsilon})$ space data structure and stabbing queries can be answered in optimal $O(1 + output)$ time.*

Proof. Using rank space reduction, each rectangle R, which is originally in an $[n]^3$ grid can be mapped to a rectangle in an $[d]^3$ grid. Now, when a 3D point p comes as a query, we can find the corresponding point p' in $[d]^3$ grid using three n-word arrays in constant time, such that the set of rectangles stabbed by p is equivalent to the set of rectangles in the rank-space reduced space stabbed by p'. Next, the rectangle stabbing query on rectangles in $[d]^3$ grid can be reduced to an equivalent range reporting problem over points in an $[d]^6$ grid. It is known that the range reporting over a set of d points in any $[d]^r$ grid, for any constant r can be answered in optimal time using $O(d^{1+\epsilon})$ words of space [23]. By combining this result with rank-space reduction, the claim follows.   ∎

Thus by combining the space time trade-offs from both phases, we obtain Theorem 2.

5.2 Achieving Succinct Space

We show how to achieve Theorem 3 in this section. We start with some notations: for any node u in $\overleftarrow{\mathcal{T}}$, let u^* be its lowest marked ancestor (i.e., a node with its path$(\cdot)$ equals to $\overleftarrow{(P_i)}$ for some $i \in [1, d]$). Similarly, for any node v in $\overrightarrow{\mathcal{T}}$, let v^* be its lowest marked ancestor (i.e., a node with its path$(\cdot)$ equals to Q_i for some $i \in [1, d]$). Using the sparsifications techniques by Hon et al. [21], both $\overleftarrow{\mathcal{T}}$ and $\overrightarrow{\mathcal{T}}$ can be replaced by an $n \log \sigma + O(d \log n) + o(n \log \sigma)$ bits structure, and whenever a text T comes as query, instead of reporting u_j and v_j for $1 \leq j \leq |T|$, we can report all u_j^*'s and v_j^*'s in $O(|T| \log^\epsilon n)$ time. Now observe that the set of marked nodes on the path to the root of $\overleftarrow{\mathcal{T}}$ from u_j and u_j^* is the same. Similarly, the set of marked nodes on the path to the root of $\overleftarrow{\mathcal{T}}$ from u_j and u_j^* is also same. Therefore, we can easily replace u_j by u_j^* and v_{j+g+1} by v_{j+g+1}^* in the solution for Problem 4.

In summary, Phase (i) can be handled using an $n \log \sigma + O(d \log n) + o(n \log \sigma)$ bits structure in $O(|T| \log^\epsilon n)$ time. Whereas Phase (ii) can be handled exactly as before, however with a new threshold $\tau = \log \lambda \log^{1+\epsilon} n$. This new threshold makes sure that the space for the structures for Phase (ii) is $o(n) + O(d \log n)$ bits. The time for long pattern case remains the same as $O(|T| \gamma \log \lambda \log d + occ_l)$, where as that of short patterns will be $O(|T| \gamma \tau \log d + occ_s)$. By combining every thing, we obtain Theorem 3.

6 Extensions and Variants

6.1 Gap Can Occur at Multiple Locations in the Pattern

First we consider a generalization where gap can be present at multiple locations in the pattern. Suppose that instead of $P_i[\alpha_i, \beta_i]Q_i$ as a one-gap-pattern (where gap location is fixed), we are given only P_i' along with parameters $[\alpha_i, \beta_i]$ and a list of positions list_i. Thus, this pattern represents a collection $\{P_i'[1..p] [\alpha_i, \beta_i] P_i'[p+1..|P_i'|] \mid p \in \text{list}_i\}$ of one-gap patterns. Then, our framework can be generalized to handle such a case, so that the total space is bounded by $O(\sum_i |P_i'|) + O(L \log L)$, where $L = \sum_i |\text{list}_i|$ denotes the total number of positions in the patterns that a gap may appear. For this, we do the following modification in our framework. Instead of indexing $\overleftarrow{P_i}$ in $\overleftarrow{\mathcal{T}}$, we use the whole string $\overleftarrow{P_i'}$ and index P_i in $\overrightarrow{\mathcal{T}}$. Now, for each position p in list_i, we create a rectangle using locus u of $\overleftarrow{P_i'[1..p]}$ in $\overleftarrow{\mathcal{T}}$ and locus v of $P_i'[p + 1..|P_i'|]$ in $\overrightarrow{\mathcal{T}}$ and the gap bounds $[\alpha_i, \beta_i]$. This increases the number of rectangle in our stabbing structure to L, giving $O(n + L \log L)$ space data structure. In the worst case, when L could be all n possible positions, this becomes $O(n \log n)$ data structure answering queries in $O(|T| \gamma \log \lambda \log n + occ)$.

6.2 Missing Substring: Induced Gap in the Text

Here, we discuss the *dictionary matching with one missing string* problem, in which a pattern may match a text substring when there is a gap of an acceptable length appearing within the text substring. In this variant, a gap of acceptable length can be created in the text, for matching puposes. More, precisely we denote missing substring pattern as $P_i[\beta_i]Q_i$, the pattern can be interpreted as P_i followed by a suffix of Q_i with anywhere between 0 to β_i beginning characters missing from Q_i.[2]

Again our framework of rectangle stabbing can be used here. For each pattern $P_i[\beta_i]Q_i$, we create $\beta_i + 1$ rectangles corresponding to loci of each valid suffix of Q_i. Now these, rectagles are 2D with no gap dimension. During the search, at each point in T we no longer have to try out all possible gaps. We simply issue one query by obtaining the stabbing point as the pair of locus of $\overleftarrow{T[...j]}$ in $\overleftarrow{T}$ and $T[j+1,...]$ in $\overrightarrow{T}$.

Since there are no more than $O(n)$ 2D rectangles, our data structures take $O(n)$ space. For query time, we get $O(|T| \log n + \mathsf{occ})$.

6.3   Missing Substring at Arbitrary Location

Here, we generalize the *dictionary matching with one missing string* problem to allow the missing substring at any arbitrary location in the pattern. Again, here we are no longer given the break up of the pattern into P_i and Q_i. We are just given a pattern P_i along with parameter k_i. Any substring of length anywhere between 0 and k_i can go missing from any location in P_i. One way to solve it would be to take P_i and create $|P_i|k_i$ number of distinct patterns, each of the form $P_i[1..j]P_i[j'+1...|P_i|]$ with $j \in [1, |P_i|]$ and $j' - j \in [0, k_i]$.

Now, when we create $\overleftarrow{T}$, we use patterns $\overleftarrow{P_i}$. Notice that every suffix of $\overleftarrow{P_i}$ finds a locus in $\overleftarrow{T}$. If seen in the reverse, it means every prefix of P_i is indexed. Similarly, when we create $\overrightarrow{T}$ we use the whole pattern P_i in forward sense. Again each suffix of P_i finds locus in $\overrightarrow{T}$. Now, for the pattern $\pi \cdot \zeta$, where π is a prefix of P_i and ζ is a suffix of P_i, let u be the node in $\overleftarrow{T}$ with $\mathsf{path}(u) = \overleftarrow{\pi}$, and v be the node in $\overrightarrow{T}$ with $\mathsf{path}(v) = \zeta$. Construct a 2D rectangle $[x, x'] \times [y, y']$, where $[x, x']$ is the preorder range of u in $\overleftarrow{T}$, and $[y, y']$ is the preorder range of v in $\overrightarrow{T}$. Finally, maintain a rectangle stabbing (point enclosure) index for this set of 2D rectangles constructed from all P_i's. The number of such rectangles is $\rho = \sum_i |P_i|k_i$.

Given this, we first find loci of each suffix of T in $\overrightarrow{T}$ and loci of the reverse of each prefix of T in $\overleftarrow{T}$. This can be done in $O(T)$ time. Then, for a particular $\ell \in [1, |T|]$, we get the locus of the reverse of $T[1..\ell]$ in $\overleftarrow{T}$, and the locus of $T[\ell+1..|T|]$ in $\overrightarrow{T}$. Let x, y respectively denote the preorder ranks of there loci. Then, all patterns that match a substring of T, with a gap inserted at position $\ell + 1$, can be found by a point enclosure query with (x, y) as input. Thus, we get $O(\rho)$ space index with query time $O(|T| \log \rho + \mathsf{occ})$.

---

[2] The lower bound α_i is redundant in this case and is set to zero. Otherwise, we can always omit first α_i characters from Q_i obtaining Q_i' and work with $P_i[\beta_i - \alpha_i]Q_i'$.

For small k_i's the space requirement ρ may be linear but in general it may be higher. We briefly sketch a space saving alternative offering trade-off. For this, note that for a particular pattern P_i all the loci corresponding to it in $\overrightarrow{T}$ appear in on a contiguous path if we consider the tree formed by following suffix links of $\overrightarrow{T}$ (i.e., the failure tree of AC automata). So now we number the nodes, in terms of preorder numbering in failure tree. For this preorder numbering, we first decompose the failure tree into centroid paths and then make sure all the vertices in a particular centroid path have contiguous numbers. Now, any pattern P_i's path only overlaps with at most $O(\log n)$ centroid paths. Thus, all the loci of suffixes of P_i form contiguous chunks of pre-order numbers, with number of chunks being at most $\log n$. Based on this idea, the number of rectangles generated by P_i can be reduced to $|P_i| + k_i \log n$. This gives us $O(n \log n)$ space. However, time goes up by a multiplicative factor $\kappa = \max_i |P_i|$. This gives us query time of $O(\kappa |T| \log n + occ)$.

7   Conclusions

We have applied Amir et al.'s framework in [6], but taking a different view to represent the dictionary, to solve a more general variant of the dictionary matching with one gap problem, thereby answering one of the open problems in [6]. We gave space-efficient and succinct space solutions for this problem. We have also proposed a new variant of the dictionary problem, in which a gap may appear in the text during a match, and showed that it can be solved using a similar framework.

Two questions are open: 1. Can we extend the techniques to handle dictionary matching with more than one gap? 2. Can we obtain succinct solutions for gap in the text variant?

References

1. Afshani, P., Arge, L., Larsen, K.G.: Higher-dimensional orthogonal range reporting and rectangle stabbing in the pointer machine model. In: Symposuim on Computational Geometry 2012, SoCG 2012, Chapel Hill, NC, USA, pp. 323–332, 17–20 June 2012
2. Aho, A.V., Corasick, M.J.: Efficient string matching: an aid to bibliographic search. Commun. ACM **18**(6), 333–340 (1975)
3. Amir, A., Farach, M.: Adaptive dictionary matching. In: 32nd Annual Symposium on Foundations of Computer Science, San Juan, Puerto Rico, pp. 760–766, 1–4 October 1991
4. Amir, A., Farach, M., Idury, R.M., Poutré, J.A.L., Schäffer, A.A.: Improved dynamic dictionary matching. Inf. Comput. **119**(2), 258–282 (1995)
5. Amir, A., Keselman, D., Landau, G.M., Lewenstein, M., Lewenstein, N., Rodeh, M.: Text indexing and dictionary matching with one error. J. Algorithms **37**(2), 309–325 (2000)
6. Amir, A., Levy, A., Porat, E., Shalom, B.R.: Dictionary matching with one gap. In: Kulikov, A.S., Kuznetsov, S.O., Pevzner, P. (eds.) CPM 2014. LNCS, vol. 8486, pp. 11–20. Springer, Heidelberg (2014)

7. Belazzougui, D.: Succinct dictionary matching with no slowdown. In: Amir, A., Parida, L. (eds.) CPM 2010. LNCS, vol. 6129, pp. 88–100. Springer, Heidelberg (2010)

8. Boyer, R.S., Moore, J.S.: A fast string searching algorithm. Commun. ACM **20**(10), 762–772 (1977)

9. Chan, T.M., Larsen, K.G., Patrascu, M.: Orthogonal range searching on the RAM, revisited. In: Proceedings of the 27th ACM Symposium on Computational Geometry, Paris, France, pp. 1–10, 13–15 June 2011

10. Chazelle, B.: Filtering search: a new approach to query-answering. SIAM J. Comput. **15**(3), 703–724 (1986)

11. Cole, R., Gottlieb, L., Lewenstein, M.: Dictionary matching and indexing with errors and don't cares. In: Proceedings of the 36th Annual ACM Symposium on Theory of Computing, Chicago, IL, USA, pp. 91–100, 13–16 June 2004

12. Feigenblat, G., Porat, E., Shiftan, A.: An improved query time for succinct dynamic dictionary matching. In: Kulikov, A.S., Kuznetsov, S.O., Pevzner, P. (eds.) CPM 2014. LNCS, vol. 8486, pp. 120–129. Springer, Heidelberg (2014)

13. Ferragina, P., Manzini, G.: Indexing compressed text. J. ACM **52**(4), 552–581 (2005)

14. Fredriksson, K., Grabowski, S.: Efficient algorithms for pattern matching with general gaps, character classes, and transposition invariance. Inf. Retr. **11**(4), 335–357 (2008)

15. Grossi, R., Vitter, J.S.: Compressed suffix arrays and suffix trees with applications to text indexing and string matching. SIAM J. Comput. **35**(2), 378–407 (2005)

16. Hofmann, K., Bucher, P., Falquet, L., Bairoch, A.: The PROSITE database, its status in 1999. Nucleic Acids Res. **27**(1), 215–219 (1999)

17. Hon, W., Ku, T., Shah, R., Thankachan, S.V., Vitter, J.S.: Compressed dictionary matching with one error. In: 2011 Data Compression Conference (DCC 2011), Snowbird, UT, USA, pp. 113–122, 29–31 March 2011

18. Hon, W., Ku, T., Shah, R., Thankachan, S.V., Vitter, J.S.: Faster compressed dictionary matching. Theor. Comput. Sci. **475**, 113–119 (2013)

19. Hon, W., Lam, T.W., Shah, R., Tam, S., Vitter, J.S.: Compressed index for dictionary matching. In: 2008 Data Compression Conference (DCC 2008), Snowbird, UT, USA, pp. 23–32, 25–27 March 2008

20. Hon, W.-K., Lam, T.-W., Shah, R., Tam, S.-L., Vitter, J.S.: Succinct index for dynamic dictionary matching. In: Dong, Y., Du, D.-Z., Ibarra, O. (eds.) ISAAC 2009. LNCS, vol. 5878, pp. 1034–1043. Springer, Heidelberg (2009)

21. Hon, W.-K., Ku, T.-H., Lam, T.-W., Shah, R., Tam, S.-L., Thankachan, S.V., Vitter, J.S.: Compressing dictionary matching index via sparsification technique. Algorithmica **72**(2), 515–538 (2015)

22. Karp, R.M., Rabin, M.O.: Efficient randomized pattern-matching algorithms. IBM J. Res. Dev. **31**(2), 249–260 (1987)

23. Karpinski, M., Nekrich, Y.: Space efficient multi-dimensional range reporting. In: Ngo, H.Q. (ed.) COCOON 2009. LNCS, vol. 5609, pp. 215–224. Springer, Heidelberg (2009)

24. Knuth, D.E., Morris Jr, J.H., Pratt, V.R.: Fast pattern matching in strings. SIAM J. Comput. **6**(2), 323–350 (1977)

25. Lewenstein, M.: Dictionary matching. In: Kao, M.-Y. (ed.) Encyclopedia of Algorithms, pp. 1–6. Springer, US (2015)

26. Manber, U., Myers, E.W.: Suffix arrays: a new method for on-line string searches. SIAM J. Comput. **22**(5), 935–948 (1993)

27. Mehldau, G., Myers, G.: A system for pattern matching applications on biosequences. Comput. Appl. Biosci. **9**(3), 299–314 (1993)
28. Navarro, G., Raffinot, M.: Fast and simple character classes and bounded gaps pattern matching, with applications to protein searching. J. Comput. Biol. **10**(6), 903–923 (2003)
29. Sleator, D.D., Tarjan, R.E.: A data structure for dynamic trees. J. Comput. Syst. Sci. **26**(3), 362–391 (1983)
30. Weiner, P.: Linear pattern matching algorithms. In: 14th Annual Symposium on Switching and Automata Theory, Iowa City, Iowa, USA, pp. 1–11, 15–17 October 1973

Partition into Heapable Sequences, Heap Tableaux and a Multiset Extension of Hammersley's Process

Gabriel Istrate[1,2(✉)] and Cosmin Bonchiş[1,2]

[1] Department of Computer Science, West University of Timişoara,
Timişoara, Romania
[2] e-Austria Research Institute, Bd. V. Pârvan 4, cam. 045 B,
300223 Timişoara, Romania
gabrielistrate@acm.org

Abstract. We investigate partitioning of integer sequences into heapable subsequences (previously defined and established by Byers et al.). We show that an extension of patience sorting computes the decomposition into a minimal number of heapable subsequences (MHS). We connect this parameter to an interactive particle system, a multiset extension of Hammersley's process, and investigate its expected value on a random permutation. In contrast with the (well studied) case of the longest increasing subsequence, we bring experimental evidence that the correct asymptotic scaling is $\frac{1+\sqrt{5}}{2} \cdot \ln(n)$. Finally we give a heap-based extension of Young tableaux, prove a hook inequality and an extension of the Robinson-Schensted correspondence.

Keywords: Heapable sequences · Hammersley process · Heap tableaux

1 Introduction

Patience sorting [16] and *the longest increasing (LIS) sequence* are well-studied topics in combinatorics. The analysis of the expected length of the LIS of a random permutation is a classical problem displaying interesting connections with the theory of interacting particle systems [2] and that of combinatorial Hopf algebras [9]. Recursive versions of patience sorting are involved (under the name of *Schensted procedure* [19]) in the theory of Young tableaux. A wonderful recent reference for the rich theory of the longest increasing sequences (and substantially more) is [18].

Recently Byers et al. [4] introduced, under the name of *heapable sequence*, an interesting variation on the concept of increasing sequences. Informally, a sequence of integers is heapable if it can be successively inserted into a (not necessarily complete) binary tree satisfying the heap property without having to resort to node rearrangements. Byers et al. showed that the longest heapable subsequence in a random permutation grows linearly (rather than asymptotically

© Springer International Publishing Switzerland 2015
F. Cicalese et al. (Eds.): CPM 2015, LNCS 9133, pp. 261–271, 2015.
DOI: 10.1007/978-3-319-19929-0_22

equal to $2\sqrt{n}$ as does LIS) and raised as an open question the issue of extending the rich theory of LIS to the case of heapable sequences.

In this paper we partly answer this open question: we define a family $MHS_k(X)$ of measures (based on decomposing the sequence into subsequences heapable into a min-heap of arity at most k) and show that a variant of patience sorting correctly computes the values of these parameters. We show that this family of measures forms an infinite hierarchy, and investigate the expected value of parameter $MHS_2[\pi]$, where π is a random permutation of order n. Unlike the case $k = 1$ where $E[MHS_1[\pi]] = E[LDS[\pi]] \sim 2\sqrt{n}$, we argue that in the case $k \geq 2$ the correct scaling is logarithmic, bringing experimental evidence that the precise scaling is $E[MHS_2[\pi]] \sim \phi \ln n$, where $\phi = \frac{1+\sqrt{5}}{2}$ is the golden ratio. The analysis exploits the connection with a new, multiset extension of the Hammersley-Aldous-Diaconis process [1], an extension that may be of independent interest. Finally, we introduce a heap-based generalization of Young tableaux. We prove (Theorem 6 below) a hook inequality related to the hook formula for Young tableaux [7] and Knuth's hook formula for heap-ordered trees [13], and (Theorem 8) an extension of the Robinson-Schensted (R-S) correspondence.

2   Preliminaries

For $k \geq 1$ define alphabet $\Sigma_k = \{1, 2, \ldots, k\}$. Define as well $\Sigma_\infty = \cup_{k \geq 1} \Sigma_k$. Given words x, y over Σ_∞ we will denote by $x \sqsubseteq y$ the fact that x is a prefix of y. The set of (non-strict) prefixes of x will be denoted by $Pref(x)$. Given words $x, y \in \Sigma_\infty^*$ define the *prefix partial order* $x \preceq_{ppo} y$ as follows: If $x \sqsubseteq y$ then $x \preceq_{ppo} y$. If $x = za$, $y = zb$, $a, b \in \Sigma_\infty$ and $a < b$ then $x \preceq_{ppo} y$. $\preceq_{ppo}$ is the transitive closure of these two constraints. Similarly, the *lexicographic partial order* $\preceq_{lex}$ is defined as follows: If $x \sqsubseteq y$ then $x \preceq_{lex} y$. If $x = za$, $y = zb$, $a, b \in \Sigma_\infty$ and $a < b$ then $x \preceq_{lex} y$. $\preceq_{lex}$ is the transitive closure of these two constraints.

A *k-ary tree* is a finite, $\preceq_{ppo}$-closed set T of words over alphabet $\Sigma_k = \{1, 2, \ldots, k\}$. That is, we impose the condition that positions on the same level in a tree are filled preferentially from left to right. The *position $pos(x)$ of node x in a k-ary tree* is the string over alphabet $\{1, 2, \ldots, k\}$ encoding the path from the root to the node (e.g. the root has position λ, its children have positions $1, 2, \ldots, k$, and so on). A *k-ary (min)-heap* is a function $f : T \to \mathbf{N}$ monotone with respect to pos, i.e. $(\forall x, y \in T), [pos(x) \sqsubseteq pos(y)] \Rightarrow [f(x) \leq f(y)]$.

A *(binary min-)heap* is a binary tree, not necessarily complete, such that $A[parent[x]] \leq A[x]$ for every non-root node x. If instead of binary we require the tree to be k-ary we get the concept of k-ary min-heap.

A sequence $X = X_0, \ldots, X_{n-1}$ is *k-heapable* if there exists some k-ary tree T whose nodes are labeled with (exactly one of) the elements of X, such that for every non-root node X_i and parent X_j, $X_j \leq X_i$ and $j < i$. In particular a 2-heapable sequence will simply be called *heapable* [4]. Given sequence of integer

- A number of individuals appear (at integer times $i \geq 1$) as random numbers X_i, uniformly distributed in the interval $[0, 1]$.
- Each individual is initially endowed with k "lifelines".
- The appearance of a new individual X_{t+1} subtracts a life from the largest individual $X_a < X_{t+1}$ (if any) still alive at moment t.

Fig. 1. HAD_k, the multiset Hammersley process with k lifelines.

numbers X, denote by $MHS_k(X)$ the smallest number of heapable (not necessarily contiguous) subsequences one can decompose X into. $MHS_1(X)$ is equal [14] to the shuffled up-sequences (SUS) measure in the theory of presortedness.

Example 1. Let $X = [2, 4, 3, 1]$. Via patience sorting $MHS_1(X)$ =SUS$(X) = 3$. $MHS_2(X) = 2$, since subsequences $[2, 3, 4]$ and $[1]$ are 2-heapable. On the other hand, for every $k \geq 1$, $MHS_k([k, k-1, \ldots, 1]) = k$.

Analyzing the behavior of LIS relies on the correspondence between longest increasing sequences and an interactive particle system [1] called the Hammersley-Aldous-Diaconis (shortly, Hammersley or HAD) process. We give it the multiset generalization displayed in Fig. 1. Technically, to recover the usual definition of Hammersley's process one should take $X_a > X_{t+1}$ (rather than $X_a < X_{t+1}$). This small difference arises since we want to capture $MHS_k(\pi)$, which generalizes $LDS(\pi)$, rather than $LIS(\pi)$ (captured by Hammersley's process). This slight difference is, of course, inconsequential: our definition is simply the flipped around the midpoint of segment [0,1] version of such a generalization, and has similar behavior).

3   A Greedy Approach to Computing MHS_k

First we show that one can combine patience sorting and the greedy approach in [4] to obtain an algorithm for computing $MHS_k(X)$. To do so, we must adapt to our purposes some notation in that paper.

A binary tree with n nodes has $n + 1$ positions (that will be called slots) where one can add a new number. We will identify a slot with the minimal value of a number that can be added to that location. For heap-ordered trees it is the value of the parent node. Slots easily generalize to forests. The number of slots of a forest with d trees and n nodes is $n + d$.

Given a binary heap forest T, the signature of T denoted $sig(T)$, is the vector of the (values of) free slots in T, in sorted (non-decreasing) order. Given two binary heap forests T_1, T_2, T_1 dominates T_2 if $|sig_{T_1}| \leq |sig_{T_2}|$ and inequality $sig_{T_1}[i] \leq sig_{T_2}[i]$ holds for all $1 \leq i \leq |sig_{T_1}|$.

Theorem 1. For every fixed $k \geq 1$ there is a polynomial time algorithm that, given sequence $X = (X_0, \ldots, X_{n-1})$ as input, computes $MHS_k(X)$.

Proof. We use the greedy approach of Algorithm 3.1.

Algorithm 3.1: GREEDY(W)

INPUT $W = (w_1, w_2, \ldots, w_n)$ a list of integers.
Start with empty heap forest $T = \emptyset$.
for i **in range(n):**
    if (there exists a slot where X_i can be inserted):
            insert X_i in the slot with the lowest value
    else :
            start a new heap consisting of X_i only.

Proving correctness of the algorithm employs the following
Lemma 1. *Let T_1, T_2 be two heap forests such that T_1 dominates T_2. Insert a new element x in both T_1 and T_2: greedily in T_1 (i.e. at the largest slot with value less or equal to x, or as the root of a new tree, if no such slot exists) and arbitrarily in T_2, obtaining forests T_1', T_2', respectively. Then T_1' dominates T_2'.*

Proof. First note that, by domination, if no slot of T_1 can accomodate x (which, thus, starts a new tree) then a similar property is true in T_2 (and thus x starts a new tree in T_2 as well).

□

Let $sig_{T_1} = (a_1, a_2, \ldots)$ and $sig_{T_2} = (b_1, b_2, \ldots)$ be the two signatures. By domination $a_i \leq b_i$ for all i. The process of inserting x can be described as adding two copies of x to the signature of $T_1(T_2)$ and (perhaps) removing a label $\leq x$ from the two signatures. The removed label is a_i, the largest label $\leq x$, in the case of greedy insertion into T_1. Let b_j be the largest value (or possibly none) in

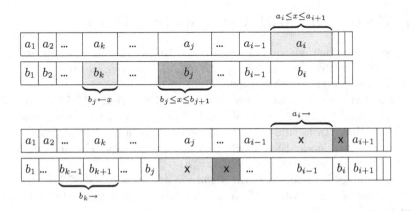

Fig. 2. The argument of Lemma 1. Pictured vectors (both initial and resulting) have equal lengths (which may not always be the case).

T_2 less or equal to x. Some b_k less or equal to b_j is replaced by two copies of x in T_2. The following are true:

- The length of $sig_{T_1'}$ is at most that of $sig_{T_2'}$.
- The element b_k (if any) deleted by x from T_2 satisfies $b_k \leq x$. Its index in T_2 is less or equal to i.
- The two x's are inserted to the left of the deleted (if any) positions in both T_1 and T_2.

Consider some position l in $sig_{T_1'}$. Our goal is to show that $a_l' \leq b_l'$. Several cases are possible (Fig. 2):

- $l < k$. Then $a_l' = a_l$ and $b_l' = b_l$.
- $k \leq l < j$. Then $a_l' = a_l$ and $b_l' = b_{l+1} \geq a_l$.
- $j \leq l \leq i + k - 1$. Then $a_l' \leq x$ and $b_l' \geq x$.
- $l > i + k - 1$. Then $a_l' = a_{l-k+1}$ and $b_l' = b_{l-k+1}$.

Let X be a sequence of integers, OPT be an optimal partition of X into k-heapable sequences and Γ be the solution produced by GREEDY. Applying Lemma 1 repeatedly we infer that whenever GREEDY adds a new heap the same thing happens in OPT. Thus the number of heaps created by Greedy is optimal, which means that the algorithm computes $MHS_k(X)$.   □

Trivially $MHS_k(X) \leq MHS_{k-1}(X)$. On the other hand

Theorem 2. *The following statements (proved in the full version [12]) are true for every $k \geq 2$: (a). there exists a sequence X such that $MHS_k(X) < MHS_{k-1}(X) < \ldots < MHS_1(X)$; (b). $\sup_X [MHS_{k-1}(X) - MHS_k(X)] = \infty$.*

4   The Connection with the Multiset Hammersley Process

Denote by $MinHAD_k(n)$ the random variable denoting *the number of times i in the evolution of process HAD_k up to time n when the newly inserted particle X_i has lower value than all the existing particles at time i*. The observation from [1,8] generalizes to:

Theorem 3. *For every fixed $k, n \geq 1$ $E_{\pi \in S_n}[MHS_k(\pi)] = E[MinHAD_k(n)]$.*

Proof Sketch. W.h.p. all X_i's are different. We will thus ignore in the sequel the opposite alternative. Informally minima correspond to new heaps and live particles to slots in these heaps (cf. also Lemma 1).   □

5   The Asymptotic Behavior of $E[MHS_2[\pi]]$

The asymptotic behavior of $E[MHS_1[\pi]]$ where π is a random permutation in S_n is a classical problem in probability theory: results in [1,8,15,22] show that it is asymptotically equal to $2\sqrt{n}$.

A simple lower bound valid for all values of $k \geq 1$ is

Theorem 4. *For every fixed* $k, n \geq 1$

$$E_{\pi \in S_n}[MHS_k(\pi)] \geq H_n, \text{ the } n'\text{th harmonic number.} \tag{1}$$

Proof. For $\pi \in S_n$ the set of its *minima* is defined as $Min(\pi) = \{j \in [n] : \pi[j] < \pi[i] \text{ for all } 1 \leq i < j\}$ (and similarly for maxima). It is easy to see that $MHS_k[\pi] \geq |Min[\pi]|$. Indeed, every minimum of π must determine the starting of a new heap, no matter what k is. Now we use the well-known formula $E_{\pi \in S_n}[|Min[\pi]|] = E_{\pi \in S_n}[|Max[\pi]|] = H_n$ [13]. □

To gain insight in the behavior of process HAD_2 we note that, rather than giving the precise values of $X_0, X_1, \ldots, X_t \in [0, 1]$, an equivalent random model inserts X_t uniformly at random in any of the $t+1$ possible positions determined by $X_0, X_1, \ldots, X_{t-1}$. This model translates into the following equivalent combinatorial description of HAD_k: word w_t over the alphabet $\{-1, 0, 1, 2\}$ describes the state of the process at time t. Each w_t conventionally starts with a -1 and continues with a sequence of 0, 1's and 2's, informally the "number of lifelines" of particles at time t. For instance $w_0 = 0$, $w_1 = 02$, w_2 is either 022 or 012, depending on $X_0 <> X_1$, and so on. At each time t a random letter of w_t is chosen (corresponding to a position for X_t) and we apply one of the following transformations, the appropriate one for the chosen position:

- *Replacing* -10^r *by* -10^r2: This is the case when X_t is the smallest particle still alive, and to its right there are $r \geq 0$ dead particles.
- *Replacing* 10^r *by* $0^{r+1}2$: Suppose that X_a is the largest live label less or equal to X_t, that the corresponding particle X_a has one lifetime at time t, and that there are r dead particles between X_a and X_t. Adding X_t (with multiplicity two) decreases multiplicity of X_a to 0.
- *Replacing* 20^r *by* 10^r2: Suppose that X_a is the largest label less or equal to X_t, its multiplicity is two, and there are $r \geq 0$ dead particles between X_a and X_t. Adding X_t removes one lifeline from particle X_a.

Simulating the (combinatorial version of the) Hammersley process with two lifelines confirms the fact that $E[MHS_2(\pi)]$ grows significantly slower than $E[MHS_1(\pi)]$: The x-axis in the figure is logarithmic. The scaling is clearly different, and is consistent (see inset) with logarithmic growth (displayed as a straight line on a plot with log-scaling on the x-axis). Experimental results (see the inset/caption of Fig. 3) suggest the following bold

Conjecture 1. We have $\lim_{n \to \infty} \frac{E[MHS_2[\pi]]}{\ln(n)} = \phi$, with $\phi = \frac{1+\sqrt{5}}{2}$ the golden ratio. More generally, for an arbitrary $k \geq 2$ the relevant scaling is

$$\lim_{n \to \infty} \frac{E[MHS_k[\pi]]}{\ln(n)} = \frac{1}{\phi_k}, \tag{2}$$

where ϕ_k is the unique root in $(0, 1)$ of equation $X^k + X^{k-1} + \ldots + X = 1$.

We plan to present the experimental evidence for the truth of equation (2) and a nonrigorous, "physics-like" justification, together with further insights on the so-called *hydrodynamic behavior* [10] of the HAD_k process in subsequent work [11]. For now we limit ourselves to showing that one can (rigorously) perform a first step in the analysis of the HAD_2 process: we prove convergence of (some of) its structural characteristics. This will likely be useful in a full rigorous proof of Conjecture 1.

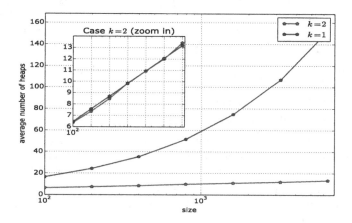

Fig. 3. Scaling of expected value of $MHS_k[\pi]$ for $k = 1, 2$. The inset shows $E[MHS_2[\pi]]$ (red) versus $\phi \cdot \ln(n) + 1$ (blue). The fit is strikingly accurate (Color figure online).

Denote by L_t the number of digits 1+2, and by C_t the number of ones in w_t. Let $l(t) = E[\frac{L(t)}{t}]$, $c(t) = E[\frac{C(t)}{t}]$. $l(t), c(t)$ always belong to $[0, 1]$.

Theorem 5. *There exist constants $l, c \in [0, 1]$ such that $l(t) \to l$, $c(t) \to c$.*

Proof Sketch. We use a standard tool, *subadditivity*: if sequence a_n satisfies $a_{m+n} \leq a_m + a_n$ for all $m, n \geq 1$ then (by Fekete's Lemma ([21] pp. 3, [20]) $\lim_{n \to \infty} a_n/n$ exists. We show in the full version [12] that this is the case for two independent linear combinations of $l(t)$ and $c(t)$.   □

Experimentally (and nonrigorously) $l = \phi - 1 = \frac{\sqrt{5}-1}{2}$ and $c = \frac{3-\sqrt{5}}{2}$. "Physics-like" nonrigorous arguments then imply the desired scaling. An additional ingredient is that digits 0/1/2 are uniformly distributed (conditional on their density) in a large w_t. This is intuitively true since for large t the behavior of the HAD_k process is described by a compound Poisson process. We defer more complete explanations to [11].

6   Heap Tableaux, a Hook Inequality and a Generalization of the Robinson-Schensted Correspondence

Finally, we present an extension of Young diagrams to heap-based tableaux. All proofs are given in the full version [12]. A *(k-)heap tableau T* is *k-ary* min-heap of integer vectors, so that for every $r \in \Sigma_k^*$, the vector V_r at address r is nondecreasing. We formally represent the tableau as a function $T : \Sigma_k^* \times \mathbf{N} \to \mathbf{N} \cup \{\bot\}$ such that (a). T has *finite support:* the set $dom(T) = \{(r, a) : T(r, a) \neq \bot\}$ of nonempty positions is finite. (b). T is $\sqsubseteq$-*nondecreasing:* if $T(r, a) \neq \bot$ and $q \sqsubset r$ then $T(q, a) \neq \bot$ and $T(q, a) \leq T(r, a)$. In other words, $T(\cdot, a)$ is a min-heap. (c). T is *columnwise increasing:* if $T(r, a) \neq \bot$ and $b < a$ then $T(r, b) \neq \bot$ and $T(r, b) < T(r, a)$. That is, each column V_r is increasing. The *shape of* T is the heap $S(T)$ where node with address r holds value $|V_r|$.

A tableau is *standard* if (e). for all $1 \leq i \leq n = |dom(T)|$, $|T^{-1}(i)| = 1$ and (f). If $x \leq_{lex} y$ and $T(y, 1) \neq \bot$ then $\bot \neq T(x, 1) \leq T(y, 1)$. I.e., labels in the first heap H_1 are increasing from left to right and top to bottom.

Example 2. A heap tableau T_1 with 9 elements is presented in Fig. 4(a) and as a Young-like diagram in Fig. 4(b). Note that: (i). Columns correspond to *rows* of T_1 (ii). Their labels are in Σ_2^*, rather than $\mathbf{N}$. (iii). Cells may contain $\bot$. (iv). Rows need not be increasing, only *min-heap ordered.*

One important drawback of our notion of heap tableaux above is that they do not reflect the evolution of the process HAD_k the way ordinary Young tableaux do (on their first line) for process HAD_1 via the Schensted procedure [19]: A generalization with this feature would seem to require that each cell contains not an integer but a *multiset* of integers. Obtaining such a notion of tableau is part of ongoing research.

However, we can motivate our definition of heap tableau by the first application below, a hook inequality for such tableaux. To explain it, note that heap tableaux generalize both heap-ordered trees and Young tableaux. In both cases there exist hook formulas that count the number of ways to fill in a structure with n cells by numbers from 1 to n: [7] for Young tableaux and [13] (Sect. 5.1.4, Ex. 20) for heap-ordered trees. It is natural to wonder whether there exists a hook formula for heap tableaux that provides a common generalization of both these results.

Theorem 6 gives a partial answer: not a formula but a *lower bound.* To state it, given $(\alpha, i) \in dom(T)$, define the *hook length* $H_{\alpha,i}$ to be the cardinal of set $\{(\beta, j) \in dom(T) : [(j = i) \wedge (\alpha \sqsubseteq \beta)] \vee [(j \geq i) \wedge (\alpha = \beta)]\}$. For example, Fig. 4(c). displays the hook lengths of cells in T_1.

Theorem 6. *Given $k \geq 2$ and a k-shape S with n free cells, the number of ways to create a heap tableau T with shape S by filling its cells with numbers $\{1, 2, \ldots, n\}$ is at least $\frac{n!}{\prod_{(\alpha,i) \in dom(T)} H_{\alpha,i}}$. The bound is tight for Young tableaux [7], heap-ordered trees [13], and infinitely many other examples, but is also* **not** *tight for infinitely many (counter)examples.*

We leave open the issue whether one can tighten up the lower bound above to a formula by modifying the definition of the hook length $H_{\alpha,i}$.

We can create k-heap tableaux from integer sequences by a version of the Schensted procedure [19]. Algorithm Schensted-HEAP$_k$ below performs *column insertions* and gives to any bumped element k choices for insertion/bumping, the children of vector V_r, with addresses $r \cdot \Sigma_k$.

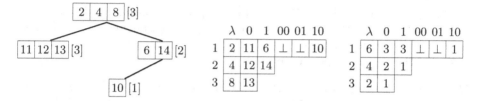

Fig. 4. (a). Heap tableau T_1 and its shape $S(T_1)$ (in brackets) (b). The equivalent Young tableau-like representation of T_1 and (c). The hook lengths.

Theorem 7. *The result of applying the Schensted-HEAP$_k$ procedure to an arbitrary permutation X is indeed a k-ary heap tableau.*

Example 3. Suppose we start with T_1 from Fig. 4(a). Then (Fig. 5) 9 is appended to vector V_λ. 7 arrives, bumping 8, which in turn bumps 11. Finally 11 starts a new vector at position 00. Modified cells are grayed.

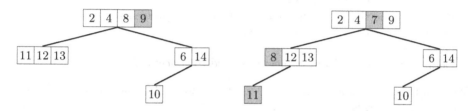

Fig. 5. Inserting 9 and 7 into T_1.

Algorithm 6.1: SCHENSTED-HEAP$_k(X = x_0, \ldots, x_{n-1})$

for i in **range**$(n) : BUMP(x_i, \lambda)$

PROCEDURE BUMP$(x, S) :$ #*S is a set of adresses.*
  - Attempt to append x to some $V_r, r \in S$ (perhaps creating it)
    (choose the first r where appending x keeps V_r increasing).
 if (this is not possible for **any** vector $V_r, r \in S$) :
    - Let B_x be the set of elements of value $> x$,
      in all vectors $V_r, r \in S$ (clearly $B_x \neq \emptyset$)
    - Let $y = min\{B_x\}$ and r the address of its vector.
    - Replace y by x into V_r
    - $BUMP(y, r \cdot \Sigma_k)$ #*bump y into some child of r*

Procedure Schensted-HEAP$_k$ does **not** help in computing the longest heapable subsequence: The complexity of computing this parameter is open [4], and we make no progress on this issue. On the other hand, we can give a $k \geq 2$ version of the R-S correspondence:

Theorem 8. *For every $k \geq 2$ there exists a bijection between permutations $\pi \in S_n$ and pairs (P, Q) of k-heap tableaux with n elements and identical shape, where Q is a standard tableau.*

Condition "Q is standard" is specific to case $k \geq 2$ where heaps simply have "too many degrees of freedom" between siblings. Schensted-HEAP$_k$ solves this problem by starting new vectors from left to right and top to bottom.

7   Conclusion and Acknowledgments

Our paper raises a large number of open issues. We briefly list a few of them:

- Rigorously justify Conjecture 1.
- Study process HAD_k and its variants [6,17].
- Reconnect the theory to the analysis of *secretary problems* [3,5].
- Determine the distribution of $MHS_k[\pi]$.
- Obtain a hook formula for heap tableaux.
- Define a version of Young tableaux related to process HAD_k.

We plan to address some of these in subsequent work. The most important open problem, however, is *the complexity of computing LHS*.

Acknowledgment. This research has been supported by CNCS IDEI Grant PN-II-ID-PCE-2011-3-0981 "Structure and computational difficulty in combinatorial optimization: an interdisciplinary approach".

References

1. Aldous, D., Diaconis, P.: Hammersley's interacting particle process and longest increasing subsequences. Probab. Theory Relat. Fields **103**(2), 199–213 (1995)
2. Aldous, D., Diaconis, P.: Longest increasing subsequences: from patience sorting to the Baik-Deift-Johansson theorem. Bull. Am. Math. Soc. **36**(4), 413–432 (1999)
3. Archibald, M., Martínez, C.: The hiring problem and permutations. DMTCS Proc. **01**, 63–76 (2009)
4. Byers, J., Heeringa, B., Mitzenmacher, M., Zervas, G.: Heapable sequences and subseqeuences. In: Proceedings of ANALCO, pp. 33–44 (2011)
5. Broder, A., Kirsch, A., Kumar, R., Mitzenmacher, M., Upfal, E., Vassilvitskii, S.: The hiring problem and Lake Wobegon strategies. SIAM J. Comput. **39**(4), 1233–1255 (2009)
6. Cator, E., Groeneboom, P.: Hammersley's process with sources and sinks. Ann. Probab. **33**(3), 879–903 (2005)

7. Frame, J.S., Robinson, G., Thrall, R.M.: The hook graphs of the symmetric group. Canad. J. Math. **6**(316), C324 (1954)
8. Hammersley, J.M.: A few seedlings of research. In: Proceedings of the Sixth Berkeley Symposium on Mathematical Statistics and Probability. Theory of Statistics, vol. 1 (1972)
9. Hivert, F.: An Introduction to Combinatorial Hopf Algebras Physics and Theoretical Computer Science: From Numbers and Languages to (Quantum) Cryptography Security. IOS Press (2007)
10. Groenenboom, P.: Hydrodynamical models for analyzing longest increasing sequences. J. Comput. Appl. Math. **142**, 83–105 (2002)
11. Istrate, G., Bonchiş, C.: Hammersley's process with multiple lifelines, and a conjecture on decomposing permutations into heapable sequences (manuscript in preparation, 2015)
12. Istrate, G., Bonchiş, C.: Partition into heapable sequences, heap tableaux and a multiset extension of Hammersley's process. http://tcs.ieat.ro/recent-papers/
13. Knuth, D.: The Art of Computer Programming. Sorting and Searching, vol. 3. Addison Wesley, Redwood City (1998)
14. Levcopoulos, C., Petersson, O.: Sorting shuffled monotone sequences. Inf. Comput. **112**(1), 37–50 (1994)
15. Logan, B.F., Shepp, L.A.: A variational problem for random Young tableaux. Adv. Math. **26**(2), 206–222 (1977)
16. Mallows, C.L.: Patience sorting. SIAM Rev. **5**(4), 375–376 (1963)
17. Montoya, L.: A rapidly mixing stochastic system of finite interacting particles on the circle. Stoch. Process. Appl. **67**(1), 69–99 (1997)
18. Romik, D.: The Surprising Mathematics of Longest Increasing Sequences. Cambridge University Press, Cambridge (2014)
19. Schensted, C.: Longest increasing and decreasing subsequences. Canad. J. Math. **13**(2), 179–191 (1961)
20. Szpankowski, W.: Average Case Analysis of Algorithms on Sequences. Wiley, New York (2001)
21. Steele, J.M.: Probability Theory and Combinatorial Optimization. SIAM, Philadelphia (1997)
22. Vershik, A., Kerov, S.: Asymptotics of Plancherel measure of symmetric group and limit form of Young tables. Dokl. Akad. Nauk SSSR **233**(6), 1024–1027 (1977)

The Approximability of Maximum Rooted Triplets Consistency with Fan Triplets and Forbidden Triplets

Jesper Jansson[1][✉], Andrzej Lingas[2], and Eva-Marta Lundell[2]

[1] Laboratory of Mathematical Bioinformatics, Institute for Chemical Research, Kyoto University, Gokasho, Uji, Kyoto 611-0011, Japan
jj@kuicr.kyoto-u.ac.jp
[2] Department of Computer Science, Lund University, Box 118, 221 00 Lund, Sweden
{Andrzej.Lingas,Eva-Marta.Lundell}@cs.lth.se

Abstract. The *maximum rooted resolved triplets consistency problem* takes as input a set $\mathcal{R}$ of resolved triplets and asks for a rooted phylogenetic tree that is consistent with the maximum number of elements in $\mathcal{R}$. This paper studies the polynomial-time approximability of a generalization of the problem where in addition to resolved triplets, the input may contain fan triplets and forbidden triplets. To begin with, we observe that the generalized problem admits a 1/4-approximation in polynomial time. Next, we present a polynomial-time approximation scheme (PTAS) for dense instances based on smooth polynomial integer programming. Finally, we generalize Wu's exact exponential-time algorithm in [19] for the original problem to also allow fan triplets, forbidden resolved triplets, and forbidden fan triplets. Forcing the algorithm to always output a k-ary phylogenetic tree for any specified $k \geq 2$ then leads to an exponential-time approximation scheme (ETAS) for the generalized, unrestricted problem.

Keywords: Bioinformatics · Approximation algorithms · Phylogenetic tree · Rooted triplet · Smooth integer program

1 Introduction

Phylogenetic trees are used by scientists to describe treelike evolutionary history for various kinds of objects such as biological species, natural languages, manuscripts, *etc.* [7]. Inferring an accurate phylogenetic tree from experimental data can be a difficult task; for example, computationally expensive methods like maximum likelihood that are known to yield good trees may be impractical for large data sets [6]. One potential remedy is the divide-and-conquer approach: first apply some expensive method to obtain a collection of highly reliable trees for small, overlapping subsets of the leaf labels, and then use a computationally cheaper method to merge these trees into a phylogenetic supertree [6,10,15].

JJ was funded by The Hakubi Project and KAKENHI grant number 26330014.

F. Cicalese et al. (Eds.): CPM 2015, LNCS 9133, pp. 272–283, 2015.
DOI: 10.1007/978-3-319-19929-0_23

A concept that captures the combinatorial aspects of the smallest meaningful building blocks of a phylogenetic supertree in the rooted case is *rooted triplets consistency*. Given a set $\mathcal{R}$ of possibly contradicting rooted phylogenetic trees with exactly three leaves each (so-called *rooted triplets*), the *maximum rooted triplets consistency problem* asks for a tree that contains as many of the rooted triplets in $\mathcal{R}$ as possible as embedded subtrees. Most previous work on the topic (e.g., [1,4,5,8,16,18,19]) has focused on the case where all the given rooted triplets are *resolved triplets*, meaning that they are binary. This paper considers a more general problem variant where $\mathcal{R}$ may also contain non-binary triplets (called *fan triplets*) that should preferably be included in the output tree as well as *forbidden triplets* that should be avoided.

1.1 Definitions

A (rooted) *phylogenetic tree* is a rooted, unordered tree with no internal nodes of degree 1 and whose leaves are distinctly labeled. To simplify the presentation, we identify each leaf in a phylogenetic tree with the unique element that labels it. The set of all leaf labels in a phylogenetic tree T is denoted by $\Lambda(T)$. For any $x, y \in \Lambda(T)$, $lca^T(x, y)$ is the lowest common ancestor in T of x and y.

Suppose that T is a phylogenetic tree. For any $x, y, z \in \Lambda(T)$, define the following four types of constraints on T:

1. $xy|z$, specifying that $lca^T(x, y)$ should be a proper descendant of $lca^T(x, z)$ (or equivalently, that $lca^T(x, y)$ should be a proper descendant of $lca^T(y, z)$).
2. $x|y|z$, specifying that $lca^T(x, y) = lca^T(x, z) = lca^T(y, z)$ should hold.
3. $\neg xy|z$, specifying that $lca^T(x, y)$ should not be a not proper descendant of $lca^T(x, z)$ (or equivalently, that $lca^T(x, y)$ should not be a proper descendant of $lca^T(y, z)$).
4. $\neg x|y|z$, specifying that the same node should not be the lowest common ancestor of a and b for all pairs $a, b \in \{x, y, z\}$.

The *maximum rooted triplets consistency problem* (MTC) is: given a set S of leaf labels and a set $\mathcal{R}$ of constraints as defined above, output a phylogenetic tree T with $\Lambda(T) = S$ that satisfies as many constraints from $\mathcal{R}$ as possible. In this paper, the special case of MTC where all constraints in $\mathcal{R}$ are of type 1 is called the *maximum rooted resolved triplets consistency problem* (MRTC), and the special case where all constraints are of type 1 or type 3 is called the *maximum mixed rooted resolved triplets consistency problem* (MMRTC).[1]

To express the size of an instance of MTC, we write $n = |S|$ and $m = |\mathcal{R}|$. An instance $(S, \mathcal{R})$ of MTC is *complete* if, for every $S' \subseteq S$ with $|S'| = 3$, $\mathcal{R}$ contains at least one constraint involving the three elements in S' only. It is called *dense* if it contains $\Omega(n^3)$ constraints. Note that any complete instance is dense.

---

[1] MRTC is called MAX-LEVEL-0 in [4], MAXRTC in [5], MILCT in [8,12], MAXCL-0-DENSE in [11], MTC in [16], and MCTT in [18,19]. MMRTC is called MMTT in [9].

Remark 1. Phylogenetic trees with exactly three leaves are commonly referred to as *rooted triplets* in the literature. A rooted triplet t is either a binary or a non-binary tree. In the former case, t is a *resolved triplet* and always satisfies a constraint of type 1, and if this constraint is also satisfied in a phylogenetic tree T then t and T are said to be *consistent*. Similarly, if t is non-binary then t is called a *fan triplet* and always satisfies a constraint of type 2; if it is also satisfied in a phylogenetic tree T then t and T are *consistent*. Thus, an equivalent formulation of MTC is: given two sets $\mathcal{C}$ and $\mathcal{F}$ of rooted triplets, output a phylogenetic tree T with $\Lambda(T) = \bigcup_{t \in \mathcal{C} \cup \mathcal{F}} \Lambda(t)$ maximizing $|T(\mathcal{C})| - |T(\mathcal{F})|$, where $T(\mathcal{X})$ for any set $\mathcal{X}$ of rooted triplets is the subset of $\mathcal{X}$ consistent with T. In analogy with this terminology, constraints of type 1, 2, 3, and 4 are called *resolved triplets*, *fan triplets*, *forbidden resolved triplets*, and *forbidden fan triplets* from here on.

1.2   Previous Results

Aho *et al.* [1] presented a polynomial-time algorithm that determines if there exists a phylogenetic tree consistent with *all* of the resolved triplets in a given set, and if so, outputs such a tree. Its time complexity was improved from $O(mn)$ to $\min\{O(n + mn^{1/2}), O(m + n^2 \log n)\}$ by Henzinger *et al.* [10]. He *et al.* [9] extended Aho *et al.*'s algorithm to the case where the input also contains forbidden resolved triplets, and the resulting running time to determine if there exists a phylogenetic tree that satisfies all the input constraints is $O((m + n)n \log n)$.

In comparison, the optimization versions of rooted triplets consistency turn out to be harder. MRTC is NP-hard [3,12,19], even if restricted to dense problem instances [11]. Furthermore, MRTC in the non-dense case is APX-complete [4]. The supplementary version of MRTC in which the objective is to remove as few elements as possible from the input $\mathcal{R}$ so that there exists a phylogenetic tree consistent with the resulting $\mathcal{R}$ is $W[2]$-hard and cannot be approximated within $c \ln n$ for some constant $c > 0$ in polynomial time, unless $P = NP$ [5]. As for positive results for MRTC, Gąsieniec *et al.* [8] presented a top-down, polynomial-time 1/3-approximation algorithm, and Wu [19] gave a bottom-up, polynomial-time heuristic that was shown experimentally to perform well in practice. Byrka *et al.* [5] later modified Wu's heuristic to guarantee that it too achieves an approximation ratio of 1/3. Other heuristics for MRTC (with unknown approximation ratios) have been published in [16,18]. An exact algorithm for MRTC running in $O(3^n(m + n^2))$ time and $O(2^n)$ space was given by Wu in [19]. Finally, we remark that the 1/3-approximation algorithm for MRTC in [8] was generalized to a polynomial-time 1/3-approximation algorithm for MMRTC in [9].

The unrooted analogue of a resolved triplet, called a *quartet* [17], is an unrooted tree with two internal nodes and four distinctly labeled leaves. The corresponding maximum quartets consistency problem is MAX SNP-hard [14,17], but the complete version of the problem admits a PTAS [14]. In an unpublished manuscript [13], we have outlined how to obtain a similar PTAS for dense MRTC.

See the survey in Sect. 2 in [5] for references to other rooted triplets consistency-related problems in the literature involving enumeration, ordered trees, phylogenetic networks, multi-labeled phylogenetic trees (MUL-trees), etc.

1.3 Our Contributions

We first show how any known polynomial-time 1/3-approximation algorithm for MRTC (e.g., [5,8]) can be applied to obtain a polynomial-time 1/4-approximation algorithm for MTC (Sect. 2).

The APX-completeness of MRTC [4] (and hence, MTC) rules out the possibility of finding a PTAS for MTC in the general case. Nevertheless, we make further progress on the approximation status of MTC by presenting a PTAS for MTC restricted to *dense* instances based on smooth polynomial integer programming, using some ideas from [14] and generalizing our unpublished work in [13] (Sect. 3).

Next, we extend Wu's exact exponential-time algorithm for MRTC [19] to MTC (Sect. 4). We let the algorithm take an additional parameter $k \geq 2$ as input and force the output to be a phylogenetic tree in which every internal node has at most k children. The resulting algorithm runs in $O(2^{(n+1)\log_2(k+1)}(m+n))$ time. This may be $\Omega(n^n)$ if k is unrestricted, but the running time is single-exponential in n when $k = O(1)$, and we use this fact to design an exponential-time approximation scheme (ETAS) for MTC with no restrictions on k.

Finally, we describe how to adapt our algorithms to the weighted case, where nonnegative weights are assigned to the triplet constraints and the objective is to construct a phylogenetic tree that maximizes the sum of the weights of the satisfied constraints (Sect. 5). In case of our PTAS and our ETAS, we have to additionally assume that the ratio between the largest and the smallest constraint weights is bounded by a constant.

2   A 1/4-Approximation Algorithm for MTC

The maximum rooted resolved triplets consistency problem (MRTC) admits a 1/3-approximation algorithm running in polynomial time [5,8]. The algorithms in [5,8] always output a binary tree, so they also yield (at least) a 1/3-approximation when in addition to resolved triplets, forbidden fan triplets are included in the input. We use this fact to design a 1/4-approximation algorithm for the maximum rooted triplets consistency problem (MTC) as follows.

Algorithm 1
Input: A set $\mathcal{R}$ of m triplet constraints over an n-element set S.
Output: A phylogenetic tree with n leaves distinctly leaf-labeled by S.

1. If $\mathcal{R}$ contains at least $m/4$ fan triplets and forbidden resolved triplets then output a tree whose root has n children, each of them a leaf with a distinct label in S, and stop.
2. Extract the set $\mathcal{R}'$ of all resolved triplets and forbidden fan triplets from $\mathcal{R}$ and apply any known polynomial-time 1/3-approximation algorithm for MRTC (e.g., [5] or [8]) to $\mathcal{R}'$. Output the tree produced by the latter.

Theorem 1. *Algorithm 1 is a polynomial-time 1/4-approximation algorithm for MTC.*

Proof. We need to show that the algorithm outputs a phylogenetic tree satisfying at least $1/4$ of the input triplet constraints. There are two cases:

If $\mathcal{R}$ contains at least $m/4$ fan triplets and forbidden resolved triplets then the star phylogenetic tree output in the first step satisfying all the fan triplets and all the forbidden resolved triplets satisfies at least $m/4$ input triplet constraints.

Otherwise, $\mathcal{R}$ contains at least $3m/4$ resolved triplets and forbidden fan triplets. The $1/3$-approximation algorithm run on them in the second step yields a phylogenetic tree satisfying at least $\frac{1}{3} \cdot \frac{3m}{4} = m/4$ input triplet constraints. □

3   A PTAS for Dense MTC

Analogously to [14] for the unrooted case, we first show that any rooted phylogenetic tree T with a leaf label set $S = \Lambda(T)$ can be represented approximately by a *decomposition tree* consisting of:

1. a bounded-size subtree (termed *kernel*) K of T on non-leaf nodes, and
2. subsets of S (forming a partition of S) in one-to-one correspondence with the leaves of K, where the elements of each subset are children of the corresponding leaf of K.

In particular, an optimal tree T_{opt} for a given instance of MTC can be approximately represented by such a decomposition tree which preserves enough of the original triplet constraints to serve as a good approximation. More precisely, the number of input triplet constraints satisfied by the approximate tree differs from that of T_{opt} by an arbitrarily small fraction, depending on the number of subsets in the partition. We find an approximate solution by enumerating all possible kernels, and for each one, finding the approximately best partition of S.

Recall that an instance of MTC is *dense* if the input set of triplet constraints has $\Omega(n^3)$ elements. The analysis of the accuracy of our approximate solution relies on the fact that for a dense instance, the number of input triplet constraints satisfied by T_{opt} is $\Omega(n^3)$, since it is at least $1/4$ of the number of the constraints by Theorem 1.

Let k be a fixed integer, and let $S_1, S_2, ..., S_k$ be a partition of the set S. A subset S_i is termed a *bin*. For each bin S_i, there is a non-leaf node of degree $|S_i| + 1$ in the decomposition tree, termed a *bin root*, connected by an edge to each element in the bin. Algorithm 2, given below, transforms an input tree into its decomposition tree by joining adjacent subtrees of T until the bin is large enough, for some given maximum bin size b. If a bin is smaller than $b/2$, and there is another bin also smaller than $b/2$ in an adjoining subtree, the two small bins may be joined into one single bin. The resulting kernel K is the subtree of the output decomposition tree induced by remaining non-leaf nodes, with the subtrees defining the bins removed. The output decomposition tree T_k consists of the kernel K, with the bin roots as leaves of K, and the elements in each bin being children of its respective bin root.

Algorithm 2. k-bin decomposition(T)
Input: A phylogenetic tree T with n leaves.
Output: A decomposition tree T_k of T.

- Traverse T, and for every node v visited, check if the size of the subtree $T(v)$ of T rooted at v is less or equal to $6n/k$. If so, v is denoted a bin root (unless v is a leaf), and all internal edges of $T(v)$ except for edges incident to a leaf are contracted, so that $T(v)$ becomes a tree of height 1. If the size of $T(v)$ is larger than $6n/k$, continue traversing T at a child of v.
- For a single leaf l that is not in a bin, the edge between l and its parent is subdivided to create a new bin root associated with l.
- A bin of size $\leq 3n/k$ is *small*. Let b be a small bin, and let v be the parent of b. If another small bin b' exists as a child of a sibling of v, b and b' are combined to a single bin.

Lemma 1. *Algorithm 2 for k-bin decomposition produces a decomposition tree T_k having at most k bins, where each bin is of size less or equal to $6n/k$.*

Proof. In Lemma 1 in [14], a proof of an analogous lemma for quartets is given. The reader is referred to this proof for more details.

As a consequence of the decomposition procedure, the number of bins will be bounded by k since the merging of small bins in the third step guarantees that there are not too many small bins. Lemma 1 in [14] shows that the number of small bins is strictly smaller than twice the number of large bins. Since a large bin has a size of at least $3n/k$, the number of large bins is at most $k/3$. Let the number of large bins be l, and the number of small bins be s. Then the total number of bins is $s + l < l + 2l = 3l < 3 \cdot k/3 = k$. So, T_k has less than k bins, each of size at most $6n/k$.   □

Let R be the input set of triplet constraints. For any phylogenetic tree T, let R_T denote the subset of triplet constraints in R that are satisfied by T.

Since the decomposition algorithm works by contracting some edges of T_{opt} and transferring leaves to neighboring bins, it follows that for any triplet $\{a, b, c\}$ where a, b and c are in different bins, $ab|c \in R_{T_k}$ if and only if $ab|c \in R_{T_{opt}}$, $\neg ab|c \in R_{T_k}$ if and only if $\neg ab|c \in R_{T_{opt}}$, and similarly, $a|b|c \in R_{T_k}$ if and only if $a|b|c \in R_{T_{opt}}$, and $\neg a|b|c \in R_{T_k}$ if and only if $\neg a|b|c \in R_{T_{opt}}$.

Lemma 2. *The tree T_k that is a k-bin decomposition of T_{opt} satisfies $|R_{T_k} \cap R| \geq |R_{T_{opt}} \cap R| - \frac{c}{k} \cdot n^3$ input triplet constraints, for some constant c.*

Proof. Any triplet topology in $R_{T_{opt}} \setminus R_{T_k}$ must have two or more leaves in the same bin. The number of such triplet topologies with three or two leaves in the same bin is at most $1/6 \left(6n/k\right)^3 k + 1/2 \left(6n/k\right)^2 nk \leq 24n^3/k$ for $k \geq 6$. Each of the above triplet topologies may contribute to at most four triplet constraints in R (one fan triplet and three forbidden resolved triplets in the worst case). Hence, assuming that $k \geq 6$, we have $|R_{T_k} \cap R| \geq |R_{T_{opt}} \cap R| - 96n^3/k$.   □

Label-to-bin Assignment: Suppose that we are given a kernel K with at most k leaves of a hypothetical phylogenetic tree distinctly leaf-labeled by S. The *Label-to-Bin Assignment* problem (LBA) for a set R of triplet constraints asks for an assignment of labels in S to at most k bins of size $\leq 6n/k$ that completes K to T_k and maximizes $|R \cap R_{T_k}|$. The supertree of K induced by such an assignment is called a *completion of K.*

Jiang *et al.* [14] showed that although the corresponding LBA problem for unrooted quartets is NP-hard, it admits a PTAS relying on a modified PTAS for smooth polynomial integer programs by Arora *et al.* [2]. We adapt this technique to our problem. First, for every resolved triplet $ab|c$ in R, define the polynomial:

$$p_{ab|c}(x) = \sum_{ij|k \in R_{T_k}} x_{ai}x_{bj}x_{ck} + x_{bi}x_{aj}x_{ck}$$

Here, the term $x_{sb} = 1$ if label s is assigned to bin b, and 0 otherwise. Next, for every fan triplet $a|b|c$ in R, define the following polynomial, where $Per(a, b, c)$ stands for the set of all one-to-one mappings from $\{a, b, c\}$ to $\{a, b, c\}$:

$$p_{a|b|c}(x) = \sum_{i|j|k \in R_{T_k}} \sum_{\delta \in Per(a,b,c)} x_{\delta(a)i} x_{\delta(b)j} x_{\delta(c)k}$$

For every forbidden resolved triplet $\neg ab|c$ in R, define the polynomial:

$$p_{\neg ab|c}(x) = p_{ac|b}(x) + p_{bc|a}(x) + p_{a|b|c}(x)$$

Similarly, for every forbidden fan triplet $\neg a|b|c$ in R, define the polynomial:

$$p_{\neg a|b|c}(x) = p_{ab|c}(x) + p_{ac|b}(x) + p_{bc|a}(x)$$

Finally, define:

$$p(x) = \sum_{ab|c \in R} p_{ab|c}(x) + \sum_{a|b|c \in R} p_{a|b|c}(x) + \sum_{\neg ab|c \in R} p_{\neg ab|c}(x) + \sum_{\neg a|b|c \in R} p_{\neg a|b|c}(x)$$

The optimization problem becomes: Maximize $p(x)$ subject to $\sum_{i=1}^{k} x_{si} = 1$ for each leaf s, and $\sum_{s=1}^{n} x_{si} \leq 6n/k$ for each bin i. (The first condition ensures that each label is assigned to exactly one bin and the second condition maintains the k-bin property.) Our polynomial integer program is an $O(1)$-*smooth degree-3 polynomial integer program* according to the following definition from [2]: An $O(1)$ - *smooth degree-d polynomial integer program* is to maximize $p(x_1, ..., x_n)$ subject to $x_i \in \{0, 1\}$, $\forall i \leq n$, where $p(x_1, ..., x_n)$ is a degree-d polynomial in which the coefficient of each degree-i monomial (term) is $O(n^{d-i})$.

Lemma 3. *(Arora* et al.*[2]) Let m be the maximum value of an $O(1)$-smooth degree-d polynomial integer program $p(x_1, ..., x_n)$. For each $\epsilon > 0$, there is a polynomial-time algorithm that finds a 0/1 assignment α for the x_i satisfying $p(\alpha(x_1), ..., \alpha(x_n)) \geq m - \epsilon n^d$.*

The PTAS of Arora *et al.* first solves the fractional version of the problem. It then rounds the obtained fractional value for each variable individually in order to obtain an integer solution. However, this is not possible in our case because of the condition $\sum_{i=1}^{k} x_{si} = 1$ for each leaf s. Instead, following [14], we set $x_{si} = 1$ and $x_{sj} = 0$ for $j \neq i$ with probability equal the fractional value for x_{si}. In effect, exactly one of the variables $x_{s1}, ..., x_{sk}$ is set to 1 and the rest to 0. In analogy to Theorem 2.6 in [14], we obtain the next lemma.

Lemma 4. *For each $\epsilon > 0$, there is a polynomial-time algorithm which, for each instance of the LBA specified by a set R of triplet constraints for dense MTC and a kernel K, produces a completion T' of K such that $|R_{T'} \cap R| \geq |R_{\hat{T}} \cap R| - \epsilon n^3$, where $\hat{T}$ is an optimal completion of K.*

T_{opt} can be decomposed into a kernel with at most k leaves and k bins of size $\leq 6n/k$ (i.e., the tree T_k) as shown in Lemmas 1 and 2. Given any input set of triplet constraints, for each kernel with k leaves, an approximate optimal assignment of leaves to bins of such size can be found in polynomial time by Lemma 4. Hence, dense MTC can be approximated in the following way:

Theorem 2. *For each $\epsilon > 0$, there is a polynomial-time algorithm which, for each instance R of dense MTC, produces a tree T_k that approximates T_{opt} in such a way that $|R_{T_k} \cap R| \geq (1 - \epsilon)|R_{opt} \cap R|$.*

Proof. Let c be the constant specified in Lemma 2. By Lemmas 2 and 4, $|R_{T_k} \cap R| \geq |R_{T_{opt}} \cap R| - (c/k + \epsilon') \cdot n^3 \geq (1 - c/(c'k) - \epsilon'/c')|R_{T_{opt}} \cap R|$, where c' is a constant satisfying $|R_{T_{opt}} \cap R| \geq c'n^3$. We estimate this constant by the density of R and Theorem 1. By picking $k \geq \frac{2c}{c'\epsilon}$ and $\epsilon' \leq \frac{c'\epsilon}{2}$, we obtain the theorem. $\square$

4   An ETAS for MTC

The following additional notation will be used. For any node u in a phylogenetic tree T with a leaf label set $S = \Lambda(T)$, let S_u be the subset of S labeling the leaves of the subtree rooted at u. For any node v of T, let P_v be the partition of S_v into $S_{v_1}, ..., S_{v_l}$, where $v_1, ..., v_l$ are the children of v.

For a partition P of $U \subseteq S$ into l subsets, let $w_2(P)$ be the number of resolved triplets $ab|c$ such that a and b belong to two distinct subsets in P and $c \notin U$. Similarly, let $w_3(P)$ be the number of fan triplets $a|b|a$ such that a, b, c belong to three different subsets in P. Next, let $w_{f2}(P)$ be the number of forbidden resolved triplets $\neg ab|c$ such that a and c belong to two distinct subsets in P and $b \notin U$, or b and c belong to two distinct subsets in P and $a \notin U$, or a, b, c belong to three different subsets in P. Finally, let $w_{f3}(P)$ be the number of forbidden fan triplets $\neg a|b|c$ such that two elements in $\{a, b, c\}$ belong to two distinct subsets in P and the remaining one does not belong to any of the subsets. We have:

Lemma 5. *Given a partition P of $U \subseteq S$ into l subsets, $w_2(P)$, $w_3(P)$, $w_{f2}(P)$ and $w_{f3}(P)$ can be computed in $O(m + n)$ time, where m is the number of input triplet constraints and n is the size of S.*

Proof. We "color" the elements in U with l colors according to P and the elements in $S \setminus U$ with another color, and then examine each input triplet constraint to check if it increases $w_2(P)$, $w_3(P)$, $w_{f2}(P)$, or $w_{f3}(P)$ by one. $\square$

Remark 2. When $l = 2$ in Lemma 5, $w_2(P)$ is the same as $w(V_1, V_2)$ in Wu's exact algorithm for MRTC [19]. Theorem 2 in [19] computes $w(V_1, V_2)$ in $O(m + n^2)$ time, so using our Lemma 5 instead slightly improves the running time of Wu's algorithm from $O(3^n(m + n^2))$ to $O(3^n(m + n))$.

Lemma 6. *For a phylogenetic tree T with leaves labeled with elements in S, the number of input triplet constraints consistent with T is equal to $\sum_{v \in T} (w_2(P_v) + w_3(P_v) + w_{f2}(P_v) + w_{f3}(P_v))$.*

We now analyze how much is lost by forcing the solution to an instance of MTC to be a k-ary *phylogenetic tree*, defined as a phylogenetic tree in which every internal node has degree at most k, where k is any integer such that $k \geq 2$:

Theorem 3. *For any phylogenetic tree T, there exists a k-ary phylogenetic tree T' with $\Lambda(T') = \Lambda(T)$ that satisfies at least a fraction of $(1 - 12/k)$ of the input triplet constraints satisfied by T.*

Proof. We shall replace each node v of T having more than k children by a subtree in which all nodes have at most k children. Let $v_1, ..., v_l$ be the children of v. Note that $l > k$. To start with, assign to each forbidden resolved triplet $\neg ab|c$ contributing to $w_{f2}(P_v)$, either the resolved triplet $ac|b$, where a and c belong to distinct S_{v_i}, S_{v_j} and $b \notin S_v$, or the resolved triplet $bc|a$, where a and c belong to distinct S_{v_i}, S_{v_j} and $a \notin S_v$, or the fan triplet $a|b|c$, where all a, b, c belong to three distinct S_{v_i}, S_{v_j}, S_{v_q}. Similarly, assign to each forbidden fan triplet $\neg a|b|c$ contributing to $w_{f3}(P_v)$, either the resolved triplet $ab|c$, where a and b belong to distinct S_{v_i}, S_{v_j} and $c \notin S_v$, or the resolved triplet $ac|b$, where a and c belong to distinct S_{v_i}, S_{v_j} and $b \notin S_v$, or the resolved triplet $bc|a$, where a and c belong to distinct S_{v_i}, S_{v_j} and $a \notin S_v$. Let $f_2(P_v)$ be the cardinality of the multiset of assigned resolved triplets and let $f_3(P_v)$ be the cardinality of the multiset of assigned fan triplets. Then $w_{f2}(P_v) + w_{f3}(P_v) = f_2(P_v) + f_3(P_v)$.

For the sake of the proof, partition the family of subsets $S_{v_1}, ..., S_{v_l}$ into k groups uniformly at random. Consider any fan triplet $a|b|c$ contributing to $w_3(P_v)$ (i.e., having each of its elements in a distinct S_{v_i}) or to $f_3(P_v)$ (i.e., being assigned to a forbidden resolved triplet). The probability that any two elements in $\{a, b, c\}$ fall into the same group is bounded from above by $1/k + 2/k \leq 3/k$. Hence, there exists a partition of the family of $S_{v_1}, ..., S_{v_l}$ into k groups such that at least a $(1 - 3/k)$ fraction of triples $a|b|c$ contributing to $w_3(P_v) + f_3(P_v)$ will have all its elements in three different groups. For each group g in the latter partition, first construct an arbitrary rooted resolved tree F_g whose leaves are labeled by the children v_i of v for which $S_{v_i} \in g$ and then replace each leaf labeled by v_i in F_g by the subtree of T rooted at v_i. Next, delete the edges in T connecting v with its children and instead connect v to the roots of the trees F_g by edges. Observe that the same fan triplet may contribute to $w_3(P_v)$ and it may also contribute up to three times to $f_3(P_v)$, (i.e., it may be assigned to up to three forbidden triplets contributing to $w_{f2}(P_v)$). It follows that the sum of the new value of $w_3(P_v) + f_3(P_v)$ (provided that we keep the same assignments if possible) is at least $(1 - 4 \cdot 3/k)$ of the sum of the previous value of $w_3(P_v) + f_3(P_v)$.

In turn, consider any resolved triplet $ab|c$ that contributes to $w_2(P_v)$ or to $f_2(P_v)$ (i.e., is assigned to a forbidden resolved triplet contributing to $w_{f2}(P_v)$ or a forbidden fan triplet contributing to $w_{f3}(P_v)$). After the transformation of T, the following holds: If the labels a and b belong to subsets in the same group g then $ab|c$ can neither contribute to $w_2(P_v)$ nor to $f_2(P_v)$ (i.e., to be assigned to a forbidden resolved triplet contributing to $w_{f2}(P_v)$ or to a forbidden fan triplet contributing to $w_{f3}(P_v)$). On the other hand, there must exist a non-leaf node u of the binary tree F_g for which $ab|c$ correspondingly contributes to $w_2(P_u)$, or it can be assigned to the same forbidden resolved triplets now contributing to $w_{f2}(P_u)$, or it can be assigned to the same forbidden fan triplets now contributing to $w_{f3}(P_u)$. Thus, by extending the notation $f_2(\ )$ to include $f_2(P_t)$, the sum of $w_2(P_t) + f_2(P_t)$ over the tree nodes t does not change. The theorem follows from Lemma 6. □

Motivated by Theorem 3, our new approximation algorithm in this section constructs a k-ary phylogenetic tree consistent with the maximum possible number of input triplet constraints for some suitable value of k. For this purpose, we generalize Wu's algorithm [19] for MRTC which always outputs a *binary* phylogenetic tree, i.e., corresponding to the special case $k = 2$. We also need to extend Wu's algorithm to allow not only resolved triplets in the input.

Our new algorithm works as follows. For each non-singleton subset U of S, define $score(U)$ recursively by $score(U) = \max_{l=2}^{k} score_l(U)$, where $score_l(U) =$

$$\max_{l-\text{partition } U_1...,U_l \text{ of } U} \sum_{i=1}^{l} score(U_i) + \sum_{j=2}^{3} w_j(U_1,..,U_l) + w_{fj}(U_1,...,U_l)$$

For a singleton U, $score(U)$ is set to 0. As in Wu's algorithm [19], $score(U)$ is evaluated in non-decreasing order of the sizes of subsets U of S. Then, the output phylogenetic tree is constructed by a traceback, starting from $score(S)$, and picking an l-partition of the current subset U that yields the maximum value of $score(U)$. The corresponding node of the constructed tree gets l children in one-to-one correspondence with the subsets of U forming the selected partition.

It follows by induction on $|U|$ and Lemma 6 that $score(U)$ equals the maximum number of input triplets that can be satisfied by a k-ary subtree leaf-labeled by U. This yields the optimality of the tree constructed during the traceback.

There are $\binom{n}{q}$ subsets U of S with q elements. The number of l-partitions of a subset U with q elements is l^q. Therefore, the total number of subsets partitions processed by our algorithm is $\sum_{q=1}^{n} \binom{n}{q} \sum_{\ell=2}^{k} \ell^q \leq \sum_{\ell=2}^{k} (\ell + 1)^n \leq (k + 1)^{n+1}$ by binomial expansion. Finally, by Lemma 5, for a given partition P of $U \subseteq S$ into l subsets, the weights $w_2(P)$, $w_3(P)$, $w_{f2}(P)$ and $w_{f3}(P)$ can be computed in $O(m + n)$ time, where m is the number of input triplet constraints and n is the size of S. We conclude that our algorithm runs in $O((k+1)^{n+1}(m+n))$ time, i.e., in $O(2^{(n+1)\log_2(k+1)}(m + n))$ time.

Theorem 4. *Let S be a set of n distinct labels and let k be an integer greater than 1. For any set R of m (resolved or forbidden resolved or fan or forbidden fan) triplet constraints on S, one can find a k-ary phylogenetic tree T with $\Lambda(T) = S$ that maximizes the number of satisfied triplet constraints in R among all k-ary phylogenetic trees in $O(2^{(n+1)\log_2(k+1)}(m + n))$ time.*

By combining Theorems 3 and 4, we obtain an exponential-time approximation scheme (ETAS) for the maximum rooted triplets consistency problem:

Theorem 5. *Let S be a set of n distinct labels and let $\epsilon > 0$ be a constant. For any set R of m (resolved or forbidden resolved or fan or forbidden fan) triplet constraints on S, one can find a phylogenetic tree T with $\Lambda(T) = S$ in $O(2^{(n+1)\log_2(\lceil 12/\epsilon\rceil+1)}(m+n))$ time satisfying at least $(1 - \epsilon)$ of the maximum number of triplet constraints in R that can be satisfied in any phylogenetic tree.*

5   Extensions to the Weighted Case

Having input triplet constraints in the form of rooted triplets and forbidden rooted triplets, it is natural to assign nonnegative real weights to them. Note that $\neg a|b|c$ is equivalent to the conjunction of $\neg ab|c$, $\neg ac|b$ and $\neg bc|a$. Consequently, MTC generalizes to the *maximum weighted rooted triplet consistency problem* (MWTC), where the objective is to construct a phylogenetic tree that maximizes the total weight of the satisfied input triplet constraints.

By Theorem 4 in [8], the 1/3-approximation algorithm for MRTC in [8] works for the weighted version of MRTC as well. Hence, the 1/4-approximation algorithm for MTC in Sect. 2 immediately generalizes to MWTC by considering sums of weights of input triplet constraints belonging to the appropriate subsets instead of just the cardinalities of the subsets. Our exact algorithm for MTC in Sect. 4 similarly generalizes to MWTC by considering sums of the weights of the respective triplet constraints instead of their numbers. However, the situation is a bit more subtle for our PTAS for dense MTC in Sect. 3 and our ETAS for MTC in Sect. 4. Because of Lemma 2 and Theorem 3, respectively, where in both cases some fraction of the triplet constraints may be lost, we need to assume that the maximum triplet constraint weight is at most $O(1)$ times larger than the minimum one in order to generalize both approximation schemes to the weighted case. Furthermore, in our PTAS, the polynomials in one-to-one correspondence with the input triplet constraints in the definition of the integer program have to be multiplied by the weight of the corresponding constraint.

6   Final Remarks

MTC is APX-complete by the APX-completeness of MRTC [4] and Theorem 1. An open problem is to improve the polynomial-time approximation ratios 1/3 and 1/4 for MRTC and MTC; by applying the technique in Sect. 2, an f-approximation for the former would give an $\frac{f}{1+f}$-approximation for the latter.

References

1. Aho, A.V., Sagiv, Y., Szymanski, T.G., Ullman, J.D.: Inferring a tree from lowest common ancestors with an application to the optimization of relational expressions. SIAM J. Comput. **10**(3), 405–421 (1981)
2. Arora, S., Karger, D., Karpinski, M.: Polynomial time approximation schemes for dense instances of NP-hard problems. J. Comput. Syst. Sci. **58**(1), 193–210 (1999)

3. Bryant, D.: Building Trees, Hunting for Trees, and Comparing Trees: Theory and Methods in Phylogenetic Analysis. Ph.D. thesis. University of Canterbury, Christchurch, New Zealand (1997)
4. Byrka, J., Gawrychowski, P., Huber, K.T., Kelk, S.: Worst-case optimal approximation algorithms for maximizing triplet consistency within phylogenetic networks. J. Discrete Algorithms **8**(1), 65–75 (2010)
5. Byrka, J., Guillemot, S., Jansson, J.: New results on optimizing rooted triplets consistency. Discrete Appl. Math. **158**(11), 1136–1147 (2010)
6. Chor, B., Hendy, M., Penny, D.: Analytic solutions for three taxon ML trees with variable rates across sites. Discrete Appl. Math. **155**(6–7), 750–758 (2007)
7. Felsenstein, J.: Inferring Phylogenies. Sinauer Associates Inc., Sunderland (2004)
8. Gąsieniec, L., Jansson, J., Lingas, A., Östlin, A.: On the complexity of constructing evolutionary trees. J. Comb. Optim. **3**(2–3), 183–197 (1999)
9. He, Y.J., Huynh, T.N.D., Jansson, J., Sung, W.-K.: Inferring phylogenetic relationships avoiding forbidden rooted triplets. J. Bioinform. Comput. Biol. **4**(1), 59–74 (2006)
10. Henzinger, M.R., King, V., Warnow, T.: Constructing a tree from homeomorphic subtrees, with applications to computational evolutionary biology. Algorithmica **24**(1), 1–13 (1999)
11. van Iersel, L., Kelk, S., Mnich, M.: Uniqueness, intractability and exact algorithms: reflections on level-k phylogenetic networks. J. Bioinform. Comput. Biol. **7**(4), 597–623 (2009)
12. Jansson, J.: On the complexity of inferring rooted evolutionary trees. In: Proceedings of the Brazilian Symposium on Graphs, Algorithms, and Combinatorics (GRACO 2001). Electronic Notes in Discrete Mathematics, vol. 7, pp. 50–53. Elsevier (2001)
13. Jansson, J., Lingas, A., Lundell, E.-M.: A triplet approach to approximations of evolutionary trees. Poster H15 presented at RECOMB 2004 (2004)
14. Jiang, T., Kearney, P., Li, M.: A polynomial time approximation scheme for inferring evolutionary trees from quartet topologies and its application. SIAM J. Comput. **30**(6), 1942–1961 (2001)
15. Kearney, P.: Phylogenetics and the quartet method. In: Jiang, T., Xu, Y., Zhang, M.Q. (eds.) Current Topics in Computational Molecular Biology, pp. 111–133. The MIT Press, Massachusetts (2002)
16. Snir, S., Rao, S.: Using max cut to enhance rooted trees consistency. IEEE/ACM Trans. Comput. Biol. Bioinf. **3**(4), 323–333 (2006)
17. Steel, M.: The complexity of reconstructing trees from qualitative characters and subtrees. J. Classif. **9**(1), 91–116 (1992)
18. Wu, B.Y.: Constructing evolutionary trees from rooted triplets. J. Inf. Sci. Eng. **20**, 181–190 (2004)
19. Wu, B.Y.: Constructing the maximum consensus tree from rooted triples. J. Comb. Optim. **8**(1), 29–39 (2004)

String Powers in Trees

Tomasz Kociumaka[✉], Jakub Radoszewski, Wojciech Rytter,
and Tomasz Waleń

Faculty of Mathematics, Informatics and Mechanics,
University of Warsaw, Warsaw, Poland
{kociumaka,jrad,rytter,walen}@mimuw.edu.pl

Abstract. We investigate the asymptotic growth of the maximal number $\mathsf{powers}_\alpha(n)$ of different α-powers (strings w with a period $|w|/\alpha$) in an edge-labeled unrooted tree of size n. The number of different powers in trees behaves much unlike in strings. In a previous work (CPM, 2012) it was proved that the number of different squares in a tree is $\mathsf{powers}_2(n) = \Theta(n^{4/3})$. We extend this result and analyze other powers. We show that there are phase-transition thresholds:

1. $\mathsf{powers}_\alpha(n) = \Theta(n^2)$ for $\alpha < 2$;
2. $\mathsf{powers}_\alpha(n) = \Theta(n^{4/3})$ for $2 \leq \alpha < 3$;
3. $\mathsf{powers}_\alpha(n) = \mathcal{O}(n \log n)$ for $3 \leq \alpha < 4$;
4. $\mathsf{powers}_\alpha(n) = \Theta(n)$ for $4 \leq \alpha$.

The difficult case is the third point, which follows from the fact that the number of different cubes in a rooted tree is linear (in this case, only cubes passing through the root are counted).

1 Introduction

Repetitions are a fundamental notion in combinatorics on words. For the first time they were studied more than a century ago by Thue [14] in the context of square-free strings, that is, strings that do not contain substrings of the form $W^2 = WW$. Since then, α-free strings, avoiding string powers of exponent α (of the form W^α), have been studied in many different contexts; see [13]. Another line of research is related to strings that are rich in string powers. It has been shown that the number of different squares in a string of length n does not exceed $2n - \Theta(\log n)$ (see [5,7,8]); stronger bounds are known for cubes [12].

Repetitions are also considered in labeled trees and graphs. In this model, a repetition corresponds to a sequence of labels of edges (or nodes) on a simple path. The origin of this study comes from a generalization of square-free

This work was supported by the Polish National Science Center, grant no 2014/13/B/ST6/00770.

T. Kociumaka—Supported by Polish budget funds for science in 2013–2017 as a research project under the 'Diamond Grant' program.

J. Radoszewski—Supported by the Polish Ministry of Science and Higher Education under the 'Iuventus Plus' program in 2015–2016 grant no 0392/IP3/2015/73. Receives financial support of Foundation for Polish Science.

© Springer International Publishing Switzerland 2015
F. Cicalese et al. (Eds.): CPM 2015, LNCS 9133, pp. 284–294, 2015.
DOI: 10.1007/978-3-319-19929-0_24

strings and α-free strings, called non-repetitive colorings of graphs. A survey by Grytczuk [6] presents several results of this kind. In particular, non-repetitive colorings of labeled trees were considered [2]. Strings related to paths in graphs have also been studied in the context of hypertexts [1].

Enumeration of squares in labeled trees has already been considered from both combinatorial [4] and algorithmic point of view [9]. Our study is a continuation of the results of [4], where it has been proved that the maximum number of different squares in a labeled tree with n nodes is of the order $\Theta(n^{4/3})$. As our main result we show a *phase transition* property: for every exponent $2 < \alpha < 3$, a tree of n nodes may contain $\Omega(n^{4/3})$ string α-powers, whereas it may only have $\mathcal{O}(n \log n)$ powers of exponent $\alpha \geq 3$.

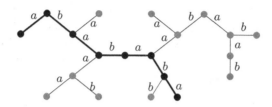

Fig. 1. There are 5 different cubic substrings in this tree: a^3, $(ab)^3$, $(ba)^3$, $(aab)^3$, $(baa)^3$. Hence, $\mathsf{powers}_3(T) = 5$. Note that the cube $(ab)^3$ occurs twice; also a^3 has multiple occurrences. The most repetitive substring, a 3.5-power $(ab)^{3.5}$, is marked in the figure.

Let T be a tree whose edges are labeled with symbols from an alphabet Σ. We denote the size of the tree, that is, the number of nodes, by $|T|$. A *substring* of T is the sequence of labels of edges on any simple path in T. We define $\mathsf{powers}_\alpha(T)$ as the number of different substrings of T which are powers of (possibly fractional) exponent α; see Fig. 1. We denote $\mathsf{powers}_\alpha(n) = \max_{|T|=n} \mathsf{powers}_\alpha(T)$. The bound $\mathsf{powers}_2(n) = \Theta(n^{4/3})$ has been shown in [4]. Here, we prove the following asymptotic bounds:

$\alpha \in (1,2)$	$\mathsf{powers}_\alpha(n) = \Theta(n^2)$
$\alpha \in [2,3)$	$\mathsf{powers}_\alpha(n) = \Theta(n^{4/3})$
$\alpha \in [3,4)$	$\mathsf{powers}_\alpha(n) = \mathcal{O}(n \log n)$
$\alpha \geq 4$	$\mathsf{powers}_\alpha(n) = \Theta(n)$

2   Preliminaries

2.1   Combinatorics of Strings

Let V be a string over an alphabet Σ. We denote its letters by $V_1, \ldots, V_m$ and its length m by $|V|$. By V^R we denote the *reverse* string $V_m \ldots V_1$. For

$1 \leq i \leq j \leq m$ a string $V[i..j] = V_i \ldots V_j$ is a *substring* of V. We say that a positive integer q is a *period* of V if $V_i = V_{i+q}$ holds for $1 \leq i \leq m - q$. In this case we also say that the prefix of V of length q is a period of V.

For an integer i, $1 \leq i \leq m$, a substring $V[1..i]$ is called a *prefix* of V, and $V[i..m]$ is called a *suffix* of V. A string U is a *border* of V if it is both a prefix and a suffix of V. It is well known that a string of length m has a border of length b if and only if it has a period $m - b$.

Fact 1 ([11]). *Let B_1, B_2 be borders of a string V. If $|B_1| < |B_2| \leq 2|B_1|$, then B_1 and B_2 have the same shortest period p, which is a divisor of $|B_2| - |B_1|$.*

We say that a string V is an *α-power* (a power of exponent α) of a string U, denoted as $V = U^\alpha$, if $|V| = \alpha|U|$ and U is a period of V. Here, $\alpha \geq 1$ may otherwise be an arbitrary rational number. Powers of exponent $\alpha = 2$ are called *squares*, and powers of exponent $\alpha = 3$ are called *cubes*. By U^* we denote the set of all integer powers of U. A string V is called *non-primitive* if $V = U^k$ for some string U and an integer $k \geq 2$. Otherwise, V is called *primitive*. Primitive strings enjoy several useful properties; see [3, 13].

Fact 2 (Synchronization Property). *If P is a primitive string, then it occurs exactly twice as a substring of P^2.*

Fact 3. *Let p be a period of a string X and P be any substring of X of length p. If p is the shortest period of X, then P is primitive. Conversely, if P is primitive and $p \leq \frac{1}{2}|X|$, then p is the shortest period of X.*

Fact 4. *Let X be a string. Suppose that an integer p is a period of a prefix Y of X and of a suffix Z of X. If $|X| \leq |Y| + |Z| - p$, then p is a period of X.*

2.2   Labeled Trees

Let T be a labeled tree. If u and v are two nodes of T, then by $val(u, v)$ we denote the sequence of labels of edges on the path from u to v. We call $val(u, v)$ a *substring* of T and (u, v) an *occurrence* of the string $val(u, v)$ in T. A *rooted tree* is a tree T with one of its nodes r designated as a root. For any two nodes u, v, by $lca(u, v)$ we denote their lowest common ancestor in T. A substring of a rooted tree is *anchored* at r if it corresponds to a path passing through r, i.e., if it has an occurrence (u, v) such that $lca(u, v) = r$. A *directed tree* T_r is a rooted tree with all its edges directed towards its root r. Every substring of a directed tree corresponds to a directed path in the tree. The following fact is a simple generalization of the upper bound of $2n$ on the number of squares in a string of length n; see [5, 7].

Lemma 5. *A directed tree with n nodes contains at most $2n$ different square substrings.*

Proof. It suffices to note that there are at most two topmost occurrences of different squares starting at each node of the tree; see [5, 7, 10]. □

3   Cubes in Rooted Trees

In this section, we show that a rooted tree T with n nodes contains $\mathcal{O}(n)$ different cubes anchored at its root r.

3.1   Cube Decompositions

For a non-empty string X, (U, V) is a *cube decomposition* of X^3 if $UV = X^3$ and there exist nodes u and v in T such that $lca(u, v) = r$, $val(u, r) = U$ and $val(r, v) = V$. A cube decomposition is called *leftist* if $|U| \geq |V|$ and *rightist* if $|U| \leq |V|$. Due to the following lemma, it suffices to consider cubes with a leftist cube decomposition.

Lemma 6. *In a rooted tree the numbers of different cubes with a leftist decomposition and with a rightist decomposition are equal.*

Proof. (U, V) is a leftist cube decomposition of a cube X^3 if and only if (V^R, U^R) is a rightist cube decomposition of a cube Y^3 where $Y = X^R$.   □

If $|U|, |V| < 2|X|$, then (U, V) is called a *balanced* cube decomposition. Otherwise, it is *unbalanced*. It turns out that the number of cubes with an unbalanced decomposition is simpler to bound.

Lemma 7. *A rooted tree with n nodes contains at most $2n$ different cubes with a leftist unbalanced cube decomposition.*

Proof. Let T be a tree rooted in r and let T_r be the corresponding directed tree. If (U, V) is an unbalanced leftist decomposition of a cube X^3, then $|U| \geq 2|X|$ and thus X^2 occurs as a square substring in T_r. By Lemma 5 there are at most $2n$ such different squares.   □

A cube X^3 is called a *p-cube* if X is primitive. Otherwise it is called an *np-cube*. A bound on the number of np-cubes also follows from Lemma 5.

Lemma 8. *A rooted tree with n nodes contains at most $4n$ different np-cubes with a leftist cube decomposition.*

Proof. Let X^3 be an np-cube with a leftist decomposition (U, V) in a tree T rooted at r. We have $X = Y^k$ for a primitive string Y and an integer $k \geq 2$. Let $\ell = \lfloor \frac{3k}{4} \rfloor$. Note that $Y^{2\ell}$ is a proper prefix of U and thus a square in the directed tree T_r. Consider an assignment $Y^{3k} \mapsto Y^{2\ell}$. Observe that a single square can be assigned this way at most two cubes: $Y^{2\ell}$ can be assigned to $Y^{4\ell}, Y^{4\ell+1}, Y^{4\ell+2}$, or $Y^{4\ell+3}$, but no more than two of these exponents may be divisible by 3.

By Lemma 5 there are at most $2n$ different squares in the directed tree T_r. Therefore the number of different np-cubes with a leftist cube decomposition is bounded by $4n$.   □

3.2   Essential Cube Decompositions

Thanks to Lemmas 6–8, from now on we only consider p-cubes in T which have a balanced leftist cube decomposition. We call such a decomposition an *essential cube decomposition*. In this section, we classify such decompositions into two types and provide a separate bound for either type.

Observation 9. *Let (U, V) be an essential cube decomposition of a p-cube X^3. Then $U = XB$ for a non-empty string B which is a border of U (and a prefix of X) and satisfies $\frac{1}{3}|U| \leq |B| < \frac{1}{2}|U|$.*

Motivated by the observation, for a string U we define

$$\mathcal{B}(U) = \{B : B \text{ is a border of } U \text{ and } \tfrac{1}{3}|U| \leq |B| < \tfrac{1}{2}|U|\}.$$

Moreover, by $\mathcal{B}'(U)$ we denote a set formed by the two longest strings in $\mathcal{B}(U)$ (we assume $\mathcal{B}'(U) = \mathcal{B}(U)$ if $|\mathcal{B}(U)| \leq 2$).

Definition 10. *Let (U, V) be an essential cube decomposition of X^3 and let $U = XB$. This decomposition is said to be of type 1 if $B \in \mathcal{B}'(U)$ and of type 2 otherwise.*

Note that the string U and its border B uniquely determine the cube X^3. Since $|\mathcal{B}'(U)| \leq 2$, the following observation follows directly from the definition above.

Observation 11 (Type-1 Reconstruction). *For every string U there are at most two strings V such that (U, V) is an essential decomposition of type 1 of some cube $X^3 = UV$.*

Below we prove a similar property of type-2 decompositions. Before that, we need to characterize them more carefully. The following lemma lists several properties of type-2 decompositions; see also Fig. 2.

Lemma 12. *Let (U, V) be a type-2 essential decomposition of a p-cube X^3. Then there exists a primitive string P such that:*

(a) $|P| \leq \frac{1}{6}|X|$,
(b) X has a prefix of the form P^* of length at least $2|X| - |V| + |P|$,
(c) X has P as a suffix, but does not have a suffix of the form P^* of length $|V| - |X|$ or more.

Proof. Let $\mathcal{B}(U) = \{B_0, \ldots, B_\ell\}$ with $|B_0| < \ldots < |B_\ell|$. Since (U, V) is a type-2 decomposition of X, we have $U = XB_k$ for some k satisfying $0 \leq k \leq \ell - 2$. In particular, this implies $\ell \geq 2$.

By Fact 1, all borders in $\mathcal{B}(U)$ share a common shortest period, whose length in particular divides $|B_{i+1}| - |B_i|$ for any i ($0 \leq i < \ell$). We denote this period by P. By Fact 3, P is primitive. Let $p = |P|$ and let $p' = |B_0| \bmod p$. Moreover, let P' be the prefix of P of length p'. Observe that $B_0 = P^j P'$ for some integer j, and in general $B_i = P^{j+i}P'$.

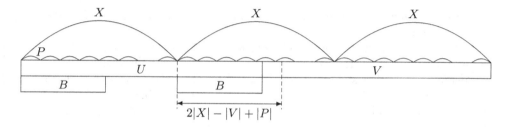

Fig. 2. Type 2 essential cube decomposition (U,V) of a cube X. Here, B is a border of U. Note that P is a period of B, but not a period of X or U.

(a) We have $\frac{1}{3}|U| \le |B_0| < |B_\ell| < \frac{1}{2}|U|$ and $|B_\ell| - |B_0| = \ell \cdot p \ge 2p$. Thus, $p \le \frac{1}{2}\left(\frac{1}{2} - \frac{1}{3}\right)|U| = \frac{1}{12}|U|$. Moreover, $|U| \le 2|X|$, and as a consequence we get $|P| = p \le \frac{2}{12}|X| = \frac{1}{6}|X|$.

(b, c) Note that $U = XB_k$ has B_ℓ as a suffix, and $B_\ell = P^{\ell-k}B_k$. Thus $P^{\ell-k}$ and, in particular, P is a suffix of X. Moreover, B_ℓ is a prefix of U, so U has $P^{j+\ell}$ as a prefix and, in particular, P is a prefix of X. Therefore, P is a border of X. Observe that P is not a period of X. Otherwise, due to synchronization property of primitive strings (Fact 2), X would be a power of P, which is a contradiction with X^3 being a p-cube.

Consequently, $|P^{j+\ell}| < |X|$, so $P^{j+\ell}$ is a prefix X. Moreover, we have $|P^{j+\ell}| \ge |B_{\ell-1}| \ge |B_k| + |P|$ since $k \le \ell - 2$, and $|B_k| = |U| - |X| = 3|X| - |V| - |X| = 2|X| - |V|$. Thus, X indeed has a prefix Y of the form P^* whose length is at least $2|X| - |V| + |P|$. Now, suppose that X has a suffix Z of the form P^* whose length is at least $|V| - |X|$. We would have $|X| \le |Y| + |Z| - |P|$, so Fact 4 would imply that P is a period of X, which we have already proved impossible. $\qquad\square$

Lemma 13. (Type-2 Reconstruction). *For every string V there is at most one string U such that (U,V) is an essential cube decomposition of type 2 of some cube $X^3 = UV$.*

Proof. Suppose there is at least one string U which satisfies the assumption of the lemma. We shall prove that U can be uniquely determined from V. Let $UV = X^3$ and let P be the primitive string obtained through Lemma 12. Our goal is to recover P and then X from V.

Recall that $|X| < |V| \le \frac{3}{2}|X|$ by the definition of essential cube decomposition. We have $X = V[i..|V|]$ for $i = |V| - |X| + 1$. Additionally, let $j = |X|$. Note that $j - i + 1 = 2|X| - |V|$, so Lemma 12(b) implies that $V[i..j'] = P^k$ for a position $j' \ge j$ and an integer exponent k. Observe that

$$i = |V| - |X| + 1 \le \tfrac{1}{3}|V| + 1 \quad \text{and} \quad j = |X| \ge \tfrac{2}{3}|V|,$$

so $p = |P|$ is a period of $V' = V[\lfloor\frac{1}{3}|V|\rfloor + 1..\lceil\frac{2}{3}|V|\rceil]$. By Lemma 12(a), $|P| \le \frac{1}{6}|X| \le \frac{1}{6}|V| \le \frac{1}{2}|V'|$ and P is primitive. Thus, by Fact 3, p can be uniquely determined as the shortest period of V'.

Once we know p, we can easily determine P: by Lemma 12(c), P is a suffix of X and thus a suffix of V. Hence, $P = V[|V| - p + 1..|V|]$.

Next, we determine the smallest position $i' > \frac{1}{3}|V|$ where P occurs in V. This occurrence must lie within $V[i..j]$, so $i \equiv i' \pmod{p}$ by the synchronization property of primitive strings (Fact 2). Let ℓ be the largest integer such that P^ℓ is a suffix of X. Then ℓ is simultaneously the largest integer such that P^ℓ is a suffix of V and the largest integer such that P^ℓ is a suffix of $V[1..i - 1]$ (since $\ell p < |V| - |X|$ by Lemma 12(c)). The former lets us uniquely determine ℓ. The latter implies that $\ell' := \ell + \frac{i' - i}{p}$ is the largest integer such that $P^{\ell'}$ is a suffix of $V[1..i' - 1]$. Since ℓ' is uniquely determined by V, so is i, and thus also $X = V[i..|V|]$. This concludes the proof that the string U can be uniquely determined from V. In particular, at most one such string exists.   □

3.3   The Upper Bound

Theorem 14. *A rooted tree with n nodes contains $\mathcal{O}(n)$ cubes anchored at its root.*

Proof. Let T be a tree with n nodes rooted in r. The whole proof reduces to proofs of the following two claims.

Claim. There are $\mathcal{O}(n)$ different cubes in T having a non-essential cube decomposition.

Proof. A non-essential decomposition of a cube is rightist, leftist unbalanced or a leftist decomposition of an np-cube. In each case, by Lemmas 6–8, there are $\mathcal{O}(n)$ different cubes with such a decomposition.   □

Claim. There are $\mathcal{O}(n)$ different p-cubes in T having an essential cube decomposition.

Proof. For each p-cube X^3 with an essential decomposition let us fix a single such decomposition UV and a single pair of nodes (u, v) that gives this decomposition.

If UV is a type-1 decomposition, we *charge* one token to the node u, otherwise we charge one token to v. By Observation 11 and Lemma 13, each node receives at most 3 tokens.   □

This concludes the proof of the theorem.   □

4   Powers in Trees

In this section we prove the announced bounds for powers$_\alpha$ for $\alpha > 1$.

Let S_m be a string $\mathsf{a}^m \mathsf{b} \mathsf{a}^m$. Note that S_m can be seen as a tree with a linear structure. Though the following fact can be treated as a folklore result, we provide its proof for completeness.

Theorem 15. *For every rational $\alpha \in (1, 2)$, we have* powers$_\alpha(S_m) = \Omega(|S_m|^2)$.

Proof. Let $\alpha = 1 + \frac{x}{y}$ where x, y are coprime positive integers. For every positive integer $c \leq \frac{m}{y}$, we construct $c(y - x)$ different powers of exponent α and length $cy\alpha$ that occur in S_m:

$$\mathsf{a}^i \mathsf{ba}^{cy-1-i} \mathsf{a}^{cx} \quad \text{for } cx \leq i < cy.$$

Note that $i < cy \leq m$ and $cy - 1 - i + cx < cy \leq m$, so they indeed occur as substrings of S_m. In total we obtain

$$\sum_{1 \leq c \leq \frac{m}{y}} c(y - x) = \Theta\left(\frac{m^2(y-x)}{y^2}\right) = \Theta(m^2)$$

different α-powers. Moreover, $|S_m| = \Theta(m)$, so this implies $\mathsf{powers}_\alpha(S_m) = \Omega(|S_m|^2)$. $\qquad\square$

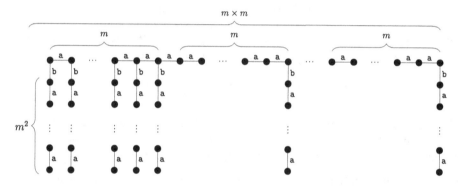

Fig. 3. Lower bound example T_m for powers of exponent $\alpha \in (2, 3)$.

Recall that for $\alpha = 2$, it has been shown that $\mathsf{powers}_2(n) = \Theta(n^{4/3})$ [4]. It turns out that the same bound applies for any $\alpha \in (2, 3)$. Moreover, the lower bound on $\mathsf{powers}_\alpha(n)$ is realized by the same family of trees called *combs*; see Fig. 3. A comb T_m consists of a path of length m^2 called the *spine*, with at most one *branch* attached to each node of the spine. Branches are located at positions $\{0, 1, 2, \ldots, m - 1, m, 2m, 3m, \ldots, m^2\}$ of the spine. All edges of the spine are labeled with letters a. Each branch is a path starting with a letter b, followed by m^2 edges labeled with letters a.

Theorem 16. *For every rational $\alpha \in (2, 3)$, we have $\mathsf{powers}_\alpha(T_m) = \Omega(|T_m|^{4/3})$.*

Proof. Let $\alpha = 2 + \frac{x}{y}$ where x, y are coprime positive integers. For every positive integer $c \leq \frac{m^2}{y}$, we construct $c(y - x)$ different α-powers of length $cy\alpha$ that occur in T_m:

$$(\mathsf{a}^i \mathsf{ba}^{cy-1-i})^2 \mathsf{a}^{cx} \quad \text{for } cx \leq i < cy.$$

Let us prove that these powers indeed occur in T_m. In [4] it was shown that for every $0 < j < m^2$ there are two branches whose starting nodes (on the spine) satisfy $distance(u, v) = j$. We apply this fact for $j = cy - 1$ and align letters b at the edges incident to u and v. Each branch contains m^2 edges labeled with a. Since $i < cy \le m^2$ and $cy - 1 - i + cx < cy \le m^2$, this is enough to extend an occurrence of $\mathsf{ba}^{cy-1}\mathsf{b}$ to an occurrence of $(\mathsf{a}^i\mathsf{ba}^{cy-1-i})^2\mathsf{a}^{cx}$. Altogether this gives $\Theta(m^4)$ different α-powers. Since $|T_m| = \Theta(m^3)$, the number of the considered powers in T_m is $\Omega(|T_m|^{4/3})$.   □

The upper bound for cubes and, consequently, for powers of rational exponent $\alpha \in (3, 4)$, is a consequence of the main result of the previous section.

Theorem 17. *For every rational $\alpha \ge 3$, we have $\mathsf{powers}_\alpha(n) = \mathcal{O}(n \log n)$.*

Proof. Recall that a *centroid* of a tree T is a node r such that each connected component of $T \setminus \{r\}$ is a tree with at most $\frac{n}{2}$ nodes. It is a well-known fact that every tree has a centroid.

We have already shown (Theorem 14) that the number of cubes in the tree T passing through a fixed node r is $\mathcal{O}(n)$. Now we need to count the remaining cubes in T. After removing the node r, the tree is partitioned into components $T_1, \ldots, T_k$. Hence, the number of cubes in T can be written as:

$$\mathsf{powers}_3(T) \le \mathcal{O}(|T|) + \sum_i \mathsf{powers}_3(T_i).$$

The components satisfy $\sum_i |T_i| = n - 1$ and $|T_i| \le \frac{n}{2}$, so a solution to this recurrence yields $\mathsf{powers}_3(n) = \mathcal{O}(n \log n)$. For every $\alpha \ge 3$, each power U^α of exponent α induces a cube U^3, so $\mathsf{powers}_\alpha(n) = \mathcal{O}(n \log n)$.   □

The final result related to the powers function may be interpreted as a generalization of the $2n$ upper bound on the number of different squares in a string.

Theorem 18. *For every $\alpha \ge 4$, $\mathsf{powers}_\alpha(n) = \Theta(n)$.*

Proof. For a string a^n, we have $\Theta(n/\alpha) = \Theta(n)$ distinct α-powers. For the proof of a linear upper bound, let T be a tree with n nodes and let r be any of its nodes. Let T_r be a directed tree obtained from T by selecting r as its root. Then any power U^α in T of exponent $\alpha \ge 4$ corresponds to square U^2 or $(U^R)^2$ in T_r. Thus, the conclusion follows from Lemma 5.   □

5   Final Remarks

We have presented an almost complete asymptotic characterization of the function powers_α specifying the maximum number of different powers of exponent α in a tree of given size. What remains is an exact asymptotic bound for powers_α, $\alpha \in [3, 4)$, for which we have shown an $\mathcal{O}(n \log n)$ upper bound.

It can be shown (see Fact 19) that a tree with n nodes contains $\mathcal{O}(n)$ different cubes of the form $(\mathsf{a}^i\mathsf{ba}^j)^3$. In comparison, the lower bound constructions for $\alpha < 3$ rely on counting powers of the form $(\mathsf{a}^i\mathsf{ba}^j)^\alpha$.

Fact 19. *A tree with n nodes with edges labeled with $\{a, b\}$ contains $\mathcal{O}(n)$ cubes of the form $(a^i b a^j)^3$.*

Proof. Let T be a tree with n nodes. Suppose that T is rooted at an arbitrary node r. Nevertheless, we bound the number of all cubes of the form $(a^i b a^j)^3$ in T, including those which are not anchored at r. We shall assign each such cube to a single node of T so that each node of T is assigned at most two cubes. For a particular occurrence of a cube $X^3 = (a^i b a^j)^3$ which starts in node u and ends in node v with $q = lca(u, v)$, we define the assignment as follows:

(A) if the string $val(u, q)$ contains at least two characters b, then the cube is assigned to node u,

(B) otherwise (in that case $val(q, v)$ contains at least two characters b) the cube is assigned to node v.

Let us prove that such procedure assigns at most one cube of type (A) and at most one cube of type (B) to a single node. If we fix the node and type of the assignment, we shall be able to uniquely recover the cube X^3 by going towards the root until we encounter the second edge labeled with b. Indeed, suppose u is a fixed node and consider the assignment of type (A). Let X_1 be the shortest prefix of $val(u, r)$ that contains exactly one character b and let X_2 be the shortest prefix of $val(u, r)$ that contains exactly two characters b. Then $X = a^{|X_1|-1} b a^{|X_2|-|X_1|-1}$. For the assignment of type (B), we use a symmetric procedure. □

We conclude with the following conjectures.

Conjecture 20 (Weak conjecture). $\mathsf{powers}_\alpha(n) = \Theta(n)$ for every $\alpha > 3$.

Conjecture 21 (Strong conjecture). $\mathsf{powers}_3(n) = \Theta(n)$.

References

1. Amir, A., Lewenstein, M., Lewenstein, N.: Pattern matching in hypertext. J. Algorithms **35**(1), 82–99 (2000)
2. Brešar, B., Grytczuk, J., Klavžar, S., Niwczyk, S., Peterin, I.: Nonrepetitive colorings of trees. Discrete Math. **307**(2), 163–172 (2007)
3. Crochemore, M., Hancart, C., Lecroq, T.: Algorithms on Strings. Cambridge University Press, Cambridge (2007)
4. Crochemore, M., Iliopoulos, C.S., Kociumaka, T., Kubica, M., Radoszewski, J., Rytter, W., Tyczyński, W., Waleń, T.: The maximum number of squares in a tree. In: Kärkkäinen, J., Stoye, J. (eds.) CPM 2012. LNCS, vol. 7354, pp. 27–40. Springer, Heidelberg (2012)
5. Fraenkel, A.S., Simpson, J.: How many squares can a string contain? J. Comb. Theory Ser. A **82**(1), 112–120 (1998)
6. Grytczuk, J.: Thue type problems for graphs, points, and numbers. Discrete Math. **308**(19), 4419–4429 (2008)
7. Ilie, L.: A simple proof that a word of length n has at most $2n$ distinct squares. J. Comb. Theory Ser. A **112**(1), 163–164 (2005)

8. Ilie, L.: A note on the number of squares in a word. Theor. Comput. Sci. **380**(3), 373–376 (2007)
9. Kociumaka, T., Pachocki, J., Radoszewski, J., Rytter, W., Waleń, T.: Efficient counting of square substrings in a tree. Theor. Comput. Sci. **544**, 60–73 (2014)
10. Kociumaka, T., Radoszewski, J., Rytter, W., Waleń, T.: Maximum number of distinct and nonequivalent nonstandard squares in a word. In: Shur, A.M., Volkov, M.V. (eds.) DLT 2014. LNCS, vol. 8633, pp. 215–226. Springer, Heidelberg (2014)
11. Kociumaka, T., Radoszewski, J., Rytter, W., Waleń, T.: Internal pattern matching queries in a text and applications. In: Indyk, P. (ed.) 26th Annual ACM-SIAM Symposium on Discrete Algorithms, SODA 2015, pp. 532–551. SIAM (2015)
12. Kubica, M., Radoszewski, J., Rytter, W., Waleń, T.: On the maximum number of cubic subwords in a word. Eur. J. Comb. **34**(1), 27–37 (2013)
13. Lothaire, M.: Combinatorics on Words. Cambridge Mathematical Library, second edn. Cambridge University Press, New York (1997)
14. Thue, A.: Über unendliche Zeichenreihen. Norske Videnskabers Selskabs Skrifter Mat.-Nat. Kl. **7**, 1–22 (1906)

Online Detection of Repetitions
with Backtracking

Dmitry Kosolobov[(✉)]

Ural Federal University, Ekaterinburg, Russia
dkosolobov@mail.ru

Abstract. In this paper we present two algorithms for the following problem: given a string and a rational $e > 1$, detect in the online fashion the earliest occurrence of a repetition of exponent $\geq e$ in the string.

1. The first algorithm supports the backtrack operation removing the last letter of the input string. This solution runs in $O(n \log m)$ time and $O(m)$ space, where m is the maximal length of a string generated during the execution of a given sequence of n read and backtrack operations.
2. The second algorithm works in $O(n \log \sigma)$ time and $O(n)$ space, where n is the length of the input string and σ is the number of distinct letters. This algorithm is relatively simple and requires much less memory than the previously known solution with the same working time and space.

Keywords: Repetition-free · Square-free · Online algorithm · Backtracking

1 Introduction

The study of algorithms analyzing different kinds of string periodicities forms an important branch of stringology. Repetitions of a given fixed order often play a central role in such investigations. We say that an integer p is a *period* of w if $w = (uv)^k u$ for some integer $k \geq 1$ and strings u and v such that $|uv| = p$. Given a rational $e > 1$, a string w such that $|w| \geq pe$ for a period p of w is called an *e-repetition*. A string is *e-repetition-free* if it does not contain an e-repetition as a substring. We consider algorithms recognizing e-repetition-free strings for any fixed $e > 1$. To be more precise, we say that an algorithm *detects e-repetitions* if it decides whether the input string is e-repetition-free. Further, we say that this algorithm detects e-repetitions *online* if it processes the input string sequentially from left to right and decides whether each prefix is e-repetition-free after reading the rightmost letter of that prefix.

In this paper we give two algorithms that detect e-repetitions online for a given fixed $e > 1$. The first one, which uses the ideas of the Apostolico-Breslauer algorithm [1], works on unordered alphabet and supports *backtracking*, the operation removing the last letter of the processed string. This solution requires $O(n \log m)$ time and $O(m)$ space, where m is the maximal length of a string generated during the execution of n given backtrack and read operations. Slightly

F. Cicalese et al. (Eds.): CPM 2015, LNCS 9133, pp. 295–306, 2015.
DOI: 10.1007/978-3-319-19929-0_25

modifying the proof from [10], one can show that this time is the best possible in the case of unordered alphabet. The second algorithm works on ordered alphabet and requires $O(n \log \sigma)$ time and linear space, where σ is the number of distinct letters in the input string and n is the length of this string. Although this result does not theoretically outperform the previously known solution [6], it is significantly less complicated and can be used in practice. Both algorithms report the position of the leftmost e-repetition.

Let us point out some previous results on the problem. Recall that a repetition of the form xx is called a *square*. A string is *square-free* if it is 2-repetition-free. Squares are, perhaps, the most extensively studied repetitions. The classical result of Thue [12] states that on a three-letter alphabet there are infinitely many square-free strings. How fast can one decide whether a string is square-free? It turns out that the orderedness of alphabet plays a crucial role here: while any algorithm detecting squares on unordered alphabet requires $\Omega(n \log n)$ time [10], it is unlikely that *any* superlinear lower bound exists in the case of ordered alphabet, in view of the recent result of the author [8]. So, we always emphasize whether an algorithm under discussion relies on order or not.

The best known offline (not online) results are the algorithm of Main and Lorentz [10] detecting e-repetitions in $O(n \log n)$ time and linear space on unordered alphabet, and Crochemore's algorithm [4] detecting e-repetitions in $O(n \log \sigma)$ time and linear space on ordered alphabets. Our interest in online algorithms detecting repetitions was partially motivated by problems in the artificial intelligence research (see [9]), where some algorithms use the online square detection. Apostolico and Breslauer [1] presented a parallel algorithm for this problem on an unordered alphabet. As a by-product, they obtained an online algorithm detecting squares in $O(n \log n)$ time and linear space, the best possible bounds as it was noted above. Later, online algorithms detecting squares in $O(n \log^2 n)$ [9] and $O(n(\log n + \sigma))$ [7] time were proposed. Apparently, their authors were unaware of the result of [1]. For ordered alphabet, Jansson and Peng [7] found an online algorithm detecting squares in $O(n \log n)$ time and Hong and Chen [6] presented an online algorithm detecting e-repetitions in $O(n \log \sigma)$ time and linear space.

An online algorithm for square detection with backtracking is in the core of the generator of random square-free strings described in [11]. Using our algorithm with backtracking, one can in a similar way construct a generator of random e-repetition-free strings for any fixed $e > 1$. This result might be useful in further studies in combinatorics on words.

The paper is organized as follows. In Sect. 2 we present some basic definitions and the key data structure, called catcher, which helps to detect repetitions. Section 3 contains an algorithm with backtracking. In Sect. 4 we describe a simpler solution without backtracking.

2 Catcher

A *string of length n* over the alphabet Σ is a map $\{1, 2, \ldots, n\} \mapsto \Sigma$, where n is referred to as the length of w, denoted by $|w|$. We write $w[i]$ for the ith letter of

w and $w[i..j]$ for $w[i]w[i+1]\ldots w[j]$. Let $w[i..j]$ be the empty string for any $i > j$. A string u is a *substring* of w if $u = w[i..j]$ for some i and j. The pair (i, j) is not necessarily unique; we say that i specifies an *occurrence* of u in w. A string can have many occurrences in another string. A substring $w[1..j]$ [resp., $w[i..n]$] is a *prefix* [resp. *suffix*] of w. For any i, j, the set $\{k \in \mathbb{Z}: i \leq k \leq j\}$ (possibly empty) is denoted by $[i..j]$; $(i..j]$ and $[i..j)$ denote $[i..j] \setminus \{i\}$ and $[i..j] \setminus \{j\}$ respectively.

We fix a rational constant $e > 1$ and use it throughout the paper. The input string is denoted by $text$ and $n = |text|$. Initially, $text$ is the empty string. We refer to the operation appending a letter to the right of $text$ as *read operation* and to the operation that cuts off the last letter of $text$ as *backtrack operation*.

Let us briefly outline the ideas behind our results. Both our algorithms utilize an auxiliary data structure based on a scheme proposed by Apostolico and Breslauer [1]. This data structure is called a *catcher*. Once a letter is appended to the end of $text$, the catcher checks whether $text$ has a suffix that is an e-repetition of length k such that $k \in [l..r]$ for some segment $[l..r]$ specific for this catcher. The segment $[l..r]$ cannot be arbitrary, so we cannot, for example, create a catcher with $l = 1$ and $r = n$. But, as it is shown in Sect. 3, we can maintain $O(\log n)$ catchers such that the union of their segments $[l..r]$ covers the whole range from 1 to n and hence these catchers "catch" each e-repetition in $text$. This construction leads to an algorithm with backtracking. In Sect. 4 we further reduce the number of catchers to a constant but this solution does not support backtracking.

In what follows we first describe an inefficient version of the read operation for catcher and show how to implement the backtrack operation; then, we improve the read operation and provide time and space bounds for the constructed catcher.

Let i and j be integers such that $1 \leq i \leq j < n$. Observe that if for some $k \leq i$, the string $text[k..n]$ is an e-repetition and $e(n - j) \geq n - k + 1$, then the string $text[i..j]$ occurs in $text[i+1..n]$ (see Fig. 1). Given this fact, the read operation works as follows. The catcher searches online occurrences of the string $text[i..j]$ in $text[i+1..n]$. If we have $text[i..j] = text[n-(j-i)..n]$, then the number $p = n - j$ is a period of $text[i..n]$. The catcher "extends" the repetition $text[i..n]$ to the left with the same period p. Then, the catcher online "extends" the repetition to the right with the same period p until an e-repetition is found. We say that *the catcher is defined by i and j*.

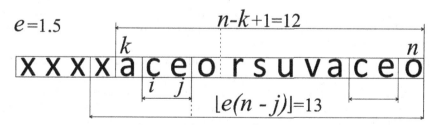

Fig. 1. An e-repetition $text[k..n]$, where $k = 5$, $n = 16$. Here $i = 6$, $j = 7$, and $text[i..j] = text[14..15]$.

Example 1. Consider $text = xxxxaceorsuv$. Denote $n = |text|$. Suppose $e = 1.5$. Let a catcher be defined by $i = 6$ and $j = 7$ (see Fig. 1). We consecutively perform the read operations that append the letters a, c, e, o to the right of $text$. The catcher online searches occurrences of the string $text[i..j] = ce$ (e.g., using the standard Boyer-Moore or Knuth-Morris-Pratt algorithm). Once we have $text = xxxxaceorsuvace$, the catcher has found an occurrence of $text[i..j]$: $text[n-1..n] = ce$. Hence, the string $text[i..n] = ceorsuvace$ has a period $p = n - j = 8$. The catcher "extends" this repetition to the left and thus obtains the repetition $text[i-1..n] = aceorsuvace$ with period p. Then the catcher online "extends" the found repetition to the right: after the next read operation, the catcher obtains the repetition $text[i-1..n] = aceorsuvaceo$ that is an e-repetition.

To support the backtrack operation, we store the states of the catcher in an array of states and when the backtracking is performed, we restore the previous state. For the described read operation, this approach has two drawbacks. First, the state does not necessarily require a fixed space, so the array of states may take a large amount of memory. Second, the catcher can spend a lot of time at some text locations (alternating backtracking with reading) and therefore the complexity of the whole algorithm can greatly increase. To solve these problems, our improved read operation performs the "extensions" of found repetitions and the searching of $text[i..j]$ simultaneously.

This approach relies on a *real-time constant-space string matching algorithm*, i.e., a constant-space algorithm that processes the input string online, spending constant time per letter; once the searched pattern occurs, the algorithm reports this occurrence. For unordered alphabet, we can use the algorithm of Galil and Seiferas [5] though in the case of ordered alphabet, it is more practical to use the algorithm of Breslauer, Grossi, and Mignosi [2].

The improved read operation works as follows. Denote $h = (j - i + 1)/2$. The real-time string matching algorithm searches for $text[i..i+\lceil h \rceil -1]$. It is easy to see that if we have $text[n-\lceil h \rceil +1..n] = text[i..i+\lceil h \rceil -1]$, then the number $p = (n - \lceil h \rceil + 1) - i$ is a period of $text[i..n]$. The catcher maintains a linked list P of pairs (p, l_p), where p is found in the described way and l_p is such that p is a period of $text[l_p+1..n]$ (initially $l_p = i - 1$). Each read operation tries to extend $text[l_p+1..n]$ with the same period p to the right and to the left. If $text[n] \neq text[n-p]$, then the catcher removes (p, l_p) from P. To extend to the left, we could assign $l_p \leftarrow \min\{l: text[l+1..n] \text{ has a period } p\}$ but the calculation of this value requires $O(n)$ time while we want to keep within the constant time on each read operation.

In order to achieve this goal, we will extend r symbols to the left after reading a letter. We choose $r = \lceil (e - 1)p/\lfloor h \rfloor \rceil$. Then one of two situations occurs at the moment when $text[i..j] = text[i+p..n]$ (i.e., an occurrence of $text[i..j]$ is found). Either we have $text[l_p] \neq text[l_p+p]$ (l_p cannot be "extended" to the left) or $text[l_p+1..n]$ is an e-repetition. Suppose $text[i..j] = text[i+p..n]$ and $text[l_p] = text[l_p+p]$. Since at this moment we have performed $\lfloor h \rfloor$ operations decreasing l_p by r, we have $l_p = i - 1 - \lfloor h \rfloor r$ and hence $n - l_p \geq p + \lfloor h \rfloor r$.

Thus, if we put $r = \lceil (e-1)p/\lfloor h \rfloor \rceil$, then $n - l_p \geq ep$ and therefore, $text[l_p+1..n]$ is an e-repetition. The following pseudocode clarifies this description.

```
 1: read a letter and append it to text (thereby incrementing n)
 2: feed the letter to the algorithm searching for text[i..i+⌈h⌉−1]
 3: if text[n−⌈h⌉+1..n] = text[i..i+⌈h⌉−1] then              ▷ found an occurrence
 4:     p ← (n − ⌈h⌉ + 1) − i;  l_p ← i − 1;         ▷ p is a period of text[l_p+1..n]
 5:     P ← P ∪ {(p, l_p)};
 6: for all (p, l_p) in P do
 7:     if text[n] ≠ text[n−p] then
 8:         P ← P \ {(p, l_p)};  ▷ text[l_p+1..n] cannot be "extended" to the right
 9:     else
10:         r ← ⌈(e − 1)p/⌊h⌋⌉;             ▷ maximal number of left "extensions"
11:         while l_p > 0 and r > 0 and text[l_p] = text[l_p+p] do
12:             l_p ← l_p − 1;  r ← r − 1;        ▷ "extend" text[l_p+1..n] to the left
13:             if n − l_p ≥ ep then           ▷ if text[l_p+1..n] is an e-repetition
14:                 detected e-repetition text[l_p+1..n]
```

A state of the catcher consists of the list P and the state of the string matching algorithm, $O(|P|+1)$ integers in total. To support the backtracking, we simply store the states of the catcher in an array of states.

Lemma 1. *Suppose that i and j define a catcher on text, n is the current length of text, and $c > 0$. If the conditions (i) $text[1..n-1]$ is e-repetition-free and (ii) $c(j - i + 1) \geq n - i$ hold, then each read or backtrack operation takes $O(c+1)$ time and the catcher occupies $O((c+1)(n-i))$ space.*

Proof. Clearly, at any time of the work, the array of states contains $n - i$ states. Each state occupies $O(|P|+1)$ integers. Hence, to estimate the required space, it suffices to show that $|P| = O(c)$. Denote $v = text[i..i+\lceil h \rceil - 1]$. It follows from the pseudocode that each $(p, l_p) \in P$ corresponds to a unique occurrence of v in $text[i+1..n]$. Thus, to prove that $|P| = O(c)$, it suffices to show that the string v has at most $O(c)$ occurrences in $text[i+1..n]$ at any time of the work of the catcher. Suppose v occurs at positions k_1 and k_2 such that $i < k_1 < k_2 < k_1 + |v|$. Hence, the number $k_2 - k_1$ is a period of v. Since $text[1..n-1]$ is e-repetition-free during the work of the catcher, we have $k_2 - k_1 > \frac{1}{e}|v|$. Therefore the string v always has at most $(n-i)/(\frac{1}{e}|v|)$ occurrences in the string $text[i+1..n]$. Finally, the inequalities $|v| \geq \frac{1}{2}(j-i+1)$ and $\frac{n-i}{j-i+1} \leq c$ imply $(n-i)/(\frac{1}{e}|v|) \leq 2ec = O(c)$.

Obviously, each backtrack operation takes $O(c)$ time. Any read operation takes at least constant time for each $(p, l_p) \in P$. But for some $(p, l_p) \in P$, the algorithm can perform $O((e-1)p/h) = O(p/h)$ iterations of the loop in lines 11–12 (see the value of r in line 10). Since $p \leq n - i$ for each $(p, l_p) \in P$, we have $p/h \leq 2(n - i)/(j - i + 1) \leq 2c$ and therefore, the loop performs at most $O(c)$ iterations. The loop is executed iff $text[l_p] = text[l_p+p]$. But since for each $(p, l_p) \in P$, the value of r is chosen in such a way that $text[l_p] = text[l_p+p]$ only if $text[i+p..n]$ is a proper prefix of $text[i..j]$ (see the discussion above), there are at most $(j - i + 1)/(\frac{1}{e}|v|) \leq 2e$ periods p for which the algorithm executes the loop. Finally, we have $O(|P| + 2ec) = O(c)$ time for each read operation. □

Lemma 2. *If for some k, the string $text[n-k+1..n]$ is an e-repetition and $n - i < k \le e(n - j)$, then a catcher defined by i and j detects this repetition.*

Proof. Let p be the minimal period of $text[n-k+1..n]$. Since $text[i..j]$ is a substring of $text[n-k+1..n]$ and $p \le \frac{k}{e} \le n-j$, the string $text[i..j]$ occurs at position $i + p$. Thus, the catcher detects this e-repetition when processes this occurrence (see Fig. 1). $\qquad\square$

We say that a *catcher covers* $[l..r]$ if the catcher is defined by integers i and j such that $n - i < n - r + 1 \le n - l + 1 \le e(n - j)$; by Lemma 2, this condition implies that if for some $k \in [l..r]$, the suffix $text[k..n]$ is an e-repetition, then the catcher detects this repetition. We also say that the catcher *covers a segment of length* $r - l + 1$. Note that if we append a letter to the end of $text$, the catcher still covers $[l..r]$. We say that a set S of catchers covers $[l..r]$ if $\bigcup_{C \in S}[l_C..r_C] \supset [l..r]$, where $[l_C..r_C]$ is a segment covered by catcher C.

3   Unordered Alphabet and Backtracking

Theorem 1. *For unordered alphabet, there is an online algorithm with backtracking that detects e-repetitions in $O(n \log m)$ time and $O(m)$ space, where m is the length of a longest string generated during the execution of a given sequence of n backtrack and read operations.*

Proof. As above, denote $n = |text|$. If $text$ is not e-repetition-free, our algorithm skips all read operations until backtrack operations make $text$ e-repetition-free. Therefore, in what follows we can assume that $text[1..n-1]$ is e-repetition-free and thus, all e-repetitions of $text$ are suffixes. In our proof we first give an algorithm without backtracking and then improve it to support the backtrack operation.

The Algorithm Without Backtracking. Our algorithm maintains $O(\log n)$ catchers that cover $[1..n-O(1)]$ and therefore "catch" almost all e-repetitions. For each $k \in [0.. \log n]$, we have a constant number of catchers covering adjacent segments of length 2^k. These segments are of the form $(l2^k..(l+1)2^k]$ for some integers $l \ge 0$ precisely defined below. Let us fix an integer constant s for which it is possible to create a catcher covering $(n-s2^k..n-(s-1)2^k]$. To show that such s exists, consider a catcher defined by $i = j = n - (s - 1)2^k$. By Lemma 2, this catcher covers $(n-s2^k..n-(s-1)2^k]$ iff $e(n - j) = e(s - 1)2^k \ge s2^k$ or, equivalently, $s \ge \lceil \frac{e}{e-1} \rceil$. As it will be clear below, to make our catchers fast, we must assume that $s > \frac{e}{e-1}$. Note that $s \ge 2$ since $e > 1$, and $s = 2$ implies $e > 2$.

Now we precisely describe the segments covered by our catchers. Denote $t_r = \max\{0, n - ((s - 1)2^r + (n \mod 2^r))\}$. For any integer $r \ge 0$, t_r is a nonnegative multiple of 2^r. Let $k \in [0.. \log n]$. The algorithm maintains catchers covering the following segments: $(t_{k+1}..t_{k+1}+2^k], (t_{k+1}+2^k..t_{k+1}+2\cdot2^k], (t_{k+1}+2\cdot2^k..t_{k+1}+3\cdot2^k],\ldots,(t_k - 2^k..t_k]$ (see Fig. 2). Thus, there are at most $\frac{1}{2^k}(t_k - t_{k+1}) \le s$ catchers for each such k. Obviously, the constructed segments cover $[1..n-s+1]$.

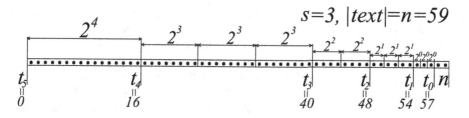

Fig. 2. A system of catchers covering $[1..n-s+1]$.

To maintain this system of catchers, the algorithm loops through all $k \in [0..\log n]$ such that $s2^k \leq n$ and, if n is a multiple of 2^k, creates a new catcher covering $(n - s2^k..n - (s-1)2^k]$; if, in addition, n is a multiple of 2^{k+1}, the algorithm removes two catchers covering $(n - s2^{k+1}..n - s2^{k+1} + 2^k]$ and $(n - s2^{k+1} + 2^k..n - (s-1)2^{k+1}]$. To prove that the derived system covers $[1..n-s+1]$, it suffices to note that if an iteration of the loop removes two catchers covering $(b_1..b_2]$ and $(b_2..b_3]$, for some b_1, b_2, b_3, then the next iteration creates a catcher covering $(b_1..b_3]$. We detect e-repetitions of lengths $2, 3, \ldots, s-1$ by a simple naive algorithm. In the following pseudocode we use the three-operand **for** loop like in the C language.

1: read a letter and append it to $text$ (thereby incrementing n)
2: check for e-repetitions of length $2, 3, \ldots, s-1$
3: **for** $(k \leftarrow 0; \; s2^k \leq n$ **and** $n \bmod 2^k = 0; \; k \leftarrow k+1)$ **do**
4:     create a catcher covering $(n - s2^k..n - (s-1)2^k]$
5:     **if** $n \bmod 2^{k+1} = 0$ **and** $n - s2^{k+1} \geq 0$ **then**
6:         remove the catcher covering $(n - s2^{k+1}..n - s2^{k+1} + 2^k]$
7:         remove the catcher covering $(n - s2^{k+1} + 2^k..n - (s-1)2^{k+1}]$

When the algorithm creates a catcher covering $(n - s2^k..n - (s-1)2^k]$, it has some freedom choosing integers i and j that define this catcher. We put $i = n - (s-1)2^k$ and $j = \max\{i, n - \lceil \frac{s}{e}2^k \rceil\}$. Indeed, in the case $j \neq i$ we have $e(n-j) = e\lceil \frac{s}{e}2^k \rceil \geq s2^k$ and, by Lemma 2, the catcher covers $(n - s2^k..n - (s-1)2^k]$; the case $j = i$ was considered above when we discussed the value of s.

Clearly, the proposed algorithm is correct. Now it remains to estimate the consumed time and space. Consider a catcher defined by integers i and j and covering a segment of length 2^k. Let us show that $j - i + 1 > \alpha 2^k$ for a constant $\alpha > 0$ depending only on e and s. We have $j - i + 1 = (s-1)2^k - \lceil \frac{s}{e}2^k \rceil + 1 > ((s-1) - \frac{s}{e})2^k$. The inequality $s > \frac{e}{e-1}$ implies $(s-1) - \frac{s}{e} > 0$ (here we use the fact that s is strictly greater than $\frac{e}{e-1}$). Hence, we can put $\alpha = (s-1) - \frac{s}{e}$.

Denote by n' the value of n at the moment of creation of the catcher. The algorithm removes this catcher when either $n' = n - s2^k$ or $n' = n - (s-1)2^k$. Thus, since $j - i + 1 > \alpha 2^k$ for some $\alpha > 0$, it follows from Lemma 1 that the catcher requires $O(1)$ time at each read operation and occupies $O(2^k)$ space. Hence, all catchers take $O(s \sum_{k=0}^{\log m} 2^k) = O(m)$ space and the algorithm requires $O(\log m)$ time at each read operation if we don't count the time for creation of catchers. We don't estimate this time in this first version of our algorithm.

The Algorithm With Backtracking. Now we modify the proposed algorithm to support the backtracking. Denote $n' = n + 1$. The backtrack operation is simply a reversed read operation: we loop through all $k \in [0..\log n']$ such that $s2^k \leq n'$ and, if n' is a multiple of 2^k, remove the catcher covering $(n' - s2^k..n' - (s-1)2^k]$; if, in addition, n' is a multiple of 2^{k+1}, the algorithm creates two catchers covering $(n - s2^{k+1}..n - s2^{k+1} + 2^k]$ and $(n - s2^{k+1} + 2^k..n - (s-1)2^{k+1}]$. Clearly, this solution is slow: if $n = 2^p$ for some integer p, then n consecutive backtrack and read operations require $O(n^2)$ time.

To solve this problem, we make the life of catchers longer. In the modified algorithm, the read and backtrack operations don't remove catchers but mark them as "removed" and the marked catchers still work some number of steps. If a backtrack or read operation tries to create a catcher that already exists but is marked as "removed", the algorithm just deletes the mark.

How long is the life of marked catcher? Consider a catcher defined by $i = n' - (s-1)2^k$ and $j = \max\{i, n' - \lceil \frac{s}{e}2^k \rceil\}$, where n' is the value of n at the moment of creation of the catcher in the corresponding read operation. The read operation marks the catcher as "removed" when either $n' = n - s2^k$ or $n' = n - (s-1)2^k$; our modified algorithm removes this marked catcher when $n' = n - (s+1)2^k$ or $n' = n - s2^k$ respectively, i.e., the catcher "lives" additional 2^k steps. The backtrack operation marks the catcher as "removed" when $n' = n+1$; we remove this catcher when $n' = n + \min\{2^k, n' - j\}$ (recall that the catcher cannot exist if $n < j$), i.e., the catcher "lives" additional $\min\{2^k, \lceil \frac{s}{e}2^k \rceil\} = \Theta(2^k)$ steps.

Let us analyze the time and space consumed by the algorithm. It is easy to see that for any $k \in [0..\log n]$, there are at most $s+2$ catchers covering segments of length 2^k. The worst case is achieved when we have s working catchers and two marked catchers. Now it is obvious that the modified algorithm, as the original one, takes $O(m)$ space and requires $O(\log m)$ time in each read or backtrack operation if we don't count the time for creation of catchers. The key property that helps us to estimate this time is that once a catcher covering a segment of length 2^k is created, it cannot be removed during any sequence of $\Theta(2^k)$ backtrack and read operations. To create this catcher, the algorithm requires $\Theta(2^k)$ time and hence, this time for creation is amortized over the sequence of $\Theta(2^k)$ backtrack and read operations. Thus, the algorithm takes $O(n \log m)$ overall time, where n is the number of read and backtrack operations.   □

4   Ordered Alphabet

It turns out that in some natural cases we can narrow the area of e-repetition search. More precisely, if $text[1..n-1]$ is e-repetition-free, then the length of any e-repetition of $text$ is close to the length of the shortest suffix v of $text$ such that v does not occur in $text[1..n-1]$. In the sequel, v is referred to as the *shortest unioccurrent suffix* of $text$. Denote $t = |v|$. Suppose u is a suffix of $text$ such that u is an e-repetition. Let us first consider some specific values of e.

Example 2. Let $e = 5$. We prove that $t \leq |u| < \frac{5}{4}t$. Denote by p a period of u such that $5p \leq |u|$. Since the suffix of length $t-1$ occurs in $text[1..n-1]$ and

$text[1..n-1]$ is 5-repetition-free, we have $|u| \geq t$. Suppose, to the contrary, $|u| \geq t + \frac{1}{4}t$. Then $t + p \leq t + \frac{1}{5}|u| \leq |u|$ and $text[n-t+1..n] = text[n-t-p+1..n-p]$ by periodicity of u (see Fig. 3a), a contradiction to the definition of t.

Example 3. Let $e = 1.5$. We show that $t \leq |u| < \frac{1.5}{0.5}t$. As above, we have $|u| \geq t$. Denote by p a period of u such that $1.5p \leq |u|$. Suppose $|u| \geq t + \frac{1}{0.5}t$ (or $t \leq \frac{0.5}{1.5}|u|$); then $t + p \leq t + \frac{1}{1.5}|u| \leq \frac{0.5}{1.5}|u| + \frac{1}{1.5}|u| = |u|$ and $text[n-t+1..n] = text[n-t-p+1..n-p]$ (see Fig. 3b), which contradicts to the definition of t.

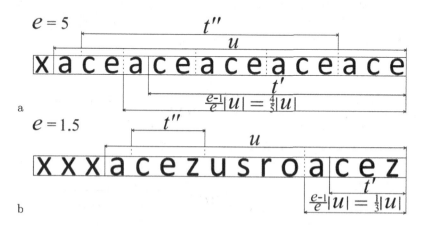

Fig. 3. (a) $n = 16$, $u = text[2..n]$, $t = 13$, $t' = 11$, $text[n-t'+1..n]=text[n-t'-2..n-3]$; (b) $n = 15$, $u = text[4..n]$, $t = 5$, $t' = 3$, $text[n-t'+1..n] = text[n-t'-7..n-8]$.

Lemma 3. *Let t be the length of the shortest unioccurrent suffix of text, and u be an e-repetition of text. If $text[1..n-1]$ is e-repetition-free, then $t \leq |u| < \frac{e}{e-1}t$.*

Proof. Clearly, u is a suffix. We have $t \leq |u|$ since the suffix of length $t-1$ occurs in $text[1..n-1]$ and $text[1..n-1]$ is e-repetition-free. Suppose, to the contrary, $|u| \geq \frac{e}{e-1}t$ (or $t \leq \frac{e-1}{e}|u|$). Denote by p the minimal period of u. We have $p \leq \frac{1}{e}|u|$. Further, we obtain $t+p \leq t+\frac{1}{e}|u| \leq \frac{e-1}{e}|u|+\frac{1}{e}|u| = |u|$, i.e., $t+p \leq |u|$. Finally, since p is a period of u, we have $text[n-t+1..n] = text[n-t-p+1..n-p]$ (see Fig. 3a,b). This contradicts to the definition of t. □

Lemma 3 describes the segment in which our algorithm must search e-repetitions. To cover this segment by catchers, we use the following technical lemma.

Lemma 4. *Let l and r be integers such that $0 \leq l \leq r < n$ and $c(n - r) > n - l$ for a constant $c > 0$. Then there is a set of catchers $\{c_k\}_{k=0}^{m}$ covering $(l..r]$ such that m is a constant depending on c and each c_k is defined by integers i_k and j_k such that $j_k - i_k + 1 \geq \frac{e-1}{2e}(n - r)$.*

Proof. Let us choose a number α such that $0 < \alpha < 1$. Denote $n-r = s$. Consider the following set of catchers $\{c_k\}_{k=0}^m$: c_k is defined by integers $i_k = n - \lceil(e\alpha)^k s\rceil$ and $j_k = n - \lceil\alpha(e\alpha)^k s\rceil$ (see Fig. 4). Denote $i'_k = n - (e\alpha)^k s$ and $j'_k = n - \alpha(e\alpha)^k s$. By Lemma 2, c_k covers $(n - e(n - j'_k)..i'_k] = (n - (e\alpha)^{k+1} s..i'_k]$. Thus, for any $k \in [0..m-1]$, the catcher c_k covers $(i'_{k+1}..i'_k]$ and therefore, the set $\{c_k\}_{k=0}^m$ covers the following segment:

$$(n - (e\alpha)^{m+1} s..i'_m] \cup (i'_m..i'_{m-1}] \cup (i'_{m-1}..i'_{m-2}] \cup \ldots \cup (i'_1..i'_0] = (n - (e\alpha)^{m+1} s..r].$$

Hence, if $e\alpha > 1$ and $(e\alpha)^{m+1} s \geq cs$, the set $\{c_k\}_{k=0}^m$ covers $(n - cs..r] \supset (l..r]$. Thus to cover $(l..r]$, we can, for example, put $\alpha = \frac{e+1}{2e}$ and $m + 1 = \lceil\frac{\log c}{\log(e\alpha)}\rceil = \lceil\frac{\log c}{\log(e+1)-1}\rceil$. Finally for $k \in [0..m]$, we have $j_k - i_k + 1 = \lceil(e\alpha)^k s\rceil - \lceil\alpha(e\alpha)^k s\rceil + 1 \geq (e\alpha)^k s - (\alpha(e\alpha)^k s + 1) + 1 = (e\alpha)^k(1 - \alpha)s \geq (1 - \alpha)s = \frac{e-1}{2e}(n - r)$. $\qquad\square$

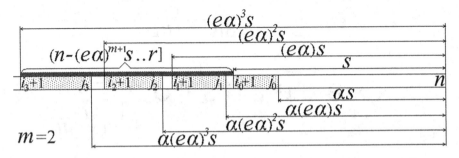

Fig. 4. The system $\{c_k\}_{k=0}^m$ with $m = 2$ (c_3 is depicted for clarity), $e \approx 1.5$, $\alpha \approx \frac{5}{6}$.

For each integer $i > 0$, denote by t_i the length of the shortest unioccurrent suffix of $text[1..i]$. We say that there is an online access to the sequence $\{t_i\}$ if any algorithm that reads the string $text$ sequentially from left to right can read t_i immediately after reading $text[i]$. The following lemma describes an online algorithm for e-repetition detection based on an online access to $\{t_i\}$. Note that the alphabet is not necessarily ordered.

Lemma 5. *If there is an online access to the sequence $\{t_i\}$, then there exists an algorithm that online detects e-repetitions in linear time and space.*

Proof. Our algorithm online reads the string $text$ while $text$ is e-repetition-free. Let $n = |text|$. Denote $l_n = \max\{0, n - \frac{e}{e-1}t_n\}$ and $r_n = n - t_n + 1$. By Lemma 3, to detect e-repetitions, it suffices to have a set of catchers covering $(l_n..r_n]$. But if the set covers only $(l_n..r_n]$, then we will have to update the catchers in each step i such that $r_{i-1} < r_i$ or $l_{i-1} > l_i$. To reduce the number of updates, we cover $(l_n..r_n]$ with significantly long left and right margins. Thus, some changes of l_n and r_n can be made without rebuilding of catchers.

We maintain two variables l and r such that $l \leq l_n \leq r_n \leq r$. Initially $l = r = 0$. To achieve linear time, we also require $n - r \leq 2(r - l)$. The following pseudocode explains how we choose l and r:

1: read a letter and append it to *text* (thereby we increment n and read t_n)
2: $l_n \leftarrow \max\{0, n - \frac{e}{e-1}t_n\}$; $r_n \leftarrow n - t_n + 1$;
3: **if** $l_n < l$ **or** $r_n > r$ **or** $n - r > 2(r - l)$ **then**
4:     $l \leftarrow \max\{0, n - \frac{2e}{e-1}t_n\}$; $r \leftarrow n - \frac{1}{2}t_n$;
5:     update catchers to cover $(l..r]$

The correctness is clear. Consider the space requirements. Since $n - r = \frac{1}{2}t_n$ and $n - l = \min\{n, \frac{2e}{e-1}t_n\}$, it follows that $c(n - r) > n - l$ for any $c > 4\frac{e}{e-1}$. Therefore, by Lemma 4, the algorithm uses a constant number of catchers and hence requires at most linear space. Denote by m the number of catchers.

Let us estimate the running time. Observe that r_n never decreases. In our analysis, we assume that to increase r_n, the algorithm performs $r_n - r_{n-1}$ increments. Obviously, our assumption does not affect the overall running time: to process any string of length k, the algorithm executes at most k increments. Also the algorithm performs k increments of n. We prove that the time required to maintain catchers is amortized over the sequence of increments of r_n and n.

Suppose the algorithm creates a set of catchers $\{c_k\}_{k=1}^m$ at some point. Denote by n' the value of n at this moment. Let us prove that it takes $O(t_{n'})$ time to create this set. For $k \in [1..m]$, let c_k be defined by i_k and j_k. By Lemma 4, for each $k \in [1..m]$, we have $j_k - i_k + 1 \geq \frac{e-1}{2e}(n' - r)$. Since $n' - r \geq \frac{4e}{4e}(n' - l) \geq \frac{e-1}{4e}(n' - i_k)$, we obtain $c(j_k - i_k + 1) \geq n' - i_k$ for any $c \geq 8e^2/(e - 1)^2$. Hence, by Lemma 1, it takes $O(n' - i_k)$ time to create the catcher c_k. Note that $n' - i_k \leq n' - l \leq \frac{2e}{e-1}t_{n'}$ and $\frac{1}{2}t_{n'} \leq n' - i_k$, i.e., $n' - i_k = \Theta(t_{n'})$. Therefore, to build the set $\{c_k\}_{k=1}^m$, the algorithm requires $O(\sum_{k=1}^m(n' - i_k)) = O(t_{n'})$ time.

Let us prove that to update the set $\{c_k\}_{k=1}^m$, the algorithm must execute $\Theta(t_{n'})$ increments of n or r_n. Consider the conditions of line 3:

1. To satisfy $l_n < l$ (clearly $l > 0$ in this case), since we have $l_{n-1} - l_n \leq \frac{e}{e-1}$ for any n, we must perform at least $(l_{n'} - l)/\frac{e}{e-1} = t_{n'}$ increments of n.
2. To satisfy $r_n > r$, we must execute $\lceil r - r_{n'} \rceil = \lceil t_n/2 \rceil$ increments of r_n.
3. To satisfy $n - r > 2(r - l)$, since $n - r = \frac{1}{2}t_{n'} + (n - n')$ and $2(r - l) \geq t_{n'}$, we must increase n by at least $\lceil \frac{1}{2}t_{n'} \rceil$.

The third condition forces us to update catchers after $\lceil \frac{4e}{e-1}t_{n'} \rceil$ increments of n. Indeed, we have $n - r = \lceil \frac{4e}{e-1}t_{n'} \rceil + n' - r \geq \frac{4e}{e-1}t_{n'} = 2(n' - l) > 2(r - l)$. Recall that for each $k \in [1..m]$, we have $n' - i_k = \Theta(t_{n'})$ and $j_k - i_k + 1 = \Theta(t_{n'})$. Hence, by Lemma 1, the catchers $\{c_k\}_{k=1}^m$ take $O(t_{n'})$ overall time. Thus the time required to maintain all catchers is amortized over the sequence of increments of n and r_n. $\square$

Theorem 2. *For ordered alphabet, there exists an algorithm that online detects e-repetitions in $O(n \log \sigma)$ time and linear space, where σ is the number of distinct letters in the input string.*

Proof. To compute the sequence $\{t_i\}$, we can use, for example, Weiner's online algorithm [13] (or its slightly optimized version [3]), which works in $O(n \log \sigma)$ time and linear space. Thus, the theorem follows from Lemma 5. $\square$

Corollary. *For constant alphabet, there exists an algorithm that online detects e-repetitions in linear time and space.*

Acknowledgement. The author would like to thank Arseny M. Shur for the help in the preparation of this paper and Gregory Kucherov for stimulating discussions.

References

1. Apostolico, A., Breslauer, D.: An optimal $O(\log \log n)$-time parallel algorithm for detecting all squares in a string. SIAM J. Comput. **25**(6), 1318–1331 (1996)
2. Breslauer, D., Grossi, R., Mignosi, F.: Simple real-time constant-space string matching. In: Giancarlo, R., Manzini, G. (eds.) CPM 2011. LNCS, vol. 6661, pp. 173–183. Springer, Heidelberg (2011)
3. Breslauer, D., Italiano, G.F.: Near real-time suffix tree construction via the fringe marked ancestor problem. J. Discrete Algorithms **18**, 32–48 (2013)
4. Crochemore, M.: Transducers and repetitions. Theor. Comput. Sci. **45**, 63–86 (1986)
5. Galil, Z., Seiferas, J.: Time-space-optimal string matching. J. Comput. Syst. Sci. **26**(3), 280–294 (1983)
6. Hong, J.J., Chen, G.H.: Efficient on-line repetition detection. Theor. Comput. Sci. **407**(1), 554–563 (2008)
7. Jansson, J., Peng, Z.: Online and dynamic recognition of squarefree strings. In: Jedrzejowicz, J., Szepietowski, A. (eds.) MFCS 2005. LNCS, vol. 3618, pp. 520–531. Springer, Heidelberg (2005)
8. Kosolobov, D.: Lempel-Ziv factorization may be harder than computing all runs. In: 32nd International Symposium on Theoretical Aspects of Computer Science (STACS 2015). Leibniz International Proceedings in Informatics (LIPIcs), vol. 30, pp. 582–593. Schloss Dagstuhl-Leibniz-Zentrum fuer Informatik (2015)
9. Leung, H., Peng, Z., Ting, H.-F.: An efficient online algorithm for square detection. In: Chwa, K.-Y., Munro, J.I. (eds.) COCOON 2004. LNCS, vol. 3106, pp. 432–439. Springer, Heidelberg (2004)
10. Main, M.G., Lorentz, R.J.: Linear time recognition of squarefree strings. In: Apostolico, A., Galil, Z. (eds.) Combinatorial Algorithms on Words, pp. 271–278. Springer, Heidelberg (1985)
11. Shur, A.M.: Generating square-free words efficiently. Accepted to WORDS 2013 Special Issue of Theoretical Computer Science (2014)
12. Thue, A.: Über unendliche zeichenreihen (1906). In: Selected Mathematical Papers of Axel Thue. Universitetsforlaget (1977)
13. Weiner, P.: Linear pattern matching algorithms. In: IEEE Conference Record of 14th Annual Symposium on Switching and Automata Theory, SWAT 2008, pp. 1–11. IEEE (1973)

Greedy Conjecture for Strings of Length 4

Alexander S. Kulikov[1]([✉]), Sergey Savinov[2], and Evgeniy Sluzhaev[1,2]

[1] St. Petersburg Department of Steklov Institute of Mathematics,
Saint Petersburg, Russia
kulikov@logic.pdmi.ras.ru
[2] St. Petersburg Academic University, Saint Petersburg, Russia

Abstract. In this short note, we prove that the greedy conjecture for the shortest common superstring problem is true for strings of length 4.

1 Introduction

In the shortest common superstring (SCS) problem one is given a set $\mathcal{S} = \{s_1, \ldots, s_n\}$ of n strings and the goal is to find a shortest string s such that each s_i is a substring of s. This is a well-known problem having applications in such areas as genome assembly and data compression.

The problem is known to be NP-hard [10] (even if the input strings have length 3 or if the alphabet is binary [2]) and APX-hard [12]. The fastest known exact solutions just reduce the problem to the Travelling salesman problem and have running time $(\sum_{i=1}^{n} |s_i|)^{O(1)} 2^n$ [1,6–8]. The currently best known approximation ratio is $2\frac{11}{23}$ [11]. Better upper bounds are known for special cases when input strings have bounded length [4,5]. A recent survey of known results (both practical and theoretical) is given in [3].

The well known greedy conjecture states that the following extremely simple greedy algorithm has approximation ratio 2 [14]: find two strings with longest mutual overlap and merge them into one string, repeat the process till only one string is left. This intriguing conjecture is open for more than 25 years already. There is a partial progress however: it is known that the conjecture is true for some orders in which the input strings are merged by the greedy algorithm [9,15].

In this short note, we consider another special case. We prove that the greedy conjecture is true if the input strings have length 4. (While for strings of length 3 the conjecture follows from the fact that the greedy algorithm achieves 2-approximation of the compression measure [13].) We do this by a careful analysis of possible overlaps produced by the greedy algorithm.

2 Preliminaries

An *overlap* $\mathrm{ov}(a, b)$ of two strings a and b is defined as the longest suffix of a which is also a prefix of b.

Let $\mathcal{S} = \{s_1, \ldots, s_n\}$ be a set of pairwise different 4-strings where by an r-string we denote just a string of length exactly r. Denote by s^{opt} and s^{gr} an

© Springer International Publishing Switzerland 2015
F. Cicalese et al. (Eds.): CPM 2015, LNCS 9133, pp. 307–315, 2015.
DOI: 10.1007/978-3-319-19929-0_26

optimal solution and a greedy solution for $\mathcal{S}$, respectively. Our goal is thus to show that

$$|s^{\mathrm{gr}}| \leq 2 \cdot |s^{\mathrm{opt}}|. \tag{1}$$

For technical reasons, we assume in this paper that in case of ties the greedy algorithm prefers strings of the form **aaaa** for $\mathbf{a} \in \Sigma$.

Let $\pi = (\pi_1, \ldots, \pi_n)$ be a permutation of $\{1, \ldots, n\}$. By overlapping n input strings in this particular order one gets a superstring of length

$$\sum_{i=1}^{n} |s_i| - \sum_{i=1}^{n-1} |\mathrm{ov}(s_{\pi_i}, s_{\pi_{i+1}})|. \tag{2}$$

The second term in the expression above is called a *compression* of $\mathcal{S}$ with respect to π. Thus, an equivalent reformulation of SCS is the following: find an order of n input strings that maximizes the compression. By c^{opt} and c^{gr} we denote the compression of the optimal solution s^{opt} and the greedy solution s^{gr}, respectively. By combining (1) with (2) we get an equivalent reformulation of what we need to prove:

$$4n - c^{\mathrm{gr}} \leq 2 \cdot (4n - c^{\mathrm{opt}}). \tag{3}$$

For a string t of length at most 3, let $\#^{\mathrm{opt}}(t)$ and $\#^{\mathrm{gr}}(t)$ be the number of overlaps that are equal to t in s^{opt} and s^{gr}, respectively. Similarly, let $\#_i^{\mathrm{opt}}$ and $\#_i^{\mathrm{gr}}$ be the number of overlaps of length exactly i. Then (3) is equivalent to

$$4n - \#_1^{\mathrm{gr}} - 2\#_2^{\mathrm{gr}} - 3\#_3^{\mathrm{gr}} \leq 2 \cdot (4n - \#_1^{\mathrm{opt}} - 2\#_2^{\mathrm{opt}} - 3\#_3^{\mathrm{opt}}) \tag{4}$$

or

$$2\#_1^{\mathrm{opt}} + 4\#_2^{\mathrm{opt}} + 6\#_3^{\mathrm{opt}} \leq 4n + \#_1^{\mathrm{gr}} + 2\#_2^{\mathrm{gr}} + 3\#_3^{\mathrm{gr}}. \tag{5}$$

Since $\#_1^{\mathrm{opt}} + \#_2^{\mathrm{opt}} + \#_3^{\mathrm{opt}} \leq n$ it suffices to prove that

$$2\#_3^{\mathrm{opt}} \leq 3\#_3^{\mathrm{gr}} + 2\#_2^{\mathrm{gr}} + \#_1^{\mathrm{gr}}. \tag{6}$$

Let $\mathcal{S}_3^{\mathrm{gr}}$ be the set of strings at the point of time when the greedy algorithm already merged all pairs of strings whose overlap is 3 and there is no more overlaps of length 3 left. In the following lemma we show that the number of overlaps equal to a 3-string t in the greedy solution cannot be much smaller than that of the optimal solution.

Lemma 1. *For any 3-string t, $\#^{\mathrm{gr}}(t) \geq \#^{\mathrm{opt}}(t) - 1$. Moreover, if $\#^{\mathrm{gr}}(t) = \#^{\mathrm{opt}}(t) - 1$ then $\mathcal{S}_3^{\mathrm{gr}}$ contains a string with prefix t and suffix t.*

Proof. Assume, for the sake of contradiction, that $\#^{\mathrm{gr}}(t) \leq \#^{\mathrm{opt}}(t) - 2$. The optimal solution contains $\#^{\mathrm{opt}}(t)$ overlaps equal to t and hence among the input n strings there are at least $\#^{\mathrm{opt}}(t)$ strings whose prefix is t and at least $\#^{\mathrm{opt}}(t)$ strings whose suffix is t. Now consider the set $\mathcal{S}_3^{\mathrm{gr}}$. Since $\#^{\mathrm{gr}}(t) \leq \#^{\mathrm{opt}}(t) - 2$, we conclude that $\mathcal{S}_3^{\mathrm{gr}}$ contains at least two strings whose suffix is t and at least two strings whose prefix is t. Hence there are two *different* strings in this set whose overlap is t which contradicts to the fact that there are no more 3-overlaps. $\square$

In the following the strings from S_3^{gr} are called *blocks*. For a 3-string t, we say that a block is *t-bad* if its suffix and its prefix are equal to t and moreover $\#^{gr}(t) = \#^{opt}(t) - 1$. We call a block *bad* if it is t-bad for a 3-string t and *good* otherwise. Let $\#_{bad}$ and $\#_{good}$ be the number of overlaps in all bad and good blocks, respectively. Then clearly $\#_{bad} + \#_{good} = \#_3^{gr}$ (recall that all the overlaps inside the blocks have length 3).

Note that if there are no bad blocks then already Lemma 1 is sufficient to prove (6): in this case, $\#^{gr}(t) \geq \#^{opt}(t)$ and therefore $\#_3^{gr} \geq \#_3^{opt}$.

Next, we consider bad blocks of fixed length: for a 3-string t, let

$$\chi_{=i}(t) = [S_3^{gr} \text{contains a } t-bad \text{ block of length exactly } i].$$

(throughout the paper, we use the standard Iverson brackets: $[P]$ is equal to 1 if P is true and is equal to 0 otherwise). Further, let

$$\chi_{=i} = \sum_{|t|=3} \chi_{=i}(t).$$

Functions $\chi_{>i}(t)$, $\chi_{\geq i}(t)$, $\chi_{>i}$, and $\chi_{\geq i}$ are defined in a similar fashion.

Note that $\chi_{=4} = 0$. Indeed a bad block of length 4 must have a form aaaa. Also, $\#^{opt}(aaaa) > 0$ and hence S contains another string starting or ending with aaa. But then the greedy algorithm must merge these two strings (as it prefers strings of the form aaaa). Hence for any 3-string t, $\chi_{\geq 5}(t)$ is exactly the number of t-bad blocks.

Lemma 2. *For any 3-string t,*

$$\min\{\#^{gr}(t), \#^{opt}(t)\} + \chi_{\geq 5}(t) = \#^{opt}(t).$$

Proof. Consider the following two cases:

1. $\#^{gr}(t) \geq \#^{opt}(t)$, then $\min\{\#^{gr}(t), \#^{opt}(t)\} = \#^{opt}(t)$ and $\chi_{\geq 5}(t) = 0$.
2. $\#^{gr}(t) < \#^{opt}(t)$, then by Lemma 1, $\min\{\#^{gr}(t), \#^{opt}(t)\} = \#^{opt}(t) - 1$. There is at least one block starting with t and ending with t. Moreover there cannot be two different such blocks as otherwise the greedy algorithm would merge them. Therefore, there is exactly one t-bad block, i.e., $\chi_{\geq 5}(t) = 1$. □

By summing up the equality from Lemma 2 over all strings t of length 3 we get the following corollary.

Corollary 1.

$$\sum_{|t|=3} \min\{\#^{gr}(t), \#^{opt}(t)\} + \chi_{\geq 5} = \#_3^{opt}.$$

Assume now that $\chi_{=5} = 0$. Then due to the fact that a bad block of length exactly i contains $i - 4$ overlaps we have that $2\chi_{>5} \leq \#_3^{gr}$. By adding twice the $\sum_{|t|=3} \min\{\#^{gr}(t), \#^{opt}(t)\}$ to both sides of this inequality and applying Corollary 1 we get

$$2\#_3^{\mathrm{opt}} \leq \#_3^{\mathrm{gr}} + 2 \sum_{|t|=3} \min\{\#^{\mathrm{gr}}(t), \#^{\mathrm{opt}}(t)\} \leq 3\#_3^{\mathrm{gr}},$$

which implies (6).

Hence the most tricky case is when there are bad blocks of length 5. The rest of the paper is devoted to the analysis of this case. Note that such blocks have the form ababa (for different letters $a, b \in \Sigma$) and therefore these are aba-bad blocks. To analyze such blocks carefully we introduce the following definitions. For a 3-string t and $1 \leq i \leq 5$, $B_i(t) = 0$ if either t is not of the form aba, or t is of the form aba and there is no block ababa. In the remaining case (i.e., t is of the form aba and there is a block ababa) B_i's are defined as follows:

$B_1(\mathsf{aba}) = [\#^{\mathrm{gr}}(\mathsf{bab}) > \#^{\mathrm{opt}}(\mathsf{bab})]$,

$B_2(\mathsf{aba}) = [\text{there exists a block with prefix ba or suffix ab}]$,

$B_3(\mathsf{aba}) = [\text{there exists a block except ababa with prefix ab or suffix ba}]$,

$B_4(\mathsf{aba}) = [\text{there exists a good block of length at least 5}$
$\qquad\qquad\qquad \text{containing aba or bab as a proper substring}]$,

$B_5(\mathsf{aba}) = [B_2(\mathsf{aba}) = 0 \text{ and } B_3(\mathsf{aba}) = 0 \text{ and there exists a bad block of}$
$\qquad\qquad\qquad \text{length at least 7 containing aba or bab as substring}]$.

Further, let for $1 \leq i \leq 5$, $B_i = \sum_{|t|=3} B_i(t)$.

Now we show B_i's provide an upper bound for the number of bad blocks of length exactly 5.

Lemma 3. $\chi_{=5} \leq \sum_{i=1}^{5} B_i$.

Proof. Note that if 3-string t is not of the form aba then $\chi_{=5}(t) = 0$ so the string t contributes nothing to the left-hand side of the inequality. We now focus on 3-strings t of the form aba. It is sufficient to prove the following inequality:

$$\chi_{=5}(\mathsf{aba}) \leq \sum_{i=1}^{5} B_i(\mathsf{aba}) \tag{7}$$

Assume that $\chi_{=5}(\mathsf{aba}) = 1$ and $B_1(\mathsf{aba}) = 0$ as otherwise the inequality holds for trivial reasons. From $B_1(\mathsf{aba}) = 0$ and Lemma 1 we have that $\#^{\mathrm{opt}}(\mathsf{bab}) - 1 \leq \#^{\mathrm{gr}}(\mathsf{bab}) \leq \#^{\mathrm{opt}}(\mathsf{bab})$. Since $\#^{\mathrm{gr}}(\mathsf{bab}) > 0$ (because S_3^{gr} contains the block ababa by definition of $\chi_{=5}(\mathsf{aba})$) we have that $\#^{\mathrm{opt}}(\mathsf{bab}) > 0$, i.e. the optimal solution has at least one overlap of the form bab. Depending of the location of this overlap in the optimal string we consider the following cases:

1. The overlap bab in the optimal solution is contained as a substring of ababa. Since $\#^{\mathrm{opt}}(\mathsf{aba}) > 0$, S contains at least one string except abab and baba containing aba as substring.
2. The overlap bab in the optimal solution is not in ababa. Hence S contains at least one string except abab and baba containing bab.

So in both cases there exists a string in S except abab and baba that contains $t' = $ aba or $t' = $ bab. This string is contained by some block $r \in S_3^{gr}$ and besides $r \neq $ ababa and $r \neq $ babab. Consider the following cases:

1. r is a good block. Then $B_4(\text{aba}) > 0$ if t' is a proper substring of r and $B_2(\text{aba}) + B_3(\text{aba}) > 0$ otherwise. Therefore (7) holds.
2. r is a bad block of length 5. Then this block has a form ababa or babab, a contradiction.
3. r is a bad block of length 6. If t' is a prefix or a suffix of r then $B_2(\text{aba}) + B_3(\text{aba}) > 0$. Otherwise either $r = r_1 t'_1 t'_2 t'_1 r_5 r_6$ or $r = r_1 r_2 t'_1 t'_2 t'_1 r_6$ where $t' = t'_1 t'_2 t'_3$. Since r is a bad block either $t'_1 t'_1 t'_2 t'_1 t'_1 t'_2$ or $t'_2 t'_1 t'_1 t'_2 t'_1 t'_1$. Finally, since either $t' = $ aba or $t' = $ bab in both these cases r has a prefix or a suffix ab or ba. Then $B_2(\text{aba}) + B_3(\text{aba}) > 0$ and (7) holds.
4. r is a bad block of length at least 7. Then $B_5(\text{aba}) > 0$ and (7) holds. $\square$

3   The Proof of the Main Theorem

In this section we prove the main result of this note: we first state auxiliary lemmas providing upper bounds on B_i's, then show how these lemmas imply the main result of the paper, and then provide the proofs of all the lemmas.

Lemma 4. $B_1 + \sum_{|t|=3} \min\{\#^{gr}(t), \#^{opt}(t)\} \leq \#_3^{gr}$.

Lemma 5. $B_2 \leq \#_2^{gr}$.

Lemma 6. $B_3 \leq \#_1^{gr} + \#_2^{gr}$.

Lemma 7. $B_4 \leq \#_{good}$.

Lemma 8. $B_5 + 2\chi_{>5} + \chi_{=5} \leq \#_{bad}$.

Theorem 1. *The greedy algorithm for strings of length 4 that prefers strings of the form* aaaa *in case of ties is 2-approximate.*

Proof. By adding the inequalities from Lemmas 5–8 to twice the inequality from Lemma 4 and applying equality $\#_{bad} + \#_{good} = \#_3^{gr}$ one gets

$$2B_1 + B_2 + B_3 + B_4 + B_5 + 2\chi_{>5} + \chi_{=5} + 2\sum_{|t|=3} \min\{\#^{gr}(t), \#^{opt}(t)\} \leq$$
$$3\#_3^{gr} + 2\#_2^{gr} + \#_1^{gr}.$$

By further adding the inequality from Lemma 3 we get

$$2\sum_{|t|=3} \min\{\#^{gr}(t), \#^{opt}(t)\} + 2\chi_{>5} + 2\chi_{=5} + B_1 \leq 3\#_3^{gr} + 2\#_2^{gr} + \#_1^{gr}.$$

Finally, applying Corollary 1 we get

$$2\#_3^{opt} + B_1 \leq 3\#_3^{gr} + 2\#_2^{gr} + \#_1^{gr}$$

which implies (6). $\square$

Proof (of Lemma 4). We have

$$B_1 + \sum_{|t|=3} \min\{\#^{\mathrm{gr}}(t), \#^{\mathrm{opt}}(t)\} = \sum_{|t|=3}(B_1(t) + \min\{\#^{\mathrm{gr}}(t), \#^{\mathrm{opt}}(t)\})$$

$$= \sum_{t \neq \mathsf{aba}}(B_1(t) + \min\{\#^{\mathrm{gr}}(t), \#^{\mathrm{opt}}(t)\}) + \sum_{\mathsf{a,b}\in\Sigma}(B_1(\mathsf{aba}) + B_1(\mathsf{bab})$$

$$+ \min\{\#^{\mathrm{gr}}(\mathsf{aba}), \#^{\mathrm{opt}}(\mathsf{aba})\} + \min\{\#^{\mathrm{gr}}(\mathsf{bab}), \#^{\mathrm{opt}}(\mathsf{bab})\})$$

To prove this lemma, we consider the following cases:

Case 1. If $t \neq \mathsf{aba}$ then $B_1(t) = 0$ and hence

$$B_1(t) + \min\{\#^{\mathrm{gr}}(t), \#^{\mathrm{opt}}(t)\} \leq \#^{\mathrm{gr}}(t)\,.$$

Case 2. If $t = \mathsf{aba}$ and $B_1(\mathsf{aba}) + B_1(\mathsf{bab}) = 0$ then

$$B_1(\mathsf{aba}) + B_1(\mathsf{bab}) + \min\{\#^{\mathrm{gr}}(\mathsf{aba}), \#^{\mathrm{opt}}(\mathsf{aba})\}$$
$$+ \min\{\#^{\mathrm{gr}}(\mathsf{bab}), \#^{\mathrm{opt}}(\mathsf{bab}))\} \leq \#^{\mathrm{gr}}(\mathsf{aba}) + \#^{\mathrm{gr}}(\mathsf{bab})$$

Case 3. If $t = \mathsf{aba}$ and $B_1(\mathsf{aba}) = 1$ then $B_1(\mathsf{bab}) = 0$ and, by definition of B_1, $\#^{\mathrm{gr}}(\mathsf{bab}) > \#^{\mathrm{opt}}(\mathsf{bab})$. Hence

$$B_1(\mathsf{aba}) + B_1(\mathsf{bab}) + \min\{\#^{\mathrm{gr}}(\mathsf{aba}), \#^{\mathrm{opt}}(\mathsf{aba})\} + \min\{\#^{\mathrm{gr}}(\mathsf{bab}), \#^{\mathrm{opt}}(\mathsf{bab})\}$$
$$= 1 + \min\{\#^{\mathrm{gr}}(\mathsf{aba}), \#^{\mathrm{opt}}(\mathsf{aba})\} + \min\{\#^{\mathrm{gr}}(\mathsf{bab}), \#^{\mathrm{opt}}(\mathsf{bab})\}$$
$$\leq 1 + \#^{\mathrm{gr}}(\mathsf{aba}) + \#^{\mathrm{opt}}(\mathsf{bab})$$
$$\leq 1 + \#^{\mathrm{gr}}(\mathsf{aba}) + \#^{\mathrm{gr}}(\mathsf{bab}) - 1 = \#^{\mathrm{gr}}(\mathsf{aba}) + \#^{\mathrm{gr}}(\mathsf{bab})$$

Case 4. If $t = \mathsf{aba}$ and $B_1(\mathsf{bab}) = 1$. This case is similar to Case 3. □

Proof (of Lemma 5). We show that $B_2 \leq \#_2^{\mathrm{gr}}$. If $B_2(t) > 0$ then t is of the form aba and there exists a block with prefix ba or suffix ab. Since $B_2(t) > 0$ there exists a pair of blocks: ababa and a block with a 2-prefix ba or a 2-suffix ab. Note that for different strings t these pairs of blocks do not intersect and cannot be merged with 2-overlaps because the sets $\{\mathsf{a}, \mathsf{b}\}$ are different. Note that at least one block in this pair must be merged with 2-overlap with some block otherwise this pair of blocks must be merged by the greedy algorithm. Thus $\sum_t B_2(t) < \#_2^{\mathrm{gr}}$

□

For Lemma 6 we need the following auxiliary definitions. Let $\mathrm{Pref}(\mathsf{a}, \mathsf{b}) = \emptyset$ if there is no block ababa and the set of blocks with prefix ab otherwise. Similarly, let $\mathrm{Suff}(\mathsf{a}, \mathsf{b}) = \emptyset$ if there is no block ababa and the set of blocks with suffix ba otherwise. Then it is easy to see that:

$$(\mathsf{a} \neq \mathsf{a}' \vee \mathsf{b} \neq \mathsf{b}') \Rightarrow (\mathrm{Pref}(\mathsf{a}, \mathsf{b}) \cap \mathrm{Pref}(\mathsf{a}', \mathsf{b}') = \emptyset \wedge \mathrm{Suff}(\mathsf{a}, \mathsf{b}) \cap \mathrm{Suff}(\mathsf{a}', \mathsf{b}') = \emptyset)$$

Let

$$\mathrm{Pref}(\mathsf{a}) = \bigcup_{\mathsf{b}\in\Sigma} \mathrm{Pref}(\mathsf{a}, \mathsf{b}) \quad \text{and} \quad \mathrm{Suff}(\mathsf{a}) = \bigcup_{\mathsf{b}\in\Sigma} \mathrm{Suff}(\mathsf{a}, \mathsf{b})\,.$$

Lemma 9. *If* a $\neq$ c *then the set of* 1- *and* 2-*suffixes of strings from* Suff(a) *does not intersect the set of* 1- *and* 2-*prefixes of strings from* Pref(c).

Proof. All 1-suffixes of strings from Suff(a) are equal to a while all 1-prefixes of strings from Pref(c) are equal to c, hence they do not intersect.

Assume that 2-suffix of $b_1 \in$ Suff(a) equals to 2-prefix of block $b_2 \in$ Pref(c). 2-suffix of block b_1 has the form xa and 2-prefix of b_2 has the form cy so $x = $ c, $y = $ a. Hence b_1 has form acaca and b_2 has form cacac, a contradiction. $\square$

Proof (of Lemma 6). $B_3(t) > 0$ only for $t = $ aba: $B_3 = \sum_t B_3(t) = \sum_a \sum_b B_3(\text{aba})$.

By Lemma 9 one can form sets X_1^a of 1-overlaps of strings from Suff(a) and Pref(a) counted in $\#_1^{\text{gr}}$. The lemma guarantees that these sets are disjoint. Similarly we can form sets X_2^a from 2-overlaps of strings from Suff(a) and Pref(a). Hence

$$\sum_a |X_1^a| \le \#_1^{\text{gr}} \text{ and } \sum_a |X_2^a| \le \#_2^{\text{gr}} . \tag{8}$$

Since for each nonzero $\chi_{=5}(\text{aba})$ there exists a block ababa we have, for each a,

$$\sum_b B_3(\text{aba}) \le \min\{|\operatorname{Pref}(a)|, |\operatorname{Suff}(a)|\} . \tag{9}$$

Since for each block ababa with $B_3(\text{aba}) > 0$ there exists by definition a string with prefix ab or suffix ba, we have:

$$\sum_b B_3(\text{aba}) < \max\{|\operatorname{Pref}(a)|, |\operatorname{Suff}(a)|\} . \tag{10}$$

Assume that $|\operatorname{Pref}(a)| \le |\operatorname{Suff}(a)|$ (the opposite case is symmetric). Let us show that

$$|X_1^a| + |X_2^a| \ge \sum_b B_3(\text{aba}) . \tag{11}$$

For this, assume the contrary. It follows from (9) and (10) that

$$|X_1^a| + |X_2^a| \le |\operatorname{Pref}(a)| - 1 \text{ and } |X_1^a| + |X_2^a| \le |\operatorname{Suff}(a)| - 2 .$$

Hence there exists at least one block from Pref(a) whose prefix is not used in overlaps and there exist at least two blocks from Suff(a) whose suffixes are not used in overlaps. But this prefix can be merged with one of these suffixes, a contradiction establishing (11).

Finally, by summing (11) for all a and applying (8) we get the required inequality:

$$\sum_{a \in \Sigma} \sum_{b \in \Sigma} B_3(\text{aba}) \le \sum_a (|X_1^a| + |X_2^a|) \le \#_1^{\text{gr}} + \#_2^{\text{gr}} . \qquad \square$$

Proof (of Lemma 7). If for some $a, b \in \Sigma$, $B_4(aba) + B_4(bab) = 1$, then either aba or bab is contained by a good block as a proper substring, so there exists at least one overlap by t in a good block. Hence

$$B_4 = \sum_{|t|=3} B_4(t) \leq \#_{\text{good}}.$$ □

Proof (of Lemma 8). Let $\#^i_{\text{bad}}$ be the number of overlaps in bad blocks of length i.

Let $B^i_5(t) = [i \geq 7 \wedge t = aba \wedge B_2(t) = B_3(t) = 0 \wedge$ there exists a block $ababa$ and a bad-block of length i which contains aba or bab as a proper substring]
By definition,

$$B_5(t) \leq \sum_{i \geq 7} B^i_5(t)$$

Since there are two 3-overlaps in bad blocks of length 6, $2\chi_{=6} = \#^6_{\text{bad}}$.

Consider bad blocks of length $i \geq 7$. Each such block contains $i - 4$ 3-overlaps. Note that overlaps aba or bab that are counted in $B^i_5(aba)$ cannot be neighbouring as otherwise B^i_5 would contain blocks $ababa$ and $babab$ (while this is only possible if the initial set S contains equal strings).

Let aba be the first overlap in a block. Then this block has prefix cab for $c \in \Sigma$. Its suffix also equals cab since this is a bad block. But in this case $B_2(aba) > 0$ and then $B_5(aba) = 0$, a contradiction. A similar contradiction arises if aba is the last overlap in a block. Thus, for $i \geq 7$ we have:

$$B^i_5 = \sum_s B^i_5(s) \leq \chi_{>5} \cdot \left\lceil \frac{i-6}{2} \right\rceil \leq \chi_{>5} \cdot (i-6).$$

Then

$$2\chi^i_{>5} + B^i_5 \leq 2\chi_{>5} + \chi_{>5} \cdot (i-6) = \chi_{>5} \cdot (i-4) \leq \#^i_{\text{bad}}.$$

Finally, we have:

$$2\chi_{>5} + \chi_{=5} + B_5 \leq \chi_{=5} + 2\chi_{=6} + \sum_{i \geq 7} (2\chi_{=i} + B^i_5)$$

$$\leq \#^5_{\text{bad}} + \#^6_{\text{bad}} + \sum_{i \geq 7} \#^i_{\text{bad}} = \#_{\text{bad}}.$$ □

4   Conclusion

We have proved that the greedy conjecture for the shortest common superstring problem is true for strings of length 4. Extending the proof to the case of 5-strings seems to be even more tedious. At the same time resolving such special cases does not seem to help to resolve the general case.

Acknowledgments. Research is partially supported by the Government of the Russian Federation (grant 14.Z50.31.0030) and Grant of the President of the Russian Federation (MK-6550.2015.1).

References

1. Bellman, R.: Dynamic programming treatment of the travelling salesman problem. J. ACM (J.ACM) **9**(1), 61–63 (1962)
2. Gallant, J., Maier, D., Storer, J.: On finding minimal length superstrings. J. Comput. Syst. Sci. **20**(1), 50–58 (1980)
3. Gevezes, T., Pitsoulis, L.: The shortest superstring problem. In: Rassias, T.M., Floudas, C.A., Butenko, S. (eds.) Optimization in Science and Engineering– In Honor of the 60th Birthday of Panos M. Pardalos, pp. 189–227. Springer, New York (2014)
4. Golovnev, A., Kulikov, A.S., Mihajlin, I.: Approximating shortest superstring problem using de bruijn graphs. In: Fischer, J., Sanders, P. (eds.) CPM 2013. LNCS, vol. 7922, pp. 120–129. Springer, Heidelberg (2013)
5. Golovnev, A., Kulikov, A.S., Mihajlin, I.: Solving SCS for bounded length strings in fewer than 2^n steps. Inf. Process. Lett. **114**(8), 421–425 (2014)
6. Held, M., Karp, R.M.: A dynamic programming approach to sequencing problems. J. Soc. Ind. Appl. Math. **10**(1), 196–210 (1962)
7. Karp, R.M.: Dynamic programming meets the principle of inclusion and exclusion. Oper. Res. Lett. **1**(2), 49–51 (1982)
8. Kohn, S., Gottlieb, A., Kohn, M.: A generating function approach to the traveling salesman problem. In: Proceedings of the 1977 annual conference. pp. 294–300. ACM (1977)
9. Laube, U., Weinard, M.: Conditional inequalities and the shortest common superstring problem. Int. J. Found. Comput. Sci. **17**(1), 247–247 (2006)
10. Maier, D., Storer, J.A.: A note on the complexity of the superstring problem. Princeton University Technical report 233 (1977)
11. Mucha, M.: Lyndon words and short superstrings. In: Proceedings of the TwentyFourth Annual ACM-SIAM Symposium on Discrete Algorithms. SODA 2013, Society for Industrial and Applied Mathematics (2013)
12. Ott, S.: Lower bounds for approximating shortest superstrings over an alphabet of size 2. In: Widmayer, P., Neyer, G., Eidenbenz, S. (eds.) WG 1999. LNCS, vol. 1665, pp. 55–64. Springer, Heidelberg (1999)
13. Tarhio, J., Ukkonen, E.: A greedy algorithm for constructing shortest common superstrings. In: Gruska, J., Rovan, B., Wiedermann, J. (eds.) MFCS 1986. LNCS, pp. 602–610. Springer, Heidelberg (1986)
14. Turner, J.: Approximation algorithms for the shortest common superstring problem. Inf. Comput. **83**(1), 1–20 (1989)
15. Weinard, M., Schnitger, G.: On the greedy superstring conjecture. In: Pandya, P.K., Radhakrishnan, J. (eds.) FSTTCS 2003. LNCS, vol. 2914, pp. 387–398. Springer, Heidelberg (2003)

Tighter Bounds for the Sum of Irreducible LCP Values

Juha Kärkkäinen[1]([✉]), Dominik Kempa[1], and Marcin Piątkowski[1,2]

[1] Helsinki Institute of Information Technology (HIIT) and
Department of Computer Science, University of Helsinki,
Helsinki, Finland
{juha.karkkainen,dominik.kempa}@cs.helsinki.fi
[2] Faculty of Mathematics and Computer Science,
Nicolaus Copernicus University, Torun, Poland
marcin.piatkowski@mat.umk.pl

Abstract. The suffix array is frequently augmented with the longest-common-prefix (LCP) array that stores the lengths of the longest common prefixes between lexicographically adjacent suffixes of a text. While the sum of the values in the LCP array can be $\Omega(n^2)$ for a text of length n, the sum of so-called irreducible LCP values was shown to be $\mathcal{O}(n \lg n)$ just a few years ago. In this paper, we improve the bound to $\mathcal{O}(n \lg r)$, where $r \leq n$ is the number of runs in the Burrows-Wheeler transform of the text. We also show that our bound is tight up to lower order terms (unlike the previous bound). Our results and the techniques used in proving them provide new insights into the combinatorics of text indexing and compression, and have immediate applications to LCP array construction algorithms.

1 Introduction

The suffix array [8], a lexicographically sorted array of the suffixes of a text, is the most important data structure in modern string processing. Modern text books spend dozens of pages in describing applications of suffix arrays, see e.g. [12]. In many of those applications, the suffix array needs to be augmented with the longest-common-prefix (LCP) array, which stores the lengths of the longest common prefixes between lexicographically adjacent suffixes (see e.g. [1,12]).

A closely related array is the Burrows–Wheeler transform (BWT) [2], which stores the characters preceding each suffix in the lexicographical order of the suffixes. The BWT was designed for text compression and is at the heart of many compressed text indexes [11]. If a text is highly repetitive (and thus highly compressible), its BWT tends to contain long runs of the same character. For example, for any string x and positive integer k, x and x^k have the same number of BWT runs [7]. Thus the number of BWT runs is a rough measure of the (in)compressibility of the text.

Partially supported by the project "Enhancing Educational Potential of Nicolaus Copernicus University" (project no. POKL.04.01.01-00-081/10).

© Springer International Publishing Switzerland 2015
F. Cicalese et al. (Eds.): CPM 2015, LNCS 9133, pp. 316–328, 2015.
DOI: 10.1007/978-3-319-19929-0_27

An entry LCP[i] in the LCP array is called reducible if BWT[i] = BWT[$i-1$], and *irreducible* otherwise. Given all the irreducible LCP values, the reducible values are easy to compute, which has been utilized in several LCP array construction algorithms [5,6,10,14]. There is also a compressed representation of the LCP array based on the fact that the number of irreducible values is one less than the number of BWT runs and thus small for repetitive texts [14].

The sum of irreducible LCP values was shown to be $\mathcal{O}(n \lg n)$ for a text of length n in [6], and there are LCP array construction algorithms relying on this bound [5,6,14]. In this paper, we improve the bound to $n \lg r + \mathcal{O}(n)$, where r is the number of BWT runs.[1] This immediately gives better time complexities for the algorithms in [6,14]. The tightness of our bound is shown by an infinite family of strings with the irreducible LCP sum of $n \lg r - \mathcal{O}(n)$.

Our proofs are derived in a setting where the suffix array, LCP array and BWT are defined for an arbitrary multiset of strings, closely related to the extended BWT introduced in [9]. This general setting offers cleaner combinatorics — for example, our upper and lower bounds match exactly in this setting — and could be useful for studying other topics in the combinatorics of text indexes.

2   Preliminaries

By $\mathcal{A}$ we denote a finite ordered set, called the *alphabet*. Elements of the alphabet are called *letters*. A finite *word* over $\mathcal{A}$ is a finite sequence of letters $w = a_0 a_1 \ldots a_{n-1}$. The length of a word w is defined as the number of its letters and denoted by $|w|$. An empty sequence of letters, called the *empty word*, is denoted by ε. The set of all finite words over $\mathcal{A}$ is denoted by $\mathcal{A}^*$ and the set of all non-empty words over $\mathcal{A}$ by $\mathcal{A}^+ = \mathcal{A}^* \setminus \{\varepsilon\}$.

For two words $x = a_0 a_1 \ldots a_{m-1}$ and $y = b_0 b_1 \ldots b_{n-1}$, their *concatenation* is $xy = x \cdot y = a_0 a_1 \ldots a_{m-1} b_0 b_1 \ldots b_{n-1}$. For a word w and an integer $k \geq 1$, we use w^k to denote the concatenation of k copies of w, also called a *power* of w. A word w is *primitive* if w is not a power of some other word. The *root* of a word w is defined as the shortest word $u = \text{root}(w)$ such that $w = u^k$ for some $k \geq 1$.

A word u is a *factor* of a word w if there exist words x and y such that $w = xuy$. Moreover, u is a *prefix* (resp. a *suffix*) of w if $x = \varepsilon$ (resp. $y = \varepsilon$). By $\text{lcp}(u, v)$ we denote the length of the *longest common prefix* of u and v. For a word $w = a_0 \ldots a_{n-1}$ and $i, j \in [0..n)$ by $w[i..j]$ we denote its factor of the form $a_i a_{i+1} \ldots a_j$. A factor/prefix/suffix u of w is *proper* if $u \neq w$. A (multi)set of words W is *prefix-free* if no word in W is a proper prefix of another word in W.

The order on letters of $\mathcal{A}$ can be extended in a natural way into the *lexicographical order* of words. For any two words x and y we have $x < y$ if x is a proper prefix of y or we have $x = uav_1$ and $y = ubv_2$, where $a, b \in \mathcal{A}$ and $a < b$.

Let $a \in \mathcal{A}$ and $x \in \mathcal{A}^*$. We define a *rotation* operator $\sigma : \mathcal{A}^+ \to \mathcal{A}^+$ as $\sigma(a \cdot x) \mapsto x \cdot a$, a *first-letter* operator $\tau : \mathcal{A}^+ \to \mathcal{A}$ as $\tau(a \cdot x) \mapsto a$ and a *reverse*

---

[1] Throughout the paper we use lg as a shorthand for $\log_2$.

operator $^- : \mathcal{A}^* \to \mathcal{A}^*$ as $\overline{\varepsilon} \mapsto \varepsilon$ and $\overline{a \cdot x} \mapsto \overline{x} \cdot a$. We say that a word w_1 is a *conjugate* of a word w_2 if $w_1 = \sigma^k(w_2)$ for some k.

The set of *infinite periodic words* is defined as $(\mathcal{A}^+)^\omega = \{w^\omega : w \in \mathcal{A}^+\}$, where $w^\omega = w \cdot w \cdot \ldots$ is the infinite power of w. We extend several of the above operators to infinite periodic words: $\mathrm{root}(w^\omega) = \mathrm{root}(w)$, $\sigma(w^\omega) = (\sigma(w))^\omega$, $\tau((a \cdot w)^\omega) = a$ and $\overline{(w^\omega)} = (\overline{w})^\omega$. Some key properties are given below:

- The operators are well defined: If $u^\omega = v^\omega$ for two words u and v, then $\mathrm{root}(u) = \mathrm{root}(v)$, $\sigma(u)^\omega = \sigma(v)^\omega$, $\tau(u) = \tau(v)$, and $\overline{u}^\omega = \overline{v}^\omega$.
- The rotation operator σ is in fact a suffix operator for infinite periodic words: $w^\omega = \tau(w^\omega)\sigma(w^\omega)$ for all $w \in \mathcal{A}^+$. However, unlike a suffix operator for finite words, σ has a well defined inverse σ^{-1}.
- The lexicographical ordering of infinite periodic words is not necessarily the same as their roots. For example, with alphabet $\{a < b\}$, $ab < aba$ but $(ab)^\omega > (aba)^\omega$. However, for two infinite periodic words with the roots u and v, either $u^\omega = v^\omega$ (and $\mathrm{lcp}(u^\omega, v^\omega) = \omega$) or $\mathrm{lcp}(u^\omega, v^\omega) \leq |u| + |v| - \gcd(|u|, |v|)$ due to properties of periodicity [3].

A *rooted tree* T is a directed graph that contains no undirected cycles and where every vertex is reachable from a single vertex called the *root*. If (u, v) is an edge in T, u is the *parent* of v and v is a *child* of u. If there is a directed path from a vertex u to vertex v, u is an *ancestor* of v and v is a *descendant* of u. A subgraph of T induced by the set of vertices that are reachable from a vertex u is called the *subtree* rooted at u.

A *compact trie* is a rooted tree, where the edges are labelled by non-empty words so that, for any vertex u with two outgoing edges (u, v_1) and (u, v_2), $\mathrm{lcp}(\mathrm{label}(u, v_1), \mathrm{label}(u, v_2)) = 0$. The edge labelling induces a vertex labelling: the label of a vertex u is the concatenation of edge labels on the path from the root to u. The compact trie $\mathrm{CTrie}(W)$ for a set W of words is the smallest compact trie that contains a vertex labelled by w for every $w \in W$. If W is prefix-free, a vertex v in $\mathrm{CTrie}(W)$ is labelled by a word in W if and only if v is a leaf. For $W \subseteq (\mathcal{A}^+)^\omega$, the leafs and the leaf edges in $\mathrm{CTrie}(W)$ are labelled by infinite periodic words, but other edges and vertices have finite labels.

3   Cyclic Suffixes

In this section, we define a generalization of the suffix array and related data structures based on the concept of cyclic suffixes.

Let $W = \{\!\{w_i\}\!\}_{i=1}^s$ be a multiset[2] of words and $n = \sum_{i=1}^s |w_i|$. The set of positions of W is defined as the set of integer pairs $\mathrm{pos}(W) := \{\langle i, p \rangle : i \in [1..s], p \in [0..|w_i|)\}$. For a position $\langle i, p \rangle \in \mathrm{pos}(W)$ we define a *cyclic suffix* $W_{\langle i, p \rangle} := (\sigma^p(w_i))^\omega \in (\mathcal{A}^+)^\omega$. The multiset of all cyclic suffixes of W is defined as $\mathrm{suf}(W) := \{\!\{W_{\langle i, p \rangle} : \langle i, p \rangle \in \mathrm{pos}(W)\}\!\}$.

---

[2] We use the double brace notation $\{\!\{\cdot\}\!\}$ to denote a multiset as opposed to a set.

We define two multisets V and W to be *cyclically equivalent* if $\mathrm{suf}(V) = \mathrm{suf}(W)$. It is easy to see that the corresponding equivalence classes are closed under conjugation of words in the multiset. Indeed, the restriction of cyclic equivalency to multisets of primitive words is the multiset conjugacy relation defined in [9]. The following lemma illustrates some further properties of our extension.

Lemma 1. *For any multiset of words* $W = \{\!\{w_i\}\!\}_{i=1}^{s}$ *there exists a multiset of primitive words* $V = \{\!\{v_i\}\!\}_{i=1}^{t}$, $t \geq s$, *such that* $\mathrm{suf}(W) = \mathrm{suf}(V)$, *and a set (not a multiset) of words* $U = \{u_i\}_{i=1}^{q}$, $q \leq s$, *such that* $\mathrm{suf}(W) = \mathrm{suf}(U)$.

Proof. To obtain V we replace each non-primitive word $w = v^k \in W$, where $v = root(w)$, with k occurrences of the primitive word v. To obtain U we replace each word w having k occurrences in W with a single word $u = w^k$.   □

Over the multiset $\mathrm{pos}(W)$, we define a total order $\preceq_W$. We say that $\langle i, p \rangle \preceq_W \langle i', p' \rangle$ if $W_{\langle i, p \rangle} < W_{\langle i', p' \rangle}$, or $W_{\langle i, p \rangle} = W_{\langle i', p' \rangle}$ and $\langle i, p \rangle \leq \langle i', p' \rangle$, where the last comparison is the usual integer pair comparison, i.e. $(i_1, j_1) < (i_2, j_2)$ if $i_1 < i_2$ or $i_1 = i_2$ and $j_1 < j_2$.

The *(cyclic) suffix array* of a multiset of words W is defined as an array $\mathrm{SA}_W[j] = \langle i_j, p_j \rangle$, where $\langle i_j, p_j \rangle \in \mathrm{pos}(W)$ for all $j \in [0..n)$ and $\langle i_{j-1}, p_{j-1} \rangle \prec_W \langle i_j, p_j \rangle$ for all $j \in [1..n)$. Note that for two cyclically equivalent multisets V and W, we may have $\mathrm{SA}_V \neq \mathrm{SA}_W$ but always $V_{\mathrm{SA}_V[j]} = W_{\mathrm{SA}_W[j]}$ for all j.

The *longest-common-prefix array* $\mathrm{LCP}_W[1..n]$ is defined as $\mathrm{LCP}_W[j] = \mathrm{lcp}\big(W_{\mathrm{SA}[j-1]}, W_{\mathrm{SA}[j]}\big)$. The *distinguishing prefix array* $\mathrm{DP}_W[1..n]$ is defined as $\mathrm{DP}_W[i] = \mathrm{LCP}_W[i] + 1$. Note that we can have $\mathrm{LCP}_W[i] = \omega = \mathrm{DP}_W[i]$.

The *Burrows-Wheeler transform* $\mathrm{BWT}_W[0..n)$ (also denoted $\mathrm{BWT}(W)$) is defined as $\mathrm{BWT}_W[j] = \tau\big(\sigma^{-1}(W_{\mathrm{SA}[j]})\big)$. This definition is a natural generalization of the original one [2] defined for a single (not necessarily primitive) word and the one in [9] defined for a multiset of primitive words. It is easy to see that if multisets V and W are cyclically equivalent, then $\mathrm{LCP}_V = \mathrm{LCP}_W$, $\mathrm{DP}_V = \mathrm{DP}_W$ and $\mathrm{BWT}_V = \mathrm{BWT}_W$.

Let v be a word of length n and $\widehat{v}$ be obtained from v by sorting its letters. The *standard permutation* [4] of v is the permutation corresponding to the *stable* sorting of the letters, i.e., it is the mapping $\Psi_v : [0..n) \rightarrow [0..n)$ such that: for each $i \in [0..n)$ we have $\widehat{v}[i] = v[\Psi_v(i)]$ and for $\widehat{v}[i] = \widehat{v}[j]$ the relation $i < j$ implies $\Psi_v(i) < \Psi_v(j)$. Let IBWT be the mapping that maps a word v into a multiset of (primitive) words W as follows. Let Ψ_v be a standard permutation of v and $C = \{c_i\}_{i=1}^{s}$ its disjoint cycle decomposition. Then $W = \{\!\{w_i\}\!\}_{i=1}^{s}$ and for each $i \in [1..s]$ and $j \in [0..|c_i|)$ we define $w_i[j] = v[\Psi_v(c_i[j])]$. The mapping IBWT is the inverse of BWT in the sense that $\mathrm{BWT}(\mathrm{IBWT}(v)) = v$ for every word v. Thus the mapping from a word v to the cyclical equivalence class of $\mathrm{IBWT}(v)$ is a bijection (see [9]).

Example 1. Let $W = \{\!\{ab, abaaba\}\!\}$. We have $v = \mathrm{BWT}(W) = bbaabaaa$ (having $r = 4$ runs) and $\Psi_v = (0,2,5)(1,3,6)(4,7)$. Then $\mathrm{IBWT}(v) = \{\!\{aab, aab, ab\}\!\}$, which is cyclically equivalent to W.

The *suffix tree* of W, denoted by $\text{STree}(W)$, is the compact trie of suffixes $\text{CTrie}(\text{suf}(W))$. If $\text{suf}(W)$ is a multiset, a single vertex in $\text{STree}(W)$ represents all copies of a suffix, and the number of leaves in $\text{STree}(W)$ is the number of *distinct* words in $\text{suf}(W)$.

4 Irreducible Sums

For a multiset of words W, we say that a value $\text{LCP}_W[i]$ is *reducible* if $\text{BWT}_W[i-1] = \text{BWT}_W[i]$ and *irreducible* otherwise. Observe that if $\text{LCP}_W[i] = \omega$, then this value is obviously reducible. We say that a value $\text{DP}_W[i]$ is irreducible if the corresponding value $\text{LCP}_W[i]$ is irreducible. Let $\Sigma\text{lcp}(W)$ denote the sum of all LCP_W values, $\Sigma\text{ilcp}(W)$ the sum of all irreducible LCP_W values, and $\Sigma\text{idp}(W)$ the sum of irreducible DP_W values. Note, that $\Sigma\text{idp} = \Sigma\text{ilcp} + r - 1$, where r is the number of runs in the BWT. For technical reasons we analyze Σidp rather than Σilcp.

We define a *lexicographically adjacent repeat (LAR)* in a multiset W as a tuple $(\langle i,p\rangle, \langle j,q\rangle, \ell)$ such that $\langle i,p\rangle, \langle j,q\rangle \in \text{pos}(W)$, ℓ is a non-negative integer, $\text{lcp}(W_{\langle i,p\rangle}, W_{\langle j,q\rangle}) \geq \ell$, $\langle i,p\rangle \prec_W \langle j,q\rangle$ and there exists no $\langle i',p'\rangle$ such that $\langle i,p\rangle \prec_W \langle i',p'\rangle \prec_W \langle j,q\rangle$ i.e., $W_{\langle i,p\rangle}$ and $W_{\langle j,q\rangle}$ are lexicographically adjacent suffixes with a common prefix of length (at least) ℓ. A LAR $(\langle i,p\rangle, \langle j,q\rangle, \ell)$ is *left-maximal* if $(\langle i,p-1\rangle, \langle j,q-1\rangle, \ell+1)$ is not a LAR.

Lemma 2. *The number of left-maximal LARs in W equals $\Sigma\text{idp}(W)$.*

Proof. Clearly, the set of all LARs is exactly $\{(\text{SA}[i-1], \text{SA}[i], \ell) : i \in [1..n], \ell \in [0..\text{DP}[i])\}$, and a LAR $(\text{SA}[i-1], \text{SA}[i], \ell)$ is left-maximal if and only if $\text{DP}[i]$ is irreducible. $\qquad\square$

Let T be a rooted tree and $\leq$ a total order over the leaves of T. Let u and v be leaves of T, and let x be the nearest common ancestor of u and v. The pair (u,v) is called a *dispersal pair* if $u < v$ and the subtree rooted at x contains no leaf w such that $u < w < v$. Let $D_x(T, \leq)$ denote the set of dispersal pairs with x as the nearest common ancestor. The *dispersal value* of T with respect to $\leq$, denoted by $d(T, \leq)$, is the number of dispersal pairs in T.

Let $\overline{\text{suf}}(W) = \{\!\{\overline{w} : w \in \text{suf}(W)\}\!\}$ be the multiset of *reverse suffixes* of a multiset W. The reverse suffix tree $\overline{\text{STree}}(W)$ of W is $\text{CTrie}(\overline{\text{suf}}(W))$. Define a total order $\leq_W$ over the leaves of $\overline{\text{STree}}(W)$ by $u \leq_W v \iff \langle i,p\rangle \preceq_W \langle j,q\rangle$, where $\overline{W_{\langle i,p\rangle}}$ is the label of u and $\overline{W_{\langle j,q\rangle}}$ is the label of v. If $\text{suf}(W)$ contains duplicates, any of the identical reverse suffixes can be used as the representative of a vertex.

Lemma 3. $d(\overline{\text{STree}}(W), \leq_W) = \Sigma\text{idp}(W)$.

Proof. Let $(\text{SA}[i-1], \text{SA}[i], \ell)$ be a left-maximal LAR, i.e., $\text{DP}[i] > \ell$ is irreducible. Let x be the length ℓ prefix of $W_{\text{SA}[i]}$, and let y and y' be infinite periodic words such that $xy = W_{\text{SA}[i-1]}$ and $xy' = W_{\text{SA}[i]}$. Then $\overline{x}$ is the longest common prefix of $\overline{y}$ and $\overline{y'}$. Let u, v and v' be the vertices of $\overline{\text{STree}}(W)$ that are labelled by $\overline{x}$, $\overline{y}$ and $\overline{y'}$, respectively. Then $v <_W v'$ and we will show that (v, v') is a dispersal pair. Suppose (v, v') is not a dispersal pair. Then there exists a leaf v'' descendant to u such that $v <_W v'' <_W v'$. If $\overline{y''}$ is the label of v'', then $xy'' \in \text{suf}(W)$ and $xy < xy'' < xy'$, which contradicts xy and xy' being adjacent in SA. Thus (v, v') is a dispersal pair. This mapping from left-maximal LARs to dispersal pairs is clearly injective, and thus $d(\overline{\text{STree}}(W), \leq_W) \geq \Sigma\text{idp}(W)$.

Let (v, v') be a dispersal pair in $\overline{\text{STree}}(W)$, and let u be the nearest common ancestor of v and v'. Let y, y' and x be words such that $\overline{x}$, $\overline{y}$ and $\overline{y'}$ are the labels of u, v and v', respectively. Then $xy, xy' \in \text{suf}(W)$, $xy < xy'$ and $\tau(\sigma^{-1}(xy)) \neq \tau(\sigma^{-1}(xy'))$. Let i be the largest integer such that $W_{\text{SA}[i]} = xy$ and i' the smallest integer such that $W_{\text{SA}[i']} = xy'$. Then $i < i'$ and we will show that $i = i' - 1$. Suppose $i < i' - 1$ and let $i'' = i' - 1$. Then we must have $W_{\text{SA}[i]} < W_{\text{SA}[i'']} < W_{\text{SA}[i']}$ and x is a prefix of $W_{\text{SA}[i'']}$. If y'' is the word such that $xy'' = W_{\text{SA}[i'']}$, then $\overline{y''} \in \overline{\text{suf}}(W)$ has $\overline{x}$ as a prefix. If v'' is the leaf in $\overline{\text{STree}}(W)$ labelled by $\overline{y''}$, then $v <_W v'' <_W v'$, which contradicts (v, v') being a dispersal pair. Thus $i = i' - 1$ and $(\text{SA}[i-1], \text{SA}[i], |x|)$ is a left-maximal LAR. This mapping from dispersal pairs to left-maximal LARs is clearly injective and thus $d(\overline{\text{STree}}(W), \leq_W) \leq \Sigma\text{idp}(W)$. $\qquad\square$

5 $n \lg n$ Upper Bound

We will now derive upper bounds on the maximum dispersal value of any tree with n leaves. By Lemma 3, these bounds are upper bounds for Σidp, too.

Define, for $n > 0$ and $k \in [1..\lfloor n/2 \rfloor]$,

$$d(1) = 0$$
$$d(n) = \max_{i \in [1..\lfloor n/2 \rfloor]} d(n, i) \quad \text{when } n > 1$$
$$d(n, k) = d(k) + d(n - k) + \min\{2k, n - 1\}$$

Lemma 4. $d(n) = \max\{d(T, \leq)\}$, where the maximum is taken over any rooted tree T with n leaves and any total order $\leq$ on the leaves of T.

Proof. We will first prove that we can restrict ourselves to proper binary trees, where every non-leaf vertex has exactly two children. Let T be a tree with a leaf order $\leq$, and let u be a vertex with at least three children v_1, v_2 and v_3. Let T' be the tree obtained from T by adding a vertex u' and replacing the edges (u, v_1) and (u, v_2) with (u, u'), (u', v_1) and (u', v_2). Let w_1, w_2 and w_3 be leaves in the subtrees rooted at v_1, v_2 and v_3, respectively. Then, (w_1, w_2) could be a dispersal pair in T' but not in T if $w_1 < w_3 < w_2$. However, any dispersal pair in T is a dispersal pair in T' too. Thus $d(T, \leq) \leq d(T', \leq)$. The above procedure can be repeated as long as the tree contains vertices with more than two children to obtain a binary tree. Furthermore, one can similarly show that unary vertices can be removed without removing any dispersal pairs to obtain a proper binary tree.

Let then T be a proper binary tree of size (number of leaves) $n \geq 2$ with a leaf order $\leq$. Let T_L and T_R be the left and right subtree of T of sizes k and $n-k$, respectively. W.l.o.g., assume that $k \leq n - k$. Let $\leq_L$ ($\leq_R$) be the leaf order $\leq$ restricted to the left (right) subtree. Let $D_{\text{root}}(T, \leq)$ be the set of dispersal pairs with the root of T as the nearest common ancestor. Then, clearly,

$$d(T, \leq) = d(T_L, \leq_L) + d(T_R, \leq_R) + |D_{\text{root}}(T, \leq)| \,.$$

If $(u, v) \in D_{\text{root}}(T, \leq)$, then u and v are adjacent in the order $\leq$, and one of u and v is in T_L and the other is in T_R. A leaf u can be involved with at most two pairs in $D_{\text{root}}(T, \leq)$, once with its immediate predecessor in $\leq$ and once with its immediate successor. Thus $|D_{\text{root}}(T, \leq)| \leq 2k$. Furthermore, if $k = n - k$, at least one of the leaves in T_L is the first in $\leq$, the last in $\leq$ or adjacent to another leaf in T_L, and thus involved in at most one pair in $D_{\text{root}}(T, \leq)$. Then $|D_{\text{root}}(T, \leq)| \leq 2k - 1 = n - 1$. It is now easy to see that $d(n)$ is an upper bound on the dispersal value over trees of size n, by induction on n:

$$\begin{aligned} d(T, \leq) &= d(T_L, \leq_L) + d(T_R, \leq_R) + |D_{\text{root}}(T, \leq)| \\ &\leq d(k) + d(n - k) + \min\{2k, n - 1\} = d(n, k) \leq d(n) \,. \end{aligned}$$

We still need to show that, for every n, there exists a tree T_n and its leaf order $\leq_n$ such that $d(T_n, \leq_n) = d(n)$. The case $n = 1$ is trivial. For $n > 1$ and $k \in [1..\lfloor n/2 \rfloor]$, let $T_{n,k}$ be a tree with T_k and T_{n-k} as the two subtrees. Define the leaf order $\leq_{n,k}$ so that the leaves at positions $2, 4, 6, \ldots, 2k$ come from T_k consistent with the order $\leq_k$ and the leaves at positions $1, 3, 5, \ldots, 2k-1, 2k+1, 2k+2, \ldots, n$ come from T_{n-k} consistent with the order $\leq_{n-k}$. Then, it is easy to see that

$$\begin{aligned} d(T_{n,k}, \leq_{n,k}) &= d(T_k, \leq_k) + d(T_{n-k}, \leq_{n-k}) + |D_{\text{root}}(T_{n,k}, \leq_{n,k})| \\ &= d(k) + d(n - k) + \min\{2k, n - 1\} = d(n, k) \end{aligned}$$

Finally, set $T_n = T_{n,k}$ and $\leq_n = \leq_{n,k}$, for $k = \text{argmax}_i \, d(n, i)$. Then $d(T_n, \leq_n) = d(n)$. $\qquad\square$

Basic properties and closed form equations for $d(n)$ are given in the following lemmas. The proofs are omitted due to lack of space.

Lemma 5. *For any* $2 \leq 2k \leq n$,

(i) $d(n, k) \leq d(n, \lfloor n/2 \rfloor)$
(ii) $d(n) - d(n - 1) = \lceil \lg n \rceil$.

Lemma 6. $d(n) = n\lceil \lg n \rceil - 2^{\lceil \lg n \rceil} + 1$.

Lemma 7. $d(n) = n \lg n - (1 - \alpha(n))n + 1$, *where* $0 \leq \alpha(n) := 1 - 2^{\lceil \lg n \rceil}/n + \lg(2^{\lceil \lg n \rceil}/n) < (1 - \lg e + \lg \lg e) < 0.0861$.

Thus we obtain an $n \lg n$ bound on the irreducible sums.

Theorem 1. *For any multiset W of words of total length $n > 0$, we have*

$$\Sigma \text{ilcp}(W) \leq \Sigma \text{idp}(W) \leq d(n) \leq n \lg n \,.$$

6 $n \lg r$ Upper Bound

We will now use the above machinery to improve the upper bound on $\Sigma \mathrm{idp}(W)$ when the number r of runs in $\mathrm{BWT}(W)$ is given.

Lemma 8. *If* $\mathrm{BWT}(W)$ *has* r *runs, then* $|D_u(\overline{\mathrm{STree}}(W), \leq_W)| < r$ *for every vertex* u *in* $\overline{\mathrm{STree}}(W)$.

Proof. Let u be a vertex in $\overline{\mathrm{STree}}(W)$ labelled by $\overline{x}$. The bijection defined in the proof of Lemma 3 maps $D_u(\overline{\mathrm{STree}}(W), \leq_W)$ into

$$\{(\mathrm{SA}[i-1], \mathrm{SA}[i], |x|) : x \text{ is prefix of } W_{\mathrm{SA}[i-1]} \text{and } W_{\mathrm{SA}[i]}, \text{ and DP}[i] \text{ is irreducible}\}.$$

Since the total number of irreducible distinguishing prefixes is $r - 1$, the size of this set cannot be more than $r - 1$, and thus $|D_u(\overline{\mathrm{STree}}(W), \leq_W)| < r$.   □

Define, for $r > 0$, $n > 0$ and $k \in [1..\lfloor n/2 \rfloor]$,

$$d_r(1) = 0$$
$$d_r(n) = \max_{i \in [1..\lfloor n/2 \rfloor]} d_r(n, i) \quad \text{when } n > 1$$
$$d_r(n, k) = d_r(k) + d_r(n - k) + \min\{2k, n - 1, r - 1\}$$

Lemma 9. $d_{r(n)} = \max\{d(T, \leq)\}$, *where the maximum is taken over any rooted tree* T *with* n *leaves and any total order* $\leq$ *on the leaves of* T *such that* $|D_u(T, \leq)| < r$ *for every vertex* u *in* T.

Proof. We will prove the claim by modifying the construction utilized in the proof of Lemma 4. First note that in the transformation from a non-binary tree to a binary tree, the new vertex u' might have $|D_{u'}(T', \leq)| \geq r$. However, we can then replace $\leq$ with $\leq'$ such that $|D_{u'}(T', \leq')| = r - 1$ and no other vertex dispersal value is changed. Then we must have $|D_u(T', \leq')| + |D_{u'}(T', \leq')| \geq |D_u(T, \leq)|$ and thus $d(T, \leq) \leq d(T' \leq')$.

The inequality $d(T, \leq) \leq d_r(n)$ follows immediately from using the bound $|D_{\mathrm{root}}(T, \leq)| \leq \min\{2k, n - 1, r - 1\}$ in place of $|D_{\mathrm{root}}(T, \leq)| \leq \min\{2k, n - 1\}$. In the construction of T_n and $\leq_n$, the only difference is in constructing $\leq_{n,k}$ when $r - 1 < \min\{2k, n - 1\}$. In that case, the interleaving of $\leq_k$ and $\leq_{n-k}$ to obtain $\leq_{n,k}$ is then chosen so that $|D_{\mathrm{root}}(T_{n,k}, \leq_{n,k})| = r - 1$. With this change, the construction shows that $d(T_n, \leq_n) = d_r(n)$.   □

Basic properties and closed form equations for $d_r(n)$ are given in the following lemmas. Again, the proofs are omitted due to lack of space.

Lemma 10. *For any* $r \geq 2$,

(i) $d_r(n) = d(n)$ *if* $n \leq r$
(ii) $d_r(n, k) \leq d_r(n, \lfloor r/2 \rfloor)$ *if* $n \geq r$ *and* $k \leq n/2$
(iii) *for* $n > \lceil r/2 \rceil$,

$$d_r(n) - d_r(n-1) = \begin{cases} \lfloor \lg r \rfloor & \text{if } n - \lceil r/2 \rceil - 1 \bmod \lfloor r/2 \rfloor \in [0..q) \\ \lceil \lg r \rceil & \text{if } n - \lceil r/2 \rceil - 1 \bmod \lfloor r/2 \rfloor \in [q..\lfloor r/2 \rfloor) \end{cases}$$

where $q = 2^{\lceil \lg r \rceil - 1} - \lceil r/2 \rceil$.

Lemma 11. *For any* $2 \le r \le n$,

$$d_r(n) = n\lceil \lg r \rceil - 2^{\lceil \lg r \rceil} + 1 - q(n - r - p)/\lfloor r/2 \rfloor - \min\{q, p\}$$
$$\le n\lceil \lg r \rceil - 2^{\lceil \lg r \rceil} + 1$$

where $q = 2^{\lceil \lg r \rceil - 1} - \lceil r/2 \rceil$ *and* $p = (n - r) \bmod \lfloor r/2 \rfloor$.

Lemma 12. *For any* $2 \le r \le n$,

$$d_r(n) = n\lg r + n(\alpha(r) + \beta(r)) - r(1 + \beta(r)) - \gamma(p, q) + 1$$
$$\le n\lg r + n\alpha(r) - r + \begin{cases} 1 & \text{if } r \text{ is even} \\ \frac{n}{r} & \text{if } r \text{ is odd} \end{cases},$$

where p *and* q *are as in Lemma 11,* $\alpha(r) \in [0, 0.0861)$ *is as in Lemma 7,*

$$\beta(r) = \begin{cases} 0 & \text{if } r \text{ is even} \\ \frac{2r - 2^{\lceil \lg r \rceil}}{r(r-1)} & \text{if } r \text{ is odd} \end{cases} \in [0, 1/r]$$

and $\gamma(p, q) = \min\{p, q\} - pq/\lfloor r/2 \rfloor \in [0, r/8)$.

Thus we obtain the following upper bound on the irreducible sums.

Theorem 2. *For any multiset* W *of words of total length* $n > 0$ *such that* BWT(W) *has* r *runs, we have*

$$\Sigma\mathrm{ilcp}(W) + r - 1 = \Sigma\mathrm{idp}(W) \le d_r(n) < n\lg r + 0.0861 \cdot n + n/r - r = n\lg r + \mathcal{O}(n).$$

7 $n \lg n$ Lower Bound

In this and the next section, we will show the tightness of the above upper bounds by constructing words and sets of words with matching irreducible sums. We will deal only with sets (not multisets) over the binary alphabet $\{a, b\}$, which allows a useful characterization of the LCP array.

Lemma 13. *For any set of words* W *of total length least two, such that* suf(W) *contains no duplicates, the sequence of the depths of internal (non-leaf) vertices in* STree(W) *listed in inorder is exactly* LCP$_W$.

Proof. Consider constructing STree(W) by inserting the suffixes into a compact trie one at a time in the lexicographical order. When inserting $W_{\mathrm{SA}[i]}$, $i > 0$, we add exactly two vertices, the leaf v_i labelled $W_{\mathrm{SA}[i]}$ and the parent u_i of v_i. Note that u_i could not have existed (or was the root and unary) before, since the tree is binary. Since the depth of u_i must be LCP[i], and the internal vertices are inserted in inorder, the claim follows.   □

A word set $W \subseteq \{a, b\}^*$ is a de Bruijn set [4] of order $k \geq 1$ if every $v \in \{a, b\}^k$ is a prefix of exactly one word in $suf(W)$ (and thus $suf(W)$ is a set).

Lemma 14. *For any de Bruijn set W of order k, $\Sigma lcp(W) = k2^k - 2^{k+1} + 2$.*

Proof. STree(W) has 2^i vertices at depth $i \in [0..k-1]$ and no internal vertices at levels $i \geq k$. By Lemma 13, $\Sigma lcp(W) = \sum_{i=1}^{k-1} i2^i = k2^k - 2^{k+1} + 2$.   □

Higgins [4] showed the following characterization of the de Bruijn sets.

Lemma 15 ([4]). *For $k \geq 1$, and any $u \in U_k = \{ab, ba\}^{2^{k-1}}$, $W = \text{IBWT}(u)$ is a de Bruijn set.*

In particular, $W_k = \text{IBWT}((ab)^{2^{k-1}})$ is a de Bruijn set. Since every entry in LCP_{W_k} is irreducible, we obtain the following result.

Theorem 3. *For any $k \geq 1$, $\Sigma ilcp(W_k) = k2^k - 2^{k+1} + 2 = n \lg n - 2n + 2$ and $\Sigma idp(W_k) = \Sigma ilcp(W_k) + n - 1 = n \lg n - n + 1$, where $n = 2^k$ is the total length of the words in W_k.*

Thus $\Sigma idp(W_k)$ matches the upper bound from Sect. 5 exactly. We still want to show that, for any $k \geq 1$, there exist a de Bruijn word, i.e., a de Bruijn set of size one, that matches the upper bound within $\mathcal{O}(n)$. First we need a bound on the size of W_k.

Lemma 16. $|W_k| \leq (2^k + (k-1)2^{k/2})/k.$

Proof. From [4, Theorem 3.8], the size of W_k is equal to the number of Lyndon words of length dividing k. Thus the claim follows by combining Eqs. (7.10) and (7.13) from [13].

Lemma 17. *Starting with $u = u_k = (ab)^{2^{k-1}}$, there exists a sequence of $|W_k| - 1$ swaps of the form $u[2i] \leftrightarrow u[2i+1]$ resulting in $u \in U_k$ such that $|\text{IBWT}(u)| = 1$.*

Proof. We will show that the following invariant is maintained during the sequence of swaps until $|\text{IBWT}(u)| = 1$: there exists i such that every value in $[0..2i]$ belongs to the same cycle in Ψ_u, $2i+1$ belongs to a different cycle, and there has been no swaps affecting $u[2i..2^k)$. Then the next swap is $u[2i] \leftrightarrow u[2i+1]$. The invariant is clearly true when $u = u_k$.

Let $j = \Psi_u^{-1}(2i)$ and $j' = \Psi_u^{-1}(2i+1)$. Since $u[2i] = a$ and $u[2i+1] = b$, after the swap we have $\Psi_u(j) = 2i+1$ and $\Psi_u(j') = 2i$, i.e., the two cycles were merged. The values of Ψ_u are not affected elsewhere. Now consider the smallest $j \in [2i+2..2^k)$ that is not in the same cycle as 0. We must have $u[j] = b$, since otherwise $\Psi_u^{-1}(j) < j$ and j would be in the same cycle. Thus j is odd, and we can choose $i = (j-1)/2$ to satisfy the invariant.   □

Theorem 4. *For any $k \geq 1$, there exists a word w of length $n = 2^k$ such that $\Sigma idp(w) = n \lg n - \mathcal{O}(n)$.*

Proof. Let $u \in U_k$ be the result of Lemma 17 and $w = \text{IBWT}(u)$. From the proof of Lemma 14, $\max \text{LCP}_w = k - 1$. Each swap reduces the number of irreducible LCP values by at most two, thus the initial $\Sigma idp(W_k) = n \lg n - \mathcal{O}(n)$ is reduced by at most $(\max \text{LCP}_w + 1)|W_k| \leq k(2^k + (k-1)2^{k/2})/k = \mathcal{O}(n)$.   □

8 $n \lg r$ Lower Bound

We will now extend the above lower bound results to cases where $r \ll n$. The following lemma shows the key idea of the construction.

Lemma 18. *Let $u \in \{a, b\}^{2k}$, $k \geq 1$ be a word containing exactly k a's and k b's. Then, for any $w = ua^{jk}$, $j \geq 0$, it holds $|\mathrm{IBWT}(w)| = |\mathrm{IBWT}(u)|$. Furthermore, $\mathrm{IBWT}(w)$ can be obtained by replacing every occurrence of b in all $\mathrm{IBWT}(u)$ with $a^j b$.*

Proof. The standard permutation of w can be expressed by the following formula:

$$\Psi_w(i) = \begin{cases} \Psi_u(i) & 0 \leq i < k \\ i + k & k \leq i < |w| - k \\ \Psi_u(i - jk) & |w| - k \leq i < |w| \end{cases}$$

Let $j > 0$ (since the case $j = 0$ is trivial), and compare the reconstructions of $\mathrm{IBWT}(u)$ and $\mathrm{IBWT}(w)$ by following Ψ_u and Ψ_w. Whenever we visit $i \in [0..k)$, $\Psi_w(i) = \Psi_u(i)$ and we append letter a to the currently decoded word in both cycles. When visiting $i \in [k..2k)$ we append b to $\mathrm{IBWT}(u)$ but a to $\mathrm{IBWT}(w)$ and $\Psi_w(i) \neq \Psi_u(i)$. However, for any such i, and any $p \in [1..j]$, we have $\Psi_w^p(i) = i + pk$, and thus $\Psi_w^{j+1}(i) = \Psi_w(i + jk) = \Psi_u(i)$. Therefore, after a detour of j extra steps in Ψ_w the cycles meet again, and where a single b was appended $\mathrm{IBWT}(u)$, $a^j b$ was appended to $\mathrm{IBWT}(w)$. $\square$

Define $U_{k,j} = \{ab, ba\}^{2^{k-1}} a^{j2^{k-1}}$ for $k \geq 1$ and $j \geq 0$. Consider arbitrary $u \in U_{k,j}$ for some k and j, and let $W = \mathrm{IBWT}(u)$. Let $S_{k,j} = S_{0,k,j} \cup S_{1,k,j} \cup \ldots \cup S_{j+1,k,j}$, where

$$S_{i,k,j} = \begin{cases} a^i ba^j \{a, ba^j\}^{k-1} & \text{if } i \leq j \\ a^{j+1} \{a, ba^j\}^{k-1} & \text{if } i > j \end{cases}.$$

Lemma 19. *Every word in $S_{k,j}$ is a prefix of exactly one word in $\mathrm{suf}(W)$.*

Proof. First observe that, since $S_{k,j}$ is prefix-free and $|S_{k,j}| = (j + 2)2^{k-1} = |\mathrm{suf}(W)|$, it is sufficient to show that every $w \in S_{k,j}$ is a prefix of at least one word in $\mathrm{suf}(W)$.

Let $u' \in \{ab, ba\}^{2^{k-1}}$ be the word such that $u = u' a^{j2^{k-1}}$, and let $W' = \mathrm{IBWT}(u')$. Since W' is a de Bruijn set, for every $v' \in \{a, b\}^k$ either $v' a^j$ or $v' a^h b$ for some $h < j$ is a prefix of a word in $\mathrm{suf}(W')$. Thus, by Lemma 18, every $v \in \{a, a^j b\}^k a^j = a^j \{a, ba^j\}^k$ is a prefix of a word in $\mathrm{suf}(W)$. Since every word $w \in S_{k,j}$ is a factor of a word $v \in a^j \{a, ba^j\}^k$, w must be a prefix of a word in $\mathrm{suf}(W)$ too. $\square$

It is easy to see from the definition of $S_{i,k,j}$, that $\mathrm{CTrie}(S_{i,k,j})$, consists of a full binary tree of height $k - 1$ connected to the root with a single edge, and that $\mathrm{CTrie}(S_{k,j})$ consists of $j + 2$ such full binary subtrees connected to the main branch labelled a^j, see Fig. 1 for examples.

Define $u_{k,j} = (ab)^{2^{k-1}} a^{j2^{k-1}} \in U_{k,j}$ and let $W_{k,j} = \mathrm{IBWT}(u_{k,j})$. Since $u_{k,j}$, $j \geq 1$ has $2^k + 1$ runs, $\Sigma \mathrm{idp}(W_{k,j}) = \Sigma \mathrm{ilcp}(W_{k,j}) + 2^k$.

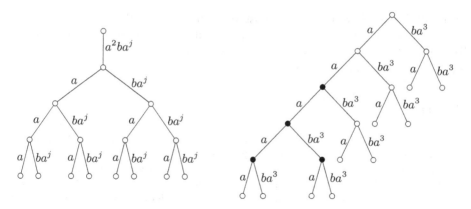

Fig. 1. Left: CTrie($S_{2,4,j}$). Right: CTrie($S_{2,3}$). Dark vertices correspond to irreducible LCP values of $W_{2,3}$.

Lemma 20. *For $j \geq 1$, Σilcp($W_{k,j}$) = $(j+2)k2^{k-1} - 2^{k+1} + j + 1$.*

Proof. By Lemma 19, STree($W_{k,j}$) is the same as CTrie($S_{k,j}$) except leaf labels have been extended to infinite words. Clearly, all LCP values in LCP$_{W_{k,j}}[1..2^k]$ (and only these) are irreducible, and by Lemma 13, they correspond to the depths of the first 2^k internal vertices in CTrie($S_{k,j}$) by inorder. These are exactly the $2^k - 1$ vertices in the subtree CTrie($\{a, ba^j\}^k$) rooted in the vertex labelled a^j plus one more vertex labelled a^{j-1} (see Fig. 1). The sum of internal vertex depths of CTrie($\{a, ba^j\}^k$) is $(j+2)(2^{k-1}(k-2)+1)$, thus Σilcp($W_{k,j}$) = $(j+2)(2^{k-1}(k-2)+1) + j(2^k - 1) + j - 1 = (j+2)k2^{k-1} - 2^{k+1} + j + 1$. $\qquad\square$

Thus we obtain the following bounds for a set of words (with the proof omitted due to lack of space) and for a single word.

Theorem 5. *For any $k \geq 1$ and $j \geq 1$, Σidp($W_{k,j}$) = $d_r(n)$ matching the upper bound in Sect. 6, where $n = (j+2)2^{k-1}$ is the total length of the words in $W_{k,j}$ and $r = 2^k + 1$ is the number of runs in BWT($W_{k,j}$).*

Theorem 6. *For any $r = 2^k + 1$, $k \geq 1$, and $n \geq r$ such that $2^{k-1} | n$, there exists a word w of length n such that BWT(w) contains $r - o(r)$ runs and Σidp(w) = $n \lg r - \mathcal{O}(n)$.*

Proof. Let $u = u_{k,j}$, where $j = n/2^{k-1} - 2 \geq 1$. From definition, u has $2^k + 1 = r$ runs. Lemmas 16 and 18 give $|\text{IBWT}(u)| = |W_{k,j}| = |W_k| = \mathcal{O}(r/\lg r)$. Note, that Lemma 17 applies also for $u \in U_{k,j}$. Let $u' \in U_{k,j}$ be the result and $w = \text{IBWT}(u')$. From Lemma 19, LCP$_w$ = LCP$_{W_{k,j}}$. Furthermore, max LCP$_w$ corresponds to the deepest internal vertex in CTrie($S_{k,j}$), i.e., max LCP$_w$ = $k(j+1) - 1 = \mathcal{O}((n \lg r)/r)$. Each swap in Lemma 17 reduces the number of irreducible LCP values by at most two, thus $u' = \mathrm{BWT}(w)$ contains $r - \mathcal{O}(r/\lg r) = r - o(r)$ runs. Altogether, the initial $\Sigma\mathrm{idp}$ is reduced by at most $|W_{k,j}|(\max \mathrm{LCP}_w + 1) = \mathcal{O}(n)$, which gives $\Sigma\mathrm{idp}(w) = \Sigma\mathrm{idp}(W_{k,j}) - \mathcal{O}(n) = n \lg r - \mathcal{O}(n)$ (see Theorem 5). $\qquad\square$

References

1. Abouelhoda, M.I., Kurtz, S., Ohlebusch, E.: Replacing suffix trees with enhanced suffix arrays. J. Discrete Algorithms **2**(1), 53–86 (2004)
2. Burrows, M., Wheeler, D.J.: A block sorting lossless data compression algorithm. Technical report 124, Digital Equipment Corporation, Palo Alto, California (1994)
3. Fine, N.J., Wilf, H.S.: Uniqueness theorems for periodic functions. Proc. Amer. Math. Soc. **16**(1), 109–114 (1965)
4. Higgins, P.M.: Burrows-Wheeler transformations and de Bruijn words. Theor. Comput. Sci. **457**, 128–136 (2012)
5. Kärkkäinen, J., Kempa, D.: LCP array construction in external memory. In: Gudmundsson, J., Katajainen, J. (eds.) SEA 2014. LNCS, vol. 8504, pp. 412–423. Springer, Heidelberg (2014)
6. Kärkkäinen, J., Manzini, G., Puglisi, S.J.: Permuted longest-common-prefix array. In: Kucherov, G., Ukkonen, E. (eds.) CPM 2009. LNCS, vol. 5577, pp. 181–192. Springer, Heidelberg (2009)
7. Mäkinen, V., Navarro, G., Sirén, J., Välimäki, N.: Storage and retrieval of highly repetitive sequence collections. J. Comp. Biol. **17**(3), 281–308 (2010)
8. Manber, U., Myers, G.W.: Suffix arrays: a new method for on-line string searches. SIAM J. Comp. **22**(5), 935–948 (1993)
9. Mantaci, S., Restivo, A., Rosone, G., Sciortino, M.: An extension of the Burrows-Wheeler transform. Theor. Comput. Sci. **387**(3), 298–312 (2007)
10. Manzini, G.: Two space saving tricks for linear time LCP array computation. In: Hagerup, T., Katajainen, J. (eds.) SWAT 2004. LNCS, vol. 3111, pp. 372–383. Springer, Heidelberg (2004)
11. Navarro, G., Mäkinen, V.: Compressed full-text indexes. ACM Comput. Surv. **39**(1), 1–61 (2007)
12. Bioinformatics Algorithms: Sequence Analysis, Genome Rearrangements, and Phylogenetic Reconstruction. Oldenbusch Verlag, Bremen, Germany (2013)
13. Ruskey, F.: Combinatorial generation, working version (1j-CSC 425/520) (2003)
14. Sirén, J.: Sampled longest common prefix array. In: Amir, A., Parida, L. (eds.) CPM 2010. LNCS, vol. 6129, pp. 227–237. Springer, Heidelberg (2010)

Parallel External Memory Suffix Sorting

Juha Kärkkäinen, Dominik Kempa[(✉)], and Simon J. Puglisi

Department of Computer Science, Helsinki Institute for Information
Technology HIIT, University of Helsinki, Helsinki, Finland
{juha.karkkainen,dominik.kempa,simon.puglisi}@cs.helsinki.fi

Abstract. Suffix sorting (or suffix array construction) is one of the most
important tasks in string processing, with dozens of applications, particularly in text indexing and data compression. Some of these applications
require the suffix array to be built for large inputs that greatly exceed the
size of RAM and so external memory must be used. However, existing
approaches for external memory suffix sorting either use debilitatingly
large amounts of disk space, or become too slow when the size of the
input data is more than a few times bigger than the size of RAM. In this
paper we address the latter problem via a non-trivial parallelization of
computation. In our experiments, the resulting algorithm is much faster
than the best prior external memory algorithms while using very little
disk space in addition to what is needed for the input and output. On
the way to this result we provide the current fastest (parallel) internal
memory algorithm for suffix sorting, which is usually around twice as
fast as previous methods, while using around one quarter of the working
space.

1 Introduction

Suffix sorting (or suffix array construction) is one of the most important tasks
in string processing. It is fundamental to building index data structures such
as suffix trees [10,32], (compressed) suffix arrays [14,24], and FM-indexes [11],
which in turn have dozens of applications in bioinformatics, including pattern
matching (i.e. read alignment [21,22]), genome assembly [30], and discovery of
repetitive structures [1]. Suffix sorting is also key to several major lossless compression transforms, such as the Burrows-Wheeler transform, Lempel-Ziv (LZ77)
parsing [16,17,34], and several grammar compressors (e.g. [4,26]). Many of these
applications deal with massive data and often suffix sorting is the computationally most demanding task.

Suffix sorting is also one of the most studied tasks in string processing [29],
but the majority of the work has focused on sequential, internal memory algorithms, which do not really scale for massive data and do not fully utilize the
resources on modern computers. There has been some research on speeding up
suffix sorting by parallel computation and on external memory suffix sorting
algorithms that escape the limits of RAM, but no really effective combination of

The work is partially supported by the Academy of Finland through grant 258308.

F. Cicalese et al. (Eds.): CPM 2015, LNCS 9133, pp. 329–342, 2015.
DOI: 10.1007/978-3-319-19929-0_28

the two approaches. This is not very surprising since external memory computation is often I/O-bound and would not benefit greatly from (internal memory) parallelism. Nevertheless, in this paper we show that the two computational paradigms can be fruitfully combined in a suffix sorting algorithm.

Our contribution. Our starting point is the recent external memory suffix sorting algorithm SAscan [15], the basic idea of which is to divide the text into blocks, construct suffix arrays for the blocks and then merge these partial suffix arrays. In this paper, we describe a parallelization of the central procedure that merges two partial suffix arrays. Using this procedure, we first design an internal memory suffix sorting algorithm that constructs several partial suffix arrays in parallel (using any sequential suffix sorter) and then merges them together. The result is the fastest internal memory algorithm that we are aware of. This internal memory suffix sorter and the parallel merging procedure are then used in designing a parallelized version of SAscan which we call pSAscan. On a machine with 12 physical cores (24 with hyper-threading), pSAscan is over four times faster than SAscan and much faster than any other external memory algorithm in all of our experiments.

The algorithms are not theoretically optimal. The internal memory algorithm needs $\Omega(n \log p)$ work on p processors, and the external memory pSAscan needs $\widetilde{\Omega}(n^2/M)$ work, where M is the size of the RAM. However, low constant factors and, crucially, space efficiency make them more scalable in practice than their competitors. The internal memory algorithm needs less than $10n$ bytes of RAM, and pSAscan needs just $7.5n$ bytes of disk space, which is nearly optimal. The best competitors use about four times as much RAM/disk space, which is likely to be a more serious limitation to their scalability than the time complexity is to our algorithms. To demonstrate the scalability, we have constructed the suffix array of a 1 TiB text in a little over 8 days.

Related work. The idea of external memory suffix sorting by merging separately constructed partial suffix arrays goes back over 20 years [13], and there has been several improvements over the years [7,12] (see also [31]). The recent incarnation SAscan [15] is one of the fastest external memory suffix sorters in practice. A different approach to merging suffix arrays in [23] is limited to merging separate files rather than blocks of the same file. The main competitor of SAscan is the eSAIS algorithm by Bingmann, Fischer and Osipov [5]. eSAIS is theoretically optimal but suffers from a large disk space usage (roughly $28n$ bytes, for an input of n symbols). SAscan needs just $7.5n$ bytes of disk space but because of its $\widetilde{O}(n^2/M)$ time complexity, it is competitive with eSAIS only when the input is less than about five times the size of RAM. The new pSAscan extends the advantage over eSAIS to much bigger inputs. Another recent external memory suffix sorter EM-SA-DS [27] appears to be slightly worse than eSAIS in practice, although a direct comparison is missing.

In contrast to the large number of algorithms for serial suffix sorting [29], results on parallel algorithms for suffix sorting are reasonably sparse. Earlier research focused on suffix tree construction (see, e.g., [3]) and was mostly of theoretical interest. More recently, research into practical algorithms has focused on either distributed [20] or GPU platforms [9,28]. Most relevant to this paper is a parallel version of DC3, a work optimal EREW-PRAM algorithm due to Kärkkäinen, Sanders and Burkhardt [19] that has been subsequently implemented by Blelloch and Shun [6]. We use their implementation as a baseline in experiments with our internal memory algorithm.

2   Preliminaries

Let $X = X[0..m)$ be a string over an integer alphabet $[0..\sigma)$. Here and elsewhere we use $[i..j)$ as a shorthand for $[i..j-1]$. For $i \in [0..m)$ we write $X[i..m)$ to denote the *suffix* of X of length $m - i$, that is $X[i..m) = X[i]X[i+1]\ldots X[m-1]$. Similarly, we write $X[0..i)$ to denote the *prefix* of X of length i and $X[i..j)$ to denote the *substring* $X[i]X[i+1]\ldots X[j-1]$ of length $j - i$. If $i = j$, the substring $X[i..j)$ is the empty string, also denoted by ε.

The suffix array SA_X of a string X contains the starting positions of the non-empty suffixes of X in the lexicographical order, i.e., it is an array $SA_X[0..m)$ which contains a permutation of the integers $[0..m)$ such that $X[SA_X[0]..m) < X[SA_X[1]..m) < \cdots < X[SA_X[m-1]..m)$. In other words, $SA_X[j] = i$ iff $X[i..m)$ is the $(j+1)^{th}$ suffix of X in ascending lexicographical order.

The Burrows-Wheeler transform $BWT_X[0..m)$ of a string X contains the characters preceding each suffix in lexicographical order: $BWT_X[i] = X[SA_X[i] - 1]$ if $SA_X[i] > 0$ and otherwise \$, a special symbol that does not appear in the text.

Partial suffix arrays. The partial suffix array $SA_{X:Y}$ is the lexicographical ordering of the suffixes of XY with a starting position in X, i.e., it is an array $SA_{X:Y}[0..m)$ that contains a permutation of the integers $[0..m)$ such that $X[SA_{X:Y}[0]..m)Y < X[SA_{X:Y}[1]..m)Y < \cdots < X[SA_{X:Y}[m-1]..m)Y$. Note that $SA_{X:\varepsilon} = SA_X$ and that $SA_{X:Y}$ is usually similar but not identical to SA_X. Also note that $SA_{X:Y}$ can be obtained from SA_{XY} by removing all entries that are larger or equal to m. The definition of the Burrows–Wheeler transform extends naturally to the partial version $BWT_{X:Y}[0..m)$.

When comparing two suffixes of XY starting in X, in most cases we only need to access characters in X, but sometimes the comparison continues beyond the end of X and may, in an extreme case, continue all the way to the end of Y. To avoid such long comparisons, we store additional information about the order of the suffixes in the form of bitvectors $gt^S_{X:Y}[0..m)$ defined as follows:

$$gt^S_{X:Y}[i] = \begin{cases} 1 & \text{if } X[i..m)Y > S \\ 0 & \text{if } X[i..m)Y \le S \end{cases}.$$

For example, for $0 \le i < j < m$, the following are equivalent:

1. $X[i..m)Y < X[j..m)Y$
2. $X[i..m) < X[j..m)Y[0..j-i)$ or $X[i..m) = X[j..m)Y[0..j-i)$ and $gt^Y_{Y:\varepsilon}[j-i] = 1$
3. $X[i..m-j+i) < X[j..m)$ or $X[i..m-j+i) = X[j..m)$ and $gt^Y_{X:Y}[m-j+i] = 0$.

3   Merging of Partial SAs

The basic building block of pSAscan is a procedure for merging two adjacent partial suffix arrays. In this section, we describe a sequential algorithm for performing the merging and then, in the next section, show how to parallelize it.

Given the partial suffix arrays $SA_{X:YZ}$ and $SA_{Y:Z}$, for some strings X, Y and Z, the task is to construct the partial suffix array $SA_{XY:Z}$. The suffixes in each input array stay in the same relative order in the output, and thus we just need to know how to interleave the input arrays. For this purpose, we compute the *gap array* $gap_{X:Y:Z}[0..|X|]$, where $gap_{X:Y:Z}[i]$ is the number of suffixes in $SA_{Y:Z}$ that are lexicographically between the suffixes $SA_{X:YZ}[i-1]$ and $SA_{X:YZ}[i]$. Formally, denoting $m = |X|$ and $n = |Y|$,

$$gap_{X:Y:Z}[0] = \left|\{j \in [0..n) : Y[j..n)Z < X[SA_{X:YZ}[0]..m)YZ\}\right|$$
$$gap_{X:Y:Z}[m] = \left|\{j \in [0..n) : X[SA_{X:YZ}[m-1]..m)YZ < Y[j..n)Z\}\right|$$

and, for $i \in [1..m)$,

$$gap_{X:Y:Z}[i] = \big|\{j \in [0..n) :$$
$$X[SA_{X:YZ}[i-1]..m)YZ < Y[j..n)Z < X[SA_{X:YZ}[i]..m)YZ\}\big|.$$

Given the gap array, the actual merging is easy; the difficult part is computing the gap array.

For a string S, let $sufrank_{X:YZ}(S)$ be the number of suffixes in $SA_{X:YZ}$ that are lexicographically smaller than S. In other words, if $sufrank_{X:YZ}(S) = k$ (and $0 < k < m$), then $X[SA_{X:YZ}[k-1]..m)YZ < S \le X[SA_{X:YZ}[k]..m)YZ$. Thus we can compute the gap array $gap_{X:Y:Z}$ by initializing all entries to zeros, and then, for all $j \in [0..n)$, computing $k = sufrank_{X:YZ}(Y[j..n)Z)$ and incrementing $gap_{X:Y:Z}[k]$. The values $sufrank_{X:YZ}(Y[j..n)Z)$ are computed starting from the end of Y using a procedure called *backward search* [11].

Backward search is based on rank operations on the Burrows–Wheeler transform $BWT_{X:YZ}$. For a character c and an integer $i \in [0..m]$, the answer to the rank query $rank_{BWT_{X:YZ}}(c, i)$ is the number of occurrences of c in $BWT_{X:YZ}[0..i)$. We preprocess $BWT_{X:YZ}[0..m)$ so that arbitrary rank queries can be answered quickly; see [15] for details. Let $C[0..\sigma)$ be an array, where $C[c]$ is the number of positions $i \in [0..m)$ such that $X[i] < c$. The following lemma shows one step of backward search.

Lemma 1. [11,15]. *Let* $k = sufrank_{X:YZ}(S)$ *for a string S. For any symbol* c,

$$sufrank_{X:YZ}(cS) = C[c] + rank_{BWT_{X:YZ}}(c, k) + \begin{cases} 1 & if\ X[m-1] = c\ and\ YZ < S \\ 0 & otherwise \end{cases}.$$

Note that when $S = Y[j..n)Z$, we can replace the comparison $YZ < S$ with $gt_{Y:Z}^{YZ}[j] = 1$. Thus, given $sufrank_{X:YZ}(Y[j..n)Z)$, we can easily compute $sufrank_{X:YZ}(Y[j-1..n)Z)$ using the lemma, and we only need to access $Y[j-1]$ and $gt_{Y:Z}^{YZ}[j]$.

Hence the whole computation of $\mathsf{gap}_{X:Y:Z}$ can be done with a single sequential pass over Y and $\mathsf{gt}_{Y:Z}^{YZ}$.

Improvements to SAscan. The procedure described above is identical to the one in the original SAscan [15], but the rest of this section describes details that differ from (and improve) the original.

First, we need $\mathsf{BWT}_{X:YZ}$ and $\mathsf{gt}_{Y:Z}^{YZ}$ for the gap array computation. In SAscan, these are computed from the strings and the partial suffix arrays as needed. This is easy and takes only linear time but is relatively expensive in practice because of frequent cache misses. We compute them differently based on the assumption that both the BWT and the bitvector are available for every partial suffix array. That is, we assume that we are given $\mathsf{BWT}_{X:YZ}$, $\mathsf{BWT}_{Y:Z}$, $\mathsf{gt}_{X:YZ}^{XYZ}$ and $\mathsf{gt}_{Y:Z}^{YZ}$ as input, and we need to compute $\mathsf{BWT}_{XY:Z}$ and $\mathsf{gt}_{XY:Z}^{XYZ}$ as output. Each BWT is stored interleaved with the corresponding SA so that the merging of the SAs produces the output BWT at almost no additional cost. The output bitvector $\mathsf{gt}_{XY:Z}^{XYZ}$ is constructed by concatenating the two bitvectors $\mathsf{gt}_{X:YZ}^{XYZ}$ and $\mathsf{gt}_{Y:Z}^{XYZ}$. The former was given as an input and the latter is computed (as in SAscan) during the backward search using the fact that $\mathsf{gt}_{Y:Z}^{XYZ}[j] = 1$ iff $\mathsf{sufrank}_{X:YZ}(Y[j..n]Z) > i_{XYZ}$, where i_{XYZ} is the position of XYZ in $\mathsf{SA}_{X:YZ}$, i.e., $\mathsf{SA}_{X:YZ}[i_{XYZ}] = 0$.

Second, we need to know $\mathsf{sufrank}_{X:YZ}(Z)$ as the starting position of the backward search. We replace the $O(m + n)$ time string range matching [18] used in SAscan by a binary search over $\mathsf{SA}_{X:YZ}$ with Z as the query. A plain binary search needs $O(\ell \log m)$ time, where ℓ is the length of the longest common prefix between Z and any suffix in $\mathsf{SA}_{X:YZ}$. This is fast enough in most cases as ℓ is typically small and the constant factors are small. However, we employ several techniques to ensure a good performance even in pathological cases. We use a string binary search algorithm with $O(\ell + \log m)$ average case time (see [24]) and $O(\ell \log_\ell m)$ worst case time (see [2] for an even better complexity); we utilize the gt-bitvectors to resolve comparisons early; and, in the full algorithm with many binary searches, we utilize the fact that all the strings are suffixes of the same text. We omit the details here due to lack of space, and because most of the advanced binary searching techniques are only used in pathological cases and have little effect on the experimental results.

The final difference to SAscan is the actual merging of SAs. In SAscan, the merging is delayed (and the gap array is stored on disk) but here we often need to perform the merging immediately. This is easily done if given a separate array for the output, but we want to do the merging *almost in-place* to reduce space usage. The basic idea, following [19, Appendix B], is to divide the SAs into small blocks, which we call *pages*, and maintain pointers to the pages in an additional array, called the *page index*. Any random access has to go through the page index, which allows us to relocate the pages independently. We assume that both the input SAs and the output SA are stored in this form. As merging proceeds and elements are moved from input to output, input pages that become empty are reused as output pages. This way merging can be performed using only a constant number of extra pages.

4   Parallel Merging of Partial SAs

In this section, we describe a parallelized implementation for the merging procedure. We assume a multicore architecture capable of running p threads simultaneously and a shared memory large enough to hold all the data structures.

The first task during merging is the construction of the rank data structure, which is easily parallelized since the data structure is naturally divided into (almost) independent blocks (see [15]).

The most expensive part of merging is the backward search, mainly because the rank queries are relatively expensive (see [15]). We parallelize it by starting the backward search in p places simultaneously. That is, we divide Y into p blocks of equal size and perform a separate backward search for each block in parallel. Each thread computes its own starting sufrank value by a binary search, and then the repeated computation of the sufrank values parallelizes trivially.

For each sufrank value computed during the backward search, we need to increment the corresponding entry in the gap array, but we cannot allow multiple threads to increment the same entry simultaneously, and guarding the gap array by locks would make the updates too expensive. Instead, each thread collects the sufrank values into a buffer. When the buffer is full, it is sorted and stored onto a queue. A separate thread takes the full buffers from the queue one at a time, divides the buffer into up to p parts and starts a thread for each part to do the corresponding gap array updates. Since the buffer is sorted, two threads can never try to increment the same gap array entry.

Once the gap array has been constructed, we still need to perform the actual merging. Recall that we assume the paged storage for the SA. We divide the output SA into p blocks, with the block boundaries always at the page boundaries, and assign a thread for each block. Each thread then finds the corresponding ranges in the input SAs using the gap array. The gap array has been preprocessed by computing cumulative sums at p equally spaced positions, so that the input ranges can be determined by scans of length $O(n/p)$ over the gap array. Next each thread performs the merging using the sequential almost-in-place procedure described in the previous section. The pages containing the beginning and the end of each input range might be shared with another thread and those pages are treated as read-only. Other input pages and all output pages are exclusive to a single thread. Thus each thread needs only four extra pages to do its part of the merging. Once all threads have finished, the extra pages can be relocated to the input boundary pages.

The whole merging procedure can be performed in $O((m + t_{\mathrm{rank}}n)/p)$ time, where t_{rank} is the time for performing one rank query. The input is overwritten by the output, and significant additional space is needed only for the rank structure, the gap array, the extra $4p$ pages and the page indexes. Using the representations from [15], the first two need about $(4.125 + 1)m$ bytes. If we choose page size $\Theta(\sqrt{n/p})$, the space needed for the latter two is $\Theta(\sqrt{np})$, which is negligible. Assuming one byte characters and five byte SA entries, the input/output itself needs about $7.125(m + n)$ bytes (text, SA, BWT and gt bitvectors). The total is $12.25m + 7.125n$ bytes (plus the $\Theta(\sqrt{np})$ bytes).

5   Parallel SA Construction

In this section, we extend the parallel merging procedure into a full parallel suffix array construction algorithm. As before, we assume a multicore architecture with p threads and a shared memory large enough for all the data structures.

The basic idea is simple: divide the input string of length n into p blocks of size $m = \lceil n/p \rceil$, construct the partial SAs for the blocks in parallel using a sequential suffix array construction algorithm, and then keep merging the partial SAs using the parallel merging procedure until the full SA is obtained.

We construct the block SAs using Yuta Mori's divsufsort [25], possibly the fastest sequential suffix sorting algorithm in practice, but we could use any other algorithm too. Let X be a block, Y the following block, and Z the full suffix starting after Y. To obtain the partial suffix array $SA_{X:YZ}$ instead of the full suffix array SA_X, we construct a string $\widehat{X}$ such that $SA_{\widehat{X}} = SA_{X:YZ}$, and for this we need the bitvector $gt^{YZ}_{X:YZ}$, which we denote by gt_X for brevity. For further details of the construction, we refer to [15], but the computation of gt_X is different. We first compute $\widetilde{gt}_X = gt^Y_{X:Y}$ in $O(m)$ time. During the computation, we identify and mark the positions i, where $X[i..m]Y[0..m-i] = Y$; we call these *undecided positions*. It is easy to see that if $\widetilde{gt}_X[i] \neq gt_X[i]$, then i must be an undecided position. Furthermore, in that case $gt_X[i] = gt_Y[i]$. Thus, if i is an undecided position in $\widetilde{gt}_X$, it depends on $\widetilde{gt}_Y[i]$. If that too is undecided, it depends on the position i in the next block and so on. Thus, given the $\widetilde{gt}$-bitvectors for all blocks, we can decide all the undecided i-positions in them in $O(p)$ time. Deciding all undecided positions requires $O(pm)$ work and $O(m + p)$ time using p threads.

Let X be a block and Z the suffix starting after the block. Given $SA_{X:Z}$, we can easily compute $BWT_{X:Z}$ and $gt^{XZ}_{X:Z}$ as well as the page index for $SA_{X:Z}$ in $O(m)$ time in preparation for the merging phase. Furthermore, we compute $O(p^2)$ sufrank values by binary searches (the suffixes starting at the block boundaries against the block SAs); these are used to ensure fast binary searches later during the merging. The worst case complexity of these binary searches is $O(np)$ work and $O(n)$ time, i.e., it does not scale with p. We have designed theoretically better ways of computing the sufrank values, but binary searching is better in practice because of small constant factors and because it is almost always much faster than the worst case. In all our experiments in Sect. 7, the binary searches never took more than 1.5 % of the total time, and even in the very worst case (a unary string) it takes less than 25 % of the total time.

To obtain the final SA from the p initial block SAs, we have to do $p-1$ pairwise merges. If we do the merges in a balanced manner, each element is involved in about $\log p$ merges, and the total time complexity is $O((t_{rank}n \log p)/p)$ for a string of length n. Surprisingly, doing the merges in a balanced manner is not necessarily the optimal way. The time for a single merge can be approximated by $a\ell + br$, where ℓ is the size of the left-hand block, r is the size of the right-hand block, and a and b are some constants. Because the merging time is dominated by the backward search phase, b is much larger than a both in theory as well as in practice. We have implemented a dynamic programming algorithm for computing the optimal merging schedule given p and the value b/a. For example, in a balanced merging with $p = 8$, a single element is involved in three merges, 1.5 times on the left-hand side and 1.5 times on the right-hand side on average. However, in an optimal merging schedule for $b/a = 4$, the averages are 2.25 times on the left-hand side and 1.125 times on the right-hand side. The optimal schedule is about 10 % faster than the balanced schedule in this case. The actual value of b/a in our experiments is about 7.

The space requirement of the algorithm is maximized during the last merge when it is about $12.25\ell + 7.125r$ bytes (see Sect. 4). The space usage can be controlled to an extent by skewing the merging to favor larger right-hand block. Thus there is a space-time tradeoff, but only for the largest merges. Smaller merges can be optimized for time only. Our dynamic program can compute a time-optimal merging schedule under a constraint on the maximal space usage.

6 Parallel SA Construction in External Memory

In this section, we combine the parallel SA construction described above and the external memory construction described in [15] to obtain a faster external memory algorithm.

The basic idea of the algorithm in [15] is:

1. Divide the text into blocks of size m that are small enough to handle in internal memory.
2. For each block X (from last to first), construct the partial suffix array $SA_{X:Z}$ and the gap array $gap_{X:Z:\varepsilon}$, where Z is the suffix starting after X.
3. After constructing all the partial SA and gap arrays, merge the SAs in one multiway merge.

The last step is dominated by I/O and does not benefit much from parallelism, but we will describe how the SA and gap array construction are parallelized.

For constructing $SA_{X:Z}$, we can use the algorithm of the previous section with minor changes required because we are constructing a *partial* SA and the tail Z is stored on disk. There are two phases affected by this: the construction of the gt bitvectors in the beginning and the computation of sufrank values before the merging phase. We assume that the bitvector $gt_{Z:\varepsilon}^Z$ is stored on disk too, which allows us to limit the access to a prefix of Z (and $gt_{Z:\varepsilon}^Z$) of length at most m.

The construction of $gap_{X:Z:\varepsilon}$ is done by backward searching Z over the rank data structure on $BWT_{X:Z}$ as described in previous sections. The only difference is that Z and $gt_{Z:\varepsilon}^Z$ are now on disk, but this is not a problem as only a sequential access is needed. For large files ($n \gg m$), this is by far the most time consuming part because the total number of backward search steps is $\Theta(n^2/m)$. Even with parallelism, the time is dominated by internal memory computation rather than I/O, because rank queries and gap array updates are expensive and the I/O volume per step is low. Thus the parallelism achieves a great speed-up compared to the sequential version.

Table 1. The memory usage of internal memory parallel suffix-sorting algorithms (in bytes). The merging schedule of pSAscan (see Sect. 5) was configured to use $10n$ bytes of RAM in all experiments.

Algor.	pDC3		divsufsort		pSAscan	
	32-bit	64-bit	32-bit	64-bit	32-bit	40-bit
RAM	$21n$	$41n$	$5n$	$9n$	$10n$	$10n$

Table 2. Dataset statistics

Name	$|X|$	σ
hg.reads	1024 GiB	6
kernel	200 GiB	229
wiki	2 GiB	210
countries	2 GiB	205
skyline	2 GiB	32
random	2 GiB	255

The block size m is chosen to fit necessary data structures in RAM. However, the gap array construction needs only about $5.2m$ bytes but the SA construction needs nearly $10m$ bytes. Therefore we add one more stage to the computation. We choose m so that $5.2m$ bytes fits in RAM, but each block X of size m is split into two halfblocks X_1 and X_2. We first compute the halfblock suffix arrays $SA_{X_1:X_2Z}$ and $SA_{X_2:Z}$ separately and write them to disk. Next we compute $gap_{X_1:X_2Z}$ and use it to merge $BWT_{X_1:X_2Z}$ and $BWT_{X_2:Z}$ into $BWT_{X:Z}$, which is then used for computing $gap_{X:Z:\varepsilon}$. This approach minimizes the total number of backward search steps. To reduce I/O, $SA_{X_1:X_2Z}$ and $SA_{X_2:Z}$ are never merged into $SA_{X:Z}$, but all halfblock SAs are merged simultaneously in the final multiway merging stage. For the final merging, we need $gap_{X_1:X_2Z:\varepsilon}$ and $gap_{X_2:Z:\varepsilon}$, which can be computed quickly and easily from $gap_{X_1:X_2Z}$ and $gap_{X:Z:\varepsilon}$.

The disk usage is less than $7.5n$ bytes consisting of the text (n bytes), SAs ($5n$), gap arrays (about n using vbyte-encoding [33]), and a gt-bitvector (n bits).

7 Experimental Results

Setup. We performed experiments on two different machines referred to as Platform S (small) and Platform L (large). Platform S was equipped with a 4-core 3.40 GHz Intel i7-3770 CPU with 8 MiB L2 cache and 16 GiB of DDR3 RAM. Platform L was equipped with two 6-core 1.9 GHz Intel Xeon E5-2420 CPUs (capable, via hyper-threading, of running 24 threads) with 15 MiB L2 cache and 120 GiB of DDR3 RAM. The machine had 7.2 TiB of disk space striped with RAID0 across four identical local disks (achieving a (combined) transfer rate of about 480 MiB/s), and an additional two-disk RAID0 which was used only for the experiment on 1 TiB input. The OS was Linux (Ubuntu 12.04, 64 bit). All programs were compiled using g++ (Cilk Plus branch) version 4.8.1 with -O2 -DNDEBUG options.

Datasets. For the experiments we used the following files varying in the number of repetitions and alphabet size (see Table 2 for some statistics):

- hg.reads: a collection of DNA reads (short fragments produced by a sequencing machine) from 40 human genomes[1] filtered from symbols other than $\{A, C, G, T, N\}$ and newline;

---
[1] http://www.1000genomes.org/.

- wiki: a prefix of English Wikipedia dump[2] (dated 20140707) in the XML format;
- kernel: a concatenation of $\sim$16.8 million source files from 510 recent versions of Linux kernel[3];
- countries: a concatenation of all versions (edit history) of four Wikipedia articles about countries in the XML format. It contains a large number of 1–5 KiB repetitions;
- skyline: an artificial, highly repetitive sequence (see [5] for details);
- random: a randomly generated sequence of bytes.

Experiments. We implemented the pSAscan algorithm in C++ using STL threads for parallelism[4]. In the first experiment we study the performance of pSAscan as a standalone internal-memory suffix sorting algorithm and compare it with the parallel implementation of DC3 algorithm [6], the fastest parallel suffix-sorter in previous studies, and the parallel version of divsufsort [25]. The latter has a parallel mode that (slightly) improves the runtime, but is mostly known as the fastest sequential suffix array construction algorithm. For each algorithm, we included two versions, one using 32-bit integers and limited to 2 GiB or 4 GiB files, and the other capable of processing larger files. The algorithms and their memory usage are summarized in Table 1. For fair comparison pSAscan produces the suffix array as a plain array (rather than in a paged form). This requires an additional permuting step and slightly slows down our algorithm. The results for Platform L are given in Fig. 1. pSAscan is clearly the fastest algorithm when using full parallelism and at least competitive when using less threads. The exception is the random input with a large alphabet (where DC3 excels due to very shallow recursion) and skyline. The poor performance of pSAscan on the skyline testfile is, however, inherited from divsufsort for which it is the worst case input. The relative performance of pDC3 and pSAscan on Platform S (see Fig. 2 for two sample graphs) is similar to Platform L.

In the second experiment we compare the EM version of pSAscan to the best EM algorithms for suffix array construction: eSAIS [5] (with the STXXL library [8] compiled in parallel mode) and SAscan [15] (sequential), using a moderate amount of RAM (3.5 GiB). Results are given in Fig. 3. For smaller files, pSAscan is several times faster than the competitors. For larger files, eSAIS approaches pSAscan and would probably overtake it somewhere around 250–300 GiB files, which coincidentally is about the size for which eSAIS would run out of disk space on the test machine. Using the full 120 GiB RAM moves the crossover point to several terabytes and allowed us to process the full 1TiB instance of hg.reads (see Table 3).

Finally, Table 4 shows that, particularly for large files, the running time of pSAscan is dominated by the gap array construction, which involves $\Theta(n^2/m)$ steps of backward searching.

---

[2] http://dumps.wikimedia.org/.

[3] http://www.kernel.org/.

[4] The implementation is available at http://www.cs.helsinki.fi/group/pads/.

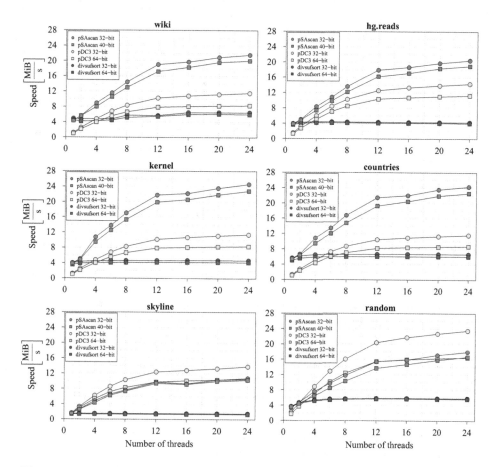

Fig. 1. Internal memory parallel suffix array construction on Platform L. All input files are of size 2 GiB (Color Figure Online).

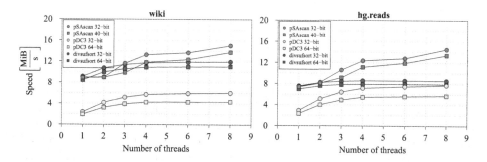

Fig. 2. Internal memory parallel suffix array construction on Platform S. All input files are of size 360 MiB (Color Figure Online).

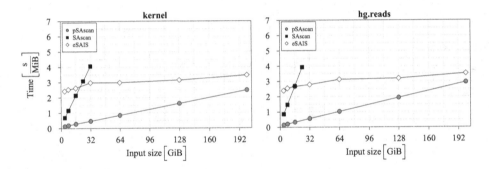

Fig. 3. Scalability of EM version of pSAscan compared to eSAIS and SAscan. All algorithms were allowed to use only 3.5 GiB of RAM for computation. pSAscan and eSAIS were allowed to use the maximum number of threads (24) (Color Figure Online).

Table 3. A performance comparison of eSAIS and pSAscan on prefixes of hg.reads testfile with varying amount of memory available to algorithms. The peak disk space usage includes input and output (which is five times the size of input).

Algorithm	Input size	RAM usage	Runtime	Peak disk usage	I/O volume
eSAIS	200 GiB	3.5 GiB	8.3 days	4.6 TiB	52.0 TiB
	200 GiB	120 GiB	4.1 days	4.6 TiB	36.1 TiB
pSAscan	200 GiB	3.5 GiB	7.0 days	1.4 TiB	43.8 TiB
	200 GiB	120 GiB	0.5 days	1.4 TiB	4.9 TiB
	1024 GiB	120 GiB	8.1 days	7.3 TiB	48.3 TiB

Table 4. A detailed runtime breakdown of external memory pSAscan on the 200GiB instance of hg.reads. The times are given in hours.

RAM usage	Internal memory suffix sort			Gap array construction	Final merge	Other
	I/O	divsufsort	Other			
3.5 GiB	0.4	0.7	2.2	132.4	29.2	1.2
120 GiB	0.6	1.1	2.6	4.3	2.4	0.9

8   Concluding Remarks

When deciding whether an algorithm scales to deal with large inputs, we are principally concerned with three values: RAM, time, and disk usage. The main advantage of pSAscan is that it measures up well on all three of these dimensions. The algorithm is also fairly versatile: for example, it would add little overhead to have it output the BWT in addition to (or instead of) the SA in order to, say, speed up construction of an FM-index.

There are many avenues for future work. Most obviously, one wonders if similar techniques for suffix sorting can be successfully applied to other parallel architectures, such as GPUs and distributed systems. We also believe our merging procedure can find other uses, such as supporting the efficient update of the suffix array when new text is appended to the underlying string.

References

1. Abouelhoda, M.I., Kurtz, S., Ohlebusch, E.: Replacing suffix trees with enhanced suffix arrays. J. Discrete Algorithms **2**(1), 53–86 (2004)
2. Andersson, A., Hagerup, T., Håstad, J., Petersson, O.: Tight bounds for searching a sorted array of strings. SIAM J. Comput. **30**(5), 1552–1578 (2000)
3. Apostolico, A., Iliopoulos, C.S., Landau, G.M., Schieber, B., Vishkin, U.: Parallel construction of a suffix tree with applications. Algorithmica **3**, 347–365 (1988)
4. Apostolico, A., Lonardi, S.: Off-line compression by greedy textual substitution. Proc. IEEE **88**(11), 1733–1744 (2000)
5. Bingmann, T., Fischer, J., Osipov, V.: Inducing suffix and LCP arrays in external memory. In: Sanders, P., Zeh, N. (eds.) ALENEX 2013. pp. 88–102. SIAM (2013)
6. Blelloch, G.E., Shun, J.: A simple parallel cartesian tree algorithm and its application to suffix tree construction. In: Müller-Hannemann, M., Werneck, R.F.F. (eds.) ALENEX 2011, pp. 48–58. SIAM (2011)
7. Crauser, A., Ferragina, P.: A theoretical and experimental study on the construction of suffix arrays in external memory. Algorithmica **32**(1), 1–35 (2002)
8. Dementiev, R., Kettner, L., Sanders, P.: STXXL: standard template library for XXL data sets. Softw. Pract. Exper. **38**(6), 589–637 (2008)
9. Deo, M., Keely, S.: Parallel suffix array and least common prefix for the GPU. In: Nicolau, A., Shen, X., Amarasinghe, S.P., Vuduc, R.W. (eds.) PPoPP 2013, pp. 197–206. ACM (2013)
10. Farach-Colton, M., Ferragina, P., Muthukrishnan, S.: On the sorting-complexity of suffix tree construction. J. ACM **47**(6), 987–1011 (2000)
11. Ferragina, P., Manzini, G.: Indexing compressed text. J. ACM **52**(4), 552–581 (2005)
12. Ferragina, P., Gagie, T., Manzini, G.: Lightweight data indexing and compression in external memory. Algorithmica **63**(3), 707–730 (2012)
13. Gonnet, G.H., Baeza-Yates, R.A., Snider, T.: New indices for text: pat trees and pat arrays. In: Frakes, W.B., Baeza-Yates, R. (eds.) Information Retrieval: Data Structures and Algorithms, pp. 66–82. Prentice-Hall, Englewood Cliffs (1992)
14. Grossi, R., Vitter, J.S.: Compressed suffix arrays and suffix trees with applications to text indexing and string matching. SIAM J. Comput. **35**(2), 378–407 (2005)
15. Kärkkäinen, J., Kempa, D.: Engineering a lightweight external memory suffix array construction algorithm. In: Iliopoulos, C.S., Langiu, A. (eds.) ICABD 2014, CEUR Workshop Proceedings, vol. 1146, pp. 53–60 (2014). CEUR-WS.org
16. Kärkkäinen, J., Kempa, D., Puglisi, S.J.: Linear time Lempel-Ziv factorization: simple, fast, small. In: Fischer, J., Sanders, P. (eds.) CPM 2013. LNCS, vol. 7922, pp. 189–200. Springer, Heidelberg (2013)
17. Kärkkäinen, J., Kempa, D., Puglisi, S.J.: Lempel-Ziv parsing in external memory. In: Bilgin, A., Marcellin, M.W., Serra-Sagristà, J., Storer, J.A. (eds.) DCC 2014, pp. 153–162. IEEE (2014)
18. Kärkkäinen, J., Kempa, D., Puglisi, S.J.: String range matching. In: Kulikov, A.S., Kuznetsov, S.O., Pevzner, P. (eds.) CPM 2014. LNCS, vol. 8486, pp. 232–241. Springer, Heidelberg (2014)

19. Kärkkäinen, J., Sanders, P., Burkhardt, S.: Linear work suffix array construction. J. ACM **53**(6), 918–936 (2006)
20. Kulla, F., Sanders, P.: Scalable parallel suffix array construction. Parallel Comput. **33**(9), 605–612 (2007)
21. Langmead, B., Trapnell, C., Pop, M., Salzberg, S.L.: Ultrafast and memory-efficient alignment of short dna sequences to the human genome. Genome Biol. **10**(3), R25 (2009)
22. Li, H., Durbin, R.: Fast and accurate short read alignment with Burrows-Wheeler transform. Bioinformatics **25**(14), 1754–1760 (2009)
23. Louza, F.A., Telles, G.P., Ciferri, C.D.D.A.: External memory generalized suffix and LCP arrays construction. In: Fischer, J., Sanders, P. (eds.) CPM 2013. LNCS, vol. 7922, pp. 201–210. Springer, Heidelberg (2013)
24. Manber, U., Myers, G.W.: Suffix arrays: a new method for on-line string searches. SIAM J. Comput. **22**(5), 935–948 (1993)
25. Mori, Y.: libdivsufsort, a C library for suffix array construction. http://code.google.com/p/libdivsufsort/
26. Nakamura, R., Inenaga, S., Bannai, H., Funamoto, T., Takeda, M., Shinohara, A.: Linear-time text compression by longest-first substitution. Algorithms **2**(4), 1429–1448 (2009)
27. Nong, G., Chan, W.H., Zhang, S., Guan, X.F.: Suffix array construction in external memory using d-critical substrings. ACM Trans. Inf. Syst. **32**(1), 1 (2014)
28. Osipov, V.: Parallel suffix array construction for shared memory architectures. In: Calderón-Benavides, L., González-Caro, C., Chávez, E., Ziviani, N. (eds.) SPIRE 2012. LNCS, vol. 7608, pp. 379–384. Springer, Heidelberg (2012)
29. Puglisi, S.J., Smyth, W.F., Turpin, A.: A taxonomy of suffix array construction algorithms. ACM Comput. Surv. **39**(2), 31 (2007). Article 4
30. Simpson, J.T., Durbin, R.: Efficient de novo assembly of large genomes using compressed data structures. Genome Res. **22**(3), 549–556 (2012)
31. Tischler, G.: Faster average case low memory semi-external construction of the Burrows-Wheeler transform. In: Iliopoulos, C.S., Langiu, A. (eds.) ICABD 2014, CEUR Workshop Proceedings, vol. 1146, pp. 61–68 (2014). CEUR-WS.org
32. Weiner, P.: Linear pattern matching algorithms. In: SWAT 1973, pp. 1–11. IEEE (1973)
33. Williams, H.E., Zobel, J.: Compressing integers for fast file access. Comput. J. **42**(3), 193–201 (1999)
34. Ziv, J., Lempel, A.: A universal algorithm for sequential data compression. IEEE Trans. Inf. Theor. **23**(3), 337–343 (1977)

On Maximal Unbordered Factors

Alexander Loptev[1], Gregory Kucherov[2], and Tatiana Starikovskaya[3](✉)

[1] Higher School of Economics, Moscow, Russia
`alexander.loptev@gmail.com`
[2] Laboratoire d'Informatique Gaspard Monge,
Université Paris-Est and CNRS, Marne-la-vallée, Paris, France
`gregory.kucherov@univ-mlv.fr`
[3] University of Bristol, Bristol, UK
`tat.starikovskaya@gmail.com`

Abstract. Given a string S of length n, its maximal unbordered factor is the longest factor which does not have a border. In this work we investigate the relationship between n and the length of the maximal unbordered factor of S. We prove that for the alphabet of size $\sigma \geq 5$ the expected length of the maximal unbordered factor of a string of length n is at least $0.99n$ (for sufficiently large values of n). As an application of this result, we propose a new algorithm for computing the maximal unbordered factor of a string.

1 Introduction

If a proper prefix of a string is simultaneously its suffix, then it is called a border of the string. Given a string S of length n, its maximal unbordered factor is the longest factor which does not have a border. The relationship between n and the length of the maximal unbordered factor of S has been a subject of interest in the literature for a long time, starting from the 1979 paper of Ehrenfeucht and Silberger [7].

Let $b(S)$ be the length of the maximal unbordered factor of S and $\pi(S)$ be the minimal period of S. Ehrenfeucht and Silberger showed that if the minimal period of S is smaller than $\frac{1}{2}n$, then $b(S) = \pi(S)$. Following this, they raised a natural question: How small $b(S)$ must be to guarantee $b(S) = \pi(S)$? Their conjecture was that $b(S)$ must be smaller than $\frac{1}{2}n$. However, this conjecture was proven false two years later by Assous and Pouzet [1]. As a counterexample they gave a string

$$S = a^m b a^{m+1} b a^m b a^{m+2} b a^m b a^{m+1} b a^m$$

of length $n = 7m+10$. The length of the maximal unbordered factor of this string is $b(S) = 3m + 6 \leq \frac{3}{7}n + 2 < \frac{1}{2}n$ (with $ba^{m+1}ba^mba^{m+2}$ and $a^{m+2}ba^mba^{m+1}b$ being unbordered), and the minimal period $\pi(S) = 4m + 7 \neq b(S)$.

The next attempt to answer the question was undertaken by Duval [3]: He improved the bound to $\frac{1}{4}n + \frac{3}{2}$. But the final answer to the question of Ehrefeucht and Silberger was given just recently by Holub and Nowotka [10]. They showed that $b(S) \leq \frac{3}{7}n$ implies $b(S) = \pi(S)$, and, as follows from the example of Assous and Pouzet, this bound is tight.

F. Cicalese et al. (Eds.): CPM 2015, LNCS 9133, pp. 343–354, 2015.
DOI: 10.1007/978-3-319-19929-0_29

Therefore, when either $b(S)$ or $\pi(S)$ is small, $b(S) = \pi(S)$. Exploiting this fact, one can even compute the maximal unbordered factor itself in linear time. The key idea is that in this case the maximal unbordered factor is an unbordered conjugate of the minimal period of S, and both the minimal period and its unbordered conjugate can be found in linear time [6,15].

The interesting cases are those where $b(S)$ (and, consequently, $\pi(S)$) is big. Yet, it is generally believed that they are the most common ones. This is supported by experimental results shown in Fig. 1 that plots the average difference between the length n of a string and the length of its maximal unbordered factor. Guided by the experimental results, we state the following conjecture:

Conjecture 1. Expected length of the maximal unbordered factor of a string of length n is $n - \mathcal{O}(1)$.

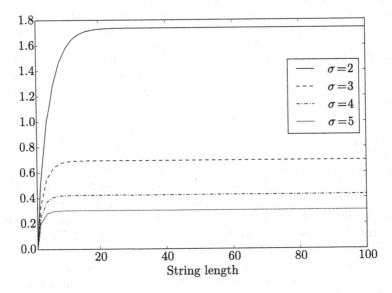

Fig. 1. Average difference between the length n of a string and the length of its maximal unbordered factor for $1 \le n \le 100$ and alphabets of size $2 \le \sigma \le 5$.

To the best of our knowledge, there have been no attempts to prove the conjecture or any lower bound at all in the literature. In Sect. 4 we address this gap and make the very first step towards proving the conjecture. We show that the expected length of the maximal unbordered factor of a string of length n over the alphabet A of size $\sigma \ge 2$ is at least $n(1 - \xi(\sigma) \cdot \sigma^{-4}) + \mathcal{O}(1)$, where $\xi(\sigma)$ is a function that converges to 2 quickly with the growth of σ. In particular, this theorem implies that for alphabets of size $\sigma \ge 5$ the expected length of the maximal unbordered factor of a string is at least $0.99n$ (for sufficiently large values of n). To prove the theorem we developed a method of generating strings with large unbordered factors which we find to be interesting on its own (see Sect. 3).

It follows that the algorithm for computing maximal unbordered factors we sketched earlier cannot be used in a majority of cases. Instead, one can consider the following algorithm. A border array of a string is an array containing the maximal length of a border of each prefix of this string. Note that a prefix of a string is unbordered exactly when the corresponding entry in the border array is zero. Therefore, to compute the maximal unbordered factor of a string S it suffices to build border arrays of all suffixes of a string. It is well-known that a single border array can be constructed in linear time, which gives quadratic time bound for the algorithm. In Sect. 5 we show how to modify this algorithm to make use of the fact that the expected length of the maximal unbordered factor is big. We give $\mathcal{O}(\frac{n^2}{\sigma^4})$ time bound for the modified algorithm, as well as confirm its efficiency experimentally.

Related work. Apart from the aforementioned results, we consider our work to be related to three areas of research.

As we have already mentioned, the maximal unbordered factor can be found by locating the rightmost zeros in the border arrays of suffixes of a string and better understanding of structure of border arrays would give more efficient algorithms for the problem. Structure of border arrays has been studied in [2,4, 5,8,9,14].

In contrast to the problem we consider in this work, one can be interested in the problem of preprocessing a string to answer online factor queries related to its borders. This problem has been considered by Kociumaka et al. [12,13]. They proposed a series of data structures which, in particular, can be used to determine if a factor is unbordered in logarithmic time.

Finally, repeating fragments in a string (borders of factors is one example of such fragments) were studied in connection with the *Longest Common Extension* problem which asks, given a pair of positions i, j in a string, to return the longest fragment that occurs both at i and j. This problem has many solutions, yet recently Ilie et al. [11] showed that the simplest solution, i.e. simply scanning the string and comparing pairs of letters starting at positions i and j, is the fastest on average. The authors also proved that the longest common extension has expected length smaller than $\frac{1}{\sigma-1}$, where σ is the size of the alphabet.

2 Preliminaries

We start by introducing some standard notation and definitions.

Power sums. We will need the following identities.

Fact 1. $S(x) = \sum_{i=1}^{k} i\, x^{i-1} = \frac{k\, x^{k+1} - (k+1)\, x^k + 1}{(x-1)^2}$ *for all* $x \neq 1$.

Proof.

$$S(x) = \left(\sum_{i=1}^{k} x^i\right)' = \left(\frac{x^{k+1} - x}{x - 1}\right)' = \frac{((k+1)x^k - 1)(x - 1) - (x^{k+1} - x)}{(x - 1)^2}$$

Simplifying, we obtain

$$S(x) = \sum_{i=1}^{k} i \, x^{i-1} = \frac{k \, x^{k+1} - (k+1) \, x^k + 1}{(x-1)^2}$$ □

Corollary 1. $S(x) = \sum_{i=1}^{k} i \, x^{i-1} = \frac{k \, x^k}{x-1} + \mathcal{O}(x^{k-2})$ *for* $x \geq 1.5$.

Strings. The alphabet A is a finite set of size σ. We refer to the elements of A as *letters*. A *string* over A is a finite ordered sequence of letters (possibly empty). Letters in a string are numbered starting from 1, that is, a string S of *length* n consists of letters $S[1], S[2], \ldots, S[n]$. The length n of S is denoted by $|S|$. A set of all strings of length n is denoted A^n.

For $1 \leq i \leq j \leq n$, $S[i..j]$ is a *factor* of S with endpoints i and j. The factor $S[1..j]$ is called a *prefix* of S, and the factor $S[i..n]$ is called a *suffix* of S. A prefix (or a suffix) different from S and the empty string is called *proper*.

If a proper prefix of a string is simultaneously its suffix, then it is called a *border*. For example, borders of a string $ababa$ are a and aba. The *maximal border* of a string is its longest border. For S we define its *border array* B (also known as the *failure function*) to contain the lengths of the maximal borders of all its prefixes, i.e. $B[i]$ is the length of the maximal border of $S[1..i]$, $i = 1..n$. The last entry in the border array, $B[n]$, contains the length of the maximal border of S. It is well-known that the border array and therefore the maximal border of S can be found in $\mathcal{O}(n)$ time and space [15].

A period of S is an integer π such that for all i, $1 \leq i \leq n - \pi$, $S[i] = S[i+\pi]$. The minimal period of a string has length $n - B[n]$, and hence can be computed in linear time as well.

Unbordered strings. A string is called *unbordered* if it has no border. Let $b(i, \sigma)$ be the number of unbordered strings in A^i. Nielsen [16] showed that unbordered strings can be constructed in a recursive manner, starting from unbordered strings of length 2 and inserting new letters in the "middle". The following theorem is a corollary of the proposed construction method:

Theorem 1 ([16]). *The sequence* $\left\{ \frac{b(i,\sigma)}{\sigma^i} \right\}_{i=1}^{\infty}$ *is monotonically nonincreasing and it converges to a constant* α*, which satisfies* $\alpha \geq 1 - \sigma^{-1} - \sigma^{-2}$.

Corollary 2 ([16]). $b(i, \sigma) \geq \sigma^i - \sigma^{i-1} - \sigma^{i-2}$ *for all* i.

This corollary immediately implies that the expected length of the maximal unbordered factor of a string of length n is at least $n(1 - \sigma^{-1} - \sigma^{-2})$. We improve this lower bound in the subsequent sections. We will make use of a lower bound on the number $b_j(i, \sigma)$ of unbordered strings such that its first letter differs from the subsequent j letters. An example of such string for $j = 2$ is $abcacbb$.

Lemma 1. $b_j(i, \sigma) \geq (\sigma - 1)^{j+1} \sigma^{i-j-1} - \sigma^{i-2}$ *for all* $i \geq j + 1$.

Proof. The number of such strings is equal to $b(i, \sigma)$ minus the number $b_j^-(i, \sigma)$ of unbordered strings of length i that do not have the property. We estimate the latter from above by the number of such strings in the set of all strings with their first letter not equal to the last letter. Hence, $b_j^-(i, \sigma) \leq (\sigma - 1)\sigma^{i-1} - (\sigma - 1)^{j+1}\sigma^{i-j-1}$. Recall that $b(i, \sigma) \geq \sigma^i - \sigma^{i-1} - \sigma^{i-2}$ by Theorem 1. The claim follows. □

Remark. *The right-hand side of the inequality of Lemma 1 is often negative for $\sigma = 2$. We will not use it for this case.*

The *maximal unbordered factor* of a string (MUF) is naturally defined to be the longest factor of the string which is unbordered.

3   Generating Strings with Large MUF

In this section we explain how to generate strings of some fixed length n with large maximal unbordered factors. To show the lower bounds we announced, we will need many of such strings. The idea is to generate them from unbordered strings.

Let S be an unbordered string of length $i \geq \lceil \frac{n}{2} \rceil$. Consider a string $SP_1 \ldots P_k$ of length n, where $P_1, \ldots, P_k$ are prefixes of S. It is not difficult to see that the maximal unbordered factor of any string of this form has length at least i. (Because S is one of its unbordered factors.) The number of such strings that can be generated from S is 2^{n-i-1}, because each of them corresponds to a composition of $n - i$, i.e. representation of $n - i$ as a sum of a sequence of strictly positive integers. But, some of these strings can be equal. Consider, for example, an unbordered string $S = aaabab$. Then the two strings $aaababaaa$ (S appended with its prefix aaa) and $aaababaaa$ (S appended with its prefixes a and aa) will be equal. However, we can show the following lemma.

Lemma 2. *Let $S_1 \neq S_2$ be two unbordered strings. Any two strings of the form above generated from S_1 and S_2 are distinct.*

Proof. Suppose that the produced strings are equal. If $|S_1| = |S_2|$, we immediately obtain $S_1 = S_2$, a contradiction. Otherwise, w.l.o.g. assume $|S_1| < |S_2|$. Then S_2 is equal to a concatenation of S_1 and some of its prefixes. The last of these prefixes is simultaneously a suffix and a prefix of S_2, i.e. S_2 is not unbordered. A contradiction. □

Our idea is to produce as many strings of the form $SP_1 \ldots P_k$ as possible, taking extra care to ensure that all strings produced from a fixed string S are distinct. From unbordered strings of length $i = n$ and $i = n - 1$ we produce just one string of length n. (For $i = n$ it is the string itself and for $i = n - 1$ it is the string appended with its first letter.) For unbordered strings of length $i \leq n - 2$ we propose a different method based on the lemma below.

Lemma 3. *Each unbordered string S of length i such that its first letter differs from the subsequent j letters, where $\lceil n/2 \rceil \leq i < n - j$, gives at least 2^j distinct strings of the form $SP_1 \ldots P_k$.*

Proof. We choose the last prefix P_k to be the prefix of S of length at least $n - i - j$. We place no restrictions on the first $k - 1$ prefixes.

Let us start by showing that all generated strings are distinct. Suppose there are two equal strings $SP_1 \ldots P_\ell$ and $SP'_1 \ldots P'_{\ell'}$. Let P_d, P'_d be the first pair of prefixes that have different lengths. W.l.o.g. assume that $|P_d| < |P'_d|$. Then $d \neq \ell$ and hence $|P_d| \leq j = n - i - (n - i - j)$. It follows that P'_d (which is a prefix of S) contains at least two occurrences of $S[1]$, one at the position 1 and one at the position $|P_d| + 1 \leq j + 1$. In other words, we have $S[1] = S[|P_d| + 1]$ and $|P_d| + 1 \leq j + 1$, which contradicts our choice of S.

If the length of the last prefix is fixed to some integer $m \geq n - i - j$, then each of the generated strings $SP_1 \ldots P_k$ is defined by the lengths of the first $k - 1$ of the appended prefixes. In other words, there is one-to-one correspondence between the generated strings and compositions of $n - i - m$. (Here we use $i \geq \lceil n/2 \rceil$ to ensure that every composition corresponds to a sequence of prefixes of S.) The number of compositions of $n - i - m$ is 1 when $m = n - i$ and $2^{n-i-m-1}$ otherwise. Summing up for all m from $n - i - j$ to $n - i$ we obtain that the number of the generated strings is 2^j. □

Let us estimate the total amount of strings produced by this method. We produce one string from each unbordered string of length i. Then, from each unbordered string of length i such that its first letter differs from the second letter, we produce $1 = 2 - 1$ more string. If the first letter differs both from the second and the third letters, we produce $2 = 2^2 - 1 - 1$ more strings. And finally, if the first letter differs from the subsequent j letters, we produce $2^{j-1} = 2^j - (1 + 1 + 2 + \ldots + 2^{j-2})$ strings. It follows that the number of strings we can produce from unbordered strings of length $i \leq n - 2$ is

$$b(i, \sigma) + \sum_{j=1}^{n-i-1} 2^{j-1} \cdot b_j(i, \sigma)$$

Recall that the maximal unbordered factor of each of the generated strings has length at least i and that none of them can be equal to a string generated from an unbordered string of different length.

4   Expected Length of MUF

In this section we prove the main result of this paper.

Theorem 2. *Expected length of the maximal unbordered factor of a string of length n over an alphabet A of size $\sigma \geq 2$ is at least*

$$n \cdot (1 - \xi(\sigma) \cdot \sigma^{-4}) + \mathcal{O}(1) \tag{1}$$

where $\xi(2) = 8$ and $\xi(\sigma) = \frac{2\sigma^3 - 2\sigma^2}{(\sigma-2)(\sigma^2 - 2\sigma + 2)}$ for $\sigma > 2$.

Before we give a proof of the theorem, let us say a few words about $\xi(\sigma)$. This function is monotonically decreasing for $\sigma \geq 2$ and quickly converges to 2. We give the first four values for $\xi(\sigma)$ (rounded up to 3 s.f.) and $1 - \xi(\sigma) \cdot \sigma^{-4}$ (rounded down to 3 s.f.) in the table below.

	$\sigma = 2$	$\sigma = 3$	$\sigma = 4$	$\sigma = 5$
$\xi(\sigma)$	8.000	7.200	4.800	3.922
$1 - \xi(\sigma) \cdot \sigma^{-4}$	0.500	0.911	0.981	0.993

Corollary 3. *Expected length of the maximal unbordered factor of a string of length n over the alphabet A of size $\sigma \geq 5$ is at least $0.99n$ (for sufficiently large values of n).*

Proof of Theorem 2. Let $\beta_i^n(\sigma)$ be the number of strings in A^n such that the length of their maximal unbordered factor is i. Expected length of the maximal unbordered factor is then equal to

$$\frac{1}{\sigma^n} \sum_{i=1}^{n} i \cdot \beta_i^n(\sigma)$$

For the sake of simplicity, we temporarily omit $\frac{1}{\sigma^n}$, and only in the very end we will add it back. Recall that in the previous section we showed how to generate a set of distinct strings of length n with maximal unbordered factors of length at least i which contains

$$b(i, \sigma) + \sum_{j=1}^{n-i-1} 2^{j-1} \cdot b_j(i, \sigma)$$

strings for all $\lceil \frac{n}{2} \rceil \leq i \leq n - 2$ and $b(i, \sigma)$ strings for $i = \{n - 1, n\}$. Then

$$\sum_{i=1}^{n} i \cdot \beta_i^n(\sigma) \geq \underbrace{\sum_{i=\lceil n/2 \rceil}^{n} i \cdot b(i, \sigma)}_{(S_1)} + \underbrace{\sum_{i=\lceil n/2 \rceil}^{n-2} \sum_{j=1}^{n-i-1} 2^{j-1} \cdot i \cdot b_j(i, \sigma)}_{(S_2)} \qquad (2)$$

We start by computing (S_1). Applying Corollary 2 and replacing $b(i, \sigma)$ with $\frac{b(n,\sigma)}{\sigma^{n-i}}$ in (S_1), we obtain:

$$(S_1) \geq \sum_{i=\lceil \frac{n}{2} \rceil}^{n} i \, \frac{b(n, \sigma)}{\sigma^{n-i}} = \frac{b(n, \sigma)}{\sigma^{n-1}} \Big(\sum_{i=\lceil \frac{n}{2} \rceil}^{n} i \, \sigma^{i-1} \Big)$$

Note that the lower limit in inner sum of (S_1) can be replaced by one because the correcting term is small:

$$\frac{b(n,\sigma)}{\sigma^{n-1}}\sum_{i=1}^{\lceil n/2\rceil-1} i\sigma^{i-1} \leq \frac{n^2\cdot b(n,\sigma)}{4\sigma^{n/2}} = \mathcal{O}(\sigma^n)$$

We finally use Corollary 1 for $x=\sigma$ and $k=n$ to compute the right-hand side of the inequality:

$$(S_1) \geq \frac{n\sigma}{\sigma-1}\cdot b(n,\sigma) + \mathcal{O}(\sigma^n) \tag{3}$$

We note that for $\sigma=2$ the right-hand side is at least $2n\cdot(2^n-2^{n-1}-2^{n-2})+\mathcal{O}(2^n) = n\cdot 2^{n-1} + \mathcal{O}(2^n)$ by Corollary 2 and $(S_2)\geq 0$. Hence, $\sum_{i=1}^{n} i\cdot\beta_i^n(2)\geq n\cdot 2^{n-1}+\mathcal{O}(2^n)$. Dividing both sides by 2^n, we obtain the theorem.

Below we assume $\sigma>2$ and for these values of σ give a better lower bound on (S_2). Recall that $b_j(i,\sigma)\geq(\sigma-1)^{j+1}\sigma^{i-j-1}-\sigma^{i-2}$ (see Lemma 1). It follows that

$$(S_2) \geq \sum_{i=\lceil n/2\rceil}^{n-2}\sum_{j=1}^{n-i-1} 2^{j-1}\cdot i\cdot\left((\sigma-1)^{j+1}\sigma^{i-j-1}-\sigma^{i-2}\right)$$

Let us change the order of summation:

$$(S_2) \geq \sum_{j=1}^{\lfloor n/2\rfloor-1} 2^{j-1}\cdot\left((\sigma-1)^{j+1}\sigma^{-j}-\sigma^{-1}\right)\sum_{i=\lceil n/2\rceil}^{n-j-1} i\cdot\sigma^{i-1}$$

We can replace the lower limit in the inner sum of (S_2) by one as it will only change the sum by $\mathcal{O}(\sigma^n)$. After replacing the lower limit, we apply Corollary 1 to compute the inner sum:

$$(S_2) \geq \sum_{j=1}^{\lfloor n/2\rfloor-1} 2^{j-1}\cdot\left((\sigma-1)^{j+1}\sigma^{-j}-\sigma^{-1}\right)\cdot(n-j-1)\frac{\sigma^{n-j-1}}{\sigma-1}+\mathcal{O}(\sigma^n)$$

We divide the sum above into positive and negative parts:

$$\underbrace{\sum_{j=1}^{\lfloor n/2\rfloor-1}(n-j-1)\,2^{j-1}(\sigma-1)^j\sigma^{n-2j-1}}_{(P)} - \underbrace{\sum_{j=1}^{\lfloor n/2\rfloor-1}(n-j-1)2^{j-1}\frac{\sigma^{n-j-2}}{\sigma-1}}_{(N)}$$

We start by computing (N). We again apply the trick with the lower limit and Fact 1, and replace $(n-j-1)$ with k.

$$(N) = \frac{2^{n-3}}{\sigma-1}\sum_{k=\lceil\frac{n}{2}\rceil}^{n-2} k\left(\frac{\sigma}{2}\right)^{k-1} = \frac{(n-2)\sigma^{n-2}}{(\sigma-1)(\sigma-2)}+\mathcal{O}(\sigma^n)$$

Computing (P) is a bit more involved. We divide it into two parts:

$$(P) = \underbrace{\frac{(n-1)\sigma^{n-1}}{2}\cdot\sum_{j=1}^{\lfloor n/2\rfloor-1}\left(\frac{2(\sigma-1)}{\sigma^2}\right)^j}_{R_1} - \underbrace{\sigma^{n-1}\sum_{j=1}^{\lfloor n/2\rfloor-1} j\,2^{j-1}(\sigma-1)^j\sigma^{-2j}}_{R_2}$$

(R_1) is a sum of a geometric progression and it is equal to

$$\frac{(n-1)\sigma^{n-1}}{2} \cdot \frac{\left(\frac{2(\sigma-1)}{\sigma^2}\right)^{\lfloor n/2 \rfloor} - \frac{2(\sigma-1)}{\sigma^2}}{\frac{2(\sigma-1)}{\sigma^2} - 1} = \frac{(n-1)\sigma^{n-1}}{2} \cdot \frac{2(\sigma-1)}{\sigma^2 - 2\sigma + 2} + \mathcal{O}(\sigma^n)$$

Lemma 4. $(R_2) = \mathcal{O}(\sigma^n)$.

Proof. We start our proof by rewriting (R_2):

$$(R_2) = \sigma^{n-3}(\sigma - 1) \cdot \sum_{j=1}^{\lfloor n/2 \rfloor - 1} j \left(\frac{2(\sigma-1)}{\sigma^2}\right)^{j-1}$$

We apply Fact 1 for $x = \frac{2(\sigma-1)}{\sigma^2}$ and $k = \lfloor n/2 \rfloor - 1$ to compute the inner sum.

$$(R_2) = \sigma^{n-3}(\sigma - 1) \cdot \frac{(\lfloor n/2 \rfloor - 1) \cdot \left(\frac{2(\sigma-1)}{\sigma^2}\right)^{\lfloor n/2 \rfloor} - \lfloor n/2 \rfloor \cdot \left(\frac{2(\sigma-1)}{\sigma^2}\right)^{\lfloor n/2 \rfloor - 1} + 1}{\left(\frac{2(\sigma-1)}{\sigma^2} - 1\right)^2}$$

The claim follows. □

We now summarize our findings. From equations for (P), (N), (R_1), and (R_2) we obtain (after simplification):

$$(S_2) \geq (P) - (N) = n \cdot \left(\frac{\sigma^n - \sigma^{n-1}}{\sigma^2 - 2\sigma + 2} - \frac{\sigma^{n-2}}{(\sigma - 1)(\sigma - 2)}\right) + \mathcal{O}(\sigma^n) \qquad (4)$$

We now return back to Eq. (2) and use our lower bounds for (S_1) and (S_2) together with Corollary 2 for $b(n, \sigma)$:

$$\sum_{i=1}^{n} i \cdot \beta_i^n(\sigma) \geq n \cdot \left(\frac{\sigma^{n+1} - \sigma^n - \sigma^{n-1}}{\sigma - 1} + \frac{\sigma^n - \sigma^{n-1}}{\sigma^2 - 2\sigma + 2} - \frac{\sigma^{n-2}}{(\sigma - 1)(\sigma - 2)}\right) + \mathcal{O}(\sigma^n)$$

We now simplify the expression above and return back $\frac{1}{\sigma^n}$ as we promised in the very beginning of the proof to obtain:

$$\frac{1}{\sigma^n} \sum_{i=1}^{n} i \cdot \beta_i^n(\sigma) \geq n \cdot (1 - \xi(\sigma) \cdot \sigma^{-4}) + \mathcal{O}(1) \qquad (5)$$

where $\xi(\sigma) = \frac{2\sigma^3 - 2\sigma^2}{(\sigma - 2)(\sigma^2 - 2\sigma + 2)}$. This completes the proof of Theorem 2. □

Remark. *Theorem 2 actually provides a lower bound on the expected length of the maximal unbordered prefix (rather than that of the maximal unbordered factor), which suggests that this bound could be improved.*

5   Computing MUF

Based on our findings we propose an algorithm for computing the maximal unbordered factor of a string S of length n and give an upper bound on its expected running time. A basic algorithm would be to compute the border arrays (see Sect. 2 for the definition) of all suffixes of S. The border arrays contain the lengths of the maximal borders of all prefixes of all suffixes of S, i.e., of all factors of S. It remains to scan the border arrays and to select the longest factor such that the length of its maximal border is zero. Since a border array can be computed in linear time, the running time of this algorithm is $\mathcal{O}(n^2)$.

The algorithm we propose is a minor modification of the basic algorithm. We build border arrays for suffixes of S starting from the longest one. After building an array B_i for $S[i..n]$ we scan it and locate the longest factor $S[i..j]$ such that the length of its maximal border stored in $B_i[j]$ is zero. We then compare $S[i..j]$ and the current maximal unbordered factor (initialized with an empty string). If $S[i..j]$ is longer, we update the maximal unbordered factor and proceed. At the moment we reach a suffix shorter than the current maximal unbordered factor, we stop.

Theorem 3. *The maximal unbordered factor of a string of length n over an alphabet A of size σ can be found in $\mathcal{O}(\frac{n^2}{\sigma^4})$ expected time.*

Proof. Let $b(S)$ be the length of the maximal unbordered factor of S. Then the running time of the algorithm is $\mathcal{O}((n - b(S)) \cdot n)$, because $b(S)$ will be a prefix of one of the first $n - b(S) + 1$ suffixes of S (starting from the longest one). Averaging this bound over all strings of length n, we obtain that the expected running time is

$$\mathcal{O}(\frac{1}{\sigma^n} \sum_{S \in A^n} (n - b(S)) \cdot n) = \mathcal{O}(n \cdot (\frac{1}{\sigma^n} \sum_{S \in A^n} (n - b(S))))$$

and $\frac{1}{\sigma^n} \sum_{S \in A^n} (n - b(S)) = \mathcal{O}(\frac{n}{\sigma^4})$ as it follows from Theorem 2 and properties of $\xi(\sigma)$.   □

We performed a series of experiments to confirm that the expected running time of the proposed algorithm is much smaller than that of the basic algorithm. We compared the time required by the algorithms for strings of length $1 \le n \le 100$ over alphabets of size $\sigma = \{2, 3, 4, 5, 10\}$. The time required by the algorithms was computed as the average time on a set of size $N = 10^6$ of randomly generated strings of given length. The experiments were performed on a PC equipped with one 2.6 GHz Intel Core i5 processor. As it can be seen in Fig. 2, the minor modification we proposed decreases the expected running time dramatically. Obtained results were similar for all considered alphabet sizes. All source files, results, and plots can be found in a repository http://github.com/avlonger/unbordered.

We note that the data structures [12,13] can be used to compute the maximal unbordered factor in a straightforward way by querying all factors in order

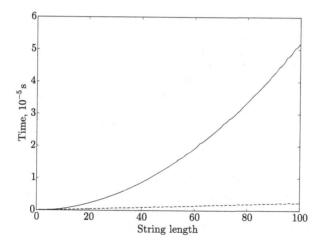

Fig. 2. Average running times of the proposed algorithm (dashed line) and the basic algorithm (solid line) for strings over the alphabet of size $\sigma = 2$.

of decreasing length. This idea seems to be very promising since these data structures need to be built just once, for the string S itself. However, the data structures are rather complex and both the theoretical bound for the expected running time, which is $\mathcal{O}(\frac{n^2}{\sigma^4} \log n)$, and our experiments show that this solution is slower than the one described above.

6   Conclusion

We consider the contributions of this work to be three-fold. We started with an explicit method of generating strings with large unbordered factors. We then used it to show that the expected length of the maximal unbordered factor and the minimal period of a string of length n is $\Omega(n)$, leaving the question raised in Conjecture 1 open. As an immediate application of our result, we gave a new algorithm for computing maximal unbordered factors and proved its efficiency both theoretically and experimentally.

Acknowledgements. The authors would like to thank the anonymous reviewers whose suggestions greatly improved the quality of this work.

References

1. Assous, R., Pouzet, M.: Une caractérisation des mots périodiques. Discrete Math. **25**(1), 1–5 (1979)
2. Clément, J., Giambruno, L.: On the number of prefix and border tables. In: Pardo, A., Viola, A. (eds.) LATIN 2014. LNCS, vol. 8392, pp. 442–453. Springer, Heidelberg (2014)

3. Duval, J.-P.: Relationship between the period of a finite word and the length of its unbordered segments. Discrete Math. **40**(1), 31–44 (1982)

4. Duval, J.-P., Lecroq, T., Lefebvre, A.: Border array on bounded alphabet. J. Autom. Lang. Comb. **10**(1), 51–60 (2005)

5. Duval, J.-P., Lecroq, T., Lefebvre, A.: Efficient validation and construction of Knuth-Morris-Pratt arrays. In: Proceedings of Conference in Honor of Donald E. Knuth (2007)

6. Duval, J.-P., Lecroq, T., Lefebvre, A.: Linear computation of unbordered conjugate on unordered alphabet. Theor. Comp. Sci. **522**, 77–84 (2014)

7. Ehrenfeucht, A., Silberger, D.M.: Periodicity and unbordered segments of words. Discrete Math. **26**(2), 101–109 (1979)

8. Franěk, F., Gao, S., Lu, W., Ryan, P.J., Smyth, W.F., Sun, Y., Yang, L.: Verifying a border array in linear time. J. Comb. Math. Comb. Comput. **42**, 223–236 (2000)

9. Gawrychowski, P., Jeż, A., Jeż, Ł.: Validating the Knuth-Morris-Pratt failure function, fast and online. Theor. Comp. Sys. **54**(2), 337–372 (2014)

10. Holub, S., Nowotka, D.: The Ehrenfeucht-Silberger problem. J. Comb. Theor., Ser. A **119**(3), 668–682 (2012)

11. Ilie, L., Navarro, G., Tinta, L.: The longest common extension problem revisited and applications to approximate string searching. J. Discrete Algorithms **8**(4), 418–428 (2010)

12. Kociumaka, T., Radoszewski, J., Rytter, W., Waleń, T.: Efficient data structures for the factor periodicity problem. In: Calderón-Benavides, L., González-Caro, C., Chávez, E., Ziviani, N. (eds.) SPIRE 2012. LNCS, vol. 7608, pp. 284–294. Springer, Heidelberg (2012)

13. Kociumaka, T., Radoszewski, J., Rytter, W., Waleń, T.: Internal pattern matching queries in a text and applications. In: Proceedings of the Twenty-Sixth Annual ACM-SIAM Symposium on Discrete Algorithms, pp. 532–551 (2014)

14. Moore, D., Smyth, W.F., Miller, D.: Counting distinct strings. Algorithmica **23**(1), 1–13 (1999)

15. Morris Jr., J.H., Pratt, V.R.: A linear pattern-matching algorithm, report 40. Technical report, University of California, Berkeley (1970)

16. Nielsen, P.: A note on bifix-free sequences. IEEE Trans. Inf. Theor. **19**(5), 704–706 (1973)

Semi-dynamic Compact Index for Short Patterns and Succinct van Emde Boas Tree

Yoshiaki Matsuoka[1]([✉]), Tomohiro I[2], Shunsuke Inenaga[1],
Hideo Bannai[1], and Masayuki Takeda[1]

[1] Department of Informatics, Kyushu University, Fukuoka, Japan
{yoshiaki.matsuoka,inenaga,bannai,takeda}@inf.kyushu-u.ac.jp
[2] Department of Computer Science, TU Dortmund, Dortmund, Germany
tomohiro.i@cs.tu-dortmund.de

Abstract. We present a compact semi-dynamic text index which allows us to find short patterns efficiently. For parameters $k \leq q \leq \log_\sigma n - \log_\sigma \log_\sigma n$ and alphabet size $\sigma = O(\text{polylog}(n))$, all *occ* occurrences of a pattern of length at most $q - k + 1$ can be obtained in $O(k \times occ + \log_\sigma n)$ time, where n is the length of the text. Adding characters to the end of the text is supported in amortized constant time. Our index requires $(n/k) \log(n/k) + n \log \sigma + o(n)$ bits of space, which is compact (i.e., $O(n \log \sigma)$) when $k = \Theta(\log_\sigma n)$. As a byproduct, we present a *succinct* van Emde Boas tree which supports insertion, deletion, predecessor, and successor on a dynamic set of integers over the universe $[0, m - 1]$ in $O(\log \log m)$ time and requires only $m + o(m)$ bits of space.

1 Introduction

A full-text index of a string T is a data structure which allows us to efficiently find all occurrences of a given pattern in T. A large number of indices have been proposed, e.g. suffix trees [15], suffix arrays [9] and FM-index [4]. An index is said to be *compact* if it occupies $O(n \log \sigma)$ bits of space, where n is the length of T and σ is the alphabet size. Suffix trees and suffix arrays are not compact since they require $O(n \log n)$ bits, while FM-index is (actually, compressed).

On the other hand, a suffix tree has an advantage that it can be constructed semi-dynamically (online), that is, a suffix tree can be modified efficiently when a character is appended to the end of the text [14]. Salson et al. [13] addressed how to modify an FM-index when the text is edited[1], rather than rebuilding it from scratch. Although their approach works well in practice, its worst-case time complexity for modification is $O(n \log n)$. There are some other related results [3,6], however, as far as we know there are no practical implementations.

In this paper, we propose a new compact full-text index for short patterns which can be constructed in linear time in an online manner. We restrict the length of patterns that can be searched for to at most around $\log_\sigma n$, and develop

---

[1] Note that they addressed more general edit operations such as insertion of a string to an arbitrary position and deletion/substitution of a substring of the text.

© Springer International Publishing Switzerland 2015
F. Cicalese et al. (Eds.): CPM 2015, LNCS 9133, pp. 355–366, 2015.
DOI: 10.1007/978-3-319-19929-0_30

novel techniques that exploit this assumption in order to achieve fast online construction using small space. More precisely, for parameters $k \leq q \leq \log_\sigma n - \log_\sigma \log_\sigma n$ and alphabet size $\sigma = O(\text{polylog}(n))$, all occ occurrences of a pattern of length at most $q - k + 1$ can be obtained in $O(k \times occ + \log_\sigma n)$ time, where n is the length of the text. Our index requires $(n/k)\log(n/k) + n\log\sigma + o(n)$ bits of space, which is compact when $k = \Theta(\log_\sigma n)$. It is also semi-dynamic, in that adding characters to the end of the text is supported in constant amortized time. Finding short patterns is useful for filtering algorithms to find exact and approximate occurrences of longer patterns [1,12]. Since our index does not need the original text T to search for pattern occurrences, we can discard it once the index is constructed. If needed, we can rebuild T from our index in $O(n)$ time.

Computational experiments on DNA sequences show that our index is practical and can be constructed much faster than suffix arrays or FM-indices. Although the query time is slower than suffix arrays, it is much faster than FM-indices. While the space requirement is larger than the static FM-index, it can be smaller than the Dynamic FM-index [13].

As a byproduct of our compact semi-dynamic index, we propose a *succinct* variant of the van Emde Boas tree [2], which we believe is of independent interest and useful for many other applications. We show that for a fully-dynamic set of integers over the universe $[0, m-1]$, we can support look-up, insertion, deletion, predecessor, and successor operations in $O(\log \log m)$ time using only $m + o(m)$ bits of space (Corollary 1). This result is valid for the Transdichotomous word RAM model with word size $\Theta(\log m)$. To our knowledge, this is the fastest succinct dynamic predecessor/successor data structure to date.

Observe that a predecessor/successor query in a dynamic set of integers over the universe $[0, m-1]$ can be replaced by a constant number of rank/select queries on a dynamic bit array B of length m. Gupta et al. [5] proposed a data structure for B which supports rank/select in $O((1/\epsilon) \log \log m)$ time for any $0 < \epsilon < 1$ and requires $m + o(m)$ bits of space, but it takes $O((1/\epsilon)m^\epsilon)$ amortized time for insertion/deletion. Navarro and Nekrich [11] proposed a data structure for B which supports rank/select and insertion/deletion in $O(\log m/\log\log m)$ time, with $m + o(m)$ bits of space. Although rank/select are more powerful than predecessor/successor, predecessor/successor are sufficient for our needs. Hence our data structure achieves $O(k \times occ + \log_\sigma n)$ pattern matching time and $O(n)$ online construction time, using our succinct van Emde boas tree.

2   Preliminaries

Let $\Sigma = \{0, \ldots, \sigma - 1\}$ be an integer alphabet of size σ. An element of Σ^* is called a *string*. The length of a string w is denoted by $|w|$. The empty string ε is a string of length 0. Let Σ^+ be the set of non-empty strings, i.e., $\Sigma^+ = \Sigma^* - \{\varepsilon\}$. For a string $w = x \cdot y \cdot z$, x, y and z are called a *prefix*, *substring*, and *suffix* of w, respectively. The position of a string starts from 0. The i-th character of a string w is denoted by $w[i]$, where $0 \leq i < |w|$. For a string w and two integers $0 \leq i \leq j < |w|$, let $w[i..j]$ denote the substring of w that begins at position i and ends at position j. For convenience, let $w[i..j] = \varepsilon$ when $i > j$.

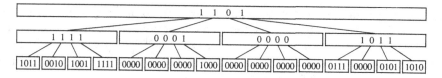

Fig. 1. Example of a complete b-ary tree where $m = 64$ and $b = 4$. The concatenation of the sequences in the leaves is equal to the bit array B maintained by the complete b-ary tree.

Throughout this paper, $T \in \Sigma^*$ denotes a string to be indexed. For convenience, we assume that we know $|T| = n$. For online processing, this restriction can be removed by adapting the standard doubling technique. The default base of logarithm is 2. We assume the unit cost Transdichotomous word RAM with the machine word size $\Theta(\log n)$ bits.

3   Semi-dynamic Bit Arrays Supporting Successor Queries

Here we present a data structure to maintain a semi-dynamic set S of integers over the universe $[0, m-1]$ on the Transdichotomous RAM model with machine word size $\Theta(\log n)$, where $m \leq n$. By semi-dynamic, we mean that a new element can be inserted to the set, but no existing elements are deleted from it. Let B be an array of length m such that $B[i] = 1$ iff $i \in S$. We propose a data structure for B which efficiently supports: $lookup(x)$ which returns $B[x]$, $successor(x)$ which returns $\min\{y \mid B[y] = 1, y > x\}$ if it exists or nil otherwise, and $insert(x)$ which sets $B[x]$ to 1. This data structure will be used in Subsect. 4.1 as a component of our compact full-text index for short patterns.

3.1   Simple $O(\log_b m)$-Time Data Structure

We begin with a simple data structure using a complete b-ary tree which supports each operation in $O(\log_b m)$ time using $m + O(\frac{m}{b} + b \log b)$ bits of space, where b is a positive integer which is not greater than the machine word size. For convenience, assume $B[i] = 0$ for any $i \geq m$. We consider a complete b-ary tree with height $h = \lceil \log_b m \rceil$. Each node v maintains a bit array $B_v[0..b-1]$ that satisfies the following conditions: (1) If v is a leaf, $B_v[0..b-1]$ is a subarray of B. More precisely, $B_v[0..b-1] = B[jb..(j+1)b-1]$ if v is the j-th leaf, and (2) If v is an internal node, for any $0 \leq j < b$, $B_v[j] = 1$ iff B_{v_j} contains 1, where v_j is the j-th child of v. Or equivalently, $B_v[j] = 1$ iff there exists a leaf v' in the complete subtree rooted at v_j s.t. $B_{v'}$ contains 1. Figure 1 shows an example of a complete b-ary tree where $m = 64$ and $b = 4$.

Since the tree is complete, we can arrange all the nodes of the tree in a single array so that given an index of any node, the index of its parent or arbitrary child can be obtained in constant time without the use of pointers. In addition, since B_v never contains 1 if it does not correspond to any region in $[0, m-1]$, we do not have to allocate memory for such out-of-bounds bit arrays. Since our tree is a complete b-ary tree, the total space is $m + \sum_{i=1}^{\log_b m} \frac{m}{b^i} = m + O(\frac{m}{b})$ bits.

For any $0 \le x < m$, $lookup(x)$ can be computed in constant time, by looking up the $(x \bmod b)$-th bit of the $\lfloor x/b \rfloor$-th leaf.

For any node v and given integer j with $0 \le j < b$, let $successor_v(j) = \min\{j' \mid B_v[j'] = 1, j' > j\}$, if it exists. This value can be calculated in constant time with the method in [7] by using a total of $O(b \log b)$ bits of space. We compute $successor(x)$ as follows. Starting from the leaf node containing x, we climb up the tree until $successor_v(x_v)$ exists at node v, where x_v is the index of the child at node v that contains x in its subtree. Then we climb down the $(successor_v(x_v))$-th child at node v. Finally, we climb down the $(successor_v(-1))$-th child at each node v, until we reach the leaf node which gives us the answer. Therefore, $successor(x)$ can be computed in a total of $O(h)$ time.

For $insert(x)$, let $(v_0, \ldots, v_h)$ be the sequence of nodes in the path from the root to the leaf representing $B[x]$, namely, v_0 is the root, v_{l-1} is the parent of v_l for each $1 \le l \le h$, and v_h is the leaf representing $B[x]$. Also, let $i_l = \lfloor x/b^{h-l} \rfloor \bmod b$ for each $0 \le l \le h$. To insert x, we climb up the path from the leaf setting $B_{v_l}[i_l] = 1$ from $l = h$ down to 0. Clearly, it takes $O(h)$ time.

Since $h = O(\log_b m)$, we get the following result.

Lemma 1. *There exists a data structure which requires $m + O(\frac{m}{b} + b \log b)$ bits of space and supports $lookup(x)$ in $O(1)$ worst-case time, and $successor(x)$, $insert(x)$ in $O(\log_b m)$ worst-case time.*

3.2 $O(\log \log m)$-Time Data Structure

In this subsection, we show the $O(\log_b m)$ time complexity for *successor* and *insert* operations can be improved to $O(\log \log m)$ by combining the b-ary trees of Lemma 1 with the van Emde Boas tree [2]. The van Emde Boas tree supports *successor* and *insert* in $O(\log \log m)$ worst-case time but requires $\Theta(m \log n)$ bits of space (i.e., $\Theta(m)$ words of space). Our underlying idea is similar to that used in the y-fast trie [16], namely, we maintain B by a top tree (a van Emde Boas tree) and bottom trees (complete b-ary trees).

Let h' be any positive constant integer and $m' = \lceil m/b^{h'+1} \rceil$. The bottom trees consist of m' complete b-ary trees of height h' such that for any $0 \le i < m' - 1$, the i-th complete b-ary tree maintains the subarray $B[ib^{h'+1}..(i+1)b^{h'+1} - 1]$, and the $(m'-1)$-th one maintains the subarray $B[(m'-1)b^{h'+1}..m-1]$. The top tree will be a van Emde Boas tree that maintains the set $R = \{\lfloor i/b^{h'+1} \rfloor \mid 0 \le i < m, B[i] = 1\}$, i.e., $j \in R$ iff there exists a set bit in the subarray of B which is maintained by the j-th bottom tree. We call this data structure the *vEBb tree*. Figure 2 shows an illustration for a vEBb tree. In this example, the top van Emde Boas tree maintains $R = \{0, 2, 3\}$ since each of the subarrays $B[0..b^{h'+1} - 1]$, $[2b^{h'+1}..3b^{h'+1} - 1]$, and $[3b^{h'+1}..4b^{h'+1} - 1]$ contains at least a 1, while the subarray $[b^{h'+1}..2b^{h'+1} - 1]$ contains no 1's.

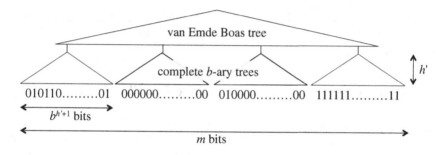

Fig. 2. Illustration for a vEBb tree.

For any $0 \le x < m$, *lookup*(x) can be computed in constant time, by looking up the $(x \bmod b)$-th bit of the $\lfloor (x \bmod b^{h'+1})/b \rfloor$-th leaf of the $\lfloor x/b^{h'+1} \rfloor$-th bottom tree.

For *successor*(x), let $i = \lfloor x/b^{h'+1} \rfloor$. If *successor*$(x)$ is in the i-th bottom tree, we can compute *successor*(x) as described previously. Otherwise, let i' be the successor of i in R. We can compute it by the top tree in $O(\log \log m')$ time. If i' exists, *successor*(x) is the minimum position that is maintained in the i'-th bottom tree.

For *insert*(x), let $i = \lfloor x/b^{h'+1} \rfloor$. We update the i-th bottom tree as described previously. Moreover, if $i \notin R$ then we add i to R, i.e., the van Emde Boas tree.

Lemma 2. *When $b = \Theta(\log n)$, the vEBb tree takes $m + O(\frac{m}{\log n} + \log n \log \log n)$ bits of space and supports lookup(x) in $O(1)$ worst-case time, and successor(x), insert(x) in $O(\log \log m)$ worst-case time.*

Proof. Since the total space for complete b-ary trees does not exceed that of Lemma 1, it is bounded by $m + O(\frac{m}{\log n} + \log n \log \log n)$ bits. The van Emde Boas tree for R needs $O(m')$ words because R is a set of integers over the universe $[0, m' - 1]$. Since $b = \Theta(\log n)$, it occupies $O(m' \log n) = O(\lceil \frac{m}{b^{h'+1}} \rceil \log n) = O(\frac{m}{\log n})$ bits of space. Hence, the total space is $m + O(\frac{m}{\log n} + \log n \log \log n)$ bits.

Since insert and successor operations on the top tree take $O(\log \log m')$ time and those operations on the bottom trees take $O(h')$ time, it takes $O(\log \log m' + h') = O(\log \log m)$ time for each operation. □

The following lemma holds for the semi-dynamic setting.

Lemma 3. *Let $b = \Theta(\log n)$. If we conduct insert operations f times on the vEBb tree of Lemma 2 in the semi-dynamic setting, then it takes $O(m + f)$ time in total.*

It is not difficult to extend the bottom b-ary trees so that deletion and predecessor operations are supported in the same time complexity using the same space. Hence, the next corollary follows from Lemma 2.

Corollary 1. *Let $m = n$ and $b = \Theta(\log m)$. Then, the vEBb tree requires $m + o(m)$ bits of space and supports lookup in $O(1)$ worst-case time, and insert, delete, predecessor, and successor in $O(\log \log m)$ worst-case time.*

Additionaly, we can hold a doubly linked list which represents R in ascending order. Then, for any i, we can compute the predecessor or successor of i in R in constant time if $i \in R$.

4   A Semi-dynamic Compact Index for Short Patterns

Let $q \leq \log_\sigma n - \log_\sigma \log_\sigma n$ and $u = \sigma^q$. Note that $u \log n \leq n \log \sigma$. We assume that $\sigma = O(\text{polylog}(n))$, thus $u \leq n/\log_\sigma n = o(n)$. We take an integer k with $k \leq q$ as a parameter, and sample the positions of q-grams of T beginning at position jk for every $0 \leq j \leq \lfloor (n-q)/k \rfloor$. These q-grams are called *sampled q-grams*. Let p_{last} denote the beginning position of the last sampled q-gram, i.e., $p_{\text{last}} = k \lfloor (n-q)/k \rfloor$. We design our data structure for patterns of length at most $q-k+1$. Given a pattern P of length at most $q-k+1$, an occurrence of P in T is completely contained in some sampled q-gram (except for the ones beginning at one of the last $O(q)$ positions of T). We can employ any linear-time string search algorithm, to compute in $O(q)$ time the occurrences of P beginning in range $[p_{\text{last}} + k, n-1]$. Hence, in what follows we only consider range $[0, p_{\text{last}} + k - 1]$. For any position $i < p_{\text{last}} + k$, i is an occurrence of P iff $i = p + d$ where p is a starting position of a sampled q-gram w and $w[d..d + |P| - 1] = P$ with $d < k$. Notice such sampled position p is unique for each occurrence i of P. We find the occurrences of P as follows: First, enumerate sampled q-grams w and offsets d such that $w[d..d + |P| - 1] = P$ (Task 1). Then, for each pair of w and d, output $p + d$ for every sampled position p of w (Task2).

Our data structure consists of the following two components (**A**) and (**B**) which respectively realize the above tasks: (**A**) to enumerate all occurrences of P in sampled q-grams; (**B**) to enumerate all sampled positions of a given q-gram.

We define $\langle w \rangle = \sum_{i=0}^{|w|-1} w[i]\sigma^{|w|-1-i}$ for any string w such that $|w| \leq q$. We call $\langle w \rangle$ as the *integer encoding* of w. For any string w and w' such that $w' = w[i..j]$ for some $0 \leq i \leq j < |w|$, $\langle w' \rangle = \lfloor \langle w \rangle /\sigma^{|w|-1-j} \rfloor \mod \sigma^{j-i+1}$. Also, for a string $w = x \cdot y$, $\langle w \rangle = \langle x \rangle \sigma^{|y|} + \langle y \rangle$. Since the number of distinct values of $j - i$ for all $0 \leq i \leq j < |w|$ is $|w| - 1$, and since we only consider concatenating substrings of w of resulting length at most q, we can precompute the powers of σ to be used in the above formulae in a total of $O(q)$ time. After this preprocessing, we can compute the integer encodings of any substring and a concatenation of two substrings in $O(1)$ time.

4.1   Data structure (A) for Task 1

Let Q be the set of q-grams occurring in T, which contains not only the sampled q-grams but *all* q-grams in T. For any integer d, let $Q(P, d) = \{w \in Q \mid w[d..d + |P| - 1] = P\}$. Our data structure (A) consists of the following two parts: (**A1**) a vEBb tree to compute $Q(P, 0)$, i.e., enumerate $w \in Q$ s.t. P is a prefix of w; (**A2**) a directed graph to update $Q(P, d)$ to $Q(P, d+1)$ for each $0 \leq d \leq k - 2$.

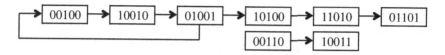

Fig. 3. Example of 5-gram transition graph where $\Sigma = \{0, 1\}$, $T = 0110100100110$.

Data Structure (A1). Let $\langle Q \rangle = \{\langle w \rangle \mid w \in Q\} \subseteq [0, u - 1]$. For a given pattern P, let $sp(P) = \langle P \rangle \sigma^{q-|P|}$ and $ep(P) = sp(P) + \sigma^{q-|P|} - 1$. We can compute $sp(P)$ and $ep(P)$ from P in $O(|P|)$ time. Here, $w \in Q(P, 0)$ iff $sp(P) \leq \langle w \rangle \leq ep(P)$ and $\langle w \rangle \in \langle Q \rangle$. That is, computing $Q(P, 0)$ is equivalent to computing $\langle Q \rangle \cap [sp(P), ep(P)]$. Using a vEBb tree that represents $\langle Q \rangle$, we can enumerate all elements of $\langle Q \rangle \cap [sp(P), ep(P)]$ in ascending order by repeating the *successor* operation starting from $sp(P)$ until we surpass $ep(P)$ or get *nil*. Note that finding the first element takes $O(\log \log n)$ time, but subsequent elements can be obtained in $O(1)$ time by using the doubly linked list. This algorithm runs in $O(\log \log n + occ + |P|) = O(occ + \log_\sigma n)$ time, where occ is the total number of occurrences of P in T. By Lemma 2 we get the following lemma.

Lemma 4. *Data structure (A1) requires* $u + O(\frac{u}{\log n} + \log n \log \log n) = o(n)$ *bits of space.*

Data Structure (A2): q-gram transition graph. We introduce a directed graph $G = (V, E)$ with $V = \Sigma^q$ and $E \subseteq Q \times Q$, called a *q-gram transition graph* of T, which satisfies the following two conditions: (1) For any $v \in V$, the indegree of v is exactly 1 iff $v \in Q$ and there exists $v' \in Q$ s.t. $v[1..q - 1] = v'[0..q - 2]$, and otherwise 0, and (2) For any $(v', v) \in E$, $v[1..q - 1] = v'[0..q - 2]$.

If $T[n - q..n - 1] \in Q$ does not occur in $T[0..n - 2]$, then the indegree of $T[n - q..n - 1]$ is 0, and this is the only case where the indegree of a node in Q is 0. Figure 3 shows an example of a q-gram transition graph where $\Sigma = \{0, 1\}$, $T = 0110100100110$ and $q = 5$. All q-grams which do not appear in T (namely, those not in Q) are omitted for simplicity. Note that $(10011, 01001) \notin E$ because the indegree of 01001 must not exceed 1.

We remark that a q-gram transition graph of T has a similar structure to a de Bruijn graph for T used in de novo assemblies. Indeed, both of them are subgraphs of the complete de Bruijn graph. An important feature of q-gram transition graphs is that the indegree of every node is at most 1. Also, for any $(v', v) \in E$, v' and v are not necessarily neighbors in T. Note that the q-gram transition graph is not necessarily unique for a given string T. Our search algorithm works fine with any graph satisfying the above conditions (see Subsect. 4.3 for the details on how to use it).

Now, we describe a space-efficient representation of the q-gram transition graph. For each $w \in V$, let $(v_0, \ldots, v_m)$ be the list of q-grams in an arbitrary order, such that $(w, v_i) \in E$ for any $0 \leq i \leq m$. We use two arrays on $\Sigma \cup \{-1\}$: $e_{\text{first}}[0..u - 1]$ and $e_{\text{next}}[0..u - 1]$ such that $e_{\text{first}}[\langle w \rangle]$ represents $(w, v_0) \in E$ and

$e_{\text{next}}[\langle v_i \rangle]$ represents $(w, v_{i+1}) \in E$ for any $0 \leq i < m$. By the definition of the q-gram transition graph, $v_i = v_i[0] \cdot w[0..q-2]$ for any $0 \leq i \leq m$. Thus, it suffices to store $v_i[0]$ to represent the edge. Precisely, if the list is empty, let $e_{\text{first}}[\langle w \rangle] = -1$. Otherwise, let $e_{\text{first}}[\langle w \rangle] = v_0[0], e_{\text{next}}[\langle v_i \rangle] = v_{i+1}[0]$ for any $0 \leq i < m$, and $e_{\text{next}}[\langle v_m \rangle] = -1$. We note that in our application sequential access to the children of v in an arbitrary order is enough as will be described in Subsect. 4.3. Hence, when a new edge (w, v) is added to E, updating the graph representation takes $O(1)$ time because at most two elements of these arrays are changed by adding the new edge at the beginning of the children's list.

Lemma 5. *The q-gram transition graph requires $o(n)$ bits of space.*

Proof. Since each element of e_{first} and e_{next} needs $\log(\sigma + 1)$ bits, these arrays use $2u \log(\sigma + 1) = O(n \log^2 \sigma / \log n) = O(n(\log \log n)^2 / \log n) = o(n)$ bits. $\square$

4.2   Data structure (B) for Task 2

We use two arrays on $\{-1, 0, \ldots, p_{\text{last}}/k\}$: $l_{\text{first}}[0..u-1]$ and $l_{\text{next}}[0..p_{\text{last}}/k]$. For each q-gram w, let $(p_0, p_1, \ldots, p_m)$ be the list of all sampled positions of sampled q-gram w in an arbitrary order. Note that each p_i $(0 \leq i \leq m)$ is divisible by k. If the list is empty, let $l_{\text{first}}[\langle w \rangle] = -1$. Otherwise, let $l_{\text{first}}[\langle w \rangle] = p_0/k$, $l_{\text{next}}[p_i/k] = p_{i+1}/k$ for any $0 \leq i < m$, and $l_{\text{next}}[p_m/k] = -1$. For example, if $\Sigma = \{0, 1\}$, $T = 0110100100110$, $q = 5$ and $k = 3$, the sampled q-grams are 01101 at position $\{0\}$ and 01001 at positions $\{3, 6\}$. Hence, we can set $l_{\text{first}}[\langle 01101 \rangle], l_{\text{first}}[\langle 01001 \rangle]$ and $l_{\text{next}}[2]$ to $0, 2$ and 1, respectively. All other elements of l_{first} and l_{next} are -1. We can regard these arrays as singly linked lists. A new starting position of a sampled q-gram w can be inserted to this list in $O(1)$ time, since we insert this edge at the beginning of this list and at most two elements of these arrays are changed.

Lemma 6. l_{first} *and* l_{next} *require at most* $(n/k) \log(n/k) + n \log \sigma$ *bits of space.*

Proof. Since each element of these arrays needs $\log(n/k)$ bits, these arrays need $(u + n/k) \log(n/k) \leq (n/k) \log(n/k) + n \log \sigma$ bits in total. $\square$

4.3   Searching Text T for Occurrences of Pattern P

Let $G = (V, E)$ be a q-gram transition graph of T. For any $v \in Q$ and integer $d \geq 0$ we define a labeled rooted tree $T_G(v, d)$ as follows: $T_G(v, 0)$ consists only of the root labeled with v. The root of $T_G(v, d)$ is labeled with v and the root connects to the root of $T_G(v', d-1)$ for every $(v, v') \in E$. The height of $T_G(v, d)$ is at most d. Given G, v, and d, we can simulate a depth-first traversal on $T_G(v, d)$ in time linear in its size, without building the tree.

Lemma 7. *Algorithm 1 enumerates all pairs of sampled q-gram w and offset d such that $w[d..d + |P| - 1] = P$ in $O(k \times occ + \log_\sigma n)$ time and $O(k)$ words of working space, where occ is the number of occurrences of P in T.*

Algorithm 1. Algorithm for Task 1.

Input: Pattern P of length at most $q - k + 1$.
Output: All pairs of sampled q-gram w and offset d s.t. $w[d..d + |P| - 1] = P$.

1   **foreach** $v \in Q(P, 0)$ **do**
2   traverse $T_G(v, k - 1)$, and when we find a sampled q-gram w at depth d
        output the pair w and d;

3   **if** *indegree of* $v' = T[n - q..n - 1]$ *is* 0 **then**
4   **foreach** *occurrence* d' *of* P *in* v' *with* $0 < d' < k$ **do**
5   traverse $T_G(v, k - 1 - d')$, and when we find a sampled q-gram w at
            depth d'' output the pair w and $d' + d''$;

Proof. Firstly we show the correctness. Take any pair of sampled q-gram w and offset d s.t. $w[d..d + |P| - 1] = P$. It follows from the definition of $G = (V, E)$ that for any $(w', w) \in E$ if $p > 0$ is an occurrence of P in w, $p - 1$ is an occurrence of P in w'. Then, when the indegree of any $w \in Q$ is 1, by traversing reversely the edges of G from w with d steps we must find $v \in Q$ s.t. $T_G(v, k - 1)$ contains a node labeled with w at depth d, and hence, it must be output at Line 2. When the indegree of $v' = T[n - q..n - 1]$ is 0, the reverse traversal may stop at v' with $0 \leq d'' < d$ steps. In this case, since $d' = d - d''$ is an occurrence of P in v', it must be output at Line 5. Therefore, all pairs of w and offset d are enumerated. Also, it is easy to see that there are no duplicates.

Let us analyze the time complexity. As in Subsect. 4.1 Line 1 takes a total of $O(occ + \log_\sigma n)$ time. Also Line 4 takes $O(q)$ time. What remains is the cost for Lines 2 and 5. Observe that, during the whole traversal in Lines 2 and 5, we visit only the nodes that contains P with offset $0 \leq d < k$, namely, the set of nodes we visit is $\bigcup_{d=0}^{k-1} Q(P, d)$. Also, we never visit a node several times with the same offset due to the definition of a q-gram transition graph. Hence the total cost for the traversal is bounded by $O(\sum_{d=0}^{k-1} |Q(P, d)|)$. Since $|Q(P, d)| \leq occ$ for any integer d, $O(\sum_{d=0}^{k-1} |Q(P, d)|) = O(k \times occ)$. Also, we use $O(k)$ words of working space for traversing $T_G(v, k - 1)$ because its depth is at most $k - 1$. □

Theorem 1. *Using data structures (A) and (B), for a given pattern P of length at most $q - k + 1$, all occurrences of P in T can be found in $O(k \times occ + \log_\sigma n)$ time.*

Proof. By Lemma 7, $Q(P, d)$ for all $0 \leq d < k$ can be computed in a total of $O(k \times occ + \log_\sigma n)$ time. Moreover, all sampled positions of a sampled q-gram including P can be computed in $O(occ)$ time, and all occurrences of P beginning in range $[p_{last} + k, n - 1]$ in $O(q) = O(\log_\sigma n)$ time. Hence we can find all occurrences of P in T in $O(k \times occ + \log_\sigma n)$ time. □

Unlike suffix arrays, in this algorithm, we can find all occurrences of P in T without T except for $T[n - q..n - 1]$, and hence, we do not have to hold $T[0..n - q - 1]$. If needed, we can retrieve whole T by data structure (B) and $T[n - q..n - 1]$ in $O(n)$ time since $T[0..n - q - 1]$ is covered by sampled q-grams.

4.4   Building Data Structures Online

We show that data structures (A) and (B) can be built in an online manner. Let $w_i = T[i..i+q-1]$ $(0 \le i \le n-q)$. First, we initialize $Q = \emptyset, E = \emptyset$ and all elements of $e_{\text{first}}, e_{\text{next}}, l_{\text{first}}, l_{\text{next}}$ to -1. For each $0 \le i \le n-q$, we update data structures (A1), (A2) and (B) as follows: (A1) Add w_i to Q by calling $insert(\langle w_i \rangle)$. (A2) If $i > 0$ and indegree of w_{i-1} is 0, then add edge (w_i, w_{i-1}) to E, i.e., let $e_{\text{next}}[\langle w_{i-1} \rangle] = e_{\text{first}}[\langle w_i \rangle]$ and $e_{\text{first}}[\langle w_i \rangle] = T[i-1]$. Note that indegree of w_{i-1} is 0 iff w_{i-1} first appears at $i-1$, and we can know it when adding w_{i-1} to Q at position $i-1$. (B) If i is divisible by k, then add a new sampled position of sampled q-gram w_i, i.e., let $l_{\text{next}}[i/k] = l_{\text{first}}[\langle w_i \rangle]$ and $l_{\text{first}}[\langle w_i \rangle] = i/k$.

Each q-gram is represented by its integer encoding. Since $\langle w_i \rangle = (\langle w_{i-1} \rangle \bmod \sigma^{q-1})\sigma + T[i+q-1]$, we can calculate $\langle w_i \rangle$ from $\langle w_{i-1} \rangle$ and $T[i+q-1]$ in $O(1)$ time if $i > 0$.

Theorem 2. *Data structures (A) and (B) for a string T of length n can be built in $O(n)$ time in an online manner.*

Proof. It is easy to see that for all $0 \le i \le n-q$, $\langle w_i \rangle$ can be computed in $O(n)$ time in total. Updating data structures (A2) and (B) takes $O(n)$ time. It follows from Lemma 3 that data structure (A1) can be updated in a total of $O(n)$ time. Hence data structures (A) and (B) can built in $O(n)$ time.   □

Theorem 3. *Data structures (A) and (B) for a string T of length n require a total of $(n/k)\log(n/k) + n\log\sigma + o(n)$ bits of space.*

Theorem 3 immediately follows from Lemmas 4, 5 and 6. When $k = \Theta(\log_\sigma n)$, our index occupies $O(n\log\sigma)$ bits, and is thus compact.

5   Computational Experiments

We implemented our algorithm in the C++ language (available at https:// github.com/ymatsuoka663/semidynamic-compact-index). For simplicity, we used a complete b-ary tree (Lemma 1) with $b = 64$ as data structure (A1), rather than the vEBb tree. We compared our algorithm with Suffix Array [9] implemented by Yuta Mori (http://code.google.com/p/libdivsufsort/), implementations of Succinct Suffix Array [8], FM-index [4] version 2, and LZ-index [10] available at the Pizza & Chili corpus web site (http://pizzachili.dcc.uchile. cl/), Dynamic FM-index [13] (http://dfmi.sourceforge.net/), Compressed q-gram index 1 (Rice, Re-Pair using blocks of 8KB in [1]) and Compressed q-gram index 2 (Rice, Plain using blocks of 8KB in [1]).

All computational experiments were conducted on a MacPro (Early 2008). For the text, we used the first n characters of DNA data taken from the Pizza & Chili corpus but removing all characters other than A,C,G,T so $\sigma = 4$. For patterns, we randomly chose 100 substrings of length 6 from the text.

Table 1 shows the construction times, average time for searching 100 patterns, and the memory usage. Our indices can be built at least 7 times faster than other

Table 1. Result of computational experiments.

n	12.5×2^{20}	25×2^{20}	50×2^{20}	100×2^{20}	200×2^{20}
Time for construction (in seconds)					
Our method ($q = 9, k = 4$)	**0.25**	**0.50**	**1.02**	**2.04**	**4.08**
Our method ($q = 11, k = 6$)	0.56	0.98	1.75	3.16	6.00
Suffix Array	1.90	4.41	9.89	21.73	46.88
Succinct Suffix Array	3.05	7.50	16.53	35.24	82.02
FM-index	2.77	6.90	15.45	33.06	78.14
LZ-index	1.95	4.53	10.17	22.59	50.05
Dynamic FM-index	47.59	123.54	313.19	763.12	1840.56
Compressed q-gram index 1	22.47	46.54	95.13	194.89	409.32
Compressed q-gram index 2	8.40	16.61	33.05	66.07	132.06
Average time for searching, using 100 patterns of length 6 (in milliseconds).					
Our method ($q = 9, k = 4$)	0.223	0.559	1.262	2.719	5.637
Our method ($q = 11, k = 6$)	0.493	0.857	1.621	3.133	6.039
Suffix Array	**0.010**	**0.015**	**0.026**	**0.046**	**0.100**
Succinct Suffix Array	26.78	66.87	149.8	317.2	668.3
FM-index	108.1	220.2	443.9	885.3	1758
LZ-index	1.747	4.546	10.37	19.67	39.75
Dynamic FM-index	281.2	741.2	1842	4434	10602
Compressed q-gram index 1	522.4	1083	2114	4309	9086
Compressed q-gram index 2	18.59	37.1	74.13	148.2	295.7
Memory usage (in megabytes).					
Our method ($q = 9, k = 4$)	9.5	18.9	38.5	79.2	163.6
Our method ($q = 11, k = 6$)	20.7	27.5	41.0	68.6	125.3
Suffix Array	62.5	125.0	250.0	500.0	1000.0
Succinct Suffix Array	9.1	18.1	36.3	72.4	144.8
FM-index	6.6	13.5	27.0	54.2	108.3
LZ-index	17.8	35.1	69.2	137.6	262.4
Dynamic FM-index	30.6	60.8	121.3	242.2	484.0
Compressed q-gram index 1	**5.3**	**10.5**	**21.0**	**41.7**	**82.6**
Compressed q-gram index 2	13.3	26.6	53.1	106.3	212.5

indices when $n = 200 \times 2^{20}$. Our indices can find pattern occurrences 3.5–7 times faster than LZ-index and at least 30 times faster than Compressed q-gram index 2. Our indices use less memory than Succinct Suffix Array when $q = 11, k = 6$ and $n = 200 \times 2^{20}$. On the other hand, when $q = 11, k = 6$ and $n = 12.5 \times 2^{20}$, our indices use more memory than other indices. This is partly because we have chosen parameter q suitable for $n = 200 \times 2^{20}$ in this experiment, which is not suitable for a smaller n. As n grows, we can choose larger q in which our indices stay compact and search longer patterns. In other words, if we fix the maximum length of patterns to be searched for, as n grows our indices could be more and more space efficient by choosing a larger q and larger k properly since the $(n/k)\log(n/k)$ bits of space for data structure (B) is a major factor dominating the memory usage. Thus our algorithm is effective when the text is long.

References

1. Claude, F., Farina, A., Martínez-Prieto, M.A., Navarro, G.: Compressed q-gram indexing for highly repetitive biological sequences. In: Proceedings of the BIBE 2010, pp. 86–91 (2010)
2. van Emde Boas, P.: Preserving order in a forest in less than logarithmic time. In: FOCS, pp. 75–84. IEEE Computer Society (1975)
3. Ferragina, P., Grossi, R.: Optimal on-line search and sublinear time update in string matching. SIAM J. Comput. **27**(3), 713–736 (1998)
4. Ferragina, P., Manzini, G.: Opportunistic data structures with applications. In: FOCS, pp. 390–398. IEEE Computer Society (2000)
5. Gupta, A., Hon, W.K., Shah, R., Vitter, J.S.: Dynamic rank/select dictionaries with applications to XML indexing. Technical report 06–014, Purdue University (2006)
6. Hon, W.K., Lam, T.W., Sadakane, K., Sung, W.K., Yiu, S.M.: Compressed index for dynamic text. In: Data Compression Conference, pp. 102–111 (2004)
7. Leiserson, C.E., Prokop, H., Randall, K.H.: Using de Bruijn sequences to index a 1 in a computer word (1998) (unpublished manuscript)
8. Mäkinen, V., Navarro, G.: Succinct suffix arrays based on run-length encoding. Nord. J. Comput. **12**(1), 40–66 (2005)
9. Manber, U., Myers, G.: Suffix arrays: a new method for on-line string searches. SIAM J. Comput. **22**(5), 935–948 (1993)
10. Navarro, G.: Indexing text using the Ziv-Lempel trie. J. Discret. Algorithms **2**(1), 87–114 (2004)
11. Navarro, G., Nekrich, Y.: Optimal dynamic sequence representations. In: Proceedings of the SODA 2013, pp. 865–876 (2013)
12. Rasmussen, K.R., Stoye, J., Myers, E.W.: Efficient q-gram filters for finding all ϵ-matches over a given length. J. Comput. Biol. **13**(2), 296–308 (2006)
13. Salson, M., Lecroq, T., Léonard, M., Mouchard, L.: Dynamic extended suffix arrays. J. Discret. Algorithms **8**(2), 241–257 (2010)
14. Ukkonen, E.: On-line construction of suffix trees. Algorithmica **14**(3), 249–260 (1995)
15. Weiner, P.: Linear pattern-matching algorithms. In: Proceedings of 14th IEEE Annual Symposium on Switching and Automata Theory, pp. 1–11 (1973)
16. Willard, D.E.: Log-logarithmic worst-case range queries are possible in space $\Theta(N)$. Inf. Process. Lett. **17**(2), 81–84 (1983)

Reporting Consecutive Substring Occurrences Under Bounded Gap Constraints

Gonzalo Navarro[1]([⊠]) and Sharma V. Thankachan[2]

[1] Center of Biotechnology and Bioengineering, Department of Computer Science,
University of Chile, Santiago, Chile
gnavarro@dcc.uchile.cl
[2] School of Computational Science and Engineering,
Georgia Institute of Technology, Atlanta, USA
sharma.thankachan@gatech.edu

Abstract. We study the problem of indexing a text $T[1 \ldots n]$ such that whenever a pattern $P[1 \ldots p]$ and an interval $[\alpha, \beta]$ comes as a query, we can report all pairs (i, j) of consecutive occurrences of P in T with $\alpha \le j - i \le \beta$. We present an $O(n \log n)$ space data structure with optimal $O(p + k)$ query time, where k is the output size.

1 Introduction

Detecting consecutive occurrences of a pattern in a text is a problem that arises, in various forms, in computational biology applications [1–3]. For example, a tandem repeat is an occurrence of the form PP of a given string $P[1 \ldots p]$ inside a sequence $T[1 \ldots n]$. Due to mutations and experimental errors, one may relax the condition that the occurrences appear exactly one after the other, and allow for a small range of distances between the two occurrences of P [1, Sect. 9.2]. Other variants of the problem are to find P closely followed by its reverse complemented version in tRNA sequences, which is useful to identify the positions where the tRNA molecule folds into a cloverleaf structure defined by stems (the two occurrences of P) and loops (the string between them) [1, Sect. 11.9, Example 42]; this process is also called RNA interference [2, Sect. 6.4].

Several related combinatorial problems stem from these motivations. For example, Iliopoulos and Rahman [4] consider the problem of finding all the k occurrences of two patterns P_1 and P_2 (of total length p) separated by a fixed distance α known at indexing time. They gave a data structure using $O(n \log^\epsilon n)$ space and query time $O(p + \log \log n + k)$, for any constant $\epsilon > 0$. Bille and Gørtz [5] retained the same space and improved the time to the optimal $O(p+k)$.[1] The problem becomes, however, much messier when we allow the distance between P_1 and P_2 to be in a range $[\alpha, \beta]$, even if these are still known at indexing time. Bille et al. [6] obtained various tradeoffs, for example $O(n)$ space and $O(p + \sigma^\beta \log \log n + k)$ time, where σ is the alphabet size; $O(n \log n \log^\beta n)$

G. Navarro—Funded with Basal Funds FB0001, Conicyt, Chile.
[1] This is optimal in the RAM model if we assume a general alphabet of size $O(n)$.

© Springer International Publishing Switzerland 2015
F. Cicalese et al. (Eds.): CPM 2015, LNCS 9133, pp. 367–373, 2015.
DOI: 10.1007/978-3-319-19929-0_31

space and $O(p + (1 + \epsilon)^\beta \log \log n + k)$ time; and $O(\sigma^{\beta^2} n \log^\beta \log n)$ space and $O((p + \beta)(\beta - \alpha) + k)$ time.

These problems, however, are more general than necessary for the applications we described, where $P_1 = P_2 = P$ (or P_2 is the reverse complement of P_1, a case that can be handled in the solution we will give). For this case, some related problems have been studied. Keller et al. [7] considered the problem of, given an occurrence of P in T, find the next one to the right. They obtained an index using $O(n \log^\epsilon n)$ space and $O(\log \log n)$ time. Another related problem they studied was to find a maximal set of nonoverlapping occurrences of P. They obtained the same space and $O(\log \log n + k)$ time. Muthukrishnan [8] considered a document-based version of the problem: T is divided into documents, and we want to report all the k documents where two occurrences of P appear at distance at most β. For β fixed at indexing time, he obtained $O(n)$ space and optimal $O(p + k)$ time; the space raises to $O(n \log n)$ when β is given as a part of the query. Finally, Brodal et al. [9] considered the related pattern mining problem: find all the z maximal patterns P that appear at least twice in T, separated by a distance in $[\alpha, \beta]$. They obtain $O(n \log n + z)$ time, within $O(n)$ space.

In this paper we focus on what is perhaps the cleanest variant of the problem, which (somewhat surprisingly) has not been considered before: find the positions in T where two occurrences of P appear, separated by a distance in the range $[\alpha, \beta]$. It is formally stated as follows.

Problem 1. *Index a text $T[1 \ldots n]$, such that whenever a pattern $P[1 \ldots p]$ and a range $[\alpha, \beta]$ comes as a query, we can report all pairs (i, j) of consecutive occurrences of P in T with $\alpha \leq j - i \leq \beta$.*

We obtain the following result.

Theorem 1. *There exists an $O(n \log n)$ space data structure with query time $O(p + k)$ for Problem 1, where k is the output size.*

Our solution makes use of heavy-path decompositions on suffix trees and geometric data structures. In the Conclusions we comment on the implications of this result on related problems.

2   Notation and Preliminaries

The ith leftmost character of T is denoted by $T[i]$, where $1 \leq i \leq n$. The substring starting at location i and ending at location j is denoted by $T[i \ldots j]$. A suffix is a substring that ends at location n and a prefix is a string that starts at location 1.

The *suffix tree* (ST) of T is a compact representation of all suffixes of $T \circ \$$, except $\$$, in the form of a compact trie [10]. Here $\$$ a special symbol that does not appear anywhere in T and $T \circ \$$ is the concatenation of T and $\$$. The number of leaves in ST is exactly n. The degree of an internal node is at least two.

We use ℓ_i to represent the ith leftmost leaf in ST. The edges are labeled with characters and the concatenation of edge labels on the path from root to a node u is denoted by $\mathsf{path}(u)$. Then, $\mathsf{path}(\ell_i)$ corresponds to the ith lexicographically smallest suffix of T, and its starting position is denoted by SA[i]. The locus of a pattern P in T, denoted by $\mathsf{locus}(P)$, is the highest node u in ST, such that P is a prefix of $\mathsf{path}(u)$. The set of occurrences of P in T is given by SA[i] over all i's, where ℓ_i is in the subtree of $\mathsf{locus}(P)$. The space occupied by ST is $O(n)$ words and the time for finding the locus of an input pattern P is $O(|P|)$. Additionally, for two nodes u and v, we shall use $\mathsf{lca}(u,v)$ to denote their lowest common ancestor.

We now describe the concept of *heavy path* and *heavy path decomposition*. The heavy path of ST is the path starting from the root, where each node u on the path is the child with the largest subtree size (ties broken arbitrary). The *heavy path decomposition* is the operation where we decompose each off-path subtree of the heavy path recursively. As a result, any $\mathsf{path}(\cdot)$ in ST will be partitioned into disjoint heavy paths. Sleator and Tarjan [11] proved the following property; we will use $\log n$ to denote logarithm in base 2.

Lemma 1. *The number of heavy paths intersected by any root to leaf path is at most $\log n$, where n is the number of leaves in the tree.*

Each node belongs to exactly one heavy path and each heavy path contains exactly one *leaf* node. The heavy path containing ℓ_i will be called the i-th heavy path (and identified simply by the number i). For an internal node u, let $\mathsf{hp}(u)$ be the unique heavy path that contains u.

Definition 1. *The set $\mathcal{H}_i$ is defined as the set of all leaf identifiers j, where the path from root to ℓ_j intersects with the i-th heavy path. That is, $\mathcal{H}_i = \{j \mid \mathsf{hp}(\mathsf{lca}(\ell_j, \ell_i)) = i\}$.*

Lemma 2. $\sum_{i=i}^{n} |\mathcal{H}_i| \leq n \log n.$

Proof. For any particular j, path from root to ℓ_j can intersect at most $\log n$ heavy paths, by Lemma 1. Therefore, j cannot be a part of more than $\log n$ sets. ∎

3   The Data Structure

The key idea is to reduce our pattern matching problem to an equivalent geometric problem. Specifically, to the *orthogonal segment intersection problem*.

Definition 2 (Orthogonal Segment Intersection). *A horizontal segment (x_i, x_i', y_i) is a line connecting the 2D points (x_i, y_i) and (x_i', y_i). A segment intersection problem asks to pre-process a given set $\mathcal{S}$ of horizontal segments into a data structure, such that whenever a vertical segment (x'', y', y'') comes as a query, we can efficiently report all the horizontal segments in $\mathcal{S}$ that intersect with the query segment. Specifically, we can output the following set: $\{(x_i, x_i', y_i) \in \mathcal{S} \mid x_i \leq x'' \leq x_i', y' \leq y_i \leq y''\}$.*

There exists an $O(|\mathcal{S}|)$ space and $O(\log|\mathcal{S}|+k)$ time solution for segment intersection problem using a persistent binary tree, where k is the output size [12]. We now proceed to describe the reduction.

3.1 Reduction

One of the main components of our data structure is the suffix tree ST of T, and is used only for finding the locus of P. Based on the heavy path on which the locus node is, we categorize the queries in different types.

Definition 3. *A query with input pattern P is type-h if $h = \mathsf{hp}(\mathsf{locus}(P))$.*

Let G_h be the data structure handling type-h queries, where G_h is a structure over a set $\mathcal{I}_h$ of horizontal segments, that can efficiently answer segment intersection queries. The set $\mathcal{I}_h$ is generated from $\mathcal{H}_h$ using the following steps for each $j \in \mathcal{H}_h$:

1. Let $P_j = \mathsf{path}(\mathsf{lca}(\ell_h, \ell_j))$.
2. Let $suc(j)$ be the first occurrence of P_j after the position SA[j] in T and let $pre(j)$ be the last occurrence of P_j before the position SA[j] in T. Clearly, neither in $[(pre(j)+1)\dots(\mathsf{SA}[j]-1)]$, nor in $[(\mathsf{SA}[j]+1)\dots(suc(j)-1)]$, P_j has an occurrence.
3. Now, obtain two segments w.r.t. j as follows:
   (a) Let P'_j be the *shortest* prefix of P_j without any occurrence in $[(pre(j)+1)\dots(\mathsf{SA}[j]-1)]$. Then, create segment $(x_i, x'_i, y_i) = (|P'_j|, |P_j|, \mathsf{SA}[j] - pre(j))$ and associate the pair $(pre(j), \mathsf{SA}[j])$ of consecutive occurrences of P_j as satellite information.
   (b) Similarly, let P''_j be the *shortest* prefix of P_j without any occurrence in $[(\mathsf{SA}[j]+1)\dots(suc(j)-1)]$. Then, create segment $(x_i, x'_i, y_i) = (|P''_j|, |P_j|, suc(j) - \mathsf{SA}[j])$ and associate it to the pair $(\mathsf{SA}[j], suc(j))$ of consecutive occurrences of P_j as satellite information.

Clearly, $|\mathcal{I}_h| = 2|\mathcal{H}_h|$. The central idea of our solution is summarized below. Figure 1 illustrates the idea.

Lemma 3. *Let P and $[\alpha, \beta]$ be the input parameters of a query in problem 1 and let $h = \mathsf{hp}(\mathsf{locus}(P))$. Then, the set of satellite information associated with all those horizontal segments in $\mathcal{I}_h$, which are stabbed by a vertical segment (p, α, β) (i.e., the segment connecting the points (p, α) and (p, β)) forms the output to Problem 1.*

Proof. First we prove that any satellite information (a, b) reported by the geometric query on G_h is an answer to the original query. Let $[s, e]$ be the x-interval corresponding to the reported satellite information (a, b). Then, $s \le p \le e$ and $\alpha \le b - a \le \beta$. Here the condition $e \ge p$ ensures that both $\ell_{\mathsf{SA}^{-1}[a]}$ and $\ell_{\mathsf{SA}^{-1}[b]}$ are leaves in the subtree of $\mathsf{locus}(P)$. Therefore a and b are occurrences of P. The condition $s \le p$ ensures that there exists no occurrence of P in any location

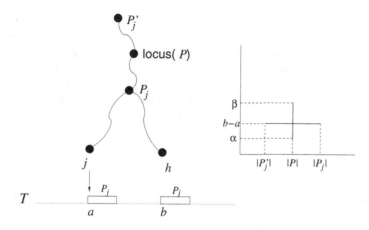

Fig. 1. Illustration of the main concepts of our data structure.

which is after a, but before b (i.e., a and b are consecutive occurrences of P). Finally the y-coordinate ensures that $\alpha \leq b - a \leq \beta$.

Now we prove that for every output (a, b) of Problem 1, there exists a segment $(s, e, b - a)$ in $\boldsymbol{\mathcal{I}}_h$ with $s \leq p \leq e$ and satellite information (a, b). Without loss of generality, let $\mathsf{lca}(\ell_h, \ell_{\mathsf{SA}^{-1}[a]})$ be either $\mathsf{lca}(\ell_h, \ell_{\mathsf{SA}^{-1}[b]})$ or an ancestor of it. Then, let $j = \mathsf{SA}^{-1}[a]$. Since P occurs at position a, the leaf j descends from the subtree of $\mathsf{locus}(P)$, and since this node belongs to the heavy path h, we have that $\mathsf{lca}(\ell_h, \ell_j)$ descends from $\mathsf{locus}(P)$, thus $e \geq p$. Since there is no occurrence of P between a and b, it holds $s \leq p$. Then, a segment of the form $(s, e, b - a)$ will indeed be created while processing $j \in \mathcal{H}_h$ during the construction of $\boldsymbol{\mathcal{I}}_h$. ∎

In the light of Lemma 3, we have the following result.

Lemma 4. *There exists an $O(n \log n)$ space and $O(p + \log n + k)$ query time solution for Problem 1, where k is the output size.*

Proof. The space of ST is $O(n)$ and the space required for maintaining the segment intersection structure over $\boldsymbol{\mathcal{I}}_h$, for all values of h, is $O(\sum_h |\boldsymbol{\mathcal{I}}_h|) = O(\sum_h |\mathcal{H}_h|) = O(n \log n)$. Thus, the total space is $O(n \log n)$ words. To answer a query, we first find the locus of P in ST in $O(p)$ time, and then query G_h, where $h = \mathsf{hp}(\mathsf{locus}(P))$, in $O(\log n + k)$ time. Therefore, the query time is $O(p + \log n + k)$. ∎

The query time in Lemma 4 is optimal if $p \geq \log n$. To handle queries where p is shorter than $\log n$, we use a different approach.

3.2   Achieving Optimal Query Time

We present an optimal query time data structure for $p < \log n$. Essentially, we associate a data structure $D(u)$ with each node u in ST, whose string depth (i.e., $|\mathsf{path}(u)|$) is at most $\log n$. Observe that the number of occurrences of $\mathsf{path}(u)$ in T is equal to $\mathsf{size}(u)$, where $\mathsf{size}(u)$ is the number of leaves in the subtree of u. Therefore, the number of consecutive occurrences (i, j) of $\mathsf{path}(u)$ is $\mathsf{size}(u) - 1$. Each such pair (i, j) can be mapped to a point $(j - i)$ in one dimension along with the pair (i, j) as an associated satellite data. We then create a one-dimensional range reporting data structure over these $(\mathsf{size}(u) - 1)$ points and call it $D(u)$. Whenever the locus of P is u, the answer can be obtained by issuing a one dimensional range reporting query on $D(u)$ with $[\alpha, \beta]$ as the input range. The satellite data associated with each reported corresponds to an answer to Problem 1.

We use the data structure summarized in Lemma 5, by which queries can be answered in optimal time and the space of $D(u)$ can be bounded by $O(\mathsf{size}(u))$ words.

Lemma 5 ([13]). *One dimensional range reporting queries over a set of m points in $\{0, 1, 2, \ldots, 2^w\}$ can be answered in optimal time using an $O(m)$ space data structure, where w is the word size.*

Note that the sum of all the $\mathsf{size}(u)$ terms for all the nodes u with the same string depth is n, and added over all the nodes with string depth up to $\log n$ is $n \log n$. Thus the space for the $D(\cdot)$ structures of all the nodes with string depth up to $\log n$ is $O(n \log n)$ words. This completes the proof of Theorem 1.

4   Conclusions

We have addressed what seems to be the cleanest variant of the problem related to finding close occurrences of a pattern $P[1 \ldots p]$ in a text $T[1 \ldots n]$: find pairs of occurrences that are within a distance range $[\alpha, \beta]$ (given at query time). Our data structure uses $O(n \log n)$ space and optimal $O(p + k)$ query time.

It is not hard to extend our result to the case where we look for the occurrence of P followed (or preceded) by some function of P, such as its reverse complemented string (as motivated in the Introduction). We can build the geometric structure at each suffix tree node v considering the function of the string represented by v, instead of the string itself. However, extending our solution to the general case of two patterns [6] seems not possible.

Our result opens several interesting questions. A first one is whether this problem is strictly harder than the restricted variant where $\alpha = \beta$. For this case, the same optimal query time has been obtained within less space, $O(n \log^\epsilon n)$ [5], even when generalizing the problem to two patterns P_1 and P_2. The significantly messier results obtained for the general case $\alpha \leq \beta$ [6] suggest that this general problem is indeed harder. Still, it is not clear whether our optimal-time result can also be obtained within $o(n \log n)$ space.

A second interesting question is whether our result can be used for pattern mining, that is, finding those P that appear twice in T separated by a distance in $[\alpha, \beta]$. A direct application of our result, which builds our structure and then traverses the suffix tree, requires $\Omega(n \log n + z)$ time, which is not better than the current result [9]. Yet, there could be harder pattern mining problems for which our result is a useful tool.

Yet a third interesting question is how our results can be extended to the document retrieval scenario, that is, listing the documents where P appears twice and separated by a distance in $[\alpha, \beta]$. The current result [8] is similar to ours in space and time, but it is restricted to the case $\alpha = 0$. It is not clear if is the problem is harder, and by how much, for an arbitrary value of α.

References

1. Gusfield, D.: Algorithms on Strings, Trees and Sequences: Computer Science and Computational Biology. Cambridge University Press, New York (1997)
2. Aluru, S. (ed.): Handbook of Computational Molecular Biology. CRC Computer and Information Science Series. Chapman and Hall, London (2005)
3. Ohlebusch, E.: Bioinformatics Algorithms: Sequence Analysis, Genome Rearrangements, and Phylogenetic Reconstruction. Oldenbusch Verlag, Bremen (2013)
4. Iliopoulos, C.S., Rahman, M.S.: Indexing factors with gaps. Algorithmica **55**, 60–70 (2009)
5. Bille, P., Gørtz, I.L.: Substring range reporting. Algorithmica **69**, 384–396 (2014)
6. Bille, P., Gørtz, I.L., Vildhøj, H.W., Vind, S.: String indexing for patterns with wildcards. Theor. Comput. Syst. **55**, 41–60 (2014)
7. Keller, O., Kopelowitz, T., Lewenstein, M.: Range non-overlapping indexing and successive list indexing. In: Dehne, F., Sack, J.-R., Zeh, N. (eds.) WADS 2007. LNCS, vol. 4619, pp. 625–636. Springer, Heidelberg (2007)
8. Muthukrishnan, S.: Efficient algorithms for document retrieval problems. In: Proceedings of 13th Annual ACM-SIAM Symposium on Discrete Algorithms (SODA), pp. 657–666 (2002)
9. Brodal, G.S., Lyngsø, R.B., Pedersen, C.N.S., Stoye, J.: Finding maximal pairs with bounded gap. In: Crochemore, M., Paterson, M. (eds.) CPM 1999. LNCS, vol. 1645, pp. 134–149. Springer, Heidelberg (1999)
10. Weiner, P.: Linear pattern matching algorithms. In: 14th Annual Symposium on Switching and Automata Theory, Iowa City, Iowa, USA, 15–17 October, pp. 1–11 (1973)
11. Sleator, D.D., Tarjan, R.E.: A data structure for dynamic trees. J. Comput. Syst. Sci. **26**, 362–391 (1983)
12. Tao, Y.: Dynamic ray stabbing. ACM Trans. Algorithms **11**, 11 (2014)
13. Alstrup, S., Brodal, G.S., Rauhe, T.: Optimal static range reporting in one dimension. In: Proceedings on 33rd Annual ACM Symposium on Theory of Computing, Heraklion, Crete, Greece, 6–8 July, pp. 476–482 (2001)

A Probabilistic Analysis of the Reduction Ratio in the Suffix-Array IS-Algorithm

Cyril Nicaud[✉]

LIGM, Université Paris-Est and CNRS,
Marne-la-Vallée Cedex 2, 77454 Paris, France
cyril.nicaud@u-pem.fr

Abstract. We show that there are asymptotically γn LMS-factors in a random word of length n, for some explicit γ that depends on the model of randomness under consideration. Our results hold for uniform distributions, memoryless sources and Markovian sources. From this analysis, we give new insight on the typical behavior of the IS-algorithm [9], which is one of the most efficient algorithms available for computing the suffix array.

1 Introduction

The suffix array of a word, is a permutation of its suffixes that orders them for the lexicographic order. Since their introduction by Manber and Meyers [7,8] in 1990, suffix arrays have been intensively studied in the literature. Nowadays, they are a fundamental, space efficient, alternative to suffix trees. They are used in many applications such as pattern matching, plagiarism detection, data compression, etc.

The first linear suffix array algorithms that do not use the suffix tree construction were proposed by Ko and Aluru [5], Kim et al. [4] and Kärkkäinen and Sanders [3] in 2003. Since then, a lot of variations or heuristics have been developed [12], motivated by the various practical uses of this fundamental data structure.

A few years ago, Ge Nong, Sen Zhang and Wai Hong Chan proposed such a linear suffix array algorithm [9], which is particularly efficient in practice. This algorithm, called the *IS-algorithm*, is a recursive algorithm, where the suffix array of a word u is deduced from the suffix array of a shorter word v. This shorter word is built using the LMS-factors of u: an LMS-position i in u is an integer such that the suffix of u that starts at position i is smaller, for the lexicographic order, than both the one that starts at position $i - 1$ and the one that starts at position $i + 1$; LMS-factors are the factors of u delimited by two consecutive LMS-positions. Once the suffix array of v is recursively calculated, the suffix array of u can be computed in linear time.

In this article we are interested in the typical reduction ratio $\frac{|v|}{|u|}$ obtained when making this recursive call. We propose a probabilistic analysis of the number of LMS-factors in a random word of length n, for classical models of random

© Springer International Publishing Switzerland 2015
F. Cicalese et al. (Eds.): CPM 2015, LNCS 9133, pp. 374–384, 2015.
DOI: 10.1007/978-3-319-19929-0_32

words: uniform distributions, memoryless sources and Markovian sources. We prove that the reduction ratio is concentrated around a constant γ, which can be explicitly computed from the parameters that describe the source.

In this extended abstract, we chose to focus on memoryless sources. After recalling the basics on words and suffix arrays in Sect. 2, we explain in Sect. 3 the steps that lead to our main statement (Theorem 2). In Sect. 4, we briefly explain how this result can be generalized to Markovian sources, and give the explicit formula for the typical reduction ratio under this model (Theorem 3). We conclude this article with some experiments, that are just intended to illustrate our theoretical results, and with a short discussion in Sect. 5.

2   Preliminaries

2.1   Definitions and Notations

Let A be a non-empty totally ordered finite alphabet. For given $n \geq 0$, we denote by A^n the set of words of length n on A. Let A^* be the set of all words on A.

If $u \in A^n$ is a word of length $n \geq 1$, let u_0 be its first letter, let u_1 be its second letter, ... and let u_{n-1} be its last letter. The *reverse* of a word $u = u_0 \cdots u_{n-1}$ is the word $\overline{u} = u_{n-1} \cdots u_0$. For given i and j such that $0 \leq i \leq j \leq n-1$, let $u[i,j]$ be the factor of u that starts at position i and ends at position j: it is the unique word w of length $j - i + 1$ such that there exists a word v of length i such that vw is a prefix of u. For given i such that $0 \leq i \leq n-1$, let $\mathtt{suff}(u,i) = u[i, n-1]$ be the suffix of u that starts at position i.

Recall that the *suffix array* of a word u of length $n \geq 1$ is the unique permutation σ of $\{0, \ldots, n-1\}$ such that, for the lexicographic order, we have

$$\mathtt{suff}(u, \sigma(0)) < \mathtt{suff}(u, \sigma(1)) < \ldots < \mathtt{suff}(u, \sigma(n-1)).$$

See [12] for a more detailed account on suffix arrays and their applications.

2.2   LMS-factors of a Word

The first step of the IS-algorithm [9] consists in marking every position in $v = u\$$, where $\$ \notin A$ is an added letter that is smaller than every letter of A. The mark of each position in v is either the letter S or the letter L. A position $i \in \{0, \ldots, n-1\}$ is marked by an S or by an L when $\mathtt{suff}(v,i) < \mathtt{suff}(v, i+1)$ or $\mathtt{suff}(v,i) > \mathtt{suff}(v, i+1)$, respectively. We also say that the position is of type S or L. By convention, the last position n of v always is of type S.

A leftmost type S position in $v = u\$$ (*LMS-position* for short) is a position $i \in \{1, \ldots, n\}$ such that i is of type S and $i-1$ is of type L. Note that with this definition, the last position of v is always an LMS-position, for a non-empty u. An *LMS-factor* of v is a factor $v[i,j]$ where $i < j$ are both LMS-positions and such that there is no LMS-position between i and j. By convention, the factor $v[n,n] = \$$ is also an LMS-factor of v.

The following notations and definitions will be used throughout this article. They are reformulations of what we just defined.

Definition 1. *Let A be a finite totally ordered non-empty alphabet. The alphabet* $\mathrm{LS}(A)$ *is defined by* $\mathrm{LS}(A) = (A \times \{L, S\}) \cup \{(\$, S)\}$. *For simplification, elements of* $\mathrm{LS}(A)$ *are written αX instead of (α, X). A letter αS of* $\mathrm{LS}(A)$ *is said to be of type S, and a letter αL is said to be of type L.*

Definition 2. *Let $u \in A^n$ for some $n \geq 1$. The LS-extension* $\mathbf{Ext}(u)$ *of u, is the word $v \in \mathrm{LS}(A)^{n+1}$ that ends with the letter $\$S$ and such that for every $i \in \{0, \dots, n-1\}$, $v_i = u_i X_i$ with $X_i = S$ if and only if $u_i < u_{i+1}$ or ($u_i = u_{i+1}$ and $X_{i+1} = S$), with the convention that $u_n = \$$.*

Observe that from its definition, $\mathbf{Ext}(u)$ is exactly the word u with an added \$ at its end, and whose positions have been marked. Thus, an LMS-position in $v = \mathbf{Ext}(u)$ is a position $i \geq 1$ such that $v_i = \alpha S$ and $v_{i-1} = \beta L$, for some $\alpha, \beta \in A \cup \{\$\}$. We extend this definition to all words of $LS(A)^*$.

Definition 3. *For any $u \in \mathrm{LS}(A)^n$, an LMS-position of u is a position $i \in \{1, \dots, n-1\}$ such that u_i is of type S and u_{i-1} is of type L.*

Example. Consider the word $u = bacbcaab$ on $A = \{a, b, c\}$. We have:

letter	b	a	c	b	c	a	a	b	\$
type	L	S	L	S	L	S	S	L	S

$\mathbf{Ext}(u) = bL \, \underline{aS} \, cL \, \underline{bS} \, cL \, \underline{aS} \, aS \, bL \, \underline{\$S},$

where the LMS-positions have been underlined.

2.3   Brief Overview of the IS-algorithm

The IS-algorithm [9] first computes the type of each position. This can be done in linear time, by scanning the word once from right to left. From this, the LMS-positions can be directly computed.

The LMS-factors are then numbered in increasing order using a radix sort (the types are kept and used for the lexicographic comparisons of these factors). This yields an array of numbers, the numbers associated with the LMS-factors, which is viewed as a word and whose suffix array σ' is recursively calculated.

The key observation is that once σ' is known, the suffixes of type L can be sorted by scanning the word once from left to right, then the suffixes of type S can be sorted by a scan from right to left. Therefore, the suffix array can be computed in linear time, once σ' is given.

For the running time analysis, if $T(n)$ is the worst case cost of the algorithm applied to a word of length n, then we have the inequality $T(n) \leq T(m) + \Theta(n)$, where m denote the number of LMS-factors. Since we always have $m \leq \frac{n}{2}$, the running time of the IS-algorithm is $\Theta(n)$. The quotient m/n is called the *reduction ratio* and it is the main focus of this article.

2.4   Distributions on Words

The *uniform distribution* on a finite set E is the probability p defined for all $e \in E$ by $p(e) = \frac{1}{|E|}$. By a slight abuse of notation, we will speak of the *uniform* *distribution on* A^* to denote the sequence $(p_n)_{n \geq 0}$ of uniform distributions on A^n. For instance, if $A = \{a, b, c\}$, then each element of A^n has probability 3^{-n} under this distribution.

An element $u \in A^n$ taken uniformly at random can also be seen as built letter by letter, from left to right or from right to left, by choosing each letter uniformly and independently in A. This is a suitable way to consider random words, which can easily be generalized to more interesting distributions. Indeed, if p is a probability on A, one can extend p to A^n by generating each letter independently following the probability p. This is called a *memoryless distribution of probability* p, and the probability of an element $u = u_0 \cdots u_{n-1} \in A^n$ is defined by $\mathbb{P}_p(u) = p(u_0)p(u_1) \cdots p(u_{n-1})$.

A further classical generalization consists in allowing some (limited) dependency from the past when generating the word. This leads to the notion of Markov chain, which we describe now. Let Q be a non-empty finite set, called the *set of states*. A sequence of Q-valued random variables $(X_n)_{n \geq 0}$ is a *homogeneous Markov chain* (or just *Markov chain* for short in this article) when for every n, every $\alpha, \beta \in Q$ and every $q_0, \ldots q_{n-1} \in Q$,

$$\mathbb{P}(X_{n+1} = \alpha \mid X_n = \beta, X_{n-1} = q_{n-1}, \ldots, X_0 = q_0) = \mathbb{P}(X_1 = \alpha \mid X_0 = \beta).$$

In the sequel, we will use the classical representation of a Markov chain by its *initial probability (row) vector* $\pi_0 \in [0, 1]^Q$ and its *transition matrix* $M \in [0, 1]^{Q \times Q}$, defined for every $i, j \in Q$ by $M(i, j) = \mathbb{P}(X_1 = j \mid X_0 = i)$. In this settings, the probability of a word $u = u_0 \cdots u_{n-1}$ on $A = Q$ is

$$\mathbb{P}_{M, \pi_0}(u) = \pi_0(u_0)\, M(u_0, u_1)\, M(u_1, u_2) \cdots M(u_{n-2}, u_{n-1}).$$

Such a Markov chain for generating words of A^* is also called a *first order* Markov chain, since the probability of a new letter only depends on the last letter. One can easily use Markov chains to allow larger dependencies from the past. For instance, a *second order* Markov chain can be defined by setting $Q = A \times A$. The probability of a word $u = u_0 \cdots u_{n-1}$, with $n \geq 2$, is now defined, for an initial probability vector $\pi_0 \in [0, 1]^Q$, by

$$\mathbb{P}_{M, \pi_0}(u) = \pi_0(u_0 u_1) M(u_0 u_1, u_1 u_2) M(u_1 u_2, u_2 u_3) \cdots M(u_{n-3} u_{n-2}, u_{n-2} u_{n-1}).$$

Higher order Markov chain are defined similarly. More general sources, such as dynamical sources [13], are also considered in the literature, but they are beyond the scope of this article.

2.5   About the Probabilistic Analysis of the Original Article

In their article [9], the authors proposed a brief analysis of the expected reduction ratio. This analysis is done under the simplistic assumption that the marks of the positions are independent and of type S or L with probability $\frac{1}{2}$ each.

We first observe that if $A = \{a, b\}$ consists of exactly two letters and if we consider the uniform distribution on A^n, then, up to the very end of the word, every a is of type S and every b is of type L. Hence, we are mostly in the model proposed in [9]. Unfortunately, if there are three or more letters, then uniform distributions, memoryless distributions and Markovian distributions failed to produce types that are i.i.d. in $\{L, S\}$. It is also the case for a binary alphabet, when the distribution under consideration is not the uniform distribution.

Their result, Theorem 3.15 page 1477, also contains a miscalculation. The average reduction ratio when the types are i.i.d. S and L with probability $\frac{1}{2}$ tends to $\frac{1}{4}$ and not to $\frac{1}{3}$ as stated. This can easily be obtained the following way: in this model, a position $i \geq 1$ is such that i is of type S and $i - 1$ is of type L with probability $\frac{1}{4}$. The result follows by linearity of the expectation.[1]

In the sequel we give formulas for the reduction ratio for alphabets of any size, and for uniform, memoryless and Markovian distributions.

3   Probabilistic Analysis for Memoryless Sources

If instead of generating a word letter by letter from left to right, we choose to perform the generation from right to left, then it is easy to compute, on the fly, the type of each position. This is a direct consequence of Definition 1. In probabilistic terms, we just defined a Markov chain, built as an extension of our random source. This is the idea developed in this section, and we will use it to compute the typical reduction ratio of the IS-algorithm.

3.1   A Markov Chain for the LS-extension

Let A be a totally ordered alphabet, with at least two letters, and let p be a probability on A such that for every $a \in A$, $p(a) > 0$. In this section we consider the memoryless distributions on A^* of probability p, as defined in Sect. 2.4. To simplify the writing, we will use p_a instead of $p(a)$ in the sequel.

Recall that if P is a property, then $[\![P]\!]$ is equal to 1 if P is true and to 0 if it is false. Let π_0 be the row vector of $[0, 1]^{\mathrm{LS}(A)}$ defined by $\pi_0(\alpha X) = [\![\alpha X =\$\ S]\!]$. Let M_p be the matrix of $[0, 1]^{\mathrm{LS}(A) \times \mathrm{LS}(A)}$ defined for every $\alpha, \beta \in A$ by

$$M_p(\alpha S, \$S) = M_p(\alpha L, \$S) = M_p(\$S, \beta S) = 0; \qquad M_p(\$S, \beta L) = p_\beta;$$
$$M_p(\alpha S, \beta S) = p_\beta \cdot [\![\beta \leq \alpha]\!]; \qquad M_p(\alpha S, \beta L) = p_\beta \cdot [\![\beta > \alpha]\!];$$
$$M_p(\alpha L, \beta S) = p_\beta \cdot [\![\beta < \alpha]\!]; \qquad M_p(\alpha L, \beta L) = p_\beta \cdot [\![\beta \geq \alpha]\!].$$

We first establish that the reverse of the LS-extension of a random word generated by a memoryless source is Markov (M_p, π_0) (Fig. 1):

Proposition 1. *Let A be a totally ordered alphabet, with at least two letters, and let p be a probability on A such that for every $a \in A$, $p_a > 0$. If u is a word on $\mathrm{LS}(A)$ such that $\mathbb{P}_{M_p, \pi_0}(u) \neq 0$, then the reverse of u is the LS-extension of a word v of A^* and $\mathbb{P}_{M_p, \pi_0}(u) = \mathbb{P}_p(v)$.*

---

[1] In their proof, they compute the mean length of an LMS-factor. The types of such a factor form a word of SS^*LL^*S. For the considered model, the mean length of an element of S^* (and of L^*) is one. Hence, the average length of an LMS-factor is 5 (and not the announced 4).

In other words, generating v using a memoryless source of probability p is the same as generating the reverse of $\mathrm{LS}(v)$ using the Markov chain (M_p, π_0) (Fig. 1).

Proposition 1 is the key observation of this article. It described a purely probabilistic way to work on LS-extensions of random words: the deterministic algorithm used to mark each position with its type is encoded into the Markov chain (M_p, π_0). We now aim at using the classical results on Markov chains to obtain some information on the number of LMS-factors.

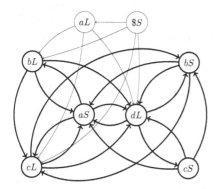

	aS	bS	bL	cS	cL	dL
aS	p_a	0	p_b	0	p_c	p_d
bS	p_a	p_b	0	0	p_c	p_d
bL	p_a	0	p_b	0	p_c	p_d
cS	p_a	p_b	0	p_c	0	p_d
cL	p_a	p_b	0	0	p_c	p_d
dL	p_a	p_b	0	p_c	0	p_d

The matrix $\underline{M}_p$

Fig. 1. On the left, the underlying graph of the Markov chain for $A = \{a, b, c, d\}$. The state dS is not depicted, as it is not reachable. Every state but $\$S$ also has a loop on itself, which is not depicted for readability. The thin states are the transient states, and the bold states are the recurrent states. For the memoryless source of probability p, the probability of each edge $\alpha X \to \beta Y$ is p_β. If we start on $\$S$ with probability 1, then this chain generates the marked words *from right to left*. On the right is presented the matrix $\underline{M}_p$, which is the restriction of M_p to its recurrent part.

3.2 Properties of the Markov Chain

From now on, except for the examples, we fix $A = \{a_1, \ldots, a_k\}$, with $k \geq 2$, and we consider the total strict order $<$ on A defined by $a_1 < a_2 \ldots < a_k$.

The *underlying graph* G_M of a Markov chain (M, π_0) of set of states Q is the directed graph whose vertices are the elements of Q and with an edge $s \to t$ whenever $M(s, t) > 0$. A state $q \in Q$ is *transient* when it is not in a terminal strongly connected component of G_M: if we start in state q, there is a non-zero probability that we will never return to q. Transient states play a minor role in our settings as with high probability they are only used during the generation of the very first letters. A state that is not transient is called *recurrent*.

Lemma 1. *The Markov chain (M_p, π_0) has three transient states: $\$S$, $a_1 L$ and $a_k S$. All other states are in the same terminal strongly connected component of its underlying graph.*

Remark 1. From the definition of M_p and π_0, a path of positive probability in the chain always starts on the state $\$S$, may pass through the state $a_1 L$, but never reaches the state $a_k S$. This property is obvious if one remember that a word generated by the chain is the reverse of the LS-extension of a word on A.

Recall that a Markov chain is *irreducible* when its underlying graph is strongly connected, and that it is *aperiodic* when the gcd of its cycles is equal to 1. Most useful theorems are stated for Markov chains that are either irreducible, or both irreducible and aperiodic. Since the chain (M_p, π_0) is not irreducible, we propose to "approximate" it with an irreducible and aperiodic one. This new Markov chain produces reversed LS-extensions where some of the types can be wrong, for a limited number of positions at the beginning. However, we will see that it does not change significantly the number of LMS-factors, the statistic we are interested in.

3.3   An Irreducible and Aperiodic Markov Chain

Let $\underline{\mathrm{LS}}(A)$ denote the alphabet $\mathrm{LS}(A)$ restricted to the recurrent states of M_p: $\underline{\mathrm{LS}}(A) = \mathrm{LS}(A) \setminus \{\$S, a_1 L, a_k S\}$. We first formalize the notion of LS-extension with errors.

Definition 4. *Let u be a word of A^n, with $n \geq 1$, and let $w = \mathbf{Ext}(u)$ be the LS-extension of u. The pseudo LS-extension $\underline{\mathbf{Ext}}(u)$ of u is the word $v \in \underline{\mathrm{LS}}(A)^n$ defined by $v_i = a_1 S$ if $w_j = a_1 L$ for all $j \in \{i, \ldots n-1\}$, and $v_i = w_i$ otherwise.*

The pseudo LS-extension of u is therefore obtained from the LS-extension w of u by first removing the last character $\$S$, and then by changing the (possibly empty) sequence of $a_1 L$'s at the end into a sequence of $a_1 S$'s. For instance, if $u = a_3 a_1 a_2 a_1 a_1 a_1$, then we have $\mathbf{Ext}(u) = a_3 L\, a_1 S\, a_2 L\, a_1 L\, a_1 L\, a_1 L\, \$S$ and $\underline{\mathbf{Ext}}(u) = a_3 L\, a_1 S\, a_2 L\, a_1 S\, a_1 S\, a_1 S$.

Lemma 2. *Let $u \in A^n$, with $n \geq 1$. If u contains at least two different letters and ends with the letter a_1, then $\mathbf{Ext}(u)$ and $\underline{\mathbf{Ext}}(u)$ have the same number of LMS-positions. Otherwise, there is exactly one more LMS-position in $\mathbf{Ext}(u)$.*

Let $\underline{M}_p$ denote the restriction of the matrix M_p to $[0,1]^{\underline{\mathrm{LS}}(A) \times \underline{\mathrm{LS}}(A)}$. This defines a stochastic matrix, since by Lemma 1, the states of $\underline{\mathrm{LS}}(A)$ form a stable subset. By construction, $\underline{M}_p$ is irreducible. It is also aperiodic, as there is a loop on every vertex of $\underline{M}_p$. Let $\underline{\pi}_0$ be the probability row vector on $\underline{\mathrm{LS}}(A)$ defined for every $\alpha \in A \setminus \{a_1\}$ by $\underline{\pi}(\alpha L) = p_\alpha$ and $\underline{\pi}(\alpha S) = 0$, and by $\underline{\pi}(a_1 S) = p_{a_1}$. We now restate Proposition 1 using the Markov chain $(\underline{M}_p, \underline{\pi}_0)$.

Proposition 2. *Let A be a totally ordered alphabet, with at least two letters, and let p be a probability on A such that for every $a \in A$, $p_a > 0$. If u is a non-empty word on $\underline{\mathrm{LS}}(A)$ such that $\mathbb{P}_{\underline{M}_p, \underline{\pi}_0}(u) \neq 0$, then the reverse of u is the pseudo LS-extension of a word v of A^* and $\mathbb{P}_{\underline{M}_p, \underline{\pi}_0}(u) = \mathbb{P}_p(v)$.*

Recall that a *stationary vector* of a Markov chain (M, π_0) is a probability row vector π that satisfies the equation $\pi \times M = \pi$. If the chain is irreducible and aperiodic, a classical theorem [10] states that there exists a unique stationary vector. Moreover, after t steps, the probability that we are on a given state q is $\pi(q) + \mathcal{O}(\lambda^t)$, for some $\lambda \in (0, 1)$ and for any choice of π_0.

For every $a \in A$, let $p_{<a} = \sum_{\alpha < a} p_\alpha$ et $p_{>a} = \sum_{\alpha > a} p_\alpha$. The following theorem gives an explicit expression for the stationary vector of $\underline{M}_p$.

Theorem 1. *Let A be a totally ordered alphabet, with at least two letters, and let p be a probability on A such that for every $a \in A$, $p_a > 0$. The unique stationary vector of $\underline{M}_p$ is the vector $\underline{\pi}$ defined on $\underline{LS}(A)$ by*

$$\underline{\pi}(\alpha S) = \frac{p_\alpha \, p_{>\alpha}}{1 - p_\alpha} \qquad and \qquad \underline{\pi}(\alpha L) = \frac{p_\alpha \, p_{<\alpha}}{1 - p_\alpha}.$$

3.4 Main Statements

Using Theorem 1 and the classical Ergodic Theorem for Markov chains (Theorem 4.16 page 58 of [6]), we get a precise estimation of the number of LMS-factors, which is also the number of LMS-positions, in a random word for the memoryless distribution of probability p. It is obtained by analyzing the number of LMS-positions in $\underline{Ext}(u)$. Indeed, by Lemma 2, counting the number of LMS-positions in u is almost the same as counting the number of LMS-positions in $\underline{Ext}(u)$.

Theorem 2. *Let A be a totally ordered alphabet, with at least two letters, and let p be a probability on A such that for every $a \in A$, $p_a > 0$. Let F_n be the random variable that counts the number of LMS-factors in a random word of length n, generated by the memoryless source of probability p. There exists a sequence $(\varepsilon_n)_{n \geq 0}$ that tends to 0 such that:*

$$\mathbb{P}_p \left(\left| \frac{1}{n} F_n - \gamma_p \right| > \varepsilon_n \right) \xrightarrow[n \to \infty]{} 0, \ with \ \gamma_p = \sum_{a \in A} \frac{p_a}{1 - p_a} p_{>a}^2. \tag{1}$$

Corollary 1. *When the input of the IS-algorithm is a random word of length n generated by the memoryless source of probability p, the expected reduction ratio tends to γ_p.*

Remark 2. The statement of Theorem 2 is more precise than a result for the expectation of F_n (as in Corollary 1). For instance, Eq. (1) also implies that the random variable $\frac{1}{n} F_n$ is concentrated around its mean.

Remark 3. It is not completely obvious from its definition, but one can rewrite γ_p as $\sum_{a \in A} \frac{p_a}{1 - p_a} p_{<a}^2$. As a consequence, if p' is the reverse of p, that is, $p'_{a_i} = p_{a_{k+1-i}}$ for every $1 \leq i \leq k$, then $\gamma_p = \gamma_{p'}$.

We conclude this section by the analysis of some specific cases. First, we simplify the formula of γ_p for uniform distributions.

Lemma 3. *If p is the uniform probability on A, i.e., $p_a = \frac{1}{k}$ for every $a \in A$, then $\gamma_p = \frac{2k-1}{6k}$. In particular, $\gamma_p \to \frac{1}{3}$ as the size of the alphabet tends to infinity.*

Observe also that if p is not uniform, then γ_p may change when one reorders the probabilities values. For instance, if $A = \{a, b, c\}$, we obtain that $\gamma_p = \frac{13}{48}$ for $(p_a, p_b, p_c) = (\frac{1}{4}, \frac{1}{4}, \frac{1}{2})$ and $\gamma_{p'} = \frac{1}{4}$ for $(p'_a, p'_b, p'_c) = (\frac{1}{4}, \frac{1}{2}, \frac{1}{4})$.

For a binary alphabet $A = \{a, b\}$, we have $p_b = 1 - p_a$ and $\gamma_p = p_a(1 - p_a)$.

4   Markovian Sources

Let (N, ν_0) be a Markov chain on A. We say that it is a *complete Markov chain* when for every $\alpha, \beta \in A$, $N(\alpha, \beta) > 0$. A complete Markov chain is always irreducible and aperiodic. The construction of Sect. 3 can readily be extended to words that are generated *backward*, i.e., from right to left, using a complete Markov chain (N, ν_0). Let π_0 be the probabilistic vector of $[0, 1]^{\underline{\mathrm{LS}}(A)}$ such that $\pi_0(a_1 S) = \nu_0(a_1)$ and for every $\alpha \neq a_1$, $\pi_0(\alpha L) = \nu_0(\alpha)$ and $\pi(\alpha S) = 0$. Let $\underline{M}_N$ be the matrix of $[0, 1]^{\underline{\mathrm{LS}}(A) \times \underline{\mathrm{LS}}(A)}$ defined for every $\alpha, \beta \in A$ by[2]

$$\underline{M}_N(\alpha S, \beta S) = N(\alpha, \beta) \cdot [\![\beta \leq \alpha]\!]; \quad \underline{M}_N(\alpha S, \beta L) = N(\alpha, \beta) \cdot [\![\beta > \alpha]\!];$$
$$\underline{M}_N(\alpha L, \beta S) = N(\alpha, \beta) \cdot [\![\beta < \alpha]\!]; \quad \underline{M}_N(\alpha L, \beta L) = N(\alpha, \beta) \cdot [\![\beta \geq \alpha]\!].$$

Proposition 2 can be generalized to first order complete Markov chains the following way:

Proposition 3. *Let A be a totally ordered alphabet, with at least two letters, and let (N, ν_0) be a complete Markov chain on A. If u is a word on $\underline{\mathrm{LS}}(A)$ such that $\mathbb{P}_{\underline{M}_N, \pi_0}(u) \neq 0$, then the reverse of u is the pseudo LS-extension of a word v of A^* and $\mathbb{P}_{\underline{M}_N, \pi_0}(u) = \mathbb{P}_{N, \nu_0}(v)$.*

Though more complicated than in the memoryless case, the stationary vector of $\underline{M}_N$ can be calculated explicitly. This yields a computable formula for the typical number of LMS-factors:

Theorem 3. *Let A be a totally ordered alphabet, with at least two letters, and let (N, ν_0) be a complete Markov chain on A of stationary vector ν. Let F_n be the random variable that counts the number of LMS-factors in a random word of length n generated backward by (N, ν_0). There exists a sequence $(\varepsilon_n)_{n \geq 0}$ that tends to 0 such that*

$$\mathbb{P}_{N, \nu_0}\left(\left|\frac{1}{n}F_n - \gamma_N\right| > \varepsilon_n\right) \xrightarrow[n \to \infty]{} 0, \text{ with } \gamma_N = \sum_{a \in A} \pi(aS) \sum_{b > a} N(a, b),$$

where $\underline{\pi}$ is the stationary vector of $\underline{M}_N$, which satisfies

$$\underline{\pi}(\alpha S) = \frac{\sum_{\beta > \alpha} \nu(\beta) N(\beta, \alpha)}{1 - N(\alpha, \alpha)} \quad \text{and} \quad \underline{\pi}(\alpha L) = \frac{\sum_{\beta < \alpha} \nu(\beta) N(\beta, \alpha)}{1 - N(\alpha, \alpha)}.$$

As a consequence, the expected reduction ratio in the first recursive call of the IS-algorithm tends to γ_N, as n tends to infinity.

Remark 4. This can be generalized to Markov chains that are not complete Markov chains, but by lack of place, we cannot describe how it works in this extended abstract. The fact that the word is generated backward is usually not an issue: if the initial distribution is equal to the stationary distribution, then there exists a Markov chain that generates the words from left to right with the same probability (see [10]). It is natural to start with the stationary distribution, as it often coincides with the empirical frequencies of the letters.

---

[2] The formulas below hold when the extended letters are in $\underline{\mathrm{LS}}(A)$ only. For instance, $\alpha L = a_1 L$ is not part of the definition, since it is not in $\underline{\mathrm{LS}}(A)$.

5   Experiments and Conclusions

Though we provide a theoretical analysis of the IS-algorithm for classical distributions on words in this article, we thought it would be interesting to include some experiments on real data, even if we are not pretending to demonstrate anything with these few tests. These results are depicted in Fig. 2. It is also not our purpose to provide a statistical analysis of this information here, but we cannot help noticing that for the human chromosome 22, a Markov chain of order 1 seems to be an accurate model for analyzing the behavior of the IS-algorithm.[3]

File	$|A|$	size	red. ratio	uniform	memoryless	Markov
bible.txt	63	4047392	**0.3113**	0.3307	0.3230	0.3251
world192.txt	93	2408281	**0.2838**	0.3315	0.3256	0.2997
Chr_22.fa	4	35033745	**0.2717**	0.2917	0.2928	0.2715

Fig. 2. In these experiments we compare the real reduction ratio with the theoretical ratios obtained when approximating the distributions by one of the models proposed in this article. The first two files are from the Canterbury corpus [11], the last one is the human chromosome 22 [2]. The real reduction ratio of the first recursive call is indicated in the column "red. ratio". The three last columns were obtained after computing a model (either uniform, memoryless or Markovian) from the file. The different values are the γ's given by Lemma 3, Theorems 2 and 3. The Markov chains of the first two files are not complete, but our results still hold, as Theorem 3 can be generalized to irreducible and aperiodic chains (see Sect. 5).

The methodology presented in Sect. 4 can be extended to Markov chains (N, ν) that are only irreducible and aperiodic; the set of recurrent states may just be strictly included in $\underline{\mathrm{LS}}(A)$. It can also be extended to Markov chains of higher order, but the formulas become more and more complicated. Lets consider, say, a Markov chain of order 3 on $A = \{a, b, c, d\}$. Observe that in the recurrent part, a state adb is necessarily of type S since $b > d$. In fact, we always know the type of the last letter, except when the state is of the form $\alpha\alpha\alpha$. We need two different states for such words, one of type S and one of type L. Furthermore, $aaaL$ is transient and $dddS$ is not reachable. There are therefore $|A|^t + |A| - 2$ recurrent states in the Markov chain $\underline{M}_N$, where t is the order.

A continuation this work would be to analyze the whole behavior of the algorithm, when the reduction ratios of all the successive recursive calls are taken into account. This is technically challenging, as the letters of a given recursive call are the LMS-factors of the word at the previous stage. The precise analysis of other algorithms that compute suffix arrays is another natural direction for further investigations.

---

[3] This may be a consequence of the well-known fact that in a vertebrate genome, a C is very rarely followed by a G. This property is well captured by a Markov chain of order 1, but invisible to a memoryless model.

References

1. Baeza-Yates, R., Chávez, E., Crochemore, M. (eds.): CPM 2003. LNCS, vol. 2676. Springer, Heidelberg (2003)
2. Dunham, I., Hunt, A., Collins, J., Bruskiewich, R., Beare, D., Clamp, M., Smink, L., Ainscough, R., Almeida, J., Babbage, A., et al.: The DNA sequence of human chromosome 22. Nature **402**(6761), 489–495 (1999)
3. Baeten, J.C.M., Lenstra, J.K., Parrow, J., Woeginger, G.J. (eds.): ICALP 2003. LNCS, vol. 2719. Springer, Heidelberg (2003)
4. Kim, D.K., Sim, J.S., Park, H., Park, K.: Linear-time construction of suffix arrays. In: Baeza-Yates et al. [1], pp. 186–199
5. Ko, P., Aluru, S.: Space efficient linear time construction of suffix arrays. In: Baeza-Yates et al. [1], pp. 200–210
6. Levin, D.A., Peres, Y., Wilmer, E.L.: Markov chains and mixing times. American Mathematical Soc., Providence (2009)
7. Manber, U., Myers, E.W.: Suffix Arrays: A New Method for On-Line String Searches. SIAM J. Comput. **22**(5), 935–948 (1993)
8. Manber, U., Myers, G.: Suffix Arrays: A New Method for On-Line String Searches. In: Johnson, D.S. (eds.) Proceedings of the First Annual ACM-SIAM Symposium on Discrete Algorithms, San Francisco, California, pp. 319–327. SIAM, 22–24 January 1990
9. Nong, G., Zhang, S., Chan, W.H.: Two efficient algorithms for linear time suffix array construction. IEEE Trans. Comput. **60**(10), 1471–1484 (2011)
10. Norris, J.R.: Markov chains. Statistical and probabilistic mathematics. Cambridge University Press, Cambridge (1998)
11. Powell, M.: The Canterbury Corpus (2001). http://www.corpus.canterbury.ac.nz/. Accessed 25 April 2002
12. Puglisi, S.J., Smyth, W.F., Turpin, A.: A taxonomy of suffix array construction algorithms. ACM Comput. Surv. **39**(2), 1–31 (2007)
13. Vallée, B.: Dynamical sources in information theory: fundamental intervals and word prefixes. Algorithmica **29**(1–2), 262–306 (2001)

Encoding Nearest Larger Values

Patrick K. Nicholson[1](✉) and Rajeev Raman[2]

[1] Max-Planck-Institut Für Informatik, Saarbrücken, Germany
pnichols@mpi-inf.mpg.de
[2] University of Leicester, Leicester, UK

Abstract. In *nearest larger value (NLV)* problems, we are given an array $A[1..n]$ of numbers, and need to preprocess A to answer queries of the following form: given any index $i \in [1, n]$, return a "nearest" index j such that $A[j] > A[i]$. We consider the variant where the values in A are distinct, and we wish to return an index j such that $A[j] > A[i]$ and $|j - i|$ is minimized, the *nondirectional NLV (NNLV)* problem. We consider NNLV in the *encoding* model, where the array A is deleted after preprocessing, and note that NNLV encoding problem has an unexpectedly rich structure: the *effective entropy* (optimal space usage) of the problem depends crucially on details in the definition of the problem. Using a new *path-compressed* representation of binary trees, that may have other applications, we encode NNLV in $1.9n + o(n)$ bits, and answer queries in $O(1)$ time.

1 Introduction

Nearest Larger Value (NLV) problems have had a long and storied history. Given an array $A[1..n]$ of values, the objective is to preprocess A to answer queries of the general form: given an index i, report the index or indices nearest to i that contain values strictly larger that $A[i]$. Berkman et al. [3] studied the parallel pre-processing for this problem and noted a number of applications, such as parenthesis matching and triangulating monotone polygons. The connection to string algorithms for both the data structuring and the pre-processing variants of this problem is since well-established.

Since the definition of "nearest" is a bit ambiguous, we propose replacing it by one of the following options in order to fully specify the problem:

- *Unidirectionally nearest*: the solution is the index $j \in [1, i - 1]$ such that $A[j] > A[i]$ and $i - j$ is minimized.
- *Bidirectionally nearest*: the solution consists of indices $j_1 \in [1, i - 1]$ and $j_2 \in [i + 1, n]$ such that $A[j_k] > A[i]$ and $|i - j_k|$ is minimized for $k \in \{1, 2\}$.
- *Nondirectionally nearest*: the solution is the index j such that $A[j] > A[i]$ and $|i - j|$ is minimized. As far as we are aware, this formulation has not been considered before.

Furthermore, the data structuring problem has different characteristics depending on whether we consider the elements of A to be distinct (Berkman et al. considered the undirectional variant when all elements in A are distinct).

© Springer International Publishing Switzerland 2015
F. Cicalese et al. (Eds.): CPM 2015, LNCS 9133, pp. 385–395, 2015.
DOI: 10.1007/978-3-319-19929-0_33

We consider the problem in the *encoding* model, where once the data structure to answer queries has been created, the array A is deleted. Since it is not possible to reconstruct A from NLV queries on A, the *effective entropy* of NLV queries [9], the log of the number of distinguishable NLV configurations, is very low and an NLV encoding of A can be much smaller than A itself. The encoding variant has several applications in space-efficient data structures for string processing, in situations where the values in A are intrinsically uninteresting:

- The bidirectional NLV when A contains distinct values boils down essentially to encoding a Cartesian tree, through which route $2n+o(n)$-bit and $O(1)$-time data structures exist [4,7].
- The unidirectional NLV when A contains non-distinct values can be encoded in $2n + o(n)$ bits and queries answered in $O(1)$ time [8,10].
- The bidirectional NLV for the case where elements in A need not be distinct was first studied by Fischer [6]. His data structure occupies $\log_2(3 + 2\sqrt{2})n + o(n) \approx 2.544n + o(n)$ bits of space, and supports queries in $O(1)$ query time.

All of the above space bounds are tight to within lower-order terms.[1]

In this paper, we consider the nondirectionally nearest larger value (NNLV) problem, in the case that all elements in A are distinct. The above results already hint at the combinatorial complexity of NLV problems. However, the NNLV problem appears to be even richer, and the space bound appears not only to depend upon whether A is distinct or not, but also upon the specific tie-breaking rule to use if there are two equidistant nearest values to the query index i.

For instance, given a location i where there is a tie, we might always select the larger value to the right of location i to be its nearest larger value. We call this *rule I*. We give an illustration in the middle panel of Fig. 1 (on page 4). Alternative tie breaking rules might be: to select the smallest of the two larger values (*rule II*), or to select the larger of the two larger values (*rule III*). Interestingly, it turns out that the tie breaking rule is important for the space bound. That is, if we count the number of distinguishable configurations of the NNLV problem for the various tie breaking rules, then we get significantly different answers. We counted the number of distinguishable configurations, for problem instances of size $n \in [1, 12]$, and got the sequences presented in Table 1.

Table 1. Number of distinguishable configurations of nearest larger value problems with the three tiebreaking rules discussed.

n	1	2	3	4	5	6	7	8	9	10	11	12
rule I	1	2	5	14	40	116	341	1010	3009	9012	27087	81658
rule II	1	2	5	14	42	126	383	1178	3640	11316	35263	110376
rule III	1	2	5	12	32	88	248	702	1998	5696	16304	46718

---

[1] For the unidirectional NLV the bound is tight even when all values are distinct.

Unfortunately, none of the above sequences appears in the Online Encyclopedia of Integer Sequences[2]. Consider the sequence generated by some arbitrary tie breaking rule. If z_i is the i-th term in this sequence, then $\lim_{n \to \infty} \lg(z_n)/n$ is the constant factor in the asymptotic space bound required to store all the answers to the NNLV problem subject to that tiebreaking rule.

Our Contributions. Our main result is the following:

Theorem 1. *Let $A[1..n]$ be an array containing distinct numbers. The array A can be processed to obtain an encoding data structure that occupies $1.9n+o(n)$ bits of space, that can answer the query NNLV(A, i) in $O(1)$ time for any $i \in [1, n]$. Ties are resolved using rule I. At no point after preprocessing does the data structure require access to the array A.*

As mentioned before, the Cartesian tree (defined later) occupies $2n + o(n)$ bits and can solve NNLV queries. In Sect. 3 we describe a novel *path-compressed* representation of a binary tree that uses $2n + O(\lg n)$ bits (but supports no operations). To get the improved space bound of Theorem 1 we prove combinatorial properties of the NNLV problem relating to long chains in the Cartesian tree. These properties allow us to compress the Cartesian tree using the representation of Sect. 3, losing some information, but still retaining the ability to answer NNLV queries. The constant factor (1.9) comes from a numeric calculation bounding the worst case structure of chains in the Cartesian tree for our compression scheme (Sect. 4). In Sect. 4.1 we show how to support operations on the "lossy" Cartesian tree, thereby proving Theorem 1.

Finally, in Sect. 5, we prove a lower bound, via exhaustive search:

Theorem 2. *Any encoding data structure that can answer the query NNLV(A, i) for any $i \in [1, n]$ (breaking ties according to rule I) must occupy at least $1.3173n - \Theta(1)$ bits, for sufficiently large values of n.*

Other Related Work: Asano et al. [1] studied the time complexity of computing all nearest larger values in an array as well as higher dimensions, and mention applications to communication protocols. Asano and Kirkpatrick [2] considered sequential time-space tradeoffs for computing the nearest larger values of all elements in the array. Finally, Jo et al. [11] and Jayapaul et al. recently studied the nearest larger value problem in two dimensional arrays.

2 Cartesian Tree Review

Given a binary tree T, let $d(v)$ denote the degree (i.e., number of children) of node v, and $p(v)$ denote the parent of v. We define the rank $r(v)$ to be the inorder rank of the node v in the binary tree T. Define the *range* of a node v to be the range $[e_1(v), e_2(v)]$, where $e_1(v)$ (resp. $e_2(v)$) is the inorder rank of the leftmost (resp. rightmost) descendant of v.

---

[2] https://oeis.org/.

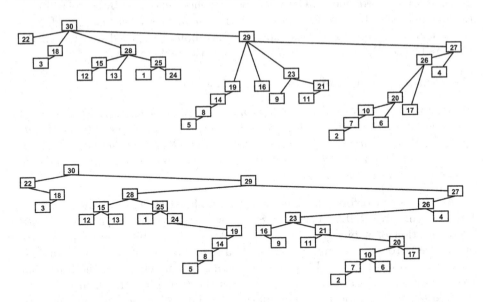

Fig. 1. Top: an array containing a permutation of $\{1, \ldots, 30\}$. Middle: The tree structure of the NNLV problem. Here the parent of a node represents its NNLV, breaking ties by selecting the element on the right (rule I). Bottom: The Cartesian tree.

Suppose we are given an array $A[1..n]$ which stores an n element permutation π, i.e., $A[i] = \pi(i)$. The Cartesian tree of $A[1..n]$ is the n node binary tree T such that the root v of T has rank $r(v) = \arg\max_i A[i]$. If $r(v) > 1$, then the left child of v is the Cartesian tree of $A[1..r(v) - 1]$, otherwise it has no left child. If $r(v) < n$ then the right child of v is the Cartesian tree of $A[r(v) + 1..n]$, otherwise it has no right child. We give an example of these definitions in Fig. 1.

We require the following technical lemma about Cartesian trees:

Lemma 1. *Consider a node v in a Cartesian tree having range $[e_1(v), e_2(v)]$. If $e_1(v) - 1 \geq 1$ then $A[e_1(v) - 1] > A[r(v)]$. Similarly, if $e_2(v) + 1 \leq n$ then $A[e_2(v) + 1] > A[r(v)]$.*

3   A Path Compressed Tree Representation

Consider an arbitrary binary tree T with n nodes. All binary trees we discuss are rooted. We next describe a path compressed encoding of such a tree that occupies no more than $2n + \Theta(\lg n)$ bits.

We identify all maximal chains $v_1, \ldots, v_\ell, v_{\ell+1}$ such that:

1. Either v_1 is the root of T, or $d(p(v_1)) = 2$;
2. $d(v_i) = 1$ for $i \in [1, \ell]$, and;
3. $d(v_{\ell+1}) \in \{0, 2\}$.

We refer to $v_{\ell+1}$ as the *terminal* of the chain. Iteratively, we remove each such maximal chain: i.e., the nodes $v_1, ..., v_\ell$ are removed from the tree. If v_1 was the root, then $v_{\ell+1}$ is set to be the new root. Otherwise, $v_{\ell+1}$ is set to be the left (resp. right) child of $p(v_1)$ iff v_1 was the left (resp. right) child of $p(v_1)$. We call the chain *left hanging* if $p(v_1)$ had v_1 as a left child, and *right hanging* otherwise. After removing all such maximal chains, the tree T' that remains is a full binary tree and has $n' \leq n$ nodes. Suppose that we have removed k nodes, for some $k \in [0, n-1]$, and so $n = n' + k$.

Suppose there are m maximal chains removed during the process just described. We now describe the representation of the original tree T.

– We store the tree T', which is a full binary tree and requires $n' + O(1)$ bits to represent.
– We store a bitvector B of length n'. Bit $B[i] = 1$ iff the node v, corresponding to the i-th node in an inorder traversal of T', is the terminal of a removed chain. This requires $\lceil \lg \binom{n'}{m} \rceil$ bits.
– Suppose we order the subset of nodes that are terminals by their inorder rank, and that v is the terminal ordered i-th. We refer to the chain having v as its terminal as C_i, and its length as c_i. We store a bitvector L of length k, which represents the lengths of each removed chain; i.e., the values $c_1, ..., c_m$. Let $p_i = \sum_{j=1}^{i} c_j$ for $i \in [1, m]$. Then $L[p_i] = 1$ for $i \in [1, m]$, and all other entries of L are 0. As L is a bit sequence of length k with m one bits, it can be stored using $\lceil \lg \binom{k}{m} \rceil$ bits.
– For each chain $C_i = \{v_1, ..., v_{c_i}\}$ having terminal node v_{c_i+1}, we store a bitvector Z_i of length c_i, in which $Z_i[j] = 0$ if v_{j+1} is the left child of v_j, and $Z_i[j] = 1$ otherwise. Let Z be the concatenation of each Z_i, $i \in [1, m]$ and is of length k. We store Z naively using k bits.

We call the above data structures, bitvectors B, L, Z and the tree T' the *path compressed* representation of T. Note that to decode this and recover the tree T, we require the value of n and n'. These can be stored using an additional $\Theta(\lg n)$ bits. By summing the above space costs, we get the following lemma.

Lemma 2. *The path compressed representation of T completely describes the combinatorial structure of T, and can be stored using $n' + \lg \binom{n'}{m} + \lg \binom{k}{m} + k + \Theta(\lg n) \leq 2n' + 2k + \Theta(\lg n) = 2n + \Theta(\lg n)$ bits.*

4   Encoding Nearest Larger Values

In this section we show how to use the path compressed tree representation to compress Cartesian trees—losing some information in the process—but still retaining the ability to answer NNLV queries. Our key observation is that chains in the Cartesian tree can be compressed to save space, as illustrated by the following lemma:

Lemma 3. *Consider the set of all possible chains with c_i deleted nodes in a path compressed representation of a Cartesian tree, excluding chains having nodes representing array elements $A[1]$ or $A[n]$. There are exactly c_i+1 combinatorially distinct chains with respect to answering nearest larger value queries, breaking ties according to rule I.*

Proof. Consider a chain with c_i deleted nodes, $\{v_1, ..., v_{c_i}\}$, where v_{c_i+1} is the terminal. Clearly, v_1 represents the maximum element in the chain, and either $r(v_j) = e_1(v_j)$ or $r(v_j) = e_2(v_j)$ for each $j \in [1, c_i]$. This follows because since v_j is in a chain it is either the left or right endpoint of the range $[e_1(v_j), e_2(v_j)]$. In turn, this implies that the range $[e_1(v_1), e_2(v_1)]$ has a *deleted prefix* and *deleted suffix* which in total contain the inorder ranks of the c_i deleted nodes.

The deleted nodes corresponding to this prefix (resp. suffix) appear contiguously in the array A, and form a decreasing (resp. increasing) run of values in A. Furthermore, by Lemma 1, and since $1, n \notin [e_1(v_1), e_2(v_1)]$ (by the assertion in the statement of the lemma), we can assert that both $A[e_1(v_1) - 1] > A[e_1(v_1)]$ and $A[e_2(v_1) + 1] > A[e_2(v_1)]$. Thus, for each k such that v_k is in the prefix we have that $A[e_1(v_k) - 1] > A[e_1(v_k)]$, and we can return the nearest larger value of $r(v_k) = e_1(v_k)$ to be $e_1(v_k) - 1$. Similarly, for each k such that v_k is in the suffix we have that $A[e_2(v_k) + 1] > A[e_2(v_k)]$, and return the nearest larger value of $r(v_k) = e_2(v_k)$ to be $e_2(v_k) + 1$.

This implies that, if we know the value c_i, then we additionally need only know how many nodes are in the prefix in order to determine the answer to a nearest larger value query for any index represented by a deleted node. There are at most $c_i + 1$ possible options: $\{0, ..., c_i\}$. Moreover, for an arbitrary index $i \in [1, n] \setminus [e_1(v_1), e_2(v_1)]$ the answer to a nearest larger value query cannot be in $[e_1(v_1), e_2(v_2)]$, since this range is sandwiched between larger values by Lemma 1. Finally, consider indices in the range $[e_1(v_{c_i+1}), e_2(v_{c_i+1})]$. Using the fact that $A[e_1(v_{c_i+1}) - 1]$ and $A[e_2(v_{c_i+1}) + 1]$ by are larger than all elements in $A[e_1(v_{c_i+1}), e_2(v_{c_i+1})]$ by Lemma 1, we can correctly answer queries for a position i in the subtree. First, we find the solution j within the subtree, and then return the nearest position to i of either j, $e_1(v_{c_i+1}) - 1$, or $e_2(v_{c_i+1}) + 1$, breaking ties according to rule I.

Recall that to recover a chain of c_i deleted nodes exactly required c_i bits in the path compressed tree representation. The previous lemma allows us to get away with $\lg(c_i+1)$ bits: an exponential improvement. Using the above lemma, we get the following upper bound for the NNLV problem (note that it does not allow queries to be performed efficiently).

Lemma 4. *The solutions to all nearest larger value queries can be encoded using $n' + \lg \binom{n'}{m} + \lg \binom{k}{m} + m\lg(\frac{k}{m} + 1) + \Theta(\lg n) \leq 1.9198n + \Theta(\lg n)$ bits.*

Proof (Sketch). We store the path compressed version of T, the Cartesian tree of A. However, we replace index Z, by an index Z' consisting of $\lceil \lg \prod_{i=1}^{m}(c_i + 1) \rceil$ bits. Z' represents, for each deleted chain—including those that contain nodes representing $A[1]$ and $A[n]$—the length of its deleted prefix. We explicitly store the answers to nearest larger value queries for $A[1]$ and $A[n]$.

The space bound for storing the data structures described is $n' + \lg \binom{n'}{m} + \lg \binom{k}{m} + \lg \prod_{i=1}^{m}(c_i + 1) + O(\lg n)$ bits. This is bounded by $n' + \lg \binom{n'}{m} + \lg \binom{k}{m} + m \lg(\frac{k}{m} + 1) + O(\lg n)$ bits using Jensen's inequality. Finally, a numerical calculation reveals that this expression is upper bounded by $1.9198n + \Theta(\lg n)$ bits.

Using a slightly more complicated analysis that bounds the space required to store L in terms of the zeroth-order empirical entropy of the sequence of chain lengths, we can improve the space bound (slightly) to $1.9n + o(n)$, resulting in Theorem 1. We defer details to the full version.

4.1 Supporting Queries

Until now we have only discussed space bounds for encoding NNLV queries, and have made no effort to actually answer them efficiently. In this section we discuss how to support NNLV queries in $O(1)$ time.

In the previous section we showed how to encode a Cartesian tree in a lossy way (losing information about the structure of chains in the tree). Thus, we can view the encoding algorithm, given an input Cartesian tree T, as mapping it to a new tree T_0, in which chains follow a path through a descending run in the prefix, then an ascending run in the suffix, and finally end at a terminal. We call T_0 the *lossy Cartesian tree* in this section. We wish to support the following tree operations on the lossy Cartesian tree T_0:

1. is_chain_prefix(i) (resp. is_chain_suffix(i)): given i, return whether the node with inorder number i in T_0 is within the prefix (resp. suffix) of a chain. To clarify what we mean by prefix or suffix, refer to Lemma 3.
2. select_inorder(i): return the node u in T_0 having inorder number i.
3. subtree_size(u): Return the size of the subtree rooted at node u in T_0.
4. left(u) (resp. right(u)): return the left (resp. right) child of node u in T_0.

Given the above operations on T_0, we can answer NNLV queries as in Algorithm 1. Correctness of the algorithm follows from the fact that the root of a subtree in T_0 is the largest value in a Cartesian tree, and Lemma 1.

Mini-micro Decomposition. All that remains is to show that we can support the operations listed above on the tree T_0. The problem is that we only have space available to store a path compressed version of T_0. Thus, we require a technical modification of the mini-micro tree decomposition presented by Farzan and Munro [5] which can be stated as follows:

Lemma 5 (Theorem 1 [5]). *For any parameter $\alpha > 1$, a tree with n nodes can be decomposed into $\Theta(\frac{n}{\alpha})$ subtrees of size at most 2α, which are pairwise disjoint aside from their roots. With the exception of edges branching from the root of a subtree, there is at most one edge from a non-root node in a subtree to a node outside the subtree.*

Algorithm 1. Computing NNLV(A, i).

1: **if** $i = 1$ or $i = n$ **then**
2:     **return** explicitly stored answer for $A[1]$ or $A[n]$.
3: **else if** is_chain_prefix(i) **then**
4:     **return** $i - 1$
5: **else if** is_chain_suffix(i) **then**
6:     **return** $i + 1$
7: **else**
8:     $\ell \leftarrow$ subtree_size(left(select_inorder(i)))
9:     $r \leftarrow$ subtree_size(right(select_inorder(i)))
10:     **if** $\ell < r$ and $i - \ell - 1 \geq 1$ **then**
11:         **return** $i - \ell - 1$
12:     **else if** $i + r + 1 \leq n$ **then**
13:         **return** $i + r + 1$
14:     **else**
15:         **return** $A[i]$ is the maximum (it has no NNLV)
16:     **end if**
17: **end if**

The binary tree structure of Davoodi et al. [4] essentially applies Lemma 5 twice to the input tree, getting a set of $O(\frac{n}{\lg^2 n})$ mini-trees of size $O(\lg^2 n)$ and $O(\frac{n}{\lg n})$ micro-trees of size $\lceil \frac{\lg n}{\gamma} \rceil$, for some $\gamma \geq 8$. Since a rooted binary tree with g nodes can be represented using $2g$ bits, we can store a *fingerprint* of size at most $\lceil \frac{2 \lg n}{\gamma} \rceil$ bits for each micro-tree. We can then perform tree operations by using these fingerprints to index into using a universal table of size $o(n)$. Overall, the space is bounded by the sum of the sizes of the fingerprints, and totals $2n + o(n)$. Their representation supports a large number of operations, which includes select_inorder, subtree_size, left, right.

The main idea of our approach is to take the lossy Cartesian tree T_0, and to decompose it using Lemma 5. We then adjust the decomposition to, roughly speaking, ensure that chains do not cross subtree boundaries. The following technical lemma captures this intution:

Lemma 6. *For any parameter $\alpha > 1$, a tree with n nodes can be decomposed into $\Theta(\frac{n}{\alpha})$ subtrees which are pairwise disjoint aside from their roots. Furthermore, we have the following properties for the subtrees:*

1. *All nodes in a chain, except possibly the terminal, are contained in the same subtree.*
2. *If a subtree contains a node of degree two, then it has size at most 2α.*
3. *Excepting edges branching from the root of a subtree, there is at most one edge from a non-root node in a subtree to a node outside the subtree.*

We apply the Lemma 6 twice to T_0. The first application has parameter $\alpha = \lceil \lg^2 n \rceil$, which gives us a set of subtrees. We change and extend the definitions of mini-trees and micro-trees slightly from the previous papers. Subtrees which have at least one degree two node are referred to as mini-trees, and are otherwise referred to as *mini-chains*. The second application of the lemma is done to each mini-tree separately with $\alpha = \lceil \frac{\lg n}{\beta} \rceil$, for $\beta \geq 16$. Similarly, the resultant subtrees are called micro-trees if they contain a degree two node, and *micro-chains* otherwise.

Next, we apply path compression to each micro-tree, micro-chain, and mini-chain. We note that micro-chains and mini-chains end up as a single node after path compression, and have degree 1. Furthermore, prior to path compression, micro-chains were chains of length at least $\lceil \frac{\lg n}{\beta} \rceil$ and at most $\Theta(\lg^2 n)$, and mini-chains were chains of length at least $\Theta(\lg^2 n)$. For micro-trees, each node (after path compression) has either degree two or zero. Each micro-tree which contains g degree two nodes can therefore be represented using g bits, rather than $2g$ bits. Recall that we used n' to represent the number of nodes in the path compressed lossy Cartesian tree. If we sum over all the micro-trees there are $\frac{n'-1}{2}$ degree two nodes in total after path compression. This means the sum of the sizes of the fingerprints for all of the micro-trees can be stored in $n' + o(n)$ bits. One technical issue is that we must mark the branching edge of each micro-tree, which can be done using an additional $\Theta(\lg \lg n)$ bits per micro-tree. Thus, this additional cost is $O(\frac{n \lg \lg n}{\lg n})$ when summed over all micro-trees. Note that the number of micro- and mini-chains is bounded by $\Theta(\frac{n}{\lg n})$, so we can also afford to mark these using a bit vector, indicating whether they are a micro-chain or a mini-chain. Recalling the encoding from the previous section, this path compressed tree we have constructed here is almost (but not quite) the path compressed version of the lossy Cartesian tree T_0: it has $\Theta(\frac{n}{\lg n})$ additional degree one nodes, but nonetheless occupies $n' + o(n)$ bits.

We call the fingerprints of the micro-trees the *path compressed fingerprints*. In the full version, we show that for an arbitrary micro-tree M we can use the path compressed fingerprint to recover the fingerprint corresponding to M the original (not path compressed) tree T_0. We have the following lemma:

Lemma 7. *We can recover the fingerprint of any micro-tree in T_0 in $O(1)$ time, using space:*

$$n' + \lg \binom{n'}{m} + \sum_{i=1}^{\sigma} \left(m_i \lg \frac{m(i+1)}{m_i} \right) + O\left(\frac{n \lg \lg n}{\lg n} \right) \quad bits.$$

Using the previous lemma, it is not hard to prove Theorem 1. The main idea is to construct the data structure of Davoodi et al. [4] using Lemma 7 as an oracle to access the fingerprints of micro-trees. This allows us to support nearly all the required query operations, except is_chain_prefix and is_chain_suffix. These two operations can be supported by considering the cases of micro-trees, micro-chains, and mini-chains separately.

5   Lower Bound

The main idea of the lower bound is to show that for a given n, there are many configurations of A that can be distinguished by NNLV queries. To do this, we define a *restricted* NNLV problem (RNNLV). The restricted problem is like the original NNLV problem on an array $A[1..n]$, except we pretend that the array has entries $A[0] = \infty$ and $A[n+1] = \infty$. Thus, an answer to the restricted NNLV query (RNNLV(A, i)) is either NNLV(A, i), 0, or $n + 1$: we choose the nearest of these three possibilities, breaking ties using rule I. This restricts the solution space, but will allow us to lower bound the unrestricted problem.

For an n element array, we use R_n to denote the number of different solutions to RNNLV, and S_n to denote the number of solutions to NNLV, both subject to tie breaking rule I. We computed the following sequences:

Table 2. Number of solutions to RNNLV problem (rule I).

n	1	2	3	4	5	6	7	8	9	10	11	12	13	14
R_n	1	2	4	9	22	55	142	378	1015	2768	7662	21340	59962	169961
S_n	1	2	5	14	40	116	341	1010	3009	9012	27087	81658	246841	747728

Next we discuss how to use Table 2 to derive a lower bound. Consider an array of length n, for n sufficiently large. Without loss of generality, we assume that a parameter $\beta \geq 1$ divides $n - 2$ and that $\frac{n-2}{\beta}$ is odd. Let D_i denote the i-th odd block, and E_i denote the i-th even block. Locations $A[1]$ and $A[n]$ are assigned values $n - 1$ and n, respectively. Odd block D_i is assigned values $[(i - 1)\beta + 1, i\beta]$, and can be arranged in one of R_β configurations, to form an instance of the RNNLV problem. Suppose there are Δ odd blocks. Even block E_i will be assigned values from $[(\Delta + i - 1)\beta + 1, (\Delta + i)\beta]$, and arranged in one of the S_β configurations of the NNLV problem.

Our claim is that each even (resp. odd) block can be assigned any of the S_β (resp. R_β) possible configurations, without interference from other blocks. To see this, consider that for each even block we have assigned values so that—with the exception of the maximum element—the nearest larger value to all elements must be within the same block. This follows since the adjacent odd blocks contain strictly smaller values than those in any even block. Moreover, for odd blocks, the values immediately to the left and right of the block are strictly larger than any values in the block. Thus, we can force the global solution to the NNLV problem on the entire array into at least $(S_\beta R_\beta)^{\frac{n-2}{2\beta}}$ distinct structures. This implies that $\lg S_n$ is at least $\frac{(n-2)}{2\beta} \lg(S_\beta R_\beta)$: selecting $\beta = 14$ yields the lower bound of Theorem 2.

6   Conclusions

We have introduced the encoding NNLV problem, and have noted its combinatorial richness. Using a novel path-compressed representation of Cartesian trees, we gave a space-efficient NNLV encoding that supports queries in $O(1)$ time. Determining the effective entropy of NNLV, and to consider the other NNLV variants, is an open problem, as is extending the path-compressed Cartesian tree representation of Sect. 4.1 to general binary trees. Finding ways to apply NNLV encodings to compressed suffix trees, as Fischer [6] did for his bidirectional NLV encoding, would also be interesting.

References

1. Asano, T., Bereg, S., Kirkpatrick, D.: Finding nearest larger neighbors. In: Albers, S., Alt, H., Näher, S. (eds.) Efficient Algorithms. LNCS, vol. 5760, pp. 249–260. Springer, Heidelberg (2009). http://dx.doi.org/10.1007/978-3-642-03456-5_17
2. Asano, T., Kirkpatrick, D.: Time-space tradeoffs for all-nearest-larger-neighbors problems. In: Dehne, F., Solis-Oba, R., Sack, J.-R. (eds.) WADS 2013. LNCS, vol. 8037, pp. 61–72. Springer, Heidelberg (2013). http://dx.doi.org/10.1007/978-3-642-40104-6_6
3. Berkman, O., Schieber, B., Vishkin, U.: Optimal doubly logarithmic parallel algorithms based on finding all nearest smaller values. J. Algorithms **14**(3), 344–370 (1993). http://dx.doi.org/10.1006/jagm.1993.1018
4. Davoodi, P., Navarro, G., Raman, R., Rao, S.: Encoding range minima and range top-2 queries. Philos. Trans. R. Soc. A **372**(2016), 1471–2962 (2014)
5. Farzan, A., Munro, J.J.: A uniform paradigm to succinctly encode various families of trees. Algorithmica **68**(1), 16–40 (2014). http://dx.doi.org/10.1007/s00453-012-9664-0
6. Fischer, J.: Combined data structure for previous- and next-smaller-values. Theor. Comput. Sci. **412**(22), 2451–2456 (2011). http://dx.doi.org/10.1016/j.tcs.2011.01.036
7. Fischer, J., Heun, V.: Space-efficient preprocessing schemes for range minimum queries on static arrays. SIAM J. Comput. **40**(2), 465–492 (2011)
8. Fischer, J., Mäkinen, V., Navarro, G.: Faster entropy-bounded compressed suffix trees. Theor. Comput. Sci. **410**(51), 5354–5364 (2009)
9. Golin, M., Iacono, J., Krizanc, D., Raman, R., Rao, S.S.: Encoding 2d range maximum queries. In: Asano, T., Nakano, S., Okamoto, Y., Watanabe, O. (eds.) ISAAC 2011. LNCS, vol. 7074, pp. 180–189. Springer, Heidelberg (2011)
10. Jayapaul, V., Jo, S., Raman, V., Satti, S.R.: Space efficient data structures for nearest larger neighbor. In: Proceedings of IWOCA 2014 (2014, to appear)
11. Jo, S., Raman, R., Rao Satti, S.: Compact encodings and indexes for the nearest larger neighbor problem. In: Rahman, M.S., Tomita, E. (eds.) WALCOM 2015. LNCS, vol. 8973, pp. 53–64. Springer, Heidelberg (2015). http://dx.doi.org/10.1007/978-3-319-15612-5_6

Sorting by Cuts, Joins and Whole Chromosome Duplications

Ron Zeira[✉] and Ron Shamir

Tel-Aviv University, 69978 Tel-Aviv, Israel
ronzeira@post.tau.ac.il, rshamir@tau.ac.il

Abstract. Genome rearrangement problems have been extensively studied due to their importance in biology. Most studied models assumed a single copy per gene. However, in reality duplicated genes are common, most notably in cancer. Here we make a step towards handling duplicated genes by considering a model that allows the atomic operations of cut, join and whole chromosome duplication. Given two linear genomes, Γ with one copy per gene, and Δ with two copies per gene, we give a linear time algorithm for computing a shortest sequence of operations transforming Γ into Δ such that all intermediate genomes are linear. We also show that computing an optimal sequence with fewest duplications is NP-hard.

Keywords: SCJ · Genome rearrangements · Computational genomics

1 Introduction

Genome organization evolves over time by undergoing rearrangement operations. Finding a shortest sequence of operations (also called a sorting scenario) between two genomes is the focus of the field of *genome rearrangements*. Such problems were studied extensively over the last two decades, due to their importance in evolution [13].

The combinatorial problems in genome rearrangements depend on the allowed operations. Hannenhalli and Pevzner showed in their seminal work that finding the minimal number of inversions that transform one signed genome into another is polynomial [15]. Many other models were studied later, allowing one or several types of operations [7–9, 11, 14, 15, 17, 18].

The *double cut and join* (*DCJ*) operation [27] models reversals, transpositions, translocations, fusions, fissions and block-interchanges as variations of one basic operation. A DCJ operation cuts the genome in two places, producing four open ends, and rejoins them in two new pairs. Finding the DCJ distance between two gene permutations can be done in linear time [4]. The *single cut or join* (*SCJ*) model [12] further simplifies the model and allows polynomial solutions to some rearrangement problems that are NP-hard under most formulations. An SCJ operation either cuts a chromosome or joins two chromosome ends. This simple model gives good results in real biological applications [5].

© Springer International Publishing Switzerland 2015
F. Cicalese et al. (Eds.): CPM 2015, LNCS 9133, pp. 396–409, 2015.
DOI: 10.1007/978-3-319-19929-0_34

Models of genomes that assume a single copy of each gene are too restrictive for many real biological problems. Duplications are frequent in cancer genomes, especially in oncogenic regions [3]. Most plant genomes contain large duplicated segments [6]. A major evolutionary event is *whole genome duplication*, wherein all chromosomes are duplicated [21].

In spite of their importance, models that allow duplications as rearrangement operations have not been the subject of extensive research to date. Ozery-Flato and Shamir [19] considered a model that includes certain duplications, deletions and SCJ operations. Under some simplifying assumptions, they provided a 3-approximation algorithm that performed well on cancer genomes. Bader [1,2] provided a heuristic for sorting by DCJs, duplications and deletions. Shao *et al.* [23] studied sorting genomes using DCJs and segmental duplications and provided an algorithm to improve an initial sorting scenario. The majority of extant models for genomes with multiple gene copies result in NP-hard problems [21, 22, 24, 25].

In this paper, we present a model that allows the operations cut, join and whole chromosome duplication. We call it the *SCJD model*. Given two linear genomes, Γ with one copy per gene, and Δ with two copies per gene, we give a linear time algorithm for computing a shortest sequence of operations transforming Γ into Δ, where all intermediate genomes must be linear too. We provide a closed form formula for that sequence length. In addition, we show that there is an optimal sequence in which all duplications are consecutive.

While cuts or joins are local events, a duplication of an entire chromosome is a more "drastic" event. We show that our algorithm actually gives an optimal scenario with a maximum number of duplications. On the other hand, we prove that finding a "conservative" optimal SCJD scenario with fewest duplications is NP-hard.

The structure of this paper is as follows. We give computational background in Sect. 2. In Sect. 3 we present the SCJD model. Section 4 gives the algorithm for the SCJD sorting problem and Sect. 5 shows the NP-hardness result. Finally, in Sect. 6, we present a brief discussion and suggest future directions. Due to lack of space, some proofs were omitted.

2   Preliminaries

Genome Representation.   We use the following standard terminology in genome rearrangements [4]. The basic entities are *genes*, denoted a, b, c etc. Gene a has *extremities:* a *head* a_h and a *tail* a_t. Gene a is assumed oriented from its tail to its head and is *positively oriented* if a_t is to the left of a_h. A *negatively oriented* gene a is denoted by $-a$. A *chromosome* is a sequence of oriented genes, e.g., $C = ab-c-d$. An *adjacency* in a chromosome is a consecutive pair of extremities from distinct neighboring genes. e.g., the adjacencies in C above are: $\{a_h, b_t\}$, $\{b_h, c_h\}, \{c_t, d_h\}$. A *telomere* is an extremity that is not adjacent to any other gene, corresponding to the end of a chromosome, e.g., $\{a_t\}, \{d_t\}$ in C. Hence, a chromosome can be equivalently represented by its set of adjacencies, where the telomeres are implicit. Note that the set of adjacencies defining a chromosome is identical to that of the reverse chromosome, where order and orientation of genes are inverted (the reverse of C is $-C = dc-b-a$). Hence, a chromosome and its reverse are equivalent.

A *genome* over gene set $\mathcal{G}$ is a collection of chromosomes. We assume for now that each gene appears once, e.g. $\Gamma = \{ab, c-d\}$. Equivalently, it can be defined by a set of adjacencies such that for each gene in $\mathcal{G}$, each extremity appears at most once. Hence $\Gamma = \{\{a_h, b_t\}, \{c_h, d_h\}\}$. The *size* of a genome Π, denoted $|\Pi|$, is the number of adjacencies in it. A chromosome is called *linear* if it starts and ends with a telomere, and *circular* if it does not contain any telomere, e.g. $D = \{\{a_h, b_t\}, \{b_h, a_t\}\}$. For a sequence of genes S, denote by S and (S) the corresponding linear and circular chromosome respectively. For example, the linear chromosome $a-b$ is defined by the set of adjacencies $\{\{a_h, b_h\}\}$ and the circular chromosome $(a-b)$ is defined by the set $\{\{a_h, b_h\}, \{b_t, a_t\}\}$. A genome is called *linear* if all its chromosomes are linear.

A gene that has several copies in the genome is called *duplicated*. We *label* different copies of the same gene by superscripts, e.g., copies a^1 and a^2 of gene a. A *duplicated genome* has exactly two copies of each gene. A genome with a single copy of each gene is called *ordinary*. The duplication of an ordinary genome Π creates a special kind of genome [26]: Each gene and each adjacency in Π is doubled, producing the genome $\Pi \oplus \Pi$. Note that in $\Pi \oplus \Pi$ the two copies of each gene are unlabeled. The set of all possible labeled genomes corresponding to $\Pi \oplus \Pi$ is denoted by 2Π. A genome $\Sigma \in 2\Pi$ is called a *perfectly duplicated genome*. Hence for Γ above, $\Gamma \oplus \Gamma = \{ab, ab, c-d, c-d\}$ and $\Sigma = \{\{a_h^2, b_t^2\}, \{c_h^2, d_h^1\}, \{a_h^1, b_t^1\}, \{c_h^1, d_h^2\}\} \in 2\Gamma$.

SCJ distance. A *cut* operation takes an adjacency $\{x, y\}$ and breaks it into two telomeres $\{x\}$ and $\{y\}$. The reverse operation, called a *join*, combines two telomeres $\{x\}$ and $\{y\}$ into an adjacency $\{x, y\}$. A *single-cut-or-join* (SCJ) operation is either a cut or a join [12]. Given two ordinary genomes Π and Σ on the same gene set, a sequence of SCJ operations that transforms Π into Σ is called a *sorting scenario*. The *SCJ distance*, denoted by $d_{SCJ}(\Pi, \Sigma)$, is the length of a shortest sorting scenario between Π and Σ. Feijão and Meidanis give the following solution for the SCJ distance:

Theorem 1. *[12]* $d_{SCJ}(\Pi, \Sigma) = |\Pi \setminus \Sigma| + |\Sigma \setminus \Pi| = |\Pi| + |\Sigma| - 2|\Pi \cap \Sigma|$. $\Pi \setminus \Sigma$ *defines the set of cuts and* $\Sigma \setminus \Pi$ *defines the set of joins in an optimal sorting scenario.*

Double Distance. The *SCJ double distance* between an ordinary genome Γ and a duplicated genome Δ is defined as

$$dd_{SCJ}(\Gamma, \Delta) \equiv \min_{\Sigma \in 2\Gamma} d_{SCJ}(\Sigma, \Delta) \qquad (1)$$

Hence, in the *double distance problem* one seeks a labeling of each gene copy in a perfectly duplicated genome $\Sigma \in 2\Gamma$ that minimizes the SCJ distance to Δ.

For a genome Σ and an adjacency $\alpha = \{x, y\}$, let Σ_α be the set of all adjacencies of the form $\{x^i, y^j\}$ in Σ. Hence $|\Sigma_\alpha|$ can be 0, 1 or 2 if Σ is duplicated, and 0 or 1 if Σ is ordinary. Let $A = \{\alpha = \{x, y\} | x \neq y\}$ be the set of all possible adjacencies with extremities belonging to distinct genes. A solution to the double distance problem is given by the following theorem:

Theorem 2. *[12] The SCJ double distance between an ordinary genome Γ and a duplicated genome Δ is*

$$dd_{SCJ}(\Gamma, \Delta) = |\Delta| + 2 \sum_{\alpha \in A} |\Gamma_\alpha|(1 - |\Delta_\alpha|).$$

A perfectly duplicated genome $\Sigma \in 2\Gamma$ realizing the distance is obtained by taking, for each adjacency $\alpha = \{x, y\} \in \Gamma$: (1) the labeled adjacencies of Δ_α, and (2) adjacencies $\{x^i, y^j\}$ with arbitrary labeling that do not conflict with (1) or among themselves.

3   The SCJD Model

In this section we generalize the SCJ model to allow duplications.

A *duplication* operation on a genome Π takes a linear chromosome C in Π and produces a new genome Π' with an additional copy of the chromosome. For example, if $\Pi = \{abcd, efg\}$ then a duplication of the first chromosome will give $\Pi' = \{abcd, abcd, efg\}$. An *SCJD operation* is either an SCJ or a duplication.

Given two linear genomes on the same gene set of size n, an ordinary one Γ and a duplicated one Δ, a sequence of SCJD operations that transforms Γ into Δ is called an *SCJD sorting scenario*. The *SCJD distance*, denoted by $d_{SCJD}(\Gamma, \Delta)$, is the number of operations in a shortest SCJD sorting scenario between Γ and Δ.

Since we focus on linear genomes we will assume from now on that all chromosomes, including intermediate ones, are linear unless specified otherwise. The following simple lemma shows that this can be satisfied when using only SCJ operations:

Lemma 1. *A sequence of SCJ operations transforming one linear genome into another linear genome can be reordered, producing another sequence with the same length, such that all intermediate genomes are linear.*

The examples below demonstrate SCJ double distances and SCJD sorting scenarios. For simplicity, we drop the braces around genomes from now on.

Example 1. $\Gamma = a$, $\Delta = a{-}a$; $dd_{SCJ}(\Gamma, \Delta) = 1$; $d_{SCJD}(\Gamma, \Delta) = 2$:

$$\Gamma \xrightarrow[dup]{} a, a \xrightarrow[join]{} \Delta$$

Example 2. $\Gamma = ab$, $\Delta = ab, ab$; $dd_{SCJ}(\Gamma, \Delta) = 0$; $d_{SCJD}(\Gamma, \Delta) = 1$:

$$\Gamma \xrightarrow[dup]{} \Delta$$

Example 3. $\Gamma = a, bc$, $\Delta = ab, abcc$; $dd_{SCJ}(\Gamma, \Delta) = 4$; $d_{SCJD}(\Gamma, \Delta) \leq 4$:

$$\Gamma \xrightarrow[join]{} abc \xrightarrow[dup]{} abc, abc \xrightarrow[cut]{} abc, ab, c \xrightarrow[join]{} \Delta$$

Example 4. $\Gamma = acb$, $\Delta = abab, cc$; $dd_{SCJ}(\Gamma, \Delta) = 8$; $d_{SCJD}(\Gamma, \Delta) \leq 7$:

$$\Gamma \xrightarrow[cut]{} a, cb \xrightarrow[cut]{} a, b, c \xrightarrow[join]{} ab, c \xrightarrow[dup]{} ab, ab, c \xrightarrow[dup]{} ab, ab, c, c \xrightarrow[join]{} abab, c, c \xrightarrow[join]{} \Delta$$

Let $\#_c \Pi$ be the number of linear chromosomes in genome Π. Let Γ be an ordinary linear genome and let Δ be a duplicated linear genome on the same gene set. A trivial upper bound for the SCJD distance between Γ and Δ is given by solving the double distance between Δ and Γ. This corresponds to first duplicating each chromosome in Γ and then computing the SCJ distance between Δ and $\Gamma \oplus \Gamma$. We get $d_{SCJD}(\Gamma, \Delta) \leq dd_{SCJ}(\Gamma, \Delta) + \#_c \Gamma$. However, Example 3 shows that this bound is not tight. It is tempting to guess that $dd_{SCJ}(\Gamma, \Delta) \leq d_{SCJD}(\Gamma, \Delta)$. Alas, Example 4 shows this conjecture is incorrect.

4 Computing the SCJD Distance

In this section we will solve the SCJD distance problem. The key idea is to show that there is an optimal scenario in which all the duplication operations are performed in sequence, one after the other. Having shown that, the sorting scenario between Γ and Δ can be presented as follows:

1. Transform Γ into another ordinary linear genome Γ' using only SCJ operations.
2. Duplicate all the chromosomes of Γ' resulting in a duplicated genome $\Gamma' \oplus \Gamma'$.
3. Solve the SCJ double distance problem between Γ' and Δ.

Let $O^* = o_1, \ldots, o_d$ be an optimal SCJD sorting scenario. Let $\Gamma_0 \equiv \Gamma$ and for every $1 \leq i \leq d$ let $\Gamma_i = o_i(\Gamma_{i-1})$ be the genome resulting from performing o_i on Γ_{i-1}. By definition, $\Gamma_d \equiv \Delta$. Let D_i be the set of duplicated genes in Γ_i. We have $D_0 = \emptyset$ and $D_d = \mathcal{G}$. Given a gene set $\mathcal{H}$, denote its extremity set by $\mathcal{E}_{\mathcal{H}} = \{a_t | a \in \mathcal{H}\} \cup \{a_h | a \in \mathcal{H}\}$.

Proposition 1. *In an optimal sorting scenario O^*, if o_i is a join operation acting on the telomeres x and y, then either $x, y \in \mathcal{E}_{D_i}$ or $x, y \notin \mathcal{E}_{D_i}$.*

Proof. Since o_i is not a duplication, we have $D_{i-1} = D_i$. Suppose by contradiction that $x \in \mathcal{E}_{D_i}$ but $y \notin \mathcal{E}_{D_i}$. Let o_j ($i < j$) be the first duplication such that $y \in \mathcal{E}_{D_j}$. The duplication operation must act on a chromosome in which all genes are not yet duplicated. Therefore, there is a cut operation o_k ($i < k < j$) that breaks the adjacency $\{x, y\}$ created by o_i.
  Let $O' = o'_1, \ldots, o'_{d-2} = o_1, \ldots, o_{i-1}, o_{i+1}, \ldots, o_{k-1}, o_{k+1}, \ldots, o_d$ be an alternative sorting sequence that results from removing o_i and o_k from O^*. Let $\Gamma'_0 \equiv \Gamma$, and denote $\Gamma'_l = o'_l(\Gamma'_{l-1})$. For every l with $1 \leq l \leq i - 1$, by definition, $o'_l = o_l$ and therefore $\Gamma'_l = \Gamma_l$.

We first show that for every l with $i \leq l \leq k-2$, $\Gamma'_l = \Gamma_{l+1} \setminus \{\{x,y\}\}$. Since o_i creates the adjacency $\{x,y\}$ we have that $\Gamma_i = \Gamma_{i-1} \cup \{\{x,y\}\}$. For every such l, $o'_l = o_{l+1}$ and since none of these operations creates a new copy of y we have that $\Gamma'_l = \Gamma_{l+1} \setminus \{\{x,y\}\}$.

Next, we show that for every l with $k-1 \leq l \leq d-2$, $\Gamma'_l = \Gamma_{l+2}$. From the previous result, and the fact that $\Gamma_k = \Gamma_{k-1} \setminus \{\{x,y\}\}$, we have $\Gamma'_{k-2} = \Gamma_k$. Now, for every such l, $o'_l = o_{l+2}$ and therefore $\Gamma'_l = \Gamma_{l+2}$.

We have established that O' is an SCJD sorting sequence of length $d-2$, contradicting the optimality of O^*. $\square$

Proposition 2. *In an optimal sorting scenario O^*, if o_i is a cut operation acting on the adjacency $\{x,y\}$, then either $x, y \in \mathcal{E}_{D_i}$ or $x, y \notin \mathcal{E}_{D_i}$.*

Corollary 1. *In an optimal sequence of SCJD operations, at the time of a cut or a join operation on the two extremities x and y, either the genes corresponding to both x and y have both already been duplicated or none of them have.* $\square$

Observe that a join operation in a sorting scenario is valid only if the two extremities it joins are not already part of any other adjacency. Similarly, a cut operation is valid only if the adjacency it breaks exists. A duplication operation is valid only if it duplicates a linear chromosome such that all its genes were not previously duplicated. A sorting scenario is valid if all its operations are valid.

Let $S = s_1, \ldots, s_m$ be a valid SCJD sorting scenario between Γ and Δ. We say the operation s_{i+1} can *preempt* the operation s_i if the sequence $S' = s_1, \ldots, s_{i+1}, s_i, \ldots, s_m$ is also a valid SCJD sorting scenario between Γ and Δ.

Proposition 3. *In a valid SCJD scenario S transforming Γ into Δ, if s_{i+1} is an SCJ operation acting on two extremities x, y that were not duplicated and s_i is a duplication, then s_{i+1} can preempt s_i.*

Proof. Suppose s_i duplicates the linear chromosome C and produces another copy of it C'. Since s_{i+1} operates on genes that are not duplicated yet, none of those genes belong to C or C'. Therefore, the sequence $s_1, \ldots, s_{i-1}, s_{i+1}$ is valid. Any operation that creates an adjacency or a telomere of C must precede s_i. Hence, $s_1, \ldots, s_{i-1}, s_{i+1}, s_i$ is valid. Finally, any s_j for $j > i+1$ that requires the results of s_i or s_{i+1} is still valid. Thus, $S' = s_1, \ldots, s_{i-1}, s_{i+1}, s_i, s_{i+2}, \ldots, s_m$ is a valid sequence.

To conclude the proof, we need to show that $\Gamma_{i+1} \equiv \Gamma'_{i+1}$. Indeed, s_{i+1} does not alter any of the adjacencies or telomeres of C or C', and therefore, $\Gamma_{i+1} = s_{i+1}(\Gamma_{i-1} \cup C') \equiv s_{i+1}(\Gamma_{i-1}) \cup C' = \Gamma'_{i+1}$. $\square$

Proposition 4. *In a valid SCJD scenario S transforming Γ into Δ, if s_{i+1} is a duplication and s_i is a cut or join acting on two duplicated extremities, then s_{i+1} can preempt s_i.*

Proposition 5. *In a valid SCJD scenario S transforming Γ into Δ, if s_{i+1} is an SCJ acting on two extremities that were not duplicated yet and s_i is an SCJ acting on two duplicated extremities, then s_{i+1} can preempt s_i.*

For a sequence of SCJ operations S, let S^D ($\overline{S}^D$, respectively) be the subsequence of operations that act on two extremities of genes that have (have not, respectively) already been duplicated at the time of the operation. By Corollary 1, for optimal S, $\overline{S}^D$ is indeed the complement of S^D.

Proposition 6. *There exists an optimal sorting scenario in which all duplication events are consecutive.*

Proof. Let $o_{i_1}, \ldots, o_{i_p}$ be the duplication events in an optimal sorting scenario. Denote by S_{i_j} the sequence of SCJ operations occurring between the duplications o_{i_j} and $o_{i_{j+1}}$. Also, denote by S_{i_0} and S_{i_p} the sequence of SCJ operations before the first duplication and after the last duplication, respectively.

   Given an optimal scenario $O^* = S_{i_0}, o_{i_1}, S_{i_1}, o_{i_2}, S_{i_2}, \ldots, S_{i_{p-1}}, o_{i_p}, S_{i_p}$ we modify it into a new sorting scenario O' as follows: Using Propositions 3 and 5, preempt SCJ operations acting on un-duplicated genes. Using Proposition 4, preempt duplication events. These steps are iterated until no preemption is possible. We get that $O' = S_{i_0}, \overline{S}_{i_1}^D, \ldots, \overline{S}_{i_p}^D, o_{i_1}, \ldots, o_{i_p}, S_{i_1}^D, \ldots, S_{i_{p-1}}^D, S_{i_p}$ is a valid SCJD optimal sequence in which all duplications are consecutive. $\qquad\square$

Corollary 2. *There exists an optimal SCJD sorting scenario, consisting, in this order, of (1) SCJ operations on single-copy genes, (2) duplications, (3) SCJ operations acting on duplicated genes.* $\qquad\square$

Denote by Γ' the intermediate (ordinary) genome after step (1). Then we can conclude:

Theorem 3. $d_{SCJD}(\Gamma, \Delta) = \min_{\Gamma'} \left(d_{SCJ}(\Gamma, \Gamma') + \#_c\Gamma' + dd_{SCJ}(\Gamma', \Delta) \right).$ $\qquad\square$

Recall that n is the number of genes in Γ. Using Theorems 1 and 2 and the fact that $\#_c\Pi = n - |\Pi|$, the distance formula can be simplified:

$$d_{SCJD} = \min_{\Gamma'} \left(|\Gamma| + |\Gamma'| - 2|\Gamma \cap \Gamma'| + n - |\Gamma'| + |\Delta| + 2\sum_{\alpha \in A} |\Gamma'_\alpha|(1 - |\Delta_\alpha|) \right)$$

$$= n + |\Delta| + |\Gamma| - 2 \max_{\Gamma'} \left(|\Gamma \cap \Gamma'| + \sum_{\alpha \in A} |\Gamma'_\alpha|(|\Delta_\alpha| - 1) \right)$$

$$= n + |\Delta| + |\Gamma| - 2 \max_{\Gamma'} \sum_{\alpha \in \Gamma'} (|\Gamma_\alpha| + |\Delta_\alpha| - 1)$$

$$= n + |\Delta| + |\Gamma| - 2 \max_{\Gamma'} \sum_{\alpha \in \Gamma'} \eta(\alpha) = n + |\Delta| + |\Gamma| - 2 \max_{\Gamma'} H(\Gamma') \qquad (2)$$

where $\eta(\alpha) = \eta(\alpha, \Gamma, \Delta) = |\Gamma_\alpha| + |\Delta_\alpha| - 1$ and $H(\Gamma') = \sum_{\alpha \in \Gamma'} \eta(\alpha)$. Since we want to maximize $H(\Gamma')$, we will focus on adjacencies with positive contribution in Eq. 2.

Lemma 2. *Let $\alpha = \{x, y\}$ be an adjacency such that $\eta(\alpha) > 0$. Then, for every extremity $z \neq y$, the conflicting adjacency $\alpha' = \{x, z\}$ has $\eta(\alpha') \leq 0$.*

Combining Lemma 2 and Theorem 3 we get a closed formula for the SCJD distance:

Theorem 4. *The genome* $\Gamma' = \{\alpha = \{x, y\} | \eta(\alpha) > 0\}$ *minimizes Equation 2. If* Γ' *is a linear genome, then the SCJD distance is given by* $d_{SCJD}(\Gamma, \Delta) = n + |\Delta| + |\Gamma| - 2H(\Gamma')$. $\qquad\square$

Let us return to the examples in Sect. 3:

- Example 1: $n = 1, |\Delta| = 1, |\Gamma| = 0, \Gamma' = \emptyset, H(\Gamma') = 0 \to d = 1+1+0-2*0 = 2$
- Example 2: $n = 2, |\Delta| = 2, |\Gamma| = 1, \Gamma' = \{\{a_h, b_t\}\}, H(\Gamma') = 2 \to d = 2 + 2 + 1 - 2 * 2 = 1$
- Example 3: $n = 3, |\Delta| = 4, |\Gamma| = 1, \Gamma' = \{\{a_h, b_t\}, \{b_h, c_t\}\}, H(\Gamma') = 1+1 \to d = 3 + 4 + 1 - 2 * 2 = 4$
- Example 4: $n = 3, |\Delta| = 4, |\Gamma| = 2, \Gamma' = \{\{a_h, b_t\}\}, H(\Gamma') = 1 \to d = 3 + 4 + 2 - 2 * 1 = 7$

Example 5. $\Gamma = abc$ and $\Delta = cab, bca$. According to Theorem 4, we get $\Gamma' = (abc)$ because $\eta(\{a_h, b_t\}) = \eta(\{b_h, c_t\}) = \eta(\{c_h, a_t\}) = 1$. The corresponding distance is $d = 3$, providing the following invalid sorting scenario:

$$\Gamma \xrightarrow[join]{} (abc) \xrightarrow[dup^*]{} (abc), (abc) \xrightarrow[cut]{} cab, (abc) \xrightarrow[cut]{} \Delta$$

dup^* indicates a duplication of a circular chromosome, an operation that is not allowed in the SCJD model (and has no cost). It is not difficult to verify that there is no valid sorting scenario with $d \leq 3$.

The reason for the discrepancy in Example 5 is that $\#_c(\Gamma') = n - |\Gamma'| = 0$ is not equal to the number of duplications if there are circular chromosomes. Therefore, in order to minimize the SCJD distance given by Eq. 2, we need to maximize $H(\Gamma')$ under the constraint that Γ' is a linear genome, i.e., $H(\Gamma') \geq H(\tilde{\Gamma})$ for every linear genome $\tilde{\Gamma}$. Lemma 3 shows that we can do so simply by removing one adjacency with $\eta = 1$ from each circular chromosome in Γ' and that such adjacency must exist.

Lemma 3. *Let* $\Gamma' = \{\alpha = \{x, y\} | \eta(\alpha) > 0\}$ *and let* Γ'' *be a genome obtained by removing one adjacency* α *with* $\eta(\alpha) = 1$ *from each circular chromosome in* Γ'. *Then,* Γ'' *is a linear genome that maximizes* $H(\cdot)$ *and the SCJD distance is given by* $d_{SCJD}(\Delta, \Gamma) = n + |\Delta| + |\Gamma| - 2H(\Gamma'')$.

Applying Lemma 3 to Example 5 we get $\Gamma'' = abc$ and $d = 5$:

$$\Gamma \xrightarrow[dup]{} abc, abc \xrightarrow[cut]{} a, bc, abc \xrightarrow[join]{} bca, abc \xrightarrow[cut]{} bca, ab, c \xrightarrow[join]{} \Delta$$

We can choose instead $\Gamma'' = cab$, which gives a different optimal sorting scenario:

$$\Gamma \xrightarrow[cut]{} ab, c \xrightarrow[join]{} cab \xrightarrow[dup]{} cab, cab \xrightarrow[cut]{} cab, a, bc \xrightarrow[join]{} \Delta$$

Algorithm 1. SCJD distance.

Input: An ordinary genome Γ and a duplicated genome Δ (both linear) on the same gene set.

Output: The SCJD distance $d_{SCJD}(\Gamma, \Delta)$ and an optimal sorting scenario $o_1, \ldots, o_d$ in which all intermediate genomes are linear.

1: $\Gamma' \leftarrow \{\alpha = \{x, y\} | \eta(\alpha) > 0\}$ (Theorem 4)
2: Create a linear genome Γ'' by removing one adjacency α with $\eta(\alpha) = 1$ from each circular chromosome in Γ' (Lemma 3)
3: $d_{SCJD}(\Gamma, \Delta) \leftarrow n + |\Delta| + |\Gamma| - 2H(\Gamma'')$ (Theorem 4, Lemma 3)
4: $o_1, \ldots, o_i \leftarrow$ Sort Γ into Γ'' (Theorem 1, Lemma 1)
5: $o_{i+1}, \ldots, o_j \leftarrow$ Duplicate all chromosomes in Γ''.
6: $o_{j+1}, \ldots, o_d \leftarrow$ Sort $2\Gamma''$ into Δ (Theorem 2, Lemma 1).
7: **return** $d, \overrightarrow{o}$

Algorithm 1 gives the full procedure for solving the SCJD distance and sorting problems. Each step of the algorithm takes $O(|\Gamma| + |\Delta|)$ time. In conclusion:

Theorem 5. *Algorithm 1 computes the SCJD distance in linear time.* □

5   Controlling the Number of Duplications

In this section we discuss how to control the number of duplications in an optimal SCJD sequence. Since the number of duplications is $n - |\Gamma''|$, selecting different intermediate genomes Γ'' that preserve the SCJD distance can produce scenarios with different number of duplications.

An optimal SCJD scenario with fewer duplications can be viewed as more conservative. The assumption behind this is that duplications are more "radical" events than breakage (cut) or fusion (join), which are local events.

Lemma 4. *Algorithm 1 gives an optimal sorting scenario with a maximum number of duplications.*

Proof. Observe first that for any sorting scenario (optimal or suboptimal) transforming Γ into Δ, we can assume w.l.o.g. that all duplications are consecutive without affecting the number of operations (Corollary 2). Call the genome right before the duplications the *last ordinary genome*. Denote by $d(\Gamma, \Pi, \Delta)$ the shortest scenario transforming Γ into Δ given that the last ordinary genome is Π. The proof of Theorem 3 implies that $d(\Gamma, \Pi, \Delta) = n + |\Delta| + |\Gamma| - 2H(\Pi)$.

Let Γ' be the last ordinary genome produced by the algorithm. Consider an optimal scenario O with a maximum number of duplications and let $\tilde{\Gamma}$ be the last ordinary linear genome in O. Since O is optimal, $H(\tilde{\Gamma})$ must be maximal. Hence, $\tilde{\Gamma}$ cannot contain adjacencies with $\eta < 0$. Moreover, it cannot contain adjacencies with $\eta = 0$, as such adjacencies increase $|\tilde{\Gamma}|$ and thus decrease the number of duplications in O. Therefore, $\tilde{\Gamma} \subseteq \Gamma'$.

We now show that $\forall \alpha \in \Gamma' \setminus \tilde{\Gamma}, \eta(\alpha) = 1$. Suppose by contradiction that there is an adjacency $\alpha \in \Gamma' \setminus \tilde{\Gamma}$ with $\eta(\alpha) > 1$ and let $\Pi = \tilde{\Gamma} \cup \{\alpha\}$. If Π is a linear genome, $d(\Gamma, \Pi, \Delta) < d(\Gamma, \tilde{\Gamma}, \Delta)$ contradicting the optimality of O. Otherwise, Π contains a circular chromosome and by Lemma 3, there is an adjacency $\beta \in \tilde{\Gamma}$ with $\eta(\beta) = 1$ such that $\Pi \setminus \{\beta\}$ is a linear genome with $H(\Pi \setminus \{\beta\}) > H(\tilde{\Gamma})$, again contradicting the optimality of O. Thus, $|\Gamma' \setminus \tilde{\Gamma}| = |\Gamma'| - |\tilde{\Gamma}| = H(\Gamma') - H(\tilde{\Gamma})$.

Γ' may contain circular chromosomes. By Lemma 3, Γ'' is produced by removing one adjacency with $\eta = 1$ from each circular chromosome in Γ'. Hence $|\Gamma' \setminus \Gamma''| = |\Gamma'| - |\Gamma''| = H(\Gamma') - H(\Gamma'')$.

Since both $\tilde{\Gamma}$ and Γ'' are last ordinary genomes in optimal SCJD scenarios, $H(\tilde{\Gamma}) = H(\Gamma'')$. Thus, $|\Gamma'| - |\tilde{\Gamma}| = H(\Gamma') - H(\tilde{\Gamma}) = H(\Gamma') - H(\Gamma'') = |\Gamma'| - |\Gamma''|$, which implies that $|\tilde{\Gamma}| = |\Gamma''|$. $\square$

One can decrease the number of duplications in an optimal SCJD scenario by adding adjacencies with $\eta(\alpha) = 0$ to Γ''. However, we need to make sure that the resulting genome is still linear. Consider the following example:

Example 6. $\Gamma = a, b, c$, $\Delta = abccba$. From Theorem 4 we have that $\Gamma' = \Gamma$ and so the SCJD distance is 8. The scenario produced by Algorithm 1 will first duplicate the three chromosomes of Γ and then perform five joins to create Δ. An alternative optimal sorting scenario is:

$$\Gamma \underset{JJ}{\longrightarrow} abc \underset{D}{\longrightarrow} abc, abc \underset{CC}{\longrightarrow} abc, a, b, c \underset{JJJ}{\longrightarrow} \Delta$$

Here, since each adjacency $\alpha \in \Delta$ has $\eta(\alpha) = 0$, we chose $\Gamma'' = abc$ and obtained an optimal scenario with a single duplication. In contrast, if we add to Γ'' the adjacencies $\{b_h, c_t\}$ and $\{c_h, b_t\}$ (which also have $\eta = 0$) we create a circular chromosome and an invalid SCJD sorting scenario.

In order to minimize the number of duplications we must add to Γ'' a maximum set of adjacencies with $\eta = 0$ such that the resulting genome is still linear. Here we show that this problem is NP-hard using a reduction similar to [16].

Theorem 6. *Given an ordinary linear genome Γ, a duplicated linear genome Δ on the same gene set, and an integer k, the problem of finding an optimal SCJD scenario with at most k duplications is NP-hard.*

Proof. Call a directed graph in which all in- and out-degrees are 2 a *2-digraph*. Deciding if a 2-digraph contains a Hamiltonian cycle is NP-hard [16,20]. This implies that the following variant is also NP-hard: Given a 2-digraph G with an edge (x, y), decide if there is a Hamiltonian path from y to x in G.

Let $G = (V, E)$ be a 2-digraph with an edge (x, y) as above. We may assume w.l.o.g. that G is strongly connected, since otherwise it would not contain a Hamiltonian path from y to x. Notice that $G \setminus (x, y)$ contains an Eulerian path from y to x [10]. Denote it by $P = e_1, e_2, \ldots, e_m$.

We construct a duplicated genome Σ as follows: for each $e_q = (u, v) \in P$ add the adjacency $\{u_h^i, v_t^j\}$ where $i = 2$ if there is an edge $e_l = (u, v')$ with

$l < q$, and $i = 1$ otherwise. Similarly, $j = 2$ if there is an edge $e_m = (u', v)$ with $m < q$ and $j = 1$ otherwise. The result is a linear chromosome created by traversing P and numbering the first occurrence of each vertex v in P as the gene copy v^1 and the second occurrence as v^2. Denote by $\overset{P}{\rightsquigarrow}$ the sequence of genes along the path P. In addition, we add two new genes w, z and the adjacencies $\{w_h^1, y_t^1\}, \{x_h^2, z_t^1\}$. Thus, Σ has three linear chromosomes: $w^1 y^1 \overset{P}{\rightsquigarrow} x^2 z^1, w^2$ and z^2. Let $\Pi = \{\{w_h, y_t\}, \{x_h, z_t\}\}$ be an ordinary genome with n chromosomes over the same set of genes. (Note that every vertex in $V \setminus \{x, y\}$ corresponds to a separate chromosome in Π.)

Let $\Sigma_{(i)}$ and $\Pi_{(i)}$ be genomes in which every gene $v \in V$ is renamed $v_{(i)}$. We define $\Delta = \bigcup_{i=1}^{k} \Sigma_{(i)}$ and $\Gamma = \bigcup_{i=1}^{k} \Pi_{(i)}$ to be the disjoint union of k different copies of Σ and Π respectively. This completes the reduction, which is clearly polynomial. We will show that there is an optimal SCJD scenario between Γ and Δ with at most k duplications iff G admits a Hamiltonian path from y to x.

For each edge $e = (u, v) \in E$ and every i, the corresponding adjacency $\alpha = \{(u_{(i)})_h^j, (v_{(i)})_t^l\}$ has $\eta(\alpha) = 1$ if there are two parallel edges from u to v, and $\eta(\alpha) = 0$ otherwise. In addition, for every i, $\eta(\{(w_{(i)})_h, (y_{(i)})_t\}) = \eta(\{(x_{(i)})_h, (z_{(i)})_t\}) = 1$, and every other adjacencies of $w_{(i)}, z_{(i)}$ have $\eta < 0$.

Suppose G contains a Hamiltonian path S from y to x. Let Γ' be the genome formed by the set of adjacencies $\{\{(w_{(i)})_h, (y_{(i)})_t\}, \{(x_{(i)})_h, (z_{(i)})_t\} | i = 1 \ldots k\}$ $\cup \{\{(u_{(i)})_h, (v_{(i)})_t\} | (u, v) \in S, i = 1 \ldots k\}$. Since S is a Hamiltonian path, Γ' is a valid ordinary linear genome with k chromosomes of the form $w_{(i)} y_{(i)} \overset{S}{\rightsquigarrow} x_{(i)} z_{(i)}$. To prove that Γ' maximizes $H(\cdot)$ we need to show it contains every adjacency with $\eta = 1$ and no adjacency with $\eta < 0$. Indeed, (suppressing the copy index i for clarity) the only adjacencies α with $\eta(\alpha) = 1$ are $\{w_h, y_t\}, \{x_h, z_t\}$ ($|\Delta_\alpha| = |\Gamma_\alpha| = 1$) and parallel edges in G ($|\Delta_\alpha| = 2, |\Gamma_\alpha| = 0$), one copy of which must be included in S. All other adjacencies in Γ' have $|\Delta_\alpha| = 1, |\Gamma_\alpha| = 0$ and $\eta(\alpha) = 0$. We conclude that Γ' is part of an optimal scenario with k duplications.

Conversely, suppose there is an optimal scenario O^* with at most k duplications and let $\tilde{\Gamma}$ be the last ordinary genome in O^*. Let $\Gamma' = \{\alpha | \eta(\alpha) > 0\}$ be a genome that minimizes the SCJD distance according to Theorem 4. First, notice that Γ' is indeed a linear genome. Otherwise, a circular chromosome of adjacencies with $\eta(\alpha) = 1$ would imply a strongly connected component without the vertices x, y, contradicting the strong connectivity of G. It follows that $\Gamma' \subseteq \tilde{\Gamma}, H(\Gamma') = H(\tilde{\Gamma})$ and $\#_c \tilde{\Gamma} \leq k$.

Since $\Sigma_{(i)}$ and $\Sigma_{(j)}$ for $i \neq j$ contain different genes, an adjacency between a gene in $\Sigma_{(i)}$ and a gene $\Sigma_{(j)}$ has negative η. Therefore, $\tilde{\Gamma}$ contains no such adjacencies. Since $\tilde{\Gamma}$ has at most k linear chromosomes, it must contain exactly k linear chromosomes, each containing all the genes of $\Sigma_{(i)}$ for one i.

Let $C = w_{(1)} y_{(1)} \ldots x_{(1)} z_{(1)}$ be the linear chromosome in $\tilde{\Gamma}$ that contains all the genes of $\Sigma_{(1)}$. Define an edge set S in G by taking for each adjacency $\{(u_{(1)})_h, (v_{(1)})_t\} \in C \setminus \{\{(w_{(1)})_h, (y_{(1)})_t\}, \{(x_{(1)})_h, (z_{(1)})_t\}\}$ the edge (u, v). Since

C is an ordinary linear chromosome containing all the genes of $\Sigma_{(1)}$, S is a Hamiltonian path in G from y to x.   □

6   Discussion

In this paper, we presented the SCJD rearrangement model, which allows the operations cut, join and whole chromosome duplication. We analyzed the problem of finding the minimum number of SCJD operations that transform an ordinary linear genome into a duplicated linear genome and provided a linear time algorithm for it. Furthermore, we showed that this algorithm gives an optimal scenario with a maximum number of duplications and that finding one with fewest duplications is NP-hard.

In the analysis, we focused on the SCJD sorting problem, which restricts the target genome to have exactly two copies of each gene. However, it is not difficult to generalize our algorithm to address the more general situation where each gene in the target genome has *at most* two copies. One can show that in this case too, an optimal solution in which all duplications are consecutive exists. In addition, each adjacency in the original genome between a gene that has two copies and a gene that has one copy in the target genome, must first be cut. This is true because duplications are defined over linear chromosomes in which every gene is unduplicated.

Our algorithm relies on the property that all duplications in the optimal solution can be clustered (Corollary 2). In this sense, the problem we study is similar to the SCJ Guided Genome Halving problem [12]. In that model the whole genome is duplicated at once, while in ours there is one duplication per chromosome, and accounting for these duplications is part of the optimization challenge.

Many aspects in the analysis of the SCJD mode require further research: How can we address the problem if there are more than two copies of each gene? Can we find the SCJD distance between two arbitrary genomes - each containing single copy and multiple copy genes? How does removing the requirement of linearity affect various SCJD problems? Moreover, duplications may be defined differently, e.g. tandem duplications [1] and segmental duplications [23]. Finally, developing a rigorous model that will allow both duplications and deletions is needed in order to analyze the full complexity of real biological data such as cancer samples.

Acknowledgments. We thank our referees for many helpful and insightful comments. This study was supported by the Israeli Science Foundation (grant 317/13) and the Dotan Hemato-Oncology Research Center at Tel Aviv University. RZ was supported in part by fellowships from the Edmond J. Safra Center for Bioinformatics at Tel Aviv University and from the Israeli Center of Research Excellence (I-CORE) Gene Regulation in Complex Human Disease (Center No 41/11).

References

1. Bader, M.: Sorting by reversals, block interchanges, tandem duplications, and deletions. BMC Bioinform. **10**(Suppl 1), S9 (2009)
2. Bader, M.: Genome rearrangements with duplications. BMC Bioinform. **11**(Suppl 1), S27 (2010)
3. Bayani, J., Selvarajah, S., Maire, G., Vukovic, B., Al-Romaih, K., Zielenska, M., Squire, J.A.: Genomic mechanisms and measurement of structural and numerical instability in cancer cells. Semin. Cancer Biol. **17**(1), 5–18 (2007)
4. Bergeron, A., Mixtacki, J., Stoye, J.: A unifying view of genome rearrangements. In: Bücher, P., Moret, B.M.E. (eds.) WABI 2006. LNCS (LNBI), vol. 4175, pp. 163–173. Springer, Heidelberg (2006)
5. Biller, P., Feijão, P., Meidanis, J.: Rearrangement-based phylogeny using the single-cut-or-join operation. IEEE/ACM Trans. Comput. Biol. Bioinform. **10**(1), 122–134 (2013)
6. Blanc, G., Barakat, A., Guyot, R., Cooke, R., Delseny, M.: Extensive duplication and reshuffling in the arabidopsis genome. Plant cell **12**(7), 1093–1101 (2000)
7. Bulteau, L., Fertin, G., Rusu, I.: Sorting by transpositions is difficult. SIAM J. Discrete Math. **26**(3), 1148–1180 (2012)
8. Caprara, A.: Sorting by reversals is difficult. In: Proceedings of the First Annual International Conference on Computational Molecular Biology (RECOMB), pp. 75–83, New York, USA (1997)
9. Christie, D.A.: Sorting permutations by block-interchanges. Inf. Process. Lett. **60**(4), 165–169 (1996)
10. Cormen, T.H., Leiserson, C.E., Rivest, R.L., Stein, C., et al.: Introduction to algorithms, vol. 2. MIT press, Cambridge (2001)
11. Dias, Z., Meidanis, J.: Genome rearrangements distance by fusion, fission, and transposition is easy. In: International Symposium on String Processing and Information Retrieval, pp. 250. IEEE Computer Society (2001)
12. Feijão, P., Meidanis, J.: SCJ: a breakpoint-like distance that simplifies several rearrangement problems. IEEE/ACM Trans. Comput. Biol. Bioinform. **8**(5), 1318–1329 (2011)
13. Fertin, G., Labarre, A., Rusu, I., Tannier, E., Vialette, S.: Combinatorics of Genome Rearrangements. MIT Press, Cambridge (2009)
14. Hannenhalli, S.: Polynomial-time algorithm for computing translocation distance between genomes. Discrete Appl. Math. **71**(1–3), 137–151 (1996)
15. Hannenhalli, S., Pevzner, P.A.: Transforming cabbage into turnip. In: Proceedings of the Twenty-Seventh Annual ACM Symposium on Theory of Computing (STOC), vol. 46, pp. 178–189, New York, USA (1995)
16. Kováč, J.: On the complexity of rearrangement problems under the breakpoint distance. J. Comput. Biol. **21**(1), 1–15 (2014)
17. Lu, C.L., Huang, Y.L., Wang, T.C., Chiu, H.-T.: Analysis of circular genome rearrangement by fusions, fissions and block-interchanges. BMC Bioinform. **7**(1), 295 (2006)
18. Mira, C.V.G., Meidanis, J.: Sorting by block-interchanges and signed reversals. ITNG **7**, 670–676 (2007)
19. Ozery-Flato, M., Shamir, R.: Sorting cancer karyotypes by elementary operations. J. Comput. Biol. **16**(10), 1445–1460 (2009)
20. Plesnik, J.: The NP-completeness of the Hamiltonian cycle problem in planar digraphs with degree bound two. Inf. Process. Lett. **8**(4), 199–201 (1979)

21. Savard, O.T., Gagnon, Y., Bertrand, D., El-Mabrouk, N.: Genome halving and double distance with losses. J. Comput. Biol. **18**(9), 1185–1199 (2011)
22. Shao, M., Lin, Y.: Approximating the edit distance for genomes with duplicate genes under DCJ, insertion and deletion. BMC Bioinform. **13**(Suppl 19), S13 (2012)
23. Shao, M., Lin, Y., Moret, B.: Sorting genomes with rearrangements and segmental duplications through trajectory graphs. BMC Bioinform. **14**(Suppl 15), S9 (2013)
24. Shao, M., Lin, Y., Moret, B.: An exact algorithm to compute the DCJ distance for genomes with duplicate genes. In: Sharan, R. (ed.) RECOMB 2014. LNCS, vol. 8394, pp. 280–292. Springer, Heidelberg (2014)
25. Tannier, E., Zheng, C., Sankoff, D.: Multichromosomal median and halving problems under different genomic distances. BMC Bioinform. **10**(1), 120 (2009)
26. Warren, R., Sankoff, D.: Genome aliquoting revisited. J. Comput. Biol. **18**(9), 1065–1075 (2011)
27. Yancopoulos, S., Attie, O., Friedberg, R.: Efficient sorting of genomic permutations by translocation, inversion and block interchange. Bioinformatics **21**(16), 3340–3346 (2005)